Modelling and Analysis of Active Biopotential Signals in Healthcare, Volume 1

IPEM–IOP Series in Physics and Engineering in Medicine and Biology

About the Series

Series in Physics and Engineering in Medicine and Biology will allow IPEM to enhance its mission to 'advance physics and engineering applied to medicine and biology for the public good.'

Focusing on key areas including, but not limited to:

- clinical engineering
- diagnostic radiology
- informatics and computing
- magnetic resonance imaging
- nuclear medicine
- physiological measurement
- radiation protection
- radiotherapy
- rehabilitation engineering
- ultrasound and non-ionising radiation.

A number of IPEM–IOP titles are published as part of the EUTEMPE Network Series for Medical Physics Experts.

Modelling and Analysis of Active Biopotential Signals in Healthcare, Volume 1

Edited by
Varun Bajaj
PDPM-Indian Institute of Information Technology, Design and Manufacturing, Jabalpur, India

G R Sinha
Myanmar Institute of Information Technology, Mandalay, Myanmar

IOP Publishing, Bristol, UK

ISBN 978-0-7503-3279-8 (ebook)
ISBN 978-0-7503-3277-4 (print)
ISBN 978-0-7503-3280-4 (myPrint)
ISBN 978-0-7503-3278-1 (mobi)

DOI 10.1088/978-0-7503-3279-8

Version: 20200801

IOP ebooks

British Library Cataloguing-in-Publication Data: A catalogue record for this book is available from the British Library.

Published by IOP Publishing, wholly owned by The Institute of Physics, London

IOP Publishing, Temple Circus, Temple Way, Bristol, BS1 6HG, UK

US Office: IOP Publishing, Inc., 190 North Independence Mall West, Suite 601, Philadelphia, PA 19106, USA

Dedicated to my father Late Mahendra Bajaj and my family members.

Varun Bajaj

Dedicated to my late grandparents, my teachers and Revered Swami Vivekananda.

G R Sinha

Contents

Preface

Bio-potential signals are often used by physicians for monitoring the pathological and physiological conditions of human organs. These signals originate from biological systems which include the nervous, cardiovascular and musculoskeletal systems, etc. Modelling and analysis of bio-potential signals involves the manipulation or transformation of the signals to enhance the relevant information for the improvement of healthcare systems.

This book, *Modelling and Analysis of Active Bio-potential Signals in Healthcare*, provides modelling of biomedical signals for understanding physiology, which can help to improve healthcare systems for the diagnosis and identification of disorders related to human organs. Therefore, this book addresses the need for a better understanding of the behaviours, concepts, fundamentals, principles, case studies, etc, of human organs for healthcare systems so that future research can be planned and carried out more effectively and the results strengthened. Biomedical signal applications are and will be used extensively in a huge number of research works and real-time applications for the improvement of healthcare systems using science, engineering and technology.

This first volume of the book provides the framework for modelling, the concepts and the applications of bio-potential signal processing. This book also emphasizes the real-time challenges in bio-potential signal processing due to the complex and non-stationary nature of the signals that are used for a variety of applications in the analysis, classification and identification of different states for the improvement of healthcare systems. Each chapter begins with a description of a biomedical example and the significance of the methods, with discussion to connect the technology with an understanding of the human organs. This book also provides information for the identification of diseases such as schizophrenia, epileptic seizures, physical action, cardiac health monitoring, etc. Moreover, the chapters can be read independently by research scholars, practising physicians, R&D engineers and graduate students who wish to explore research in the field of biomedical engineering.

Acknowledgements

Dr Bajaj expresses his heartfelt appreciation to his mother Prabha, wife Anuja and daughter Avadhi for their wonderful support and encouragement throughout the completion of this important book. His deepest gratitude goes to his mother-in-law and father-in-law for their constant motivation. This book is the outcome of sincere effort that could only be achieved due to the great support of his family. He also extends his thanks to Professor Sanjeev Jain, Director of PDPM IIITDM, Jabalpur, for his support and encouragement.

Dr Sinha expresses his gratitude and sincere thanks to his family members, wife Shubhra, daughter Samprati, parents and teachers.

We would like to thank all our friends, well-wishers and all those who keep us motivated to do more and more, better and better. We sincerely thank all the contributors for their writing on the relevant theoretical background and real-time applications of bio-potential signals for healthcare. We are also deeply grateful to many whose names are not mentioned here but whose help during this work we appreciate and wish to acknowledge.

We express our humble thanks to Michael Slaughter (Senior Commissioning Editor), Sarah Armstrong (Editorial Assistant) and the staff of IOP Publishing for their great support, necessary help, appreciation and quick responses. We also wish to thank Jessica Fricchione and Emily Tapp for their support during the review and approval of the book proposal. We also wish to thank IOP Publishing for giving us this opportunity to contribute on a relevant topic with a reputed publisher. Finally, we want to thank everyone who, in one way or another, helped us in editing this book.

Dr Bajaj, in particular, thanks his family who provided encouragement throughout the editing of this book. This book is whole-heartedly dedicated to his father who took the lead to heaven before the completion of this book.

Last, but not least, we would also like to thank God for showering us with his blessings and strength to perform novel and quality work of this type.

Varun Bajaj
G R Sinha

Editor biographies

Varun Bajaj

Varun Bajaj has been working as a faculty in the discipline of Electronics and Communication Engineering, at Indian Institute of Information Technology, Design and Manufacturing (IIITDM) Jabalpur, India since 2014. He worked as a visiting faculty in IIITDM Jabalpur from September 2013 to March 2014. He served as an Assistant Professor at Department of Electronics and Instrumentation, Shri Vaishnav Institute of Technology and Science, Indore, India during 2009–2010. He received B.E. degree in Electronics and Communication Engineering from Rajiv Gandhi Technological University, Bhopal, India in 2006, M.Tech. Degree with Honors in Microelectronics and VLSI design from Shri Govindram Seksaria Institute of Technology & Science, Indore, India in 2009. He received his PhD degree in the Discipline of Electrical Engineering, at Indian Institute of Technology Indore, India in 2014.

He is also serving as a Subject Editor-in-Chief of *IET Electronics Letters*. He served as a Subject Editor of *IET Electronics Letters* November 2018 to June 2020. He is Senior Member IEEE June 2020, MIEEE 16-20, and also contributing as active technical reviewer of leading International journals of IEEE, IET, and Elsevier, etc. He has authored more than 100 research papers in various reputed international journals/conferences like *IEEE Transactions*, Elsevier, Springer, IOP etc. He has edited *Modelling and Analysis of Active Biopotential Signals in Healthcare* Volumes 1 and 2 published by IOP Publishing. He also edited a book by CRC Press. The citation impact of his publications is around 1800 citations, an h-index of 19, and i10 index of 40 (Google Scholar July 2020). He has guided Six (03 Competed 3 Ongoing) PhD Scholars, 5 M. Tech. Scholars. He is a recipient of various reputed national and international awards. His research interests include biomedical signal processing, image processing, time-frequency analysis, and computer-aided medical diagnosis.

G R Sinha

G R Sinha is an Adjunct Professor at the International Institute of Information Technology Bangalore (IIITB) and currently deputed as a Professor at the Myanmar Institute of Information Technology (MIIT), Mandalay, Myanmar. He obtained his BE in Electronics Engineering and MTech in Computer Technology with a gold medal from the National Institute of Technology Raipur, India. He received his PhD in Electronics and Telecommunication Engineering from Chhattisgarh Swami Vivekanand Technical University (CSVTU), Bhilai, India. He is a Visiting Professor (Honorary) at the Sri Lanka Technological Campus Colombo for the year 2019–20. He has published 254 research papers, book chapters and books at the international and national level, that include *Biometrics* (published by Wiley India, a subsidiary of John Wiley), *Medical Image Processing* (published by Prentice Hall of India), and five edited books with IOP, Elsevier and Springer. He is an active reviewer and editorial member of more than 12 reputed international journals published by IEEE, IOP, Springer, Elsevier, etc. He has teaching and research experience of 21 years. He has been the Dean of Faculty and an Executive Council Member of CSVTU, and is currently a member of the Senate of MIIT. Dr Sinha has been delivering ACM lectures across the world as an ACM Distinguished Speaker in the field of DSP since 2017. A few of his more important roles include the Expert Member for Vocational Training Programme by Tata Institute of Social Sciences (TISS) for two years (2017–19), Chhattisgarh Representative of IEEE MP Sub-Section Executive Council (2016–19) and Distinguished Speaker in the field of Digital Image Processing by the Computer Society of India (2015). He is the recipient of many awards and recognitions, such as the TCS Award 2014 for Outstanding Contributions in the Campus Commune of TCS, Rajaram Bapu Patil ISTE National Award 2013 for Promising Teacher in Technical Education by ISTE New Delhi, Emerging Chhattisgarh Award 2013, Engineer of the Year Award 2011, Young Engineer Award 2008, Young Scientist Award 2005, IEI Expert Engineer Award 2007, ISCA Young Scientist Award 2006 Nomination and Deshbandhu Merit Scholarship for five years. He served as a Distinguished IEEE Lecturer in the IEEE India Council for the Bombay section. He is a Senior Member of IEEE, a Fellow of the Institute of Engineers India and a Fellow of IETE India.

He has delivered more than 50 keynote/invited talks and has chaired many technical sessions at international conferences across the world. His Special Session on 'Deep Learning in Biometrics' was included in the IEEE International Conference on Image Processing 2017. He is also a member of many national professional bodies such as ISTE, CSI, ISCA and IEI. He is a member of various committees of the University and has been Vice President of the Computer Society of India for the Bhilai chapter for two consecutive years. He is a consultant for various skill development initiatives of NSDC, the Government of India. He is a regular referee of project grants under the DST-EMR scheme and several other schemes of the Government of India. He has received important consultancy support, such as grants and travel support. Dr Sinha has supervised eight PhD scholars, 15 MTech scholars and is currently supervising another PhD scholar. His research interests include biometrics, cognitive science, medical image processing, computer vision, outcome based education (OBE) and ICT tools for developing employability skills.

Contributor list

Smith K Khare
Electronics and Communication Department, PDPM-Indian Institute of Information Technology, Design and Manufacturing, Jabalpur, India

Varun Bajaj
Electronics and Communication Department, PDPM-Indian Institute of Information Technology, Design and Manufacturing, Jabalpur, India

Siuly Siuly
Institute for Sustainable Industries and Liveable Cities, Footscray Park Victoria University, Australia

G R Sinha
Myanmar Institute of Information Technology (MIIT), Mandalay, Myanmar

Zahra Ghanbari
Amirkabir University of Technology, Tehran, Iran

Mohammad Hassan Moradi
Amirkabir University of Technology, Tehran, Iran

Supriya Supriya
Institute for Sustainable Industries and Liveable Cities, Footscray Park Victoria University, Australia

Hua Wang
Institute for Sustainable Industries and Liveable Cities, Footscray Park Victoria University, Australia

Yanchun Zhang
Institute for Sustainable Industries and Liveable Cities, Footscray Park Victoria University, Australia

Hamid Reza Marateb
Biomedical Engineering Department, Engineering Faculty, University of Isfahan, Isfahan, Iran

Carolina Migliorelli Falcone
CIBER-BBN, UPC, Escola Tècnica Superior d'Enginyeria Industrial de Barcelona (ETSEIB), Barcelona, Spain

Alejandro Bachiller Matarranz
Universitat Politècnica de Catalunya—BarcelonaTech, Escola Tècnica Superior d'Enginyeria Industrial de Barcelona (ETSEIB), Barcelona, Spain

Tayebe Azimi
Biomedical Engineering Department, Engineering Faculty, University of Isfahan, Isfahan, Iran

Farzad Ziaie Nezhad
Biomedical Engineering Department, Engineering Faculty, University of Isfahan, Isfahan, Iran

Marjan Mansourian
Biostatistics and Epidemiology Department, Faculty of Health, Isfahan University of Medical Sciences, Isfahan, Iran

Joan Francesc Alonso López
Universitat Politècnica de Catalunya—BarcelonaTech, Escola Tècnica Superior d'Enginyeria Industrial de Barcelona (ETSEIB), Barcelona, Spain

Javier Aparicio
Epilepsy Unit, Department of Neuropediatrics, Universitary Hospital Sant Joan de Déu Barcelona, Spain

Maria Victoria San Antonio Arce
Epilepsy Unit, Department of Neuropediatrics, Universitary Hospital Sant Joan de Déu Barcelona, Spain
Freiburg Epilepsy Centre, Medical Center—University of Freiburg, Germany

Sergio Romero Lafuente
Universitat Politècnica de Catalunya—BarcelonaTech, Escola Tècnica Superior d'Enginyeria Industrial de Barcelona (ETSEIB), CREB, Barcelona, Spain

Miguel Angel Mañanas Villanueva
Universitat Politècnica de Catalunya—BarcelonaTech, Escola Tècnica Superior d'Enginyeria Industrial de Barcelona (ETSEIB), Barcelona, Spain

Virender Mehla
Bennett University, Greater Noida, Uttar Pradesh, India

Ashish Singh
Bennett University, Greater Noida, Uttar Pradesh, India

Pushpendra Singh
National Institute of Technology Hamirpur, Himachal Pradesh, India

Amit Singhal
Bennett University, Greater Noida, Uttar Pradesh, India

Mislav Jordanić
Universitat Politècnica de Catalunya—BarcelonaTech, Escola Tècnica Superior d'Enginyeria Industrial de Barcelona (ETSEIB), Barcelona, Spain

Mónica Rojas-Martínez
Universidad El Bosque, Programa de Bioingeniería, Universidad El Bosque, Bogotá, Colombia

Leidy Yanet Serna Higuita
Universitat Politècnica de Catalunya—BarcelonaTech, Escola Tècnica Superior d'Enginyeria Industrial de Barcelona (ETSEIB), Barcelona, Spain

Mehdi Shirzadi
Universitat Politècnica de Catalunya—BarcelonaTech, Escola Tècnica Superior d'Enginyeria Industrial de Barcelona (ETSEIB), Barcelona, Spain

Marjan Nosouhi
Biomedical Engineering Department, Engineering Faculty, University of Isfahan, Isfahan, Iran

Kevin McGill
US Department of Veterans Affairs, United States

Ömer Faruk Alçin
Malatya Turgut Ozal University, Faculty of Engineering and Natural Sciences, Department of Electrical Engineering, Malatya, Turkey

Ümit Budak
Bitlis Eren University, Engineering Faculty, Electrical and Electronics Engineering Department, Bitlis, Turkey

Muzaffer Aslan
Bingol University, Engineering Faculty, Electrical-Electronics Engineering Department, Bingol, Turkey

Yaman Akbulut
Firat University, Informatics Department, Elazig, Turkey

Zafer Cömert
Samsun University, Engineering Faculty, Software Engineering Department, Samsun, Turkey

Muhammed H Akpınar
Firat University, Technology Faculty, Electrical and Electronics Engineering Department, Elazig, Turkey

Abdulkadir Şengür
Firat University, Technology Faculty, Electrical and Electronics Engineering Department, Elazig, Turkey

Anurag Nishad
EEE Department, BITS Pilani, KK Birla Goa Campus, India, India

Abhay Upadhyay
IET, Bundelkhand University Jhansi, UP, India, India

Ajay Somkuwar
Department of Electronics and Communication, MANIT Bhopal MP, INDIA

Vandana Somkuwar
Department of Mechanical Engineering, NITTTR Bhopal MP, INDIA

Abdulhamit Subasi
Effat University, College of Engineering, Jeddah, Saudi Arabia

Saeed Mian Qaisar
Effat University, College of Engineering, Jeddah, Saudi Arabia

Siddharth Bhalerao
Electronics and Communication Department, PDPM-Indian Institute of Information Technology, Design and Manufacturing, Jabalpur, India

Irshad Ahmad Ansari
Electronics and Communication Department, PDPM-Indian Institute of Information Technology, Design and Manufacturing, Jabalpur, India

Anil Kumar
Electronics and Communication Department, PDPM-Indian Institute of Information Technology, Design and Manufacturing, Jabalpur, India

Ranjeet Kumar
School of Electronics Engineering (SENSE), VIT University, Tamilnadu, India

Shiwangi Mishra
Electronics and Communication Department, PDPM-Indian Institute of Information Technology, Design and Manufacturing, Jabalpur, India

Pritee Khanna
Electronics and Communication Department, PDPM-Indian Institute of Information Technology, Design and Manufacturing, Jabalpur, India

IOP Publishing

Modelling and Analysis of Active Biopotential Signals in Healthcare, Volume 1

Varun Bajaj and G R Sinha

Chapter 1

Classification of schizophrenia patients through empirical wavelet transformation using electroencephalogram signals

Smith K Khare[1], Varun Bajaj[1], Siuly Siuly[2], G R Sinha[3]

[1]*PDPM-Indian Institute of Information Technology, Design and Manufacturing, Jabalpur, India*

[2]*Institute for Sustainable Industries & Liveable Cities, Victoria University, Australia*

[3]*Myanmar Institute of Information Technology, Mandalay, Myanmar*

Schizophrenia is a chronic and complex mental health disorder characterized by symptoms such as delusions, disorganized speech or behavior, hallucinations and impaired cognitive ability. Electroencephalogram (EEG) signals can provide detailed information about the brain activity associated with the behavioral changes associated with schizophrenia. Accurate and timely detection of this disease can help in diagnosis. In this chapter, empirical wavelet transformation is used to decompose the highly non-stationary EEG signals into modes in a Fourier spectrum. Linear and non-linear time domain features are extracted from the modes. Highly discriminant features are selected using the Kruskal–Wallis test. Different types of classification techniques are employed to classify the healthy and patients with schizophrenia. The effectiveness of the system is measured by evaluating various performance parameters such as accuracy, sensitivity, precision and specificity. Accuracy, precision, sensitivity and specificity of 88.7% 83.78%, 91.13% and 89.29%, respectively, are obtained.

1.1 Introduction

Schizophrenia is a mental disorder which mostly occurs during adulthood, causing deficits such as interpersonal engagement and relationships, etc. About 1% of the global population is affected by schizophrenia. Patients with schizophrenia show symptoms such as disorganized speech, hallucinations or delusions, according to the

doi:10.1088/978-0-7503-3279-8ch1

American Psychiatry Association [1]. Schizophrenia treatment involves long-term medication and is a great burden on healthcare systems and families [2]. The early prediction of schizophrenia involves a large number of aspects [3]. The reliability and comparability of studies have increased dramatically due to the introduction of standardized tools for evaluating symptoms and diagnosis. However, the problems of selecting proper methods and evaluation tools, and repeatability, etc, remain. Electroencephalogram (EEG) signals have gained much attention in the diagnosis of schizophrenia due to their non-invasive nature and ease of use [4–7]. EEG signals are electrical measures of the brain activity of billions of neurons connected together to form a network. An EEG signal is acquired from the scalp and has played a key role in clinical diagnosis and the dynamics of brain research. EEG signals provide increased coherence that reflects the presence of anomalous cortical organization in schizophrenics rather than transient states or medication effects related to severe clinical disturbance [8]. The temporal, occipital, frontal and parietal portions of the scalp play significant roles in analysing the changes during schizophrenia [9].

To date, researchers have proposed various methods for the detection and diagnosis of schizophrenia using EEG signals. The detection of schizophrenia by clustering the EEG signals with the help of the *k*-means method has been proposed [10]. The psychopharmacological and physiological changes occurring in the EEGs of schizophrenic and healthy patients have been monitored [11]. The biological and clinical association of the alpha and gamma frequency bands and power has been studied to separate schizophrenic and normal patients [12–14]. The identification of schizophrenic and normal patients has been carried out using spectral analysis [15, 16]. A rhythm based risk rate evaluation of healthy and schizophrenic patients is used in [17]. The alpha, delta, beta and gamma rhythms of occipital, central and frontal sites have been classified using the support vector machine (SVM) [18]. The separation of rhythm based features using filtering methods, multilayer back-propagation and self-organizing maps has been used [19]. Rhythm based features using a band pass filtering method with SVM, Sammon map and deep neural network classification techniques have been utilized for the identification of schizophrenia [20–23] as has the evaluation and classification of frequency based features using linear discriminant analysis [24]. The detection of schizophrenic patients using matched filtering and the fast Fourier transform (FFT) is proposed in [25]. The separation of rhythms using the Grey Walter passive filter has also been used to identify schizophrenic patients.

Positive and negative schizophrenia have been separated using the FFT of EEG signals of the frontal, temporal, parietal and occipital regions [26]. The brain activities of schizophrenic patients have been detected by evaluation of the spectral energy using the FFT [27]. The utility of the FFT along with principal component analysis, the Wilcoxon method and Welch's averaged periodogram method has been demonstrated in identifying the changes in the delta, beta, alpha and gamma bands of schizophrenic patients [28–33]. The different spikes, namely the focal, paroxysmal and independent spikes, occurring in the EEGs of schizophrenic patients have been analysed [34]. Post-imperative negative variation and contingent negative variation analysis have been used for identifying schizophrenic patients [35]. The steady-state visual evoked potential (SSVEP) and Fisher score have been used to evaluate different features classified using quadratic discriminant analysis (QDA), linear discriminant analysis (LDA), SVM with k-nearest neighbors (k-NN), second order polynomial kernels and logistic regression analysis (LRA) [36]. Features based on band power, autoregressive coefficients, Lampel–Ziv complexity (LZC), fractal dimensions (FDs) and entropies have been classified using SVM, adaptive boosting and LDA to detect schizophrenia [37–40]. The Welch periodogram technique for spectral estimation has been used to detect schizophrenia using the Kolmogorov–Smirnov test [41]. A genetic algorithm with a Butterworth filter and SVM has also been used to identify schizophrenia [42].

Ensemble synchronization measurement and Hilbert phase synchronization based methods have been used for classification using a logistic regression classifier [43]. The Hilbert–Huang transform, PCA, ICA and local discriminant bases have been used to extract the features of schizophrenic patients [44]. The utility of LZC for the identification of patients with schizophrenia is described in [45]. Power analysis of the alpha and delta bands has been carried out to distinguish control and affected patients [46]. The Higuchi, Katz and Petrosian methods have been used to extract classification features using LDA [47]. The weighted nearest neighbor, band power, FDs and autoregressive methods have been used to classify schizophrenic patients and control patients [48]. An auto-correlation and autoregressive coefficient have been classified using the independent t-test and neural networks [49–54]. LZC and correlation have been explored for measuring the alpha band activity of schizophrenic patients [55]. Multi-set canonical correlation analysis (MCCA) and SVM with recursive feature elimination (SVM-RFE) have been used for discriminating schizophrenia [56]. Entropy measurements and mean coherence with SVM have been used to discriminate schizophrenia [57]. Hurst exponent and FDs have been used to differentiate schizophrenic and control patients [22]. Kolmogorov complexity (KC), entropy and LZC methods have been used to find a useful discriminative tool for diagnostic purposes [58]. Feature vectors based on LZC and ANN have also been used to identify schizophrenic patients [59].

Autoregressive (AR), band power and FD coefficient based features extracted after preprocessing have been classified using LDA, multi-LDA (MLDA) and adaptive boosting (Adaboost) [60]. FDs and Pearson's correlation coefficient have been used to apply the brief psychiatric rating scale (BPRS) for the detection of schizophrenia [61]. Power spectral density based features have been classified using a combination of factor analysis based on maximum likelihood theory [62, 63]. Spectral features extracted from combinatorial analysis have been classified using the Kora-N algorithm [64]. The weighted minimum distance to mean and Riemannian geometric mean have been used for the classification of schizophrenia [65]. The equivalent current dipole power and asymmetry coefficient have been used for the analysis of the positive symptoms of schizophrenia [66]. Factor analysis and Kaiser's criteria have been used to identify patients suffering from schizophrenia [67]. Eigenvector power spectrum estimation and SVM have been used for the identification of schizophrenia [68]. Higuchi's method of computation of FDs has also been used to detect schizophrenia [69]. Energy and power based features have been classified using a high order pattern discovery algorithm [70]. The ε-complexity of continuous vector functions with RFs and an SVM have been used for binary classification of healthy and schizophrenic patients [71]. A fuzzy accuracy based classifier system has been used to generate fuzzy rules for discriminating healthy and schizophrenic subjects [72]. Autoregression based directed connectivity (DC) and graph-theoretical complex network (CN) based features have been classified using deep neural networks [73]. Inherent spatial pattern of network (SPN) features have been classified using LDA and an SVM to separate healthy patients from schizophrenic patients [74]. The analysis of entropy has been used for the evocation of emotions from visual cues in schizophrenia [75]. Mutual information (MI) has used to construct functional brain networks for analysis using graph theory [76]. Statistic significance probability maps based on the BPRS and a scale for the assessment of negative symptoms have been used for morphological findings in schizophrenia [77]. The spectral power of 192-channel resting EEG has been analysed using the Pearson correlation coefficient [78]. Spectral, complexity and variability measures evaluated from EEG signals have been classified using k-NN [79]. The long-term replicability of EEG spectra and auditory evoked potentials have been analysed to identify patients suffering from schizophrenia [80]. Sample covariance matrix and linear eigenvalue statistics have been used to classify schizophrenic patients using decision tree, random forest, SVM and naïve Bayes classifiers [81]. The Lyapunov exponent and Kolmogorov entropy have been evaluated to identify the classification accuracy of schizophrenic and controlled patients [82]. Average reference potential maps corresponding to global field power peaks in rhythms have been used to classify patients with schizophrenia [83]. Higuchi's FD, entropy and Kolmogorov complexity based features have been classified using SVM [84].

Independent component analysis (ICA) and time–frequency representation using the Stockwell transform have been used to find the most significant rhythms in schizophrenic patients [85]. ICA, spectral analysis and analysis of variance (ANOVA) have been carried out on the frequency bands to identify control patients from schizophrenic patients [86]. Filtering, ICA and Fisher analysis have been used to classify patients using connectivity maps [87]. ICA for the spectral analysis of 200 bands and RFs have been used for accurately detecting schizophrenia based on one-minute EEG recordings [88]. Fourier statistical analysis, evaluative power spectra, averaged power spectra and spectral variance have been used to identify the traits of schizophrenia among patients [89]. Time–frequency distributions (TFDs), FFTs, eigenvector methods, the wavelet transform (WT) and AR method have been used for the extraction of features, with advantages and disadvantages [90]. Filtering, FFT, STFT and entropy based features have been used for the classification of schizophrenia using an SVM and multilayer perceptron (MLP) [91]. Short-time Fourier transforms with a sliding window have been used to distinguish schizophrenic patients [92]. Feature extraction based on wavelet filtering with a genetic algorithm and SVM has been used to identify control patients [93]. Classification based on PCA, wavelet transform and k-NNs is proposed in [94]. Time–frequency analysis has been used, with a Morlet wavelet having a Gaussian shape in time and frequency, for the detection of schizophrenia [95]. Analysis of schizophrenic patients has been carried out using wavelet decomposition and Welch power spectral density (PSD) methods [96]. Analysis of alpha band frequencies has been carried out to detect the activity in schizophrenic patients [97]. Discriminant analysis (DA) has also been employed for classifying schizophrenic patients [98]. Features evaluated using Kolmogorov entropy, permutation entropy, correlation dimensions and spectral entropy have been selected using the Fisher criterion and classified using k-NN, SVM and back-propagation neural networks [99]. The phase lock value and phase coherence value of the intrinsic mode functions of empirical mode decomposition have been used to differentiate schizophrenia [100]. Multi-domain convolutional neural networks have been used for the classification of EEG based brain connectivity networks in schizophrenia [101].

Various methods for the detection of schizophrenia have been proposed in the literature. The FFT suffers from time–frequency localization. Other rigid methods such as STFT, wavelet transform and filtering use a basis which is independent of the processed signal. Moreover, the majority of these methods involve the direct evaluation of features from the raw EEG signals. The empirical wavelet transform (EWT) is capable of building an adaptive wavelet to extract the AM–FM components of a signal. The adaptive selection of the wavelet can capture useful hidden information from non-stationary EEG signals. In this chapter, a wavelet based decomposition method is employed to decompose the signal into AM–FM components. Dominant time domain features evaluated from the AM–FM components are selected using the Kruskal–Wallis test. The selected features are given as the input for the classifiers to distinguish patients with schizophrenia from control patients. The performance of the system is tested by evaluating four performance parameters and the receiver operating characteristics curve. The remainder of the chapter is organized as follows: section 1.2 presents the methodology, the results and discussion are provided in section 1.3, and section 1.4 concludes the chapter.

1.2 Methodology

This section includes descriptions of the dataset, empirical wavelet transform, features and classification techniques. The EEG signals are decomposed into AM–FM components using the EWT. Multiple time domain features are extracted from the obtained AM–FM components. Highly discriminant features are selected using the Kruskal–Wallis test and are classified using different classification techniques. The flowchart of the proposed methodology is shown in figure 1.1.

1.2.1 Dataset

The dataset used in this chapter contains EEG recordings of 14 female and 67 male patients. The average age and years of education are 39 years and 14.5 years, respectively. The details of the dataset are available online [102]. Three press button tasks were performed by the subjects, namely (1) pressing a button to immediately generate a tone, (2) passively listening to the same tone and (3) pressing a button without generating a tone to study the corollary discharge in people with

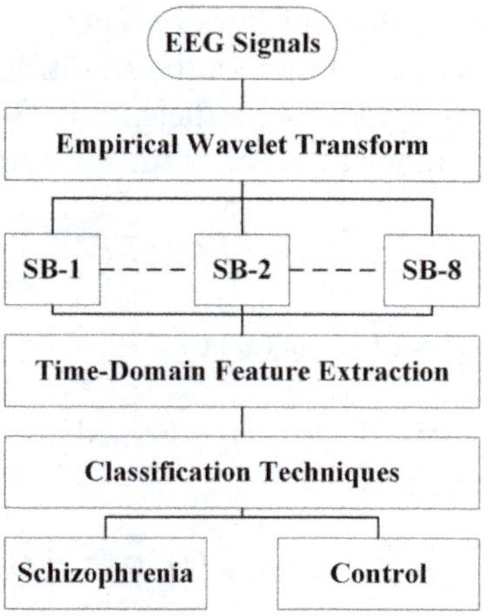

Figure 1.1. Flowchart of the proposed methodology.

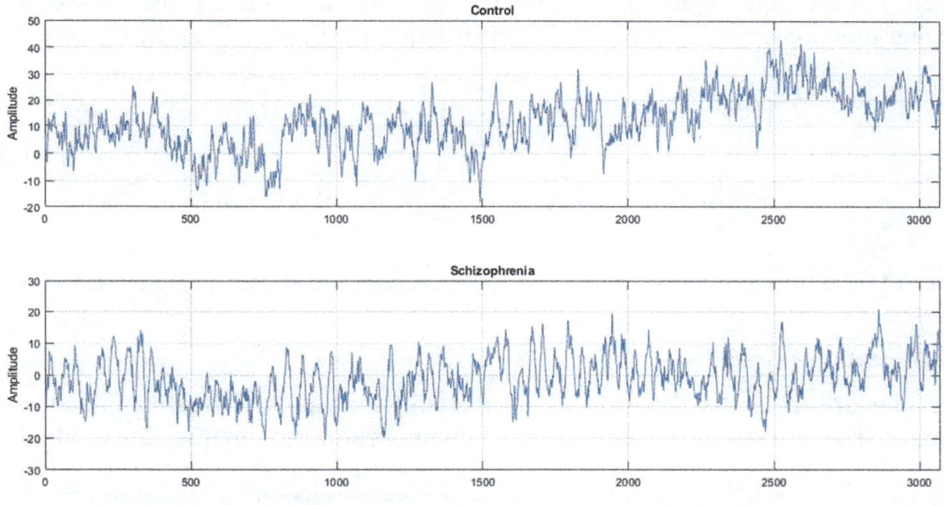

Figure 1.2. EEG signals of a healthy control and a schizophrenia patient.

schizophrenia and comparison controls. The healthy controls generated a press button tone while the schizophrenia patients did not. Hence, only condition one is tested to classify healthy and schizophrenia patients. The data are acquired from 64 sites on the scalp. The EEG data are sampled at a rate of 1024 Hz. EEG recordings of control and schizophrenic patients are shown in figure 1.2.

1.2.2 Empirical wavelet transform

To extract information from the highly complex EEG signal, the signal is split into multiple components. The empirical wavelet transform (EWT) is one such adaptive mechanism to split the signal into multiple components. The EWT is capable of extracting some components from the signal by building adaptive wavelets. Each component obtained by the EWT has a compact support Fourier spectrum. The separation of these modes is similar to Fourier spectrum segmentation and filtering. The EWT is defined in the same manner as the traditional wavelet transform. The EWT has detailed and approximation coefficients [103]. The detailed coefficients ($W_f^\alpha(n, t)$) are defined as the inner products with the empirical wavelet given as

$$\begin{aligned} W_f^\alpha(n, t) &= \langle f, \phi_n \rangle = \int f(\tau)\overline{\phi_n(\tau - t)}\mathrm{d}\tau \\ &= \left(\hat{f}(\omega)\overline{\hat{\phi}_n(\omega)}\right)^\nu. \end{aligned} \tag{1.1}$$

The approximation coefficients ($W_f^\alpha(0, t)$), defined as the inner product with a scaling function, can be written as

$$\begin{aligned} W_f^\alpha(0, t) &= \langle f, \varphi_1 \rangle = \int f(\tau)\overline{\varphi_1(\tau - t)}\mathrm{d}\tau \\ &= (\hat{f}(\omega)\overline{\hat{\varphi}_1(\omega)})^\nu, \end{aligned} \tag{1.2}$$

where f is the input signal in the time domain, and φ and ϕ are the wavelet and scaling functions, respectively. The reconstruction of the signal can be denoted as

$$\begin{aligned} f(t) &= W_f^\alpha(0, t) \times \varphi_1(t) + \sum_{n=1}^{N} W_f^\alpha(\omega, t) \times \phi_n(t) \\ &= \left(\hat{W}_f^\alpha(0, \omega) \times \hat{\varphi}_1(\omega) + \sum_{n=1}^{N} \hat{W}_f^\alpha(n, \omega) \times \hat{\phi}_n(\omega) \right). \end{aligned} \tag{1.3}$$

The AM–FM components obtained from the signal are shown in figure 1.3.

1.2.3 Feature extraction

Features are the statistical measures evaluated from the AM–FM components of signals. These statistical measures play an important role in the dimensionality

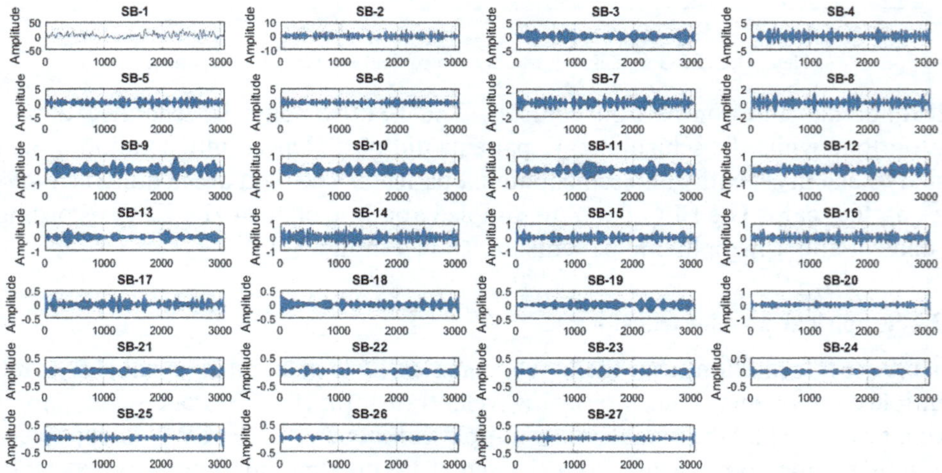

Figure 1.3. Modes obtained from EWT.

reduction and classification of signals. In this chapter, various statistical measures have been evaluated. Based on the Kruskal–Wallis analysis, five statistical measures are selected as features, which are kurtosis, covariance, root mean square, minimum and mean.

1.2.3.1 Kurtosis

Kurtosis measures the thickness along the tail of a given distribution for a given random variable. Kurtosis can be mathematically expressed as

$$\text{Kurtosis} = \frac{\sum_{n=1}^{N}(f_n - \bar{f})^4/N}{\sigma^4}, \tag{1.4}$$

where N is the number of signals, $\bar{f}$ is the mean and σ is the standard deviation.

1.2.3.2 Variance

Variance measures the spread of numbers from its mean value. It is the expectation of a squared deviation from the mean. The variance can be expressed as

$$\text{Variance} = \frac{1}{N}\sum_{n=1}^{N}(f_n - \bar{f})^2. \tag{1.5}$$

1.2.3.3 Root mean square

The root mean square (RMS) is the quadratic mean of the variable that measures the magnitude of varying quantity:

$$\text{RMS} = \frac{1}{N}\sum_{n=1}^{N}(f_n)^2. \tag{1.6}$$

1.2.3.4 Mean

The mean is the average value of all the samples in the variable and is expressed as

$$\text{Mean} = \frac{1}{N}\sum_{n=1}^{N}f_n. \tag{1.7}$$

1.2.3.5 Minimum

The minimum value of the variable is expressed as

$$\text{Minimum} = \min_{n=1}^{N}(f_n). \tag{1.8}$$

1.2.4 Classification techniques

The purpose of classifiers is to classify input data into two or more classes. The feature matrix is given as an input for classifiers. In this chapter, five different classification techniques are employed to classify binary class data. The k-NN, DA, ensemble method, SVM and decision tree based classifiers are used. To classify the input signals different kernels are employed. In the case of k-NN, six kernels are used, namely the fine, medium, coarse, cosine, cubic and weighted kernels. Linear and quadratic kernels are employed with discriminant analysis. Four kernels, namely the bagged tree, boosted tree, subspace k-NN (SS-k-NN) and subspace discriminant (SS-D) kernels are used with an ensemble based classifier. Linear, medium Gaussian and coarse Gaussian kernels are used with the SVM. For the decision tree based classifiers, simple tree, complex tree and medium tree are used. The details of the classification methods can be found in [104–108]. The process of k-NN is denoted as

$$d(y_i, y_t) = \sqrt{(y_{i1} - y_{t1})^2 + (y_{i2} - y_{t2})^2 + \dots + (y_{ip} - y_{tp})^2}$$
$$= \sqrt{\sum\nolimits_{i=1}^{n} (y_{1i} - y_{2i})^2} \tag{1.9}$$
$$V^1 = \frac{V - \min A}{\max A - \min A},$$

where d is the distance, y_i is an input with p features, n is the total inputs and p is the total number of features. V^1 is the max–min normalization matrix. In this chapter the total number of neighbors is selected as 5.

The mathematical modeling of the SVM is formed by minimizing the objective function $K(w)$, by taking the constraint $z_i(w_i^T y + b) \geqslant 1 (i = 1, 2, \dots, N)$:

$$K(w) = \min\left(\frac{1}{2}||w||^2\right). \tag{1.10}$$

By augmenting the objective function, the Lagrangian function for the SVM thus formed is denoted by

$$\Psi(w, b, \lambda) = \frac{1}{2}w^t w - \sum_{i=1}^{N} \lambda_i [z_i(w_i^T y + b) - 1], \tag{1.11}$$

where, $K(w)$ is the kernel, w is the weight matrix, b is the bias and y is the input.

The ensemble method for classification is mathematically represented by

$$\hat{G}_{\text{ens}}(\bullet) = \sum_{i=1}^{N} c_i G_i(\bullet), \tag{1.12}$$

where $\hat{G}_{\text{ens}}(\bullet)$ is the ensemble based function estimator, $G_i(\bullet)$ is the reweighted original data and c_i is the averaging weights.

The discriminant analysis classifier can be represented as

$$S_{W_i} = \sum_{y \in w} (y_i - \mu_j)(y_i - \mu_j)^T$$
$$W_i = S_{\mathrm{wi}}^{-1} S_{\mathrm{B}}, \tag{1.13}$$

where S_{B} and S_{wi} are the variance between the classes and the variance within the class, respectively.

1.2.5 Performance parameters

The performance of the system is tested by evaluating four performance parameters. In this chapter, accuracy, sensitivity, specificity and precision are measured. In the following, true positive (TP) is the number of true positives correctly identified from the positive class, true negative (TN) is the number of true negatives correctly identified from the negative class, false positive (FP) is the number of data points classified into the positive class that actually belong to the negative class and false negative (FN) is the number of data points classified into the negative class that belong to the positive class. ACC, SEN, SPE and PRE denote the accuracy, sensitivity, specificity and precision, respectively. Accuracy is defined as the ratio of the total number of correctly identified instances to the total number of instances. The mathematical formulation of accuracy is given as

$$\mathrm{ACC} = \frac{\mathrm{TP} + \mathrm{TN}}{\mathrm{TP} + \mathrm{FP} + \mathrm{TN} + \mathrm{FN}}. \tag{1.14}$$

The sensitivity or probability of detection is defined as the ability to correctly identify positive results. Sensitivity is represented by

$$\mathrm{SEN} = \frac{\mathrm{TP}}{\mathrm{TP} + \mathrm{FN}}. \tag{1.15}$$

The specificity or true negative rate is the ability to correctly identify actual negatives. The specificity is denoted by

$$\mathrm{SPE} = \frac{\mathrm{TN}}{\mathrm{TN} + \mathrm{FP}}. \tag{1.16}$$

The precision is the ratio of the total number of true positives to the total number of true positives and false positives. The precision is represented by

$$\mathrm{PRE} = \frac{\mathrm{TP}}{\mathrm{TP} + \mathrm{FP}}. \tag{1.17}$$

1.3 Results and discussion

This methodology uses the empirical wavelet transform and different classification techniques to separate schizophrenic patients from normal patients. There are 4108 signals available for schizophrenia patients and 3608 control signals. Every signal of each class has a total of 3072 samples. N-100 channels play an important role [102] in the detection of schizophrenia. Based on this, the first ten N-100 channels are considered for the evaluation. To maintain uniformity, a common experimental platform is used for both classes. Each signal is given as an input to the EWT. The boundary conditions are kept the same for both classes. The boundaries are chosen as {4 8 12 30}. Different numbers of AM–FM components are obtained from the signals using EWT. The minimum number of AM–FM components is eight and the maximum is 28. However, to maintain the synchronism between all signals for further computation, the number of AM–FM components is considered to be eight. Various statistical parameters are evaluated from the AM–FM components. Highly discriminable features are selected based on the results of the Kruskal–Wallis (KW) test. The KW test is a non-parameterized analysis of variance. It is used to find the discrimination ability of features by evaluating the probability of χ. A probability value $\leqslant 0.05$ is considered to be significant for classification. A total of five features are selected based on the KW test. These are kurtosis, variance, root mean square, minima and maxima, respectively. The probabilistic values of all the features are shown in tables 1.1–1.5, respectively. It is evident from these tables that most of the AM–FM components and channels are highly discriminable.

Inspired by the obtained results presented in tables 1.1–1.5, the selected features are given as the input for different classifiers. All the channels of every feature of each AM–FM component are combined. For every AM–FM component, the feature matrices obtained for schizophrenia and normal patients are 4108 × 50 and 3608 × 50, respectively. In this methodology, the ten-fold cross-validation method is employed for classification. Here, the input data are partitioned randomly into ten disjoint sets. Nine sets are used for training the input data and the remaining set is utilized for testing. The patients with schizophrenia are separated from the normal patients using five types of classification techniques.

Table 1.6 shows the classification accuracy obtained by the *k*-NN classifier. Six kernels are used for the classification. The classification accuracies obtained with the fine, medium, coarse, cosine, cubic and weighted kernels are, respectively, 81%, 84.1%, 82.1%, 84%, 83.3% and 84.3% for M-1, 72.1%, 75.6%, 71.5%, 75.2%, 72.4% and 75.9% for M-2, 66%, 70.2%, 68.2%, 69.7%, 67.5% and 71.1% for M-3, 66%, 69.5%, 67.6%, 68.3%, 66.2% and 70% for M-4, 65%, 69.6%, 70%, 68.6%, 65.7% and 70.7% for M-5, 61.3%, 65.4%, 67.9%, 65%, 62.8% and 66.4% for M-6, 59.3%, 61.8%, 64.8%, 61.5%, 61.2% and 63.8% for M-7, and 60.4%, 63.4%, 65.6%, 63.9%, 62.4% and 64.6% for M-8. The maximum accuracies obtained with the fine, medium, coarse, cosine, cubic and weighted kernel are, respectively, 81%, 84.1%, 82.1%, 84%, 83.3% and 84.3% for M-1.

Table 1.7 shows the accuracy of four classifiers, namely the discriminant analysis, SVM, ensemble and decision tree classifiers. The classification accuracies with the linear kernel are 74.9%, 59.3%, 56.9%, 56.4%, 55.8%, 55.7%, 56.3% and 57.1% for, respectively, M-1, M-2, M-3, M-4, M-5, M-6, M-7 and M-8, and with QDA the accuracies are 59%, 60.4%, 58.6%, 58.2%, 58.2%, 58.3%, 58.6% and 58.8%, respectively. The maximum accuracies for LDA and QDA are 74.9% and 60.4%

Table 1.1. Kruskal–Wallis test of kurtosis.

SB\C	C-1	C-2	C-3	C-4	C-5	C-6	C-7	C-8	C-9	C-10
M-1	3.25×10^{-1}	1.23×10^{-2}	3.74×10^{-3}	2.53×10^{-3}	9.64×10^{-6}	7.03×10^{-5}	1.52×10^{-1}	1.80×10^{-2}	1.24×10^{-2}	4.22×10^{-2}
M-2	1.13×10^{-3}	5.02×10^{-2}	5.34×10^{-7}	6.40×10^{-1}	2.27×10^{-2}	3.63×10^{-1}	3.44×10^{-2}	1.80×10^{-1}	5.97×10^{-1}	3.27×10^{-1}
M-3	5.15×10^{-11}	8.32×10^{-3}	4.06×10^{-5}	8.68×10^{-1}	1.19×10^{-2}	4.11×10^{-2}	2.31×10^{-2}	4.57×10^{-2}	5.14×10^{-3}	1.88×10^{-2}
M-4	3.21×10^{-4}	6.99×10^{-4}	6.27×10^{-5}	3.92×10^{-1}	1.26×10^{-3}	1.47×10^{-2}	4.67×10^{-2}	8.59×10^{-2}	1.64×10^{-2}	2.97×10^{-2}
M-5	3.58×10^{-2}	4.97×10^{-4}	6.70×10^{-3}	1.43×10^{-2}	1.91×10^{-2}	4.70×10^{-5}	3.08×10^{-2}	1.16×10^{-2}	6.09×10^{-3}	1.90×10^{-2}
M-6	4.53×10^{-3}	8.24×10^{-5}	1.77×10^{-3}	2.24×10^{-3}	1.55×10^{-2}	9.11×10^{-2}	9.22×10^{-3}	2.71×10^{-2}	1.05×10^{-4}	1.17×10^{-1}
M-7	6.88×10^{-3}	1.22×10^{-3}	1.75×10^{-2}	3.05×10^{-2}	2.24×10^{-3}	4.92×10^{-2}	2.22×10^{-3}	1.08×10^{-2}	4.18×10^{-3}	6.71×10^{-2}
M-8	2.31×10^{-7}	1.22×10^{-4}	4.63×10^{-3}	5.03×10^{-1}	5.79×10^{-2}	3.19×10^{-2}	6.06×10^{-4}	4.83×10^{-2}	7.39×10^{-3}	2.93×10^{-3}

Table 1.2. Kruskal–Wallis test of variance.

SB\C	C-1	C-2	C-3	C-4	C-5	C-6	C-7	C-8	C-9	C-10
M-1	7.48×10^{-43}	7.36×10^{-57}	2.07×10^{-42}	2.52×10^{-2}	1.10×10^{-12}	6.28×10^{-29}	6.66×10^{-13}	2.54×10^{-6}	5.76×10^{-16}	1.03×10^{-12}
M-2	5.55×10^{-9}	3.49×10^{-1}	7.43×10^{-9}	2.13×10^{-2}	1.48×10^{-5}	9.66×10^{-1}	1.30×10^{-5}	8.51×10^{-3}	3.53×10^{-2}	4.04×10^{-2}
M-3	5.11×10^{-11}	9.80×10^{-1}	5.67×10^{-8}	6.03×10^{-9}	1.33×10^{-4}	6.48×10^{-1}	5.08×10^{-4}	4.12×10^{-6}	8.38×10^{-3}	1.80×10^{-4}
M-4	3.18×10^{-6}	2.07×10^{-2}	1.15×10^{-2}	1.14×10^{-3}	1.36×10^{-3}	8.44×10^{-1}	1.94×10^{-4}	1.08×10^{-2}	4.13×10^{-2}	8.06×10^{-2}
M-5	3.21×10^{-2}	5.39×10^{-5}	2.52×10^{-1}	1.02×10^{-1}	7.85×10^{-1}	2.84×10^{-3}	2.35×10^{-1}	2.13×10^{-1}	5.61×10^{-1}	7.94×10^{-2}
M-6	3.88×10^{-2}	5.01×10^{-10}	1.90×10^{-4}	5.73×10^{-2}	7.41×10^{-4}	4.43×10^{-8}	3.58×10^{-2}	9.47×10^{-1}	9.06×10^{-2}	1.90×10^{-2}
M-7	7.70×10^{-3}	1.24×10^{-12}	2.52×10^{-9}	2.76×10^{-2}	8.84×10^{-4}	3.74×10^{-7}	7.44×10^{-1}	3.75×10^{-2}	1.26×10^{-4}	1.65×10^{-5}
M-8	1.08×10^{-5}	1.51×10^{-16}	7.61×10^{-14}	1.21×10^{-2}	1.10×10^{-6}	9.06×10^{-9}	3.81×10^{-1}	2.36×10^{-1}	3.41×10^{-5}	1.01×10^{-4}

Table 1.3. Kruskal–Wallis test of RMS.

SB\C	C-1	C-2	C-3	C-4	C-5	C-6	C-7	C-8	C-9	C-10
M-1	5.28×10^{-48}	2.33×10^{-51}	1.13×10^{-33}	2.75×10^{-5}	2.54×10^{-17}	1.00×10^{-22}	3.27×10^{-21}	1.65×10^{-12}	9.28×10^{-21}	9.28×10^{-18}
M-2	2.92×10^{-4}	7.71×10^{-1}	1.99×10^{-3}	2.12×10^{-1}	4.25×10^{-3}	9.52×10^{-1}	1.26×10^{-3}	3.37×10^{-3}	9.71×10^{-2}	2.17×10^{-1}
M-3	2.66×10^{-5}	4.79×10^{-2}	3.24×10^{-4}	1.08×10^{-5}	1.30×10^{-2}	7.32×10^{-1}	8.76×10^{-3}	2.72×10^{-8}	2.17×10^{-2}	1.90×10^{-2}
M-4	3.28×10^{-3}	4.11×10^{-2}	4.61×10^{-2}	2.80×10^{-2}	2.52×10^{-2}	6.49×10^{-1}	1.06×10^{-3}	9.36×10^{-4}	1.04×10^{-2}	3.70×10^{-2}
M-5	9.94×10^{-1}	3.07×10^{-6}	4.18×10^{-2}	1.82×10^{-1}	5.64×10^{-1}	1.73×10^{-3}	3.18×10^{-1}	7.21×10^{-2}	4.06×10^{-1}	3.72×10^{-1}
M-6	1.05×10^{-2}	2.91×10^{-11}	1.09×10^{-4}	2.84×10^{-2}	2.48×10^{-4}	2.35×10^{-7}	6.45×10^{-1}	4.57×10^{-2}	5.48×10^{-2}	1.30×10^{-2}
M-7	2.59×10^{-3}	1.50×10^{-13}	2.56×10^{-10}	7.41×10^{-3}	4.44×10^{-4}	1.97×10^{-6}	4.98×10^{-1}	9.96×10^{-1}	2.43×10^{-4}	4.38×10^{-6}
M-8	1.58×10^{-7}	5.41×10^{-18}	5.50×10^{-15}	3.11×10^{-2}	1.23×10^{-6}	7.29×10^{-8}	2.27×10^{-1}	2.84×10^{-1}	1.05×10^{-4}	2.16×10^{-5}

Table 1.4. Kruskal–Wallis test of mean.

SB\C	C-1	C-2	C-3	C-4	C-5	C-6	C-7	C-8	C-9	C-10
M-1	1.76×10^{-38}	2.47×10^{-54}	2.07×10^{-41}	1.58×10^{-2}	2.57×10^{-12}	5.38×10^{-26}	8.75×10^{-11}	4.00×10^{-5}	2.24×10^{-13}	1.72×10^{-10}
M-2	5.55×10^{-9}	3.49×10^{-1}	7.43×10^{-9}	2.13×10^{-2}	1.48×10^{-5}	9.66×10^{-1}	1.30×10^{-5}	8.51×10^{-3}	3.53×10^{-2}	4.04×10^{-2}
M-3	5.11×10^{-11}	9.80×10^{-1}	5.67×10^{-8}	6.03×10^{-9}	1.33×10^{-4}	6.48×10^{-1}	5.08×10^{-4}	4.12×10^{-6}	8.38×10^{-3}	1.80×10^{-4}
M-4	3.18×10^{-6}	2.07×10^{-1}	1.15×10^{-2}	1.14×10^{-3}	1.36×10^{-3}	8.44×10^{-1}	1.94×10^{-4}	1.08×10^{-2}	8.65×10^{-2}	8.06×10^{-2}
M-5	3.21×10^{-2}	5.39×10^{-5}	2.52×10^{-1}	1.02×10^{-2}	7.85×10^{-1}	2.84×10^{-3}	2.35×10^{-1}	2.13×10^{-1}	5.61×10^{-1}	7.94×10^{-2}
M-6	3.88×10^{-1}	5.01×10^{-10}	1.90×10^{-4}	5.73×10^{-2}	7.41×10^{-4}	4.43×10^{-8}	3.58×10^{-1}	9.47×10^{-1}	9.06×10^{-2}	1.90×10^{-1}
M-7	7.70×10^{-3}	1.24×10^{-12}	2.52×10^{-9}	2.76×10^{-2}	8.84×10^{-4}	3.74×10^{-7}	7.44×10^{-1}	7.04×10^{-1}	1.26×10^{-4}	1.65×10^{-5}
M-8	1.08×10^{-5}	1.51×10^{-16}	7.61×10^{-14}	1.21×10^{-2}	1.10×10^{-6}	9.06×10^{-9}	3.81×10^{-1}	2.36×10^{-1}	3.41×10^{-5}	1.01×10^{-4}

Table 1.5. Kruskal–Wallis test of minima.

SB\C	C-1	C-2	C-3	C-4	C-5	C-6	C-7	C-8	C-9	C-10
M-1	2.15×10^{-12}	1.23×10^{-16}	3.32×10^{-13}	4.20×10^{-1}	3.74×10^{-4}	5.62×10^{-9}	5.48×10^{-4}	1.34×10^{-2}	2.16×10^{-4}	4.09×10^{-3}
M-2	1.58×10^{-5}	3.95×10^{-2}	5.61×10^{-4}	1.14×10^{-1}	8.64×10^{-2}	9.38×10^{-1}	3.80×10^{-2}	3.75×10^{-3}	9.29×10^{-3}	1.85×10^{-1}
M-3	1.24×10^{-4}	2.71×10^{-1}	1.38×10^{-2}	6.77×10^{-2}	1.24×10^{-3}	1.35×10^{-1}	2.65×10^{-3}	4.76×10^{-2}	9.39×10^{-2}	4.74×10^{-2}
M-4	5.84×10^{-3}	2.61×10^{-1}	7.80×10^{-2}	5.03×10^{-2}	4.23×10^{-2}	6.87×10^{-1}	1.37×10^{-2}	1.75×10^{-1}	2.03×10^{-1}	3.41×10^{-1}
M-5	1.43×10^{-2}	3.40×10^{-2}	7.36×10^{-1}	3.35×10^{-2}	8.60×10^{-1}	1.32×10^{-1}	1.87×10^{-1}	3.37×10^{-3}	5.23×10^{-1}	1.23×10^{-2}
M-6	9.76×10^{-1}	1.17×10^{-2}	1.50×10^{-1}	7.92×10^{-1}	7.01×10^{-2}	1.77×10^{-6}	4.44×10^{-1}	3.39×10^{-1}	9.05×10^{-1}	6.17×10^{-1}
M-7	5.91×10^{-1}	1.82×10^{-2}	1.85×10^{-3}	3.01×10^{-2}	7.64×10^{-1}	2.16×10^{-1}	5.12×10^{-1}	5.72×10^{-2}	2.63×10^{-2}	3.01×10^{-2}
M-8	1.09×10^{-2}	1.84×10^{-3}	4.25×10^{-2}	7.42×10^{-1}	2.11×10^{-2}	3.57×10^{-2}	2.32×10^{-2}	5.83×10^{-2}	2.41×10^{-1}	3.39×10^{-2}

Table 1.6. Classification accuracy of *k*-NN.

	k-nearest neighbors					
SB	Fine	Medium	Coarse	Cosine	Cubic	Weighted
M-1	81	84.1	82.1	84	83.3	84.3
M-2	72.1	75.6	71.5	75.2	72.4	75.9
M-3	66	70.2	68.2	69.7	67.5	71.1
M-4	66	69.5	67.6	68.3	66.2	70
M-5	65	69.6	70	68.6	65.7	70.7
M-6	61.3	65.4	67.9	65	62.8	66.4
M-7	59.3	61.8	64.8	61.5	61.2	63.8
M-8	60.4	63.4	65.6	63.9	62.4	64.6

Table 1.7. The classification accuracy of the DA, ensemble, SVM and decision tree classifiers.

	Discriminant analysis		Ensemble				Support vector machine			Decision tree classifier		
SB	L	Q	Bag-T	BT	SS-*k*-NN	SS-D	Linear	Medium	Coarse	CT	ST	MT
M-1	74.9	59	87.5	85.6	72.2	81.4	86.4	88.7	83.4	78.9	75.6	68.7
M-2	59.3	60.4	87.7	85.5	69.9	62	69	88.2	62.8	80.1	78.8	76.7
M-3	56.9	58.6	83.4	80.8	66.1	60.8	63.4	86.1	61.5	76.3	73.9	72.2
M-4	56.4	58.2	82.2	80.3	65.7	61	61.6	82.9	61.3	73.5	70.9	69.6
M-5	55.8	58.2	83	81.4	63.8	60.9	61.2	82.3	61.3	74.2	71.6	68.5
M-6	55.7	58.3	76.5	74.9	63.7	61	61.1	69.5	61.4	68.1	66.7	62.5
M-7	56.3	58.6	66.8	65.1	63.4	61.2	61.4	61.6	61.4	61.4	62.6	60.8
M-8	57.1	58.8	68.2	67.4	63.1	61.2	63.9	67.5	61.4	62.6	62.5	60.7

for M-1 and M-2, respectively. The accuracies obtained with the ensemble classifier for M-1, M-2, M-3, M-4, M-5, M-6, M-7 and M-8, respectively, are 87.5%, 87.7%, 83.4%, 82.2%, 83%, 76.5%, 66.8% and 68.2% for bagged tree (Bag-T), 85.6%, 85.5%, 80.8%, 80.3%, 81.4%, 74.9%, 65.1% and 67.4% for boosted tree (BT), 72.2%, 69.9%, 66.1%, 65.7%, 63.8%, 63.7%, 63.4% and 63.1% for SS-*k*-NN and 81.4%, 62%, 60.8%, 61%, 60.9%, 61%, 61.2% and 61.2% for SS-D. The maximum accuracies obtained with Bag-T, BT, SS-*k*-NN and SS-D are 87.7%, 85.6%, 72.2% and 81.4% for M-1 and M-2. Linear, medium and coarse kernels of SVM are used to test the accuracy. For M-1, M-2, M-3, M-4, M-5, M-6, M-7 and M-8, respectively, accuracies are achieved of 86.4%, 69%, 63.4%, 61.6%, 61.2%, 61.1%, 61.4% and 63.9% for the linear kernel, 88.7%, 88.2%, 86.1%, 82.9%, 82.3, 69.5%, 61.6% and 67.5% for the medium kernel, and 83.4%, 62.8%, 61.5%, 61.3%, 61.3%, 61.4%, 61.4% and 61.4% for the coarse kernel. M-1 provides maximum accuracies of 86.4%, 88.7% and 83.4% for the linear, medium and coarse kernels, respectively. For M-1,

Table 1.8. The performance parameters of the medium kernel SVM.

Performance parameters				
SB	CC	SEN	SPE	PRE
M-1	88.70	91.13	87.37	79.70
M-2	88.20	86.51	89.29	83.78
M-3	86.10	85.87	86.24	78.41
M-4	82.90	83.44	82.54	71.65
M-5	82.30	83.73	81.65	69.72
M-6	69.50	76.44	67.89	34.97
M-7	61.60	72.19	61.10	7.43
M-8	67.50	63.49	69.16	45.59

M-2, M-3, M-4, M-5, M-6, M-7 and M-8, respectively, the accuracies are 78.9%, 80.1%, 76.3%, 73.5%, 74.2%, 68.1%, 61.4% and 62.6% for the complex tree (CT), 75.6%, 78.8%, 73.9%, 70.9%, 71.6%, 66.7%, 62.6% and 62.5% for the simple tree (ST), and 68.7%, 76.7%, 72.2%, 69.6%, 68.5%, 62.5%, 60.8% and 60.7% for the medium tree (MT). The decision tree type classifier provides the maximum accuracy for M-2 for CT, ST and MT with values of 80.1%, 78.8% and 76.7%, respectively.

Among all the classifiers, the SVM produces the highest accuracy with the medium kernel. Hence, the performance parameters of the medium kernel are shown in table 1.8. Four performance parameters are evaluated, namely accuracy (ACC), sensitivity (SEN), specificity (SPE) and precision (PRE). For M-1, M-2, M-3, M-4, M-5, M-6, M-7 and M-8, respectively, we find a sensitivity of 91.3%, 86.51%, 85.87%, 83.44%, 83.73%, 76.44%, 72.19% and 63.49%, a specificity 7.37%, 89.29%, 86.24%, 82.54%, 8165%, 67.89%, 61.10% and 69.16%, and a precision of 79.7%, 83.78%, 78.41%, 71.65%, 69.72%, 34.97%, 7.43% and 45.59%. The highest accuracy and sensitivity are obtained as 88.70% and 91.13%, respectively, for M-1. Maximum specificity and precision are obtained as 89.29% and 83.78%, respectively, for M-2.

The receiver operating characteristic (ROC) curve shows the performance of a classification model for all classification thresholds. The ROC curve of all the AM–FM components for a medium Gaussian SVM is shown in figure 1.4. As evident from figures 1.4(a) and (b), the area under curve (AUC) is 94%. The change in classifier characteristics is identified at a true positive rate (TPR) of 80% and a false positive rate (FPR) of 5% for M-1. The change in classifier characteristics is identified at 84% TPR and 9% FPR for M-2. Figures 1.4(c) and (d) represent the ROC curves of M-3 and M-4. The AUC for M-3 is 93% while for M-4 it is 91%. The change in classifier characteristics for M-3 and M-4 are obtained at FPRs of 9% and 10% and TPRs of 78% and 72%, respectively. The ROCs of M-5, M-6, M-7 and M-8 are shown in figures 1.4(e), (f), (g) and (h). The AUCs are 91%, 84%, 69% and 72% for M-5, M-6, M-7 and M-8, respectively. The change is observed at TPRs of 70% and 35% and FPRs of 9% and 7% for M-5 and M-6, respectively, while for M-7 and M-8 it is observed at TPRs of 2% and 18% and FPRs of 7% and 46%.

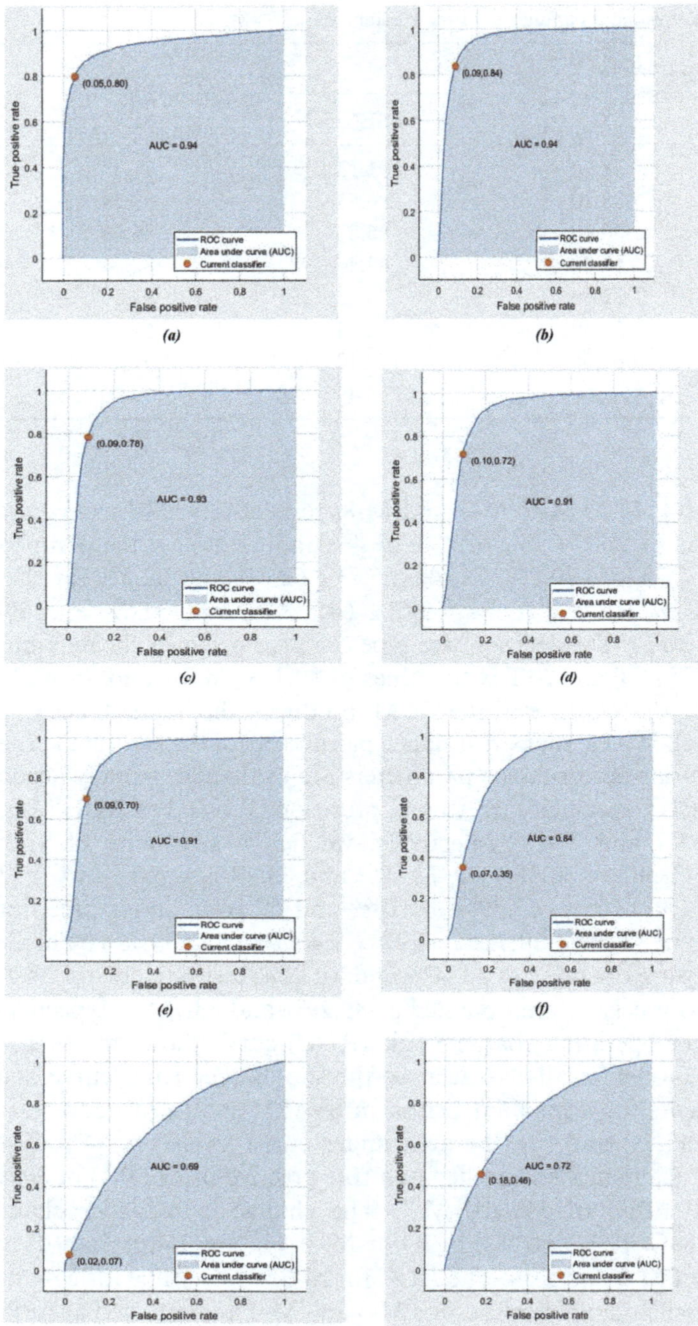

Figure 1.4. Receiver operator characteristics for a medium SVM of the first eight subbands M-1–M-8, (a) to (h), respectively.

1.4 Conclusion

Patients with schizophrenia cannot be easily identified through visual inspection. Doctors recommend a number of neurological tests to identify the symptoms of schizophrenia. However, these tests are not always effective. Electroencephalogram signals provide vital information about the neurological changes that happen in the schizophrenic state. In this chapter, a novel method based on the empirical wavelet transform is proposed for the identification of schizophrenia. The majority of the information resides in the first two AM–FM components, as these provide the highest correct classifications of schizophrenic patients and normal patients. The classification abilities of different classification techniques are tested. It is found that SVM is the best method, followed by the ensemble, *k*-nearest neighbors, decision tree and, finally, discriminant analysis classifier. The medium kernel of the SVM provides the best performance parameters with an accuracy of 88.7%, a sensitivity of 91.13%, a specificity of 89.29% and a precision of 83.78%.

References

[1] Samuel B G 1995 Diagnostic and statistical manual of mental disorders 4th edn (DSM-IV) *Am. J. Psychiatry* **152** 1228

[2] McGlashan T H 1998 Early detection and intervention of schizophrenia: rationale and research *Br. J. Psychiatry* **172** 3–6

[3] Lett T A, Voineskos A N, Kennedy J L, Levine B and Daskalakis Z J 2014 Treating working memory deficits in schizophrenia: a review of the neurobiology *Biol. Psychiatry* **75** 361–70

[4] Hee N S, Jin S H, Kim S Y and Ham B J 2002 EEG in schizophrenic patients: mutual information analysis *Clin. Neurophysiol.* **113** 1954–60

[5] Knox H F and Macfie C C 1941 Electroencephalography in schizophrenia *Am. J. Psychiatry* **98** 374–81

[6] Koukkou M, Lehmann D, Wackermann J, Dvorak I and Henggeler B 1993 Dimensional complexity of EEG brain mechanisms in untreated schizophrenia *Biol. Psychiatry* **33** 397–407

[7] Karson C N, Coppola R, Daniel D G and Weinberger D R 1988 Computerized EEG in schizophrenia *Schizophr. Bull.* **14** 193–7

[8] Merrin E L, Floyd T C and Fein G 1989 EEG coherence in unmedicated schizophrenic patients *Biol. Psychiatry* **25** 60–6

[9] Davis P A 1940 Evaluation of the electroencephalograms of schizophrenic patients *Am. J. Psychiatry* **96** 851–60

[10] Lehmann D, Faber P L, Galderisi S, Herrmann W M and Koenig T 2005 EEG microstate duration and syntax in acute, medication-naïve, first-episode schizophrenia: a multi-center study *Psychiatry Res.: Neuroimaging* **138** 141–56

[11] Kornetsky C and Mirsky A F 1965 On certain psychopharmacological and physiological differences between schizophrenic and normal persons *Psychopharmacologia* **8** 309–18

[12] Sponheim S R, Clementz B A, Iacono W G and Beiser M 2000 Clinical and biological concomitants of resting state EEG power abnormalities in schizophrenia *Biol. Psychiatry* **48** 1088–97

[13] Roach B J and Mathalon D H 2008 Event-related EEG time-frequency analysis: an overview of measures and an analysis of early gamma band phase locking in schizophrenia *Schizophr. Bull.* **34** 907–26

[14] Moran L and Hong L 2011 High vs low frequency neural oscillations in schizophrenia *Schizophr. Bull.* **37** 659–63

[15] Giannitrapani D and Kayton L 1974 Schizophrenia and EEG spectral analysis *Electroencephalogr. Clin. Neurophysiol.* **36** 377–86

[16] Boutros N N, Arfken C, Galderisi S, Warrick J, Pratt G and Iacono W 2008 The status of spectral EEG abnormality as a diagnostic test for schizophrenia *Schizophr. Res.* **99** 225–37

[17] Clementez B A, Sponheim S R, Iacono W G and Beiser M 1994 Resting EEG in first-episode schizophrenia patients, bipolar psychosis patients, and their first-degree relatives *Psychophysiology* **31** 486–94

[18] Johannesen J, Bi J, Jiang R, Kenney J and Chen C M 2016 Machine learning identification of EEG features predicting working memory performance in schizophrenia and healthy adults *Neuropsychiatr. Electrophysiol.* **2** 3

[19] Li Y and Fan F 2005 Classification of schizophrenia and depression by EEG with ANNs *2005 IEEE Engineering in Medicine and Biology 27th Annual Conf.* pp 2679–82

[20] Shim M, Hwang H, Kim D, Lee S and Im C 2016 Machine-learning-based diagnosis of schizophrenia using combined sensor-level and source-level EEG features *Schizophr. Res.* **176** 314–9

[21] Winterer G, Ziller M, Dor H, Frick K, Mulert C, Wuebben Y and Herrmann W M 2000 Frontal dysfunction in schizophrenia—a new electrophysiological classifier for research and clinical applications *Eur. Arch. Psychiatry Clin. Neurosci.* **250** 207–14

[22] Chu L, Qiu R, Liu H, Ling Z and Shi X 2017 Individual recognition in schizophrenia using deep learning methods with random forest and voting classifiers: insights from resting state EEG streams, arXiv:1707.03467

[23] Bose T, Sivakumar S and Kesavamurthy B 2016 Identification of schizophrenia using EEG alpha band power during hyperventilation and post-hyperventilation *J. Med. Biol. Eng.* **36** 901–11

[24] Ince N F, Pellizzer G, Tewfik A H, Nelson K, Leuthold A, McClannahan K and Stephane M 2009 Classification of schizophrenia with spectro-temporo-spatial MEG patterns in working memory *Clin. Neurophysiol.* **120** 1123–34

[25] Flor H P, Lind J C and Koles Z J 2004 A source-imaging (low-resolution electromagnetic tomography) study of the EEGs from unmedicated males with depression *Psychiatry Res.: Neuroimaging* **130** 191–207

[26] Begić D, Hotujac L and Jokić-Begić N 2000 Quantitative EEG in 'positive' and 'negative' schizophrenia *Acta Psychiatr. Scand.* **101** 307–11

[27] Morstyn R, Duffy F H and McCarley R W 1983 Altered topography of EEG spectral content in schizophrenia *Electroencephalogr. Clin. Neurophysiol.* **56** 263–71

[28] Michelogiannis S, Paritsis N and Trikas P 1991 EEG coherence during hemispheric activation in schizophrenics *Eur. Arch. Psychiatry Clin. Neurosci.* **241** 31–4

[29] Dierks T K, Maurer R and Ihl A 1989 Schmidtke *Topographic Brain Mapping of EEG and Evoked Potentials* (Berlin: Springer) pp 507–17

[30] Michel C M, Koukkou M and Lehmann D 1993 EEG reactivity in high and low symptomatic schizophrenics, using source modelling in the frequency domain *Brain Topogr.* **5** 389–94

[31] Umesh S, Nizamie S H, Goyal N, Tikka S and Bose S 2016 Social anhedonia and gamma band abnormalities as a composite/multivariate endophenotype for schizophrenia: a dense array EEG study *Early Interv. Psychiatry* **12** 362–71

[32] Alfimova M and Uvarova L 2003 Cognitive peculiarities in relatives of schizophrenic and schizoaffective patients: heritability and resting EEG-correlates *Int. J. Psychophysiol.* **49** 201–16

[33] Lacroix D, Chaput Y, Rodriguez J P, Filion M, Morrison D, St. Denis P and Albert J-M 1995 Quantified EEG changes associated with a positive clinical response to clozapine in schizophrenia *Progr. Neuro-Psychopharmacol. Biol. Psychiatry* **19** 861–76

[34] White P T, Demyer W and Demyer M K 1964 EEG abnormalities in early childhood schizophrenia: a double-blind study of psychiatrically disturbed and normal children during promazine sedation *Am. J. Psychiatry* **120** 950–8

[35] Verleger R, Wascher E, Arolt V, Daase C, Strohm A and Kömpf D 1999 Slow EEG potentials (contingent negative variation and post-imperative negative variation) in schizophrenia: their association to the present state and to Parkinsonian medication effects *Clin. Neurophysiol.* **110** 1175–92

[36] Alimardani F, Cho J H, Boostani R and Hwang H J 2018 Classification of bipolar disorder and schizophrenia using steady-state visual evoked potential based features *IEEE Access* **6** 40379–88

[37] Sabeti M, Katebi S and Boostani R 2009 Entropy and complexity measures for EEG signal classification of schizophrenic and control participants *Artif. Intell. Med.* **47** 263–74

[38] Sabeti M, Boostani R, Katebi S D and Price G W 2007 Selection of relevant features for EEG signals classification of schizophrenic patients *Biomed. Signal Process. Control* **2** 122–34

[39] Sabeti M, Katebi S D, Boostani R and Price G W 2011 A new approach for EEG signal classification of schizophrenic and control participants *Expert Syst. Appl.* **38** 2063–71

[40] Taghavi M, Boostani R, Sabeti M and Taghavi S M 2011 Usefulness of approximate entropy in the diagnosis of schizophrenia *Iran. J. Psychiatry Behav. Sci.* **5**

[41] Akbar Y, Khotimah S N and Haryanto F 2016 Spectral and brain mapping analysis of EEG based on Pwelch in schizophrenic patients *J. Phys. Conf. Ser.* **694** 012070

[42] Sabeti M, Boostani R and Katebi S D 2007 A new approach to classify the schizophrenic and normal subjects by finding the best channels and frequency bands *2007 15th Int. Conf. on Digital Signal Processing*

[43] Prasad P D, Halahalli H N, John J P and Majumdar K K 2014 Single-trial EEG classification using logistic regression based on ensemble synchronization *IEEE J. Biomed. Health Inform.* **18** 1074–80

[44] Azlan W A and Low Y F 2014 Feature extraction of electroencephalogram (EEG) signal—a review *2014 IEEE Conf. on Biomedical Engineering and Sciences (IECBES)*

[45] Li Y, Tong S, Liu D, Gai Y, Wang X, Wang J, Qiu Y and Zhu Y 2008 Abnormal EEG complexity in patients with schizophrenia and depression *Clin. Neurophysiol.* **119** 1232–41

[46] Fenton G W, Fenwick P B, Dollimore J, Dunn T L and Hirsch S R 1980 EEG spectral analysis in schizophrenia *Br. J. Psychiatry* **136** 445–55

[47] Boostani R, Khadijeh S and Sabeti M 2009 An efficient classifier to diagnose of schizophrenia based on the EEG signals *Expert Syst. Appl.* **36** 6492–9

[48] Parvinnia E, Sabeti M, Zolghadri M J and Boostani R 2014 Classification of EEG signals using adaptive weighted distance nearest neighbor algorithm *J. King Saud Univ. Comput. Inf. Sci.* **26** 1–6

[49] Abrams R and Taylor M A 1979 Differential EEG patterns in affective disorder and schizophrenia *Arch. Gen. Psychiatry* **36** 1355–8

[50] Saletu B, Anderer P, Saletu Z G M, Arnold O and Pascual-Marqui R D 2002 Classification and evaluation of the pharmacodynamics of psychotropic drugs by single-lead pharmaco-EEG, EEG mapping and tomography (LORETA) *Methods Find. Exp. Clin. Pharmacol.* **24** 97–120

[51] Streit M, Wölwer W, Brinkmeyer J, Ihl R and Gaebel W 2001 EEG-correlates of facial affect recognition and categorisation of blurred faces in schizophrenic patients and healthy volunteers *Schizophr. Res.* **49** 145–55

[52] Jeong J H, Kim D J, Chae J H, Kim S Y, Ko H J and Paik I H 1998 Nonlinear analysis of the EEG of schizophrenics with optimal embedding dimension *Med. Eng. Phys.* **20** 669–76

[53] Bob P 2012 Consciousness, schizophrenia and complexity *Cogn. Syst. Res.* **13** 87–94

[54] Vasios C E, Papageorgiou C, Matsopoulos G K, Nikita K S and Uzunoglu N 2003 A decision support system of evoked potentials for the classification of patients with first-episode schizophrenia *Ger. J. Psychiatry* **5** 05

[55] Cerquera A, Gjini K, Bowyer S M and Boutros N 2017 Comparing EEG nonlinearity in deficit and nondeficit schizophrenia patients: preliminary data *Clin. EEG Neurosci.* **48** 376–82

[56] Sui J, Castro E, He H, Bridwell D A, Du Y, Pearlson G D, Jiang T and Calhoun V D 2014 Combination of FMRI-SMRI-EEG data improves discrimination of schizophrenia patients by ensemble feature selection *2014 36th Annual Int. Conf. of the IEEE Engineering in Medicine and Biology Society* pp 3889–92

[57] Aydın S, Arica N, Ergul E and Tan O 2015 Classification of obsessive compulsive disorder by EEG complexity and hemispheric dependency measurements *Int. J. Neural Syst.* **25** 1550010

[58] Akar S A, Kara S, Latifoğlu F and Bilgiç V 2016 Analysis of the complexity measures in the EEG of schizophrenia patients *Int. J. Neural Syst.* **26** 1650008

[59] Fan F-Y, Li Y-J, Qiu Y-H and Zhu Y-S 2005 Use of ANN and complexity measures in cognitive EEG discrimination *2005 IEEE Engineering in Medicine and Biology 27th Annu. Conf.*

[60] Abhishekh H A, Thirthalli J, Manjegowda A, Phutane V H, Muralidharan K and Gangadhar B N 2013 Ictal EEG fractal dimension in ECT predicts outcome at 2 weeks in schizophrenia *Psychiatry Res.* **209** 155–9

[61] Abhishekh H A, Thirthalli J, Manjegowda A, Phutane V H, Muralidharan K and Gangadhar B N 2013 Ictal EEG fractal dimension in ECT predicts outcome at 2 weeks in schizophrenia *Psychiatry Res.* **209** 155–9

[62] Khodayari-Rostamabad A, Reilly J P, Hasey G, deBruin H and MacCrimmon D 2010 Diagnosis of psychiatric disorders using EEG data and employing a statistical decision model *2010 Annual Int. Conf. of the IEEE Engineering in Medicine and Biology* pp 4006–9

[63] Julieta R, Luis F, Cerdán M A, Guevara C A and Araceli S 2001 Abnormal EEG patterns in treatment-resistant schizophrenic patients *Int. J. Neurosci.* **109** 47–59

[64] Kaplan A Y, Borisov S V and Zheligovskiy V A 2005 Classification of the adolescent EEG by the spectral and segmental characteristics for normals *Zh. Vyssh. Nerv. Deiat. Im. I P Pavlova* **55** 478–86

[65] AliMardani F, Boostani R and Blankertz B 2016 Presenting a spatial-geometric EEG feature to classify BMD and schizophrenic patients *Int. J. Adv. Telecommun. Electrotech. Signals Syst.* **5** 79–85

[66] Youn T, Park H-J, Kim J-J, Kim M S and Kwon J S 2003 Altered hemispheric asymmetry and positive symptoms in schizophrenia: equivalent current dipole of auditory mismatch negativity *Schizophr. Res.* **59** 253–60

[67] Shagass C, Roemer R A, Straumanis J J and Josiassen R C 1984 Psychiatric diagnostic discriminations with combinations of quantitative EEG variables *Br. J. Psychiatry* **144** 581–92

[68] Lou E, Zhang S and Qiao S 2009 Application of eigenvector estimation and SVM for EEG signals classification *2009 9th Int. Conf. on Electronic Measurement & Instruments*

[69] Raghavendra B S, Dutt D N, Halahalli H N and John J P 2009 Complexity analysis of EEG in patients with schizophrenia using fractal dimension *Physiol. Meas.* **30** 795–808

[70] Zhang S, Qiao S and Wang W 2010 Classification of schizophrenia's EEG based on high order pattern discovery *2010 IEEE Fifth Int. Conf. on Bio-Inspired Computing: Theories and Applications (BIC-TA)*

[71] Piryatinska A, Darkhovsky B and Kaplan A 2017 Binary classification of multichannel-EEG records based on the ϵ-complexity of continuous vector functions *Comput. Methods Programs Biomed.* **152** 131–9

[72] Sabeti M, Mohammad H S and Pric G W 2007 Fuzzy accuracy-based classifier systems for EEG classification of schizophrenic patients *First Joint Congress on Fuzzy and Intelligent Systems*

[73] Phang C, Ting S, Samdin B and Ombao H 2019 Classification of EEG-based effective brain connectivity in schizophrenia using deep neural networks *9th Int. IEEE/EMBS Conf. on Neural Engineering (NER)*

[74] Li F *et al* 2019 Differentiation of schizophrenia by combining the spatial EEG brain network patterns of rest and task P300 *IEEE Trans. Neural Syst. Rehabil. Eng.* **27** 594–602

[75] Chu W, Huang M, Jian B and Cheng K 2017 Analysis of EEG entropy during visual evocation of emotion in schizophrenia *Ann. Gener. Psychiatry* **16** 34

[76] Yin Z, Li J, Zhang Y, Ren A, Von Meneen K M and Huang L 2017 Functional brain network analysis of schizophrenic patients with positive and negative syndrome based on mutual information of EEG time series *Biomed. Signal Process. Control* **31** 331–8

[77] Kirino E and Inoue R 1999 The relationship of mismatch negativity to quantitative EEG and morphological findings in schizophrenia *J. Psychiatr. Res.* **33** 445–56

[78] Tikka S K, Yadav S, Nizamie S H, Das B, Tikka D L and Goyal N 2014 Schneiderian first rank symptoms and gamma oscillatory activity in neuroleptic naïve first episode schizophrenia: a 192 channel EEG study *Psychiatry Invest.* **11** 465–75

[79] Sabeti M, Behroozi R and Moradi E 2016 Analysing complexity, variability and spectral measures of schizophrenic EEG signal *Int. J. Biomed. Eng. Technol.* **21** 109

[80] Lifshitz K, Lee K L and Susswein S 1987 Long-term replicability of EEG spectra and auditory evoked potentials in schizophrenic and normal subjects *Neuropsychobiology* **18** 205–11

[81] Liu H, Zhang T, Ye Y, Pan C, Yang G, Wang J and Qiu R C 2017 A data driven approach for resting-state EEG signal classification of schizophrenia with control participants using random matrix theory, arXiv:1712.05289

[82] Röschke J, Fell J and Beckmann P 1995 Nonlinear analysis of sleep EEG data in schizophrenia: calculation of the principal Lyapunov exponent *Psychiatry Res.* **56** 257–69

[83] Merrin E L, Meek P, Floyd T C and Callaway E 1990 Topographic segmentation of waking EEG in medication-free schizophrenic patients *Int. J. Psychophysiol.* **9** 231–6

[84] Thilakvathi B, Shenbagadevi S, Bhanu K S and Malaippan M 2017 EEG signal complexity analysis for schizophrenia during rest and mental activity *Biomed. Res.* **28** 1–9

[85] Dvey-Aharon Z, Fogelson N, Peled A and Intrator N 2015 Schizophrenia detection and classification by advanced analysis of EEG recordings using a single electrode approach *PLoS One* **10** e0123033

[86] Kim J W, Lee Y S, Han D H, Min K J and Lee K 2015 Diagnostic utility of quantitative EEG in un-medicated schizophrenia *Neurosci. Lett.* **589** 126–31

[87] Dvey-A Z, Fogelson N, Peled A and Intrator N 2017 Connectivity maps based analysis of EEG for the advanced diagnosis of schizophrenia attributes *PLoS One* **12** e0185852

[88] Buettner R, Beil D, Scholtz S and Djemai A 2020 Development of a machine learning based algorithm to accurately detect schizophrenia based on one-minute EEG recordings *HICSS-53 Proc.: 53nd Hawaii Int. Conf. on System Sciences*

[89] Etevenon P 1984 Intra and inter-hemispheric changes in alpha intensities in EEGs of schizophrenic patients versus matched controls *Biol. Psychol.* **19** 247–56

[90] Al-Fahoum A S and Al-Fraihat A 2014 Methods of EEG signal features extraction using linear analysis in frequency and time-frequency domains *ISRN Neurosci.* **2014** 730218

[91] Khaleghi A, Sheikhani A, Mohammadi M R, Nasrabadi A M, Vand S R, Zarafshan H and Moeini M 2015 EEG classification of adolescents with type I and type II of bipolar disorder *Australas. Phys. Eng. Sci. Med.* **38** 551–9

[92] Bachiller A *et al* 2014 Decreased entropy modulation of EEG response to novelty and relevance in schizophrenia during a P300 task *Eur. Arch. Psychiatry Clin. Neurosci.* **265** 525–35

[93] Hiesh M, Lam A Y, Shen C, Chen W, Lin F, Sung H, Lin J, Chiu M and Lai F 2013 Classification of schizophrenia using genetic algorithm-support vector machine (GA-SVM) *2013 35th Annual Int. Conf. of the IEEE Engineering in Medicine and Biology Society (EMBC)* pp 6047–50

[94] Nasehi S and Pourghassem H 2011 A novel effective feature selection algorithm based on S-PCA and wavelet transform features in EEG signal classification *2011 IEEE 3rd Int. Conf. on Communication Software and Networks*

[95] Doege K, Bates A T, White T P, Das D, Boks M P and Liddle P F 2009 Reduced event-related low frequency EEG activity in schizophrenia during an auditory oddball task *Psychophysiology* **46** 566–77

[96] Akar S A, Kara S, Latifoglu F and Bilgic V 2012 Wavelet–Welch methodology for analysis of EEG signals of schizophrenia patients *2012 Cairo Int. Biomedical Engineering Conf. (CIBEC)*

[97] Pauline A D 1942 Comparative study of the EEGs of schizophrenic and manic-depressive patients *Am. J. Psychiatry* **99** 210–7

[98] Haring C, Neudorfer C, Schwitzer J, Hummer M, Saria A, Hinterhuber H and Fleischhacker W W 1994 EEG alterations in patients treated with clozapine in relation to plasma levels *Psychopharmacology* **114** 97–100

[99] Zhao Q, Hu B J, Li Y, Peng H, Li L, Liu Q, Li Y D, Shi Q and Feng J 2013 An alpha resting EEG study on nonlinear dynamic analysis for schizophrenia *2013 6th Int. IEEE/EMBS Conf. on Neural Engineering (NER)* pp 484–8

[100] Ziqiang Z and Puthusserypady S 2007 Analysis of schizophrenic EEG synchrony using empirical mode decomposition *2007 15th Int. Conf. on Digital Signal Processing*

[101] Phang C R, Ting C M, Numan F and Ombao H 2019 Classification of EEG-based brain connectivity networks in schizophrenia using a multi-domain connectome convolutional neural network, arXiv:1903.08858

[102] Ford J M, Palzes V A, Roach B J and Mathalon D H 2014 Did I do that? Abnormal predictive processes in schizophrenia when button pressing to deliver a tone *Schizophr. Bull.* **40** 804–12

[103] Gilles J 2013 Empirical wavelet transform *IEEE Trans. Signal Process.* **61** 3999

[104] Song Y Y and Lu Y 2015 Decision tree methods: applications for classification and prediction *Shanghai Arch. Psychiatry* **27** 130–5

[105] Awad M and Khanna R 2015 Support vector machines for classification *Efficient Learning Machines* (Berkeley, CA: Apress)

[106] Adeniyi D A, Wei Z and Yong-quan Y 2016 Automated web usage data mining and recommendation system using *k*-nearest neighbor (KNN) classification method *Appl. Comput. Inform.* **12** 1

[107] Bühlmann P 2012 Bagging, boosting and ensemble methods *Handbook of Computational Statistics* (Berlin: Springer)

[108] Tharwat A 2016 Linear vs quadratic discriminant analysis classifier: a tutorial *Int. J. Appl. Pattern Recogn.* **3** 145

IOP Publishing

Modelling and Analysis of Active Biopotential Signals in Healthcare, Volume 1

Varun Bajaj and G R Sinha

Chapter 2

Fuzzy scale invariant feature transform phase locking value and its application to PTSD EEG data

Zahra Ghanbari, Mohammad H Moradi

Biomedical Department, Amirkabir University of Technology, Tehran, Iran

In this study we introduce a novel powerful phase synchrony index called the scale invariant feature transform phase locking value (FSIFT-PLV). This index benefits from robust SIFT descriptors, a phase estimation approach based on reduced interference Rihaczek distribution and a proposed fuzzy framework. FSIFT-PLV can detect bivariate phase synchrony properly in the presence of noise, volume conductance, common reference and sample size bias, as well as small shifts in time. It is applied to electroencephalogram (EEG) signals recorded from combat related post-traumatic stress disorder (PTSD) veterans and two control groups, including trauma exposed non-PTSD veterans and healthy controls who have not experienced any trauma including war. PTSD is a chronic debilitating disorder which may occur as a result of life-threatening mental trauma. The EEG signals are recorded in two resting states ('eyes-open' and 'eyes-closed'). FSIFT-PLV is used to generate functional connectivity matrices. ANOVA is applied to extracted features at a confidence level of 99%.

Our study possesses some unique properties: (i) investigating patients who have experienced PTSD symptoms for more than 30 years, (ii) considering trauma exposed non-PTSD veterans as the second control group, (iii) studying the resting state in both the eyes-open and eyes-closed conditions, and (iv) introducing a novel powerful phase synchrony and applying it to EEG signals.

doi:10.1088/978-0-7503-3279-8ch2

2.1 Introduction

This section provides a brief introduction to the background of post-traumatic stress disorder (PTSD), as well as resting state eyes-closed and eyes-open EEG signals. In the following, we will mention some points about phase synchrony calculation.

According to the American Psychiatric Association (APA), post-traumatic stress disorder (PTSD) is defined as a psychiatric disorder which may follow a traumatic event [1]. PTSD is a chronic debilitating anxiety condition which is characterized by unremitting distressing repetition of the traumatic experience, avoidance, hyper-arousal, hyper-vigilance, dissociation, emotional numbing and negative alteration in cognition [2].

fMRI studies have reported functional abnormalities in the cortical and sub-cortical regions of PTSD patients' brains [3, 4]. Functional connectivity is a powerful approach for investigating the brain as a sophisticated network. Many disorders have been studied using functional connectivity, including PTSD [5–8]. Although studying abnormal patterns in the presence of emotional elicitation is of great importance, dysregulated patterns of the resting state functional connectivity may provide valuable knowledge about PTSD pathology [9, 10]. Correlations embedded in haemodynamic activity levels among various brain regions are addressed by resting state functional connectivity. Functional connectivity uses synchronization of the neural activation of the aforementioned regions [11], which can be calculated using different modalities, based on various methods.

EEG is a low cost, commonly available modality with high temporal resolution. It can provide optimal observation of brain activities [12, 13]. Only a few studies have focused on the resting state functional connectivity of PTSD patients based on EEG signals. Resting state EEG can be recorded in two forms, including resting state eyes-closed (REC), and resting state eyes-open (REO). REC and REO have different basic properties. For example, the alpha band EEG is defined using synchronization in REC but desynchronization in REO as the feature [14]. Comparing REC and REO demonstrates differences in the delta, theta, alpha and beta sub-bands, both in adult and young participants [15]. One hypothesis states that the differences between REC and REO can be translated as a signal which reflects the brain's activity in response to visual stimuli. To test this hypothesis, REC and REO EEG signals have been recorded in a completely dark environment. The reported results imply significantly different spectral powers and coherence values in the delta, theta, alpha1, alpha2, beta1, beta2 and gamma sub-bands. These findings suggest that the differences between REC and REO are independent of external stimuli to the visual system. This study also proposes that such differences actually reflect externally directed attention and, in contrast, internally directed attention, specific to REO and REC, respectively [16].

In this study we aim to examine the REC and REO signals recorded from combat related PTSD participants, and two control groups including combat trauma exposed non-PTSD participants and healthy controls who have not experienced any serious trauma including war. For this purpose, we will use functional connectivity. To make the connectivity matrices, a reliable measure of synchrony is needed.

Synchronization is defined as the coordination among coupled oscillators [17]. In chaotic systems synchronization is of great importance and it has found applications in many fields, including signal processing [18]. Synchronization can be quantified using linear or nonlinear indices between simultaneously recorded signals [19]. Phase synchrony, as a nonlinear measure, has received much attention in neuroscience as it can be detected in the brain as a complex system [20–22]. Synchronization is defined as the temporal adjustment of rhythms between two oscillators [22]. Phase estimation is the first step in calculating phase synchrony, which can be a function of time or time and frequency. Although there are many methods for extracting time and frequency dependent phases, we will use a phase estimation method based on the reduced interference Rihaczek distribution [23]. The phase locking value (PLV) is a measure for evaluating bivariate phase synchrony [24]. Four main issues can rise in quantifying phase synchrony, due to noise, using a common reference, sample size bias and volume conductance of source activity [25]. Many efforts have been made to solve these problems, which have resulted in state-of-the-art phase synchrony measures. The imaginary part of coherency (ImC) was proposed as a conservative phase synchronization measure [25–27]. However, it is highly affected by the phase of coherency which results in weak performance in the case of two in-phase or in-phase-opposition sources. The phase lag index (PLI) was introduced to solve the issue. The PLI reveals higher performance in detecting phase synchronization and in the presence of volume conduction, compared to ImC. However, its discontinuity causes some problems. Therefore, the weighted phase lag index (wPLI) and the debiased wPLI (dwPLI) were proposed as modified versions [25]. wPLI uses the imaginary part of the cross spectrum as the weights corresponding to the contribution of the observed phase leads and lags. wPLI outperforms PLI, in particular for additional uncorrelated sources. The imaginary part of PLV (iPLV) and its corrected counterpart (ciPLV) were proposed as faster versions of PLV obtained via reformulation. Not only are they 100 times faster, they possess some additional properties, including insensitivity to zero lag synchronization and robustness against source leakage effects and volume conduction [28].

To modify the detected phase synchrony, in particular in the presence of the four main known issues (noise, volume conductance, common reference and sample size bias), a novel phase synchrony measure is proposed, called the fuzzy scale invariant feature transform phase locking value (FSIFT-PLV) [29]. FSIFT-PLV benefits from a powerful phase estimation method called the reduced interference Rihaczek distribution, in addition to the scale invariant feature transform (SIFT). The SIFT provides distinctive invariant features of images [30, 31]. Despite its strengths, SIFT has shortcomings. FSIFT-PLV uses a fuzzy framework to compensate for these shortcomings, in addition to adapting it for our particular purpose.

In this chapter FSIFT-PLV is introduced. Then, its ability in capturing actual phase synchrony values will be assessed using synthetic signals. Rossler oscillators are used for generating synthetic EEG signals. The next step is studying the robustness of the proposed measure against the four underlying problems in quantifying phase synchrony: noise, volume conduction, common reference and sample size bias. Moreover, the problem of the presence of small time shifts and the robustness of FSIFT-PLV against this will be addressed. After studying the proposed measure in comparison with state-of-the-art measures, it will be applied to real EEG data.

The rest of this chapter is organized as follows. The next section will cover the methods. Section 2.3 is dedicated to the EEG data. The results are reported in section 2.4. Finally, our conclusion is provided in the last section.

2.2 Method

Our proposed measure, FSIFT-PLV, will be discussed in this section. The following sub-sections are dedicated to introducing some graph features and the statistical approach which will be used in the assessment of the results.

2.2.1 FSIFT-PLV

As FSIFT-PLV is a combination of different parts, in this section each part will be discussed briefly. The following sub-sections include phase estimation, the scale invariant feature transform and the combination method, as well as the proposed fuzzy framework.

2.2.1.1 Phase estimation

The phase locking value (PLV) is an important approach for calculating phase synchrony between pairwise signals. Moreover, it is part of our proposed method. Thus, it will be reviewed briefly here. PLV is defined as

$$\mathrm{PLV}(t,\ \omega) = \frac{1}{N}\left|\sum\nolimits_{k=1}^{N} \mathrm{e}^{j\varphi_{1,2}^{k}(t,\ \omega)}\right|, \tag{2.1}$$

where N stands for the number of trials and $\varphi_{1,\ 2}^{k}(t,\ \omega)$ represents the estimated phase difference between pairwise signals of the kth trial. In perfect synchronization this measure is equal to 1, and it will approach zero in the case of independent oscillators [24]. We use the complex reduced inference distribution (RID)–Rihaczek distribution based method for phase synchrony estimation, which is formulated as [32]

$$C(t,\ \omega) = \int_{-\infty}^{\infty}\int_{-\infty}^{\infty} \mathrm{e}^{-\frac{(\theta\tau)^2}{\sigma}}\mathrm{e}^{\frac{j\theta\tau}{2}} A(\theta,\ \tau)\mathrm{e}^{-j(\theta t+\tau\omega)}\mathrm{d}\tau\mathrm{d}\theta, \tag{2.2}$$

where $\mathrm{e}^{-\frac{(\theta\tau)^2}{\sigma}}$ and $\mathrm{e}^{\frac{j\theta\tau}{2}}$ represent the Choi–Williams and Rihaczek kernels, respectively. $A(\theta,\ \tau)$ is the ambiguity function of the signal and is defined as

$$A(\theta,\ \tau) = \int_{-\infty}^{\infty} s\left(u + \frac{\tau}{2}\right)s^{*}\left(u - \frac{\tau}{2}\right)\mathrm{e}^{j\theta u}\mathrm{d}u. \tag{2.3}$$

Based on the above formulation, the time–frequency dependent phase will be

$$\Phi(t, \omega) = \left[\frac{C(t, \omega)}{| C(t, \omega)|}\right]. \tag{2.4}$$

2.2.1.2 Scale invariant feature transform (SIFT)

The scale invariant feature transform (SIFT) as a powerful method for feature extraction from images is discussed in this section. Extracting features from images is done using the following steps: (1) selecting scale space peaks, (2) localizing key-points, (3) assigning orientations and (4) generating key-point descriptors.

The scale space of an image, $L(x, y, \sigma)$, corresponding to a given image, $I(x, y)$, is as follows:

$$L(x, y, \sigma) = G(x, y, \sigma) * I(x, y), \tag{2.5}$$

where G is a variable-scale Gaussian kernel and is equal to

$$G(x, y, \sigma) = \frac{1}{2\pi\sigma^2} e^{-\frac{x^2+y^2}{2\sigma^2}}. \tag{2.6}$$

For finding stable key-points in the scale space, the difference of Gaussian (DoG) is used. The DoG is generated from the nearby scales,

$$D(x, y, \sigma) = I(x, y) * (G(x, y, k\sigma) - G(x, y, \sigma)) = L(x, y, k\sigma) - L(x, y, \sigma), \tag{2.7}$$

where k is a constant. As the most stable features of the image are provided by maxima and minima of $\sigma^2\nabla^2 G$, the following approximation is considered:

$$G(x, y, k\sigma) - G(x, y, \sigma) \approx (k - 1)\sigma^2\nabla^2 G. \tag{2.8}$$

The scale space peaks of DoG can be located by searching 3×3 regions including 26 pixels. According to the above formulation, this is equal to finding local maxima and minima of $D(x, y, \sigma)$. For accurate localization of key-points, it is essential to apply a detailed fit to nearby data corresponding to each peak candidate:

$$D(\mathbf{x}) = D + \frac{\partial D^T}{\partial \mathrm{x}} + \frac{1}{2}\frac{\mathrm{x}^T \partial^2 D}{\partial \mathrm{x}^2}\mathbf{x}, \tag{2.9}$$

where $\mathbf{x} = (x, y, \sigma)^T$ represents the offset from the sample point. By setting the derivative of D, which is calculated with respect to $\mathbf{x}$, to zero, the extremum ($\hat{\mathbf{x}}$) will be obtained. Differences of adjacent sample points are used to approximate the derivative and Hessian of D. Cosequently, we have a 3×3 linear system which can be easily solved with minimal cost.

Unstable peaks with low contrast are rejected by checking the function value at peaks. On the other hand, rejecting key-points which have low contrast will not be sufficient for stability. Since DoG shows a strong response along the edges, the principle curvatures will be computed. For this purpose $\mathbf{H}$, which is a 2×2 Hessian matrix, will be computed at the scale and location of key-points. A threshold value, r, is employed to filter inappropriate values. r is defind as the ratio of the largest and the smallest eigen-values. If

$$\frac{\mathrm{Tr}(\mathbf{H})^2}{\mathrm{Det}(\mathbf{H})} < \frac{(r+1)^2}{r} \tag{2.10}$$

holds, it automatically means that the ratio of the principle curvature is below r.

A consistent orientation is assigned to each key-point, based on the local image properties. Thus the descriptor corresponding to a key-point can be represented based on this orientation. Consequently, invariance to image rotation will be achieved. Based on pixel differences, corresponding to each image sample, $L_{x,y}$, orientation, θ is computed as

$$\theta = \tan^{-1}\left(\frac{L_{x,\,y+1} - L_{x,y-1}}{L_{x+1,\,y} - L_{x-1,y}}\right). \tag{2.11}$$

Similarly, the gradient magnitude, m, is obtained from

$$m = \sqrt{(L_{x+1,y} - L_{x-1,y})^2 + (L_{x,y+1} - L_{x,y-1})^2}\,. \tag{2.12}$$

An orientation histogram will be generated from the gradient orientations, associated with each key-point. It contains 36 bins, thus covering the 360° range of orientations. This will be done within a circular area around each sample point, and for all of them. Each sample will be added to the histogram with a weight of its gradient magnitude, in addition to a Gaussian-weighted circular window. In the orientation histogram, peaks are associated with the local gradients' dominant directions. The highest local peak will be found. At the next step, a key-point with the mentioned orientation will be created using other local peaks within 80% of the highest peak. To achieve higher accuracies, a parabola is fitted to the three histogram values around each peak. This will help to interpolate the peak position.

The last step is generating the key-point descriptor as well as representing it. The orientations and magnitudes of the image gradient are sampled around a key-point. A Gaussian weighting function is used to assign a proper weight to the magnitude of each sample point. By creating the orientation histogram over 4×4 sample regions, the descriptor allows for a significant shift in gradient positions. Eight directions exist corresponding to each orientation histogram. A weight is assigned to each histogram entry for the purpose of preventing boundary effects. Further descriptions can be found in [30].

2.2.1.3 FSIFT-PLV

FSIFT-PLV is brought up in this section. As the first step, phase estimation is done using the reduced interference distribution (RID)–Rihaczek distribution. This will provide phases as a function of time and frequency. Therefore, we have time–frequency maps which are considered as images. The SIFT is applied to the obtained maps to extract key-points and generate their corresponding descriptors. Although,

SIFT features are highly distinctive and robust against noise, they have properties which are not applicable here. SIFT descriptors are invariant to scale and rotation. Moreover, they are robust against illumination changes and affine distortion. More importantly, as a weakness, the SIFT may find matched key-points which belong to different regions of two images [3]. Here we are dealing with images which are actually time–frequency planes corresponding to the estimated phase of simultaneously recorded EEG signals. Therefore, it is necessary to restrict matched key-points so that key-points belonging to far away locations in time or frequency are not accepted as matches. Consequently, a fuzzy framework is added to the FSIFT-PLV calculating procedure. The proposed fuzzy framework has three inputs associated with each pair of key-points. The first input is a similarity measure achieved using SIFT matching, which is actually a distance index between descriptors of pairwise key-points. An index of Euclidean distance is used as the second input. This is the distance between locations of key-points in two maps. Although using the fuzzy framework with these two entries improves the results, we did not achieve the ultimate goal, i.e. generating a phase synchrony measure which reflects the actual values of phase synchrony. Therefore, another entry is added to our system. The third input is the initial estimation of similarity between two planes. Our proposed formulation for calculating this measure is as follows. First, standard SIFT matching is applied. Then, far away matched key-points are removed according to a chosen threshold. Two points should be emphasized about choosing the threshold. (i) This is a very primary inaccurate filtering which does not need a precise value as the threshold. This threshold is chosen based on the data. As here we are working on simultaneously recorded EEG signals, two matched key-points could not have distance of more than 500 ms in time. After removing far away match key-points the third input will be defined as

$$\frac{\text{remaining matched key-points}}{\text{minimum No. of key-points founded in two planes}}. \tag{2.13}$$

Since our fuzzy framework is designed with inputs in [0, 1], the above inputs should be normalized in [0, 1] if needed. The membership functions associated with the three aforementioned inputs are Gaussians, indicating low and high for the second and third inputs, and low, medium and high for the first input. For low and high values the [0, 1] interval is divided into two equal parts. Similarly, for the low, medium and high values the [0, 1] interval is divided into three equal parts. 'Mamdani' is the fuzzy system used. In addition, the 'and' method used is product. Additionally, the centroid approach is applied for de-fuzzification.

Similar to the entries, outputs are defined using Gaussian functions. The output of the fuzzy system is in [0, 1]. Therefore, for the output value [0, 1] is equally divided into 12 parts. The fuzzy system rules are summarized in table 2.1. L, M and H indicate low, medium and high, respectively, corresponding to the input values. For the output values, regions 1–12 are associated with the first to twelfth parts of the [0, 1] interval, respectively.

Table 2.1. Fuzzy system rules [29].

Values												
Input 1	L	L	L	L	M	M	M	M	H	H	H	H
Input 2	L	L	H	H	L	L	H	H	L	L	H	H
Input 3	L	H	L	H	L	H	L	H	L	H	L	H
Output	reg. 1	reg. 12	reg. 1	reg. 4	reg. 3	reg. 12	reg. 1	reg. 8	reg. 5	reg. 12	reg. 1	reg. 10

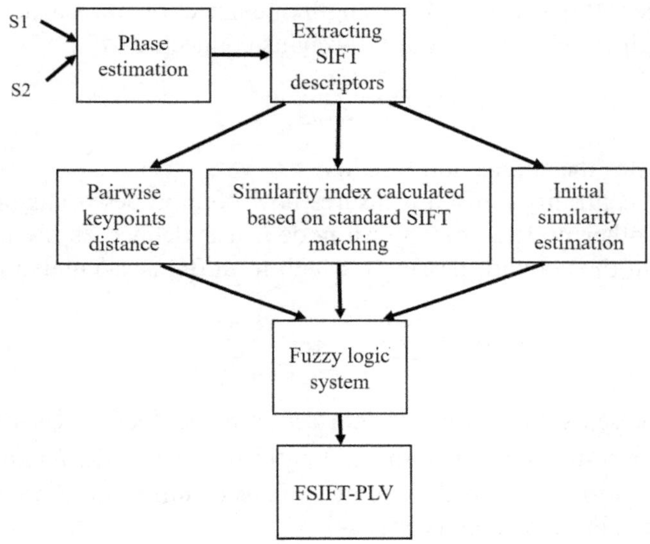

Figure 2.1. The overall block diagram of FSIFT-PLV [29].

To have more reliable parameters, the output parameters are optimized using particle swarm optimization (PSO) as a powerful evolutionary algorithm [33]. The centers of the Gaussian functions of the output are optimized, as well as the variances. The overall block diagram of FSIFT-PLV is illustrated in figure 2.1.

The final output of FSIFT-PLV is a bivariate phase synchrony measure in [0, 1], which is equal to 1 in the case of perfect synchronization and approaches 0 in the case of lacking synchrony.

2.2.2 Functional connectivity graph indices

Calculated phase synchronization between pairwise channels are used to generate connectivity graphs. To assess the obtained graphs, the three following nodal centrality measures are extracted from networks: the nodal efficiency (a measure of communication efficiency), nodal degree and betweenness centrality [34]. These features are defined as follows.

For a given node i, nodal efficiency (NE), is defined as the inverse of the harmonic mean of the shortest path length between i and all other nodes of the network. It is calculated as

$$\mathrm{NE}(i) = \frac{1}{n-1}\sum_{j\in N,\ j\neq i}\frac{1}{\sum_{wst\in g^{w}_{i\to j}} f(wst)}, \tag{2.14}$$

where f is length to weight, and $g^{w}_{i\to j}$ represents the path with the shortest weight between i and j.

Nodal degree (ND) measures the strength of connection at a particular node. For a given node i, it is defined as the total weights connected to it:

$$\mathrm{ND}(i) = \sum_{j\in N} w_{ij}, \tag{2.15}$$

where w_{ij} indicates the connection weights.

Betweenness centrality (unBC) measures how often nodes occur on the shortest paths between other nodes. For a given node i, it is defined as the fraction of the shortest paths in the network passing through it. unBC is calculated as

$$\mathrm{unBC} = \sum_{\substack{h,\, j\in N \\ h\neq j,\ h\neq i,\ j\neq i}} \frac{g_{hj}(i)}{g_{hj}}, \tag{2.16}$$

where $g_{hj}(i)$ represents the number of shortest paths between nodes h and j passing through i. g_{hj} is also the number of shortest paths between nodes h and j. The above measure can be normalized to the average values of unBCs in a network. We will denote this normalized version as BC [34].

2.3 Data

As mentioned before, FSIFT-PLV is first applied to the synthetic data to assess its ability to accurately capture phase synchrony values. In addition, its robustness will be studied. Then, it will be applied to real EEG data. The synthetic and real EEG data will be presented in this section.

2.3.1 Synthetic data

Time series obtained from numerical simulation using the Rossler system are used as the synthetic EEG data. The irregular sine-like waves generated by the Rossler system are similar to EEG signals. Therefore, they are used in our study. The Rossler oscillator is defined by the following equations [35]:

$$\begin{cases} \dfrac{dx}{dt} = -\omega y - z \\ \dfrac{dy}{dt} = \omega x + ay \\ \dfrac{dz}{dt} = d + z(x - c), \end{cases} \tag{2.17}$$

where (x, y, z) represents the oscillator state, ω stands for the oscillator natural frequency, and a, c and d are its characterizing parameters. In a given network, nodes stand for individual oscillators, and their pairwise coupling are represented by links. Couplings are considered through the x-state. The following equations define the motion of the coupled oscillators:

$$\begin{cases} \dfrac{dx_i}{dt} = -\omega_i y_i - z_i + \sigma \sum_{j=1}^{N} A_{ij}(x_j - x_i) + R \\ \dfrac{dy_i}{dt} = \omega_i x_i + a_i y_i \\ \dfrac{dz_i}{dt} = d_i + z_i(x_i - c_i), \end{cases} \tag{2.18}$$

$$i = 1, \ldots, N,$$

where σ represents the unified coupling strength between pairwise oscillators, A is the adjacency matrix associated with the connection graph and R stands for the noise. The nominal parameters of the oscillators are set to $\omega = 1, a = 0.165, c = 10$ and $d = 0.2$. Furthermore, a 10% mismatch is considered between oscillator parameters. Thus we will have non-identical oscillators. It should be emphasized that for Rossler oscillators, increasing coupling strength over a particular threshold may result in a de-synchronized regime. Therefore, parameter selection is done such that the gradient of the Lyapunov exponent remains negative while increasing the coupling strength [36]. This will guarantee that the system remains in the synchronized regime. Consequently, increasing the coupling strength just increases the synchronization. The connection structure is generated using Erdos–Renyi (ER) networks with $N = 25$ [37]. Each pair of nodes in the generated networks is connected by a link with a probability of $P = 0.1$. The networks' differential equations are iterated starting from random initial conditions using the Heun algorithm [11]. $T_f = 300$ time units is considered as the simulation time. In addition, $d_T = 0.01$ is used as the integration time step. Time series associated with the x-component are used as synthetic EEG signals. A number of initial iterations are neglected for the purpose of transient elimination. Finally, the remaining signals are segmented into 25 equal length parts [38].

2.3.2 EEG data

This sub-section is dedicated to introducing EEG data which are recorded in the resting state with eyes-open and eyes-closed. In the following, the participants, EEG recording and pre-processing will be explained in detail.

2.3.2.1 Participants

Forty five right-handed men were involved in our study. Fifteen participants were diagnosed with chronic PTSD. Fifteen participants were trauma exposed but non-PTSD. All 30 had participated the Iraq–Iran war (September 1980–August 1988). The rest of the participants were healthy non-trauma exposed controls.

The exclusion criteria for our study included psychosis, current substance abuse or dependence, and a history or current symptoms of a diagnosed neurological disorder (i.e. head injury, stroke, etc). The participants of the two control groups (trauma exposed non-PTSD participants and healthy non-trauma exposed participants) were recruited from the local community (i.e. Amirkabir University of Technology and Sadr Hospital, Tehran, Iran). The participants of the healthy non-trauma exposed group did not express any history of major trauma, including physical injury or surgery, serious disease, sexual assault, car accident or combat experience. Moreover, the participants of the two control groups were not taking medications with potential psychoactive effects.

All participants underwent a physical examination by a psychiatrist and a psychologist to classify them as PTSD or non-PTSD participants. Moreover, they were asked to answer a structured clinical interview from the *Diagnostic and Statistical Manual of Mental Disorder (*fifth edition). In addition, participants were asked to answer a 20 item self-reported questionnaire PTSD check-list (PCL). It monitors any post-traumatic intrusion symptoms, avoidance and/or hyper-arousal experienced in the previous week associated with a particular trauma, and measures their severity. The PCL questionnaire rates each item on a scale of 0–4. Participants are also asked to answer the depression, anxiety stress scales (dass-21) questionnaire. Dass-21 is a 21 item self-reported questionnaire. It measures anxiety, stress and depressive severity by monitoring the last two weeks. It rates each item on a scale of 0–3. The clinical and demographic characteristics for the three groups of participants are summarized in table 2.2. For the sake of simplicity, in the following the combat related PTSD, combat exposed non-PTSD and non-trauma exposed healthy controls will be described as PTSD, non-PTSD and control, respectively.

Ethical approval for our study was provided by Amirkabir University of Technology, and Shahid Beheshti University on behalf of Sadr Hospital, Tehran, Iran. After a full description of the study procedure, a written consent form was signed by all participants. Furthermore, they received financial compensation for their participation at the end of the session.

2.3.2.2 EEG recording and pre-processing

The EEG signals were recorded in a quiet room with normal lighting. Data acquisition was performed in 16 channels, with a 1200 Hz sampling rate, using an active electrode g.Tec system. The recording sites were AF3, AF4, F3, Pz, F4, F8, FC3, FC4, T8, CP3, P5, P6, PO3, PO4, O1 and O2 according to the standard 10/20 system. Fz and the left ear were used as the ground and reference, respectively. For the sake of simplicity, in the following the above recording sites will be described as channels 1–16, respectively. The system filters have cut-off frequencies of 0.1 Hz and

Table 2.2. The clinical and demographic characteristics of participants.

Characteristics	PTSD (n = 15) Mean (± SD)	Non-PTSD (n = 15) Mean (± SD)	Control (n = 15) Mean (± SD)
Age (years)	52.818 (±4.283)	53.111 (±2.0276)	50.867 (±3.870)
Marital state (1: married; 0: divorced)	0.818 (±0.395)	1 (±0)	1 (±0)
Number of children	2.500 (±1.012)	2.444 (±0.7265)	1.800 (±0.775)
Years of education	9.000 (±3.532)	14.778 (±1.856)	11.867 (±2.475)
Age of trauma	18.591 (±3.202)	17.556 (±1.236)	—
Depression and anxiety	50.046 (±6.897)	8.667 (±6.801)	9.133 (±3.829)
PTSD	55.476 (±12.734)	7.650 (±4.012)	—

100 Hz. Additionally, a 50 Hz notch filter is used during data acquisition. The impedance of the electrodes was less than 5 kΩ while recording the data. The EEG signals were recorded for 5 min during which the participants were asked to relax and try not to think (or to not think about a particular topic). Moreover, they were asked to avoid moving their heads up and down or turning their heads around. A neurologist checked the spontaneous signals. No focal abnormalities, epileptic discharges or any other relevant modifications were reported.

In order to avoid unwanted bias due to using the left ear as the reference, re-referencing was applied. Gross movement artifacts were removed visually. Then, the signals were de-noised using independent component analysis (ICA), in particular in the case of blinking artifacts. After pre-processing, signals were filtered to the delta, theta, alpha, beta and gamma sub-bands, which were considered to be 1–4 Hz, 4–8 Hz, 8–13 Hz, 13–25 Hz and 25–40 Hz, respectively. After filtering, the signals were divided into 1 s epochs. It was applied to the whole band signal as well.

2.4 Results

This section is dedicated to the results of applying FSIFT-PLV to the synthetic and real EEG data. In the first sub-section our goal is evaluating the performance of FSIFT-PLV in detecting the actual phase synchrony between pairwise synthetic signals. Moreover, we will evaluate FSIFT-PLV in the four above-mentioned situations, i.e. in the presence of noise, volume conductance issues, sampling size variation and common reference issue. Additionally, the case of existing time shifts between signals will be studied. The results are reported in terms of root mean square error (RMSE) corresponding to synchrony levels of 0, 0.1, 0.2, 0.3, … , 1. Moreover, the results are compared to other state-of-the-art approaches including PLV (based on reduced interference Rihaczek distribution), ImC, PLI, wPLI, dwPLI and ciPLV in terms of RMSE. The real EEG data are investigated in the second sub-section.

Statistical analysis is done using analysis of variance (ANOVA). The significance level is applied after a Bonferroni correction. All statistical calculations are done using IBM SPSS Statistics 25. Other calculations are performed in MATLAB 2018b.

Table 2.3. Phase synchrony detection error [29].

Method	PLV	PLI	WPLI	dwPLI	ImC	ciPLV	FSIFT-PLV
Error (RMSE)	0.194	0.319	0.225	0.198	0.473	0.206	0.058

2.4.1 Synthetic EEG data

In order to study the ability of FSIFT-PLV in capturing synchrony values in the following situations, simulated time series with known parameters will be used. The average values of the synchrony index associated with each epoch are reported. Results are obtained by running numerical simulations 200 times. It should be emphasized that each run is performed using a different random network with the same N and P, but independent initial conditions.

2.4.1.1 Without interference

First of all, the performance of FSIFT-PLV in capturing phase synchrony values, without any disturbing factor will be studied. Table 2.3 illustrates the results in comparison with other approaches, in terms of RMSE. FSIFT-PLV shows the least amount of error.

The following experiments are designed to investigate the robustness of FSIFT-PLV in comparison with the state-of-the-art approaches. It should be emphasized that PLV is calculated based on the reduced interference Rihaczek distribution.

2.4.1.2 Amplitude noise

The first interfering item is the amplitude noise, which is considered as the additive white Gaussian noise. By changing the signal-to-noise ratio (SNR) in the range of $-20\text{ dB} \leqslant \text{SNR} \leqslant 30\text{ dB}$, RMSE values associated with capturing phase synchrony are obtained. Figure 2.2 illustrates the results for all approaches. As can be seen, FSIFT-PLV presents the minimum RMSE values associated with each SNR and on average.

2.4.1.3 Common reference and volume conduction issues

As mentioned previously, common reference and volume conduction are two underlying concerns in calculating the phase synchrony between EEG signals. Dealing with EEG and LFP signals, a common reference can be modeled as a transition from the case of an ideal sensor to the case of adding a single source to the linear mixture model. Using an ideal sensor is equivalent to recording with electro-neutral or no reference. Consequently, considering uncorrelated noise sources whose activity is collected by the reference sensor would automatically address the common reference problem. Additionally, the volume conductance effect can be modeled as noise on the phases of the signals [25]. According to the above reasoning, studying the phase noise can address the volume conduction and common reference issues. Thus, phase noise with a white additive Gaussian model is studied in this part. The SNR values will be changed in the range of $-20\text{ dB} \leqslant \text{SNR} \leqslant 30\text{ dB}$.

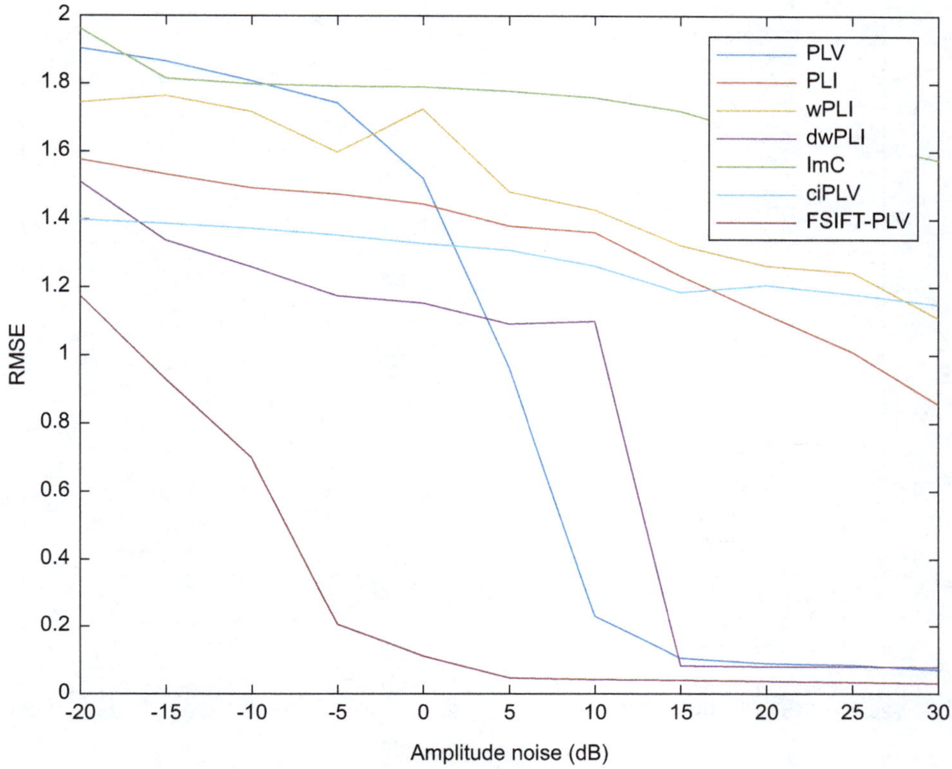

Figure 2.2. RMSE values of capturing phase synchrony values in the presence of amplitude noise, corresponding to $-20\text{ dB} \leqslant \text{SNR} \leqslant 30\text{ dB}$.

Corresponding RMSE values of capturing phase synchrony are summarized in figure 2.3 for all approaches. As can be seen, FSIFT-PLV presents the minimum RMSE vales associated with each SNR, and also on average.

2.4.1.4 Sample size bias

One of the main issues in quantifying phase synchrony between EEG signals occurs due to the smallness of the sample size. To study the robustness of FSIFT-PLV compared to other methods, sample sizes of 10–200 are studied. Figure 2.4 demonstrates the RMSE values achieved for each method, corresponding to each sample size. FSIFT-PLV shows higher performance in comparison to other approaches.

2.4.1.5 Shift in time

Assume that an approach for quantifying phase synchrony between pairwise signals uses estimated phases belonging to the same time point. Now, a constant shift in time may cause considerable changes in the calculated phase synchrony value, even in the case of identical signals. In this section, the robustness of FSIFT-PLV to time shifts will be examined. In our designed experiment, signals with the actual phase

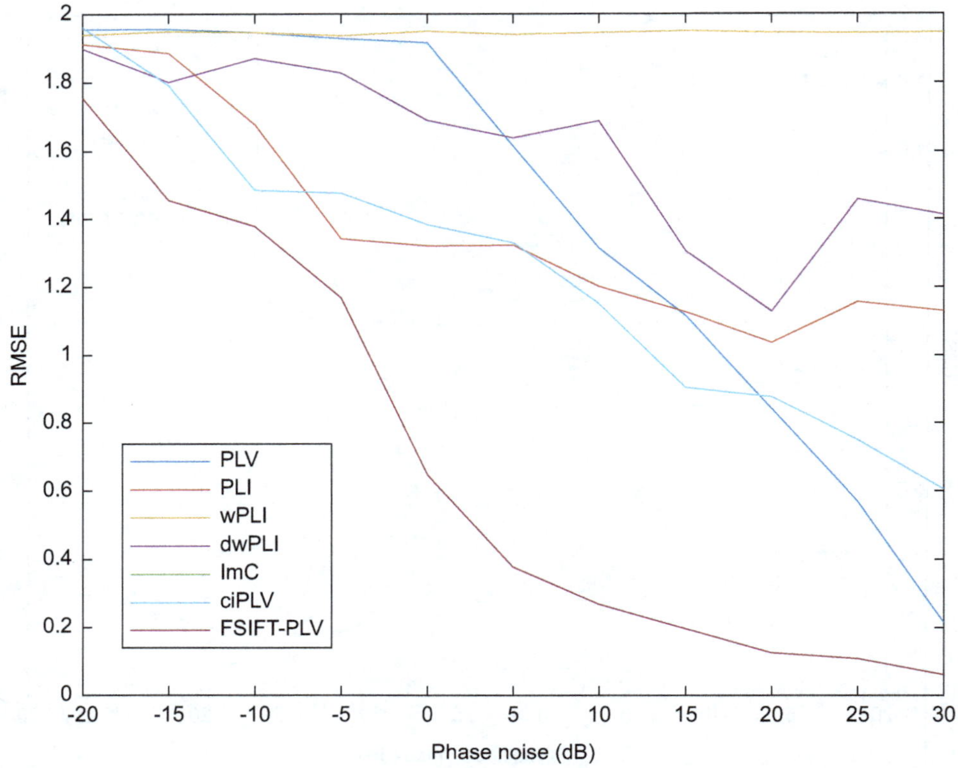

Figure 2.3. RMSE values of capturing phase synchrony values in the presence of phase noise, corresponding to -20 dB $\leqslant$ SNR $\leqslant$ 30 dB..

synchrony of 0, 0.1, 0.2, 0.3, … , 1 are shifted in time versus each other, from 1 to 10 points. Such a shift is plausible in the case of EEG signals. Errors are reported in terms of RMSE values in figure 2.5. The results confirm the superiority of FSIFT-PLV over the other methods.

2.4.2 Real EEG data

In this part we will focus on the real EEG data which are recorded in the REO and REC states for 5 min, as described in section 2.3. Three graph features (ND, NE and BC) are calculated corresponding to each of the 16 channels in five sub-bands as well as the whole band. After presenting the FSFIT-PLV accuracy and robustness in previous sections, it will be applied to real EEG data. As mentioned above, FSIFT-PLV provides a synchrony measure in [0, 1]. These pairwise values will be used for generating a 16 × 16 connectivity matrix which corresponds to an undirected graph. Nodes represent surface electrodes and links are phase synchrony values obtained by FSIFT-PLV. Consequently, corresponding to each epoch, we will have a connectivity matrix with the above properties. These steps are applied to both REO and REC signals. In the following sections REC and REO are discussed.

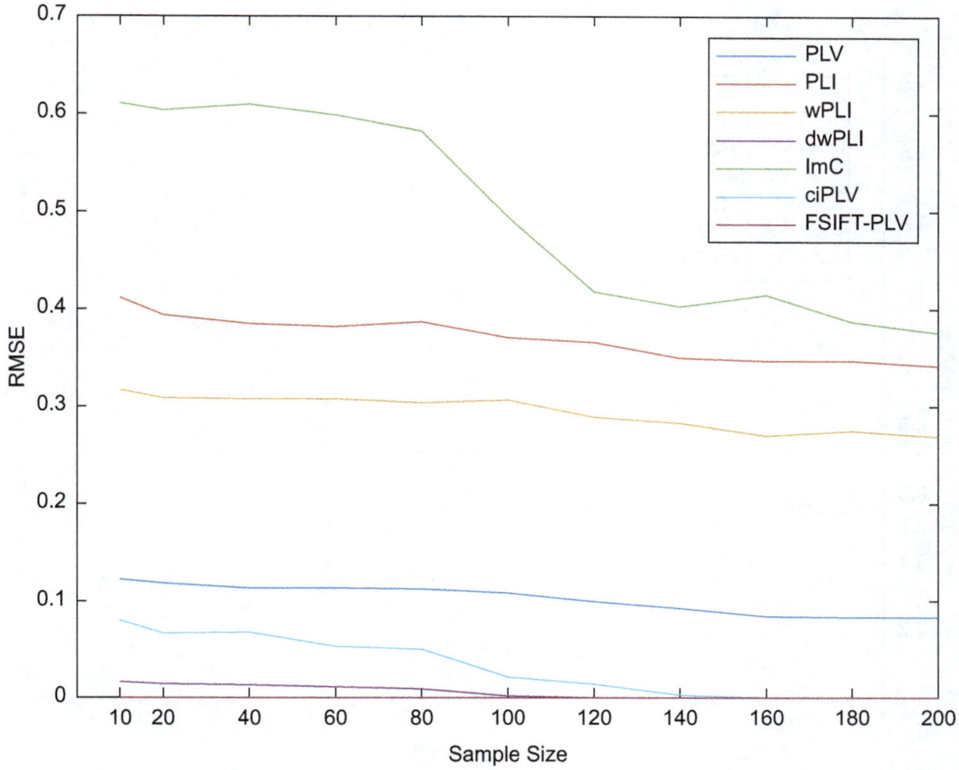

Figure 2.4. RMSE values corresponding to different sample sizes.

2.4.2.1 Resting state eyes-closed

As mentioned earlier, we have three groups in our study including PTSD, non-PTSD and control. In this part we investigate intergroup differences among each pair of groups. First, the graph features which were defined in the methods section (NE, ND and BC) are calculated for all channels. For each of these features, the average difference is obtained. It is done for each channel, each sub-band and the whole band. Figure 2.6 presents this difference for the control and non-PTSD groups. In other words, the topo-maps of figure 2.6 illustrate $BC_{non\text{-}PTSD}-BC_{control}$, $NE_{non\text{-}PTSD}-NE_{control}$ and $ND_{non\text{-}PTSD}-ND_{control}$.

As can be seen, ND and NE present quite similar behavior. Positive values (represented by warm colors) in subtraction mean greater values for NE and ND in the non-PTSD group compared to the control group. This can be found in all channels, the whole band signal and all sub-bands (although in the gamma band it is weaker than in the other sub-bands). In contrast, BC neither shows a particular trend, nor behavior similar to NE and ND.

Similarly, figure 2.7 indicates the average differences of the aforementioned features for the non-PTSD and PTSD groups. Actually, it shows $BC_{PTSD}-BC_{non\text{-}PTSD}$, $NE_{PTSD}-NE_{non\text{-}PTSD}$ and $ND_{PTSD}-ND_{non\text{-}PTSD}$.

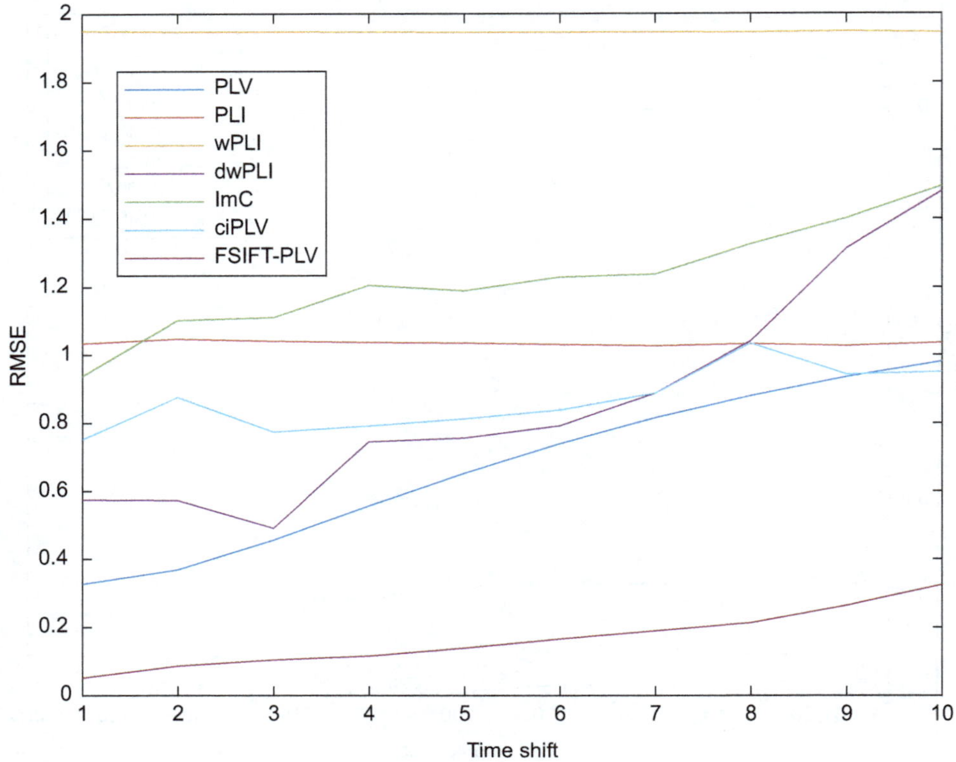

Figure 2.5. Performance in the presence of time shifts.

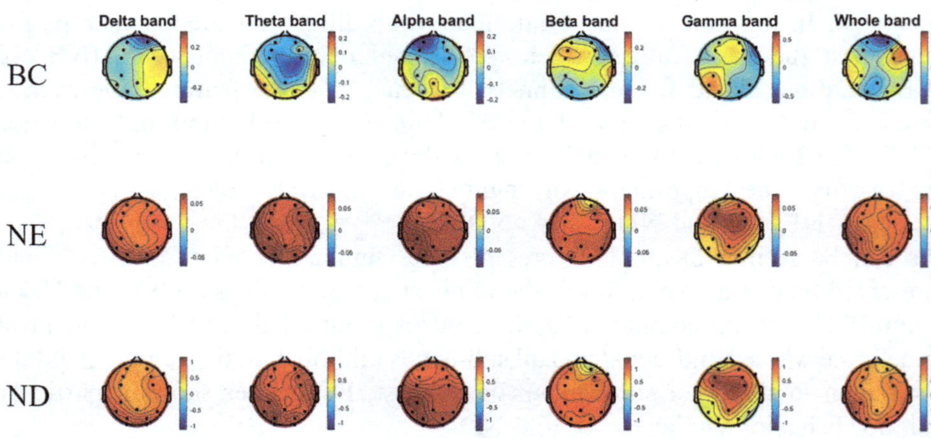

Figure 2.6. Intergroup differences between the control and non-PTSD groups, based on BC, NE and ND graph features, in sub-bands and the whole band.

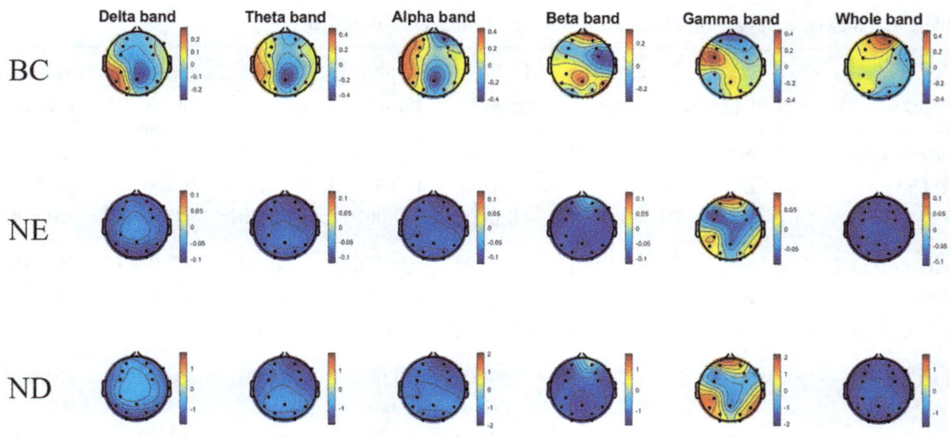

Figure 2.7. Intergroup differences between the non-PTSD and PTSD groups, based on BC, NE and ND graph features, in sub-bands and the whole band.

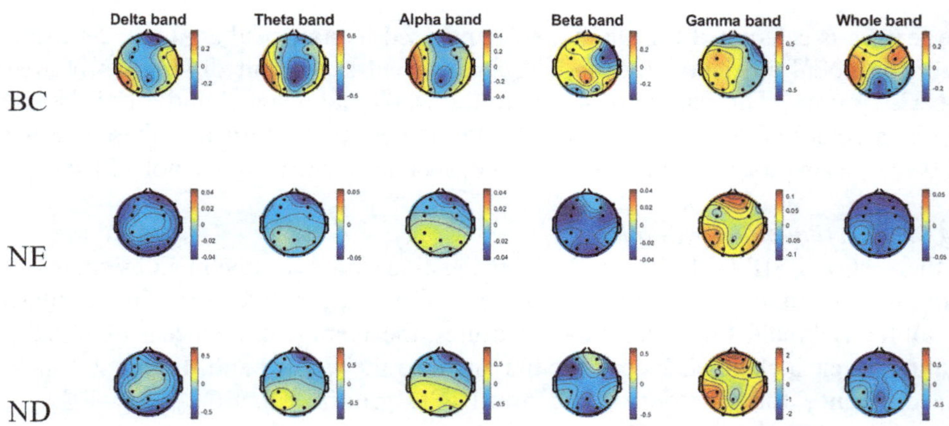

Figure 2.8. Intergroup differences between the control and PTSD groups, based on BC, NE and ND graph features, in sub-bands and the whole band.

As this figure demonstrates, ND and NE present a similar trend in all sub-bands except the gamma band. In the delta, theta and beta sub-bands as well as the whole band, ND and NE have greater average values in the non-PTSD group compared to the PTSD group. This holds for all channels. In contrast, BC does not show such a trend.

Finally, figure 2.8 represents intergroup differences between the control and PTSD group, i.e. $BC_{PTSD} - BC_{control}$, $NE_{PTSD} - NE_{control}$ and $ND_{PTSD} - ND_{control}$ as topo-maps. As can be seen NE and ND have negative average values in the delta and beta sub-bands as well as the whole band, for all channels.

To have more reliable results we need to perform a statistical analysis. As mentioned before, ANOVA is used at a 99% confidence interval. Table 2.4 illustrates the *p*-values associated with the comparison of each pair of groups (control–non-PTSD, non-PTSD–PTSD and control–PTSD). For each comparison,

Table 2.4. Pairwise comparison of three groups based on ND, NE and BC.

Classes	Feature	Delta band	Theta band	Alpha band	Beta band	Gamma band	Whole band
Control–non-PTSD	**NE**	<0.0001	<0.0001	<0.0001	<0.0001	<0.0001	<0.0001
	BC	0.081	0.867	0.390	0.211	0.724	0.286
	ND	<0.0001	<0.0001	<0.0001	<0.0001	<0.0001	<0.0001
Control–PTSD	**NE**	<0.0001	<0.0001	<0.0001	<0.0001	<0.0001	<0.0001
	BC	<0.0001	0.322	0.713	0.390	0.420	0.096
	ND	<0.0001	<0.0001	0.002	<0.0001	<0.0001	<0.0001
Non-PTSD–PTSD	**NE**	<0.0001	<0.0001	<0.0001	<0.0001	<0.0001	<0.0001
	BC	0.035	0.971	0.959	0.407	0.593	0.273
	ND	<0.0001	<0.0001	<0.0001	<0.0001	<0.0001	<0.0001

the feature is calculated for all channels and used in statistical analysis. As can be seen, the nodal efficiency and nodal degree reveal significant differences between pairwise groups. This can be observed in the whole band signal and all of the sub-bands. *p*-values of less than 0.0001 indicate the high performance of these features. However, as can be seen, the results for betweenness centrality are not satisfactory.

2.4.2.2 Resting state eyes-open

In this section FSIFT-PLV is applied to the resting state eyes-open EEG signals. The procedure is similar to the previous sub-section. First, graph features are calculated for all the channels. For each of these features, the average difference is obtained. It is done for each channel, each sub-band and also the whole band. Figure 2.9 shows this difference for the control and non-PTSD groups (i.e. $BC_{non\text{-}PTSD}-BC_{control}$, $NE_{non\text{-}PTSD}-NE_{control}$ and $ND_{non\text{-}PTSD}-ND_{control}$, respectively).

As can be seen, ND and NE present quite similar behavior. Despite the REC, here we have negative values which indicate higher values corresponding to the control group compared to the non-PTSD group. It should be emphasized that these features are greater in all channels in the delta band. However, they do not show this trend for all channels in the gamma band and the whole band signal. In the theta, alpha and beta sub-bands, we cannot find such a trend. Similarly to REC, BC shows different behavior compared to NE and ND.

Figure 2.10 illustrates the average differences of the features between the non-PTSD and PTSD groups (i.e. $BC_{PTSD}-BC_{non\text{-}PTSD}$, $NE_{PTSD}-NE_{non\text{-}PTSD}$ and $ND_{PTSD}-ND_{non\text{-}PTSD}$). Similarly to REC, here we can find that the average values of the nodal degree and nodal efficiency in all channels are higher in the non-PTSD group in comparison with the PTSD one, except in the gamma sub-band. BC does not reveal a particular trend.

Figure 2.11 demonstrates the average differences of features between the control and PTSD groups (i.e. $BC_{PTSD}-BC_{control}$, $NE_{PTSD}-NE_{control}$ and $ND_{PTSD}-ND_{control}$).

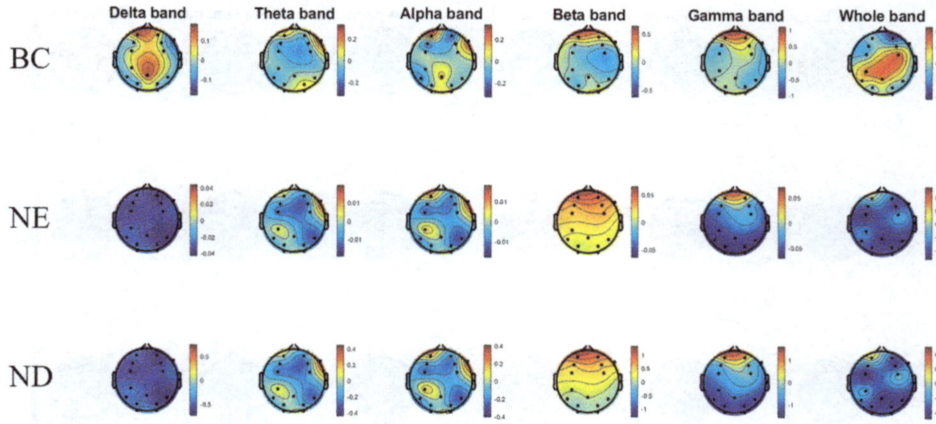

Figure 2.9. Intergroup differences between the control and non-PTSD groups, based on BC, NE and ND graph features, in sub-bands and the whole band.

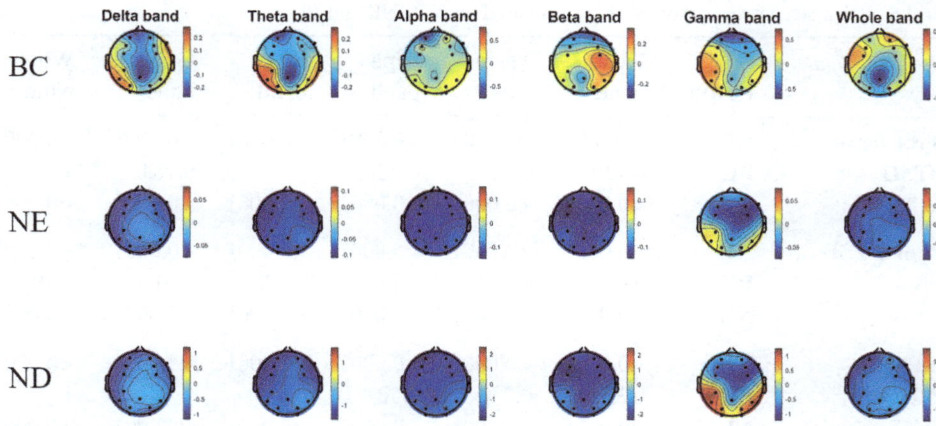

Figure 2.10. Intergroup differences between non-PTSD and PTSD groups, based on BC, NE and ND graph features, in sub-bands and the whole band.

As indicated in these topo-maps, ND and NE have higher values in the control group. This can be found in all channels, the whole band signal and all sub-bands except the gamma band. However, this trend cannot be found in REC. Here, BC does not show any particular trend, as in REO and REC previously.

Again, to have a more precise comparison we use statistical analysis of the aforementioned features, which are calculated corresponding to each channel. The results are summarized in table 2.5.

According to these results, NE and ND show proper performance. As mentioned before, ANOVA is used at a 99% confidence interval. Table 2.5 illustrates *p*-values associated with the comparison of each pair of groups (control–non-PTSD, non-PTSD–PTSD and control–PTSD). For each comparison, the feature is calculated for all channels and used in statistical analysis. As can be seen, nodal efficiency and

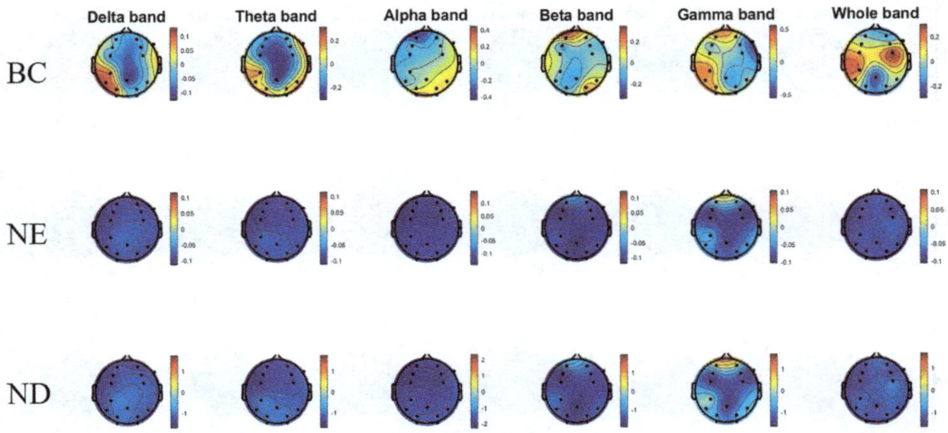

Figure 2.11. Intergroup differences between the control and PTSD groups, based on BC, NE and ND graph features, in sub-bands and the whole band.

Table 2.5. Pairwise comparison of three groups based on ND, NE and BC.

Classes	Feature	Delta band	Theta band	Alpha band	Beta band	Gamma band	Whole band
Control–non-PTSD	**NE**	<0.0001	<0.0001	<0.0001	<0.0001	<0.0001	<0.0001
	BC	0.139	0.467	0.851	1	0.284	1
	ND	<0.0001	<0.0001	<0.0001	<0.0001	<0.0001	<0.0001
Control–PTSD	**NE**	<0.0001	<0.0001	<0.0001	<0.0001	<0.0001	<0.0001
	BC	0.471	0.948	0.170	0.729	0.318	0.192
	ND	<0.0001	<0.0001	<0.0001	<0.0001	<0.0001	<0.0001
Non-PTSD–PTSD	**NE**	<0.0001	<0.0001	<0.0001	<0.0001	<0.0001	<0.0001
	BC	0.325	0.730	0.527	0.628	1	1
	ND	<0.0001	<0.0001	<0.0001	<0.0001	<0.0001	<0.0001

nodal degree reveal significant differences between pairwise groups. This can be observed in the whole band signal and all of the sub-bands. p-values of less than 0.0001 indicate the high performance of these features. However, as can be seen, the results of betweenness centrality are not satisfactory.

2.5 Conclusion

In this chapter a novel bivariate phase synchrony index called the fuzzy scale invariant phase locking value (FSIFT-PLV) is introduced. This measure benefits from a powerful phase estimation method based on the reduced interference Rihaczek distribution, in addition to a robust method for extracting features from images called the scale invariant feature transform (SIFT). For two given signals, the phases of the signals are first estimated as a function of time and frequency.

This provides time–frequency maps which will be considered as images. Then the SIFT is applied to the time–frequency maps. Therefore, key-points will be obtained in two maps. For certain reasons, using standard SIFT matching may not be useful for the purpose of finding phase synchrony. First, some SIFTs may provide matched key-points which do not belong to the same region of comparison images. This is a general issue. Second, we are using the SIFT in a particular application, on EEG signals. Therefore, due to the nature of EEG signals, the matched key-points cannot be located far away in time or frequency. To address this issue, we proposed a fuzzy framework with three inputs, including distance of key-point descriptors (used in SIFT matching), a Euclidean distance between the locations of pairwise key-points, in addition to an initial similarity estimation. The first input has three values: low, medium and high. The next two inputs have two values: low and high. The output of the fuzzy system is a phase synchrony measure in [0, 1].

To assess the ability of FSIFT-PLV and study its robustness, some experiments are designed and applied to synthetic data which are generated using Rossler oscillators. We investigated four underlying issues of calculating phase synchrony: noise, volume conduction, common reference and sample size. In addition, the presence of a time shift between signals is studied. The results are compared to state-of-the-art methods including reduced interference Rihaczek distribution based-PLV, ImC, PLI, wPLI, dwPLI and ciPLV. The results indicate the superiority of FSIFT-PLV to other approaches which was expected due to the nature of our proposed method.

In the next part FSIFT-PLV is applied to real EEG signals. Signals are recorded in resting states eyes-closed and eyes-open, for 5 min. There are three groups: combat related PTSD, combat trauma exposed non-PTSD (non-PTSD) and healthy controls. After denoising and pre-processing, the signals are filtered into sub-bands. For each 1000 ms epoch, connectivity matrices are generated using FSIFT-PLV. Three graph features, i.e. nodal degree, nodal efficiency and betweenness centrality, are extracted from the networks. These features are calculated for all channels.

For REC signals, our statistical analysis supports that NE and ND, as graph features, are able to reveal significant differences between each pair of groups. Our findings demonstrate that in the whole band and delta sub-band, the ND and NE features are decreased in PTSD participants compared to the control group in all the channels. In contrast, these two features are greater than the control group in all the channels for the non-PTSD participants. In other words in the lower and higher frequencies we have $ND_{\text{non-PTSD}} > ND_{\text{control}} > ND_{\text{PTSD}}$ in all channels. Similarly, we have $NE_{\text{non-PTSD}} > NE_{\text{control}} > NE_{\text{PTSD}}$. In the theta, alpha and beta sub-bands, the ND and NE of all the channels are greater in non-PTSD participants compared to the PTSD and control groups. However, there are some channels in which ND, NE or both have lower values in the PTSD group compared to the control group. In other words, our research demonstrates that functional connectivity decreases in the PTSD group compared to the healthy controls. In contrast, it is increased in the non-PTSD trauma exposed group in comparison to the healthy controls.

For REO signals, the results demonstrate that BC is not a proper feature for this application, which was previously found in REC signals. However, ND and NE demonstrate proper performance. It can be inferred that $NE_{control} > NE_{PTSD}$ and $NE_{control} > NE_{non\text{-}PTSD}$ except in the gamma band. In a similar manner, $ND_{control} > ND_{PTSD}$ and $ND_{control} > ND_{non\text{-}PTSD}$. However, the expressions $NE_{control} > NE_{non\text{-}PTSD} > NE_{PTSD}$ and $ND_{control} > ND_{non\text{-}PTSD} > ND_{PTSD}$ only hold in the delta sub-band. Of course, this conclusion is based on the average values and needs further studies to be confirmed.

Acknowledgments

The authors wish to express their kind appreciation to Professor Ali Reza Moradi (Kharazmi University), Dr Jafar Mirzaei (Sadr Hospital), Professor Soltani, Mrs Dehghan (Amirkabir University of Technology), Dr Rahele Davoodi and the personnel of Sadr Hospital for their valuable help.

References

[1] Vahia V N 2013 Diagnostic and statistical manual of mental disorders 5: a quick glance *Indian J. Psychiatry* **55** 220–3

[2] Kessler R C 2000 Posttraumatic stress disorder: the burden to the individual and to society *J. Clin. Psychiatry* **61** 4–12

[3] Pitman R K *et al* 2012 Biological studies of post-traumatic stress disorder *Nat. Rev. Neurosci.* **13** 769–87

[4] Shin L M and Liberzon I 2010 The neurocircuitry of fear, stress, and anxiety disorders *Neuropsychopharmacology* **35** 169–91

[5] Hata M *et al* 2016 Functional connectivity assessed by resting state EEG correlates with cognitive decline of Alzheimer's disease—an eLORETA study *Clin. Neurophysiol.* **127** 1269–78

[6] Hatz F and Fuhr P 2018 Functional EEG connectivity alterations in Alzheimer's disease *Biomarkers for Preclinical Alzheimer's Disease* ed R Perneczky (New York: Springer) pp 157–66

[7] Engels M M A, Stam C J, van der Flier W M, Scheltens P, de Waal H and van Straaten E C W 2015 Declining functional connectivity and changing hub locations in Alzheimer's disease: an EEG study *BMC Neurol.* **15** 145

[8] Rubinov M *et al* 2009 Small-world properties of nonlinear brain activity in schizophrenia *Hum. Brain Mapp.* **30** 403–16

[9] Yuan H *et al* 2018 Tracking resting state connectivity dynamics in veterans with PTSD *NeuroImage: Clin.* **19** 260–70

[10] Lee S H, Yoon S, Kim J I, Jin S H and Chung C K 2014 Functional connectivity of resting state EEG and symptom severity in patients with post-traumatic stress disorder *Progress Neuro-psychopharmacol. Biol. Psychiatry* **51** 51–7

[11] Greicius M D, Menon V, Supekar K and Dougherty R F 2008 Resting-state functional connectivity reflects structural connectivity in the default mode network *Cereb. Cortex* **19** 72–8

[12] Liu Y-H, Lin L-F, Chou C-W, Chang Y, Hsiao Y-T and Hsu W-C 2019 Analysis of electroencephalography event-related desynchronisation and synchronisation induced by lower-limb stepping motor imagery *J. Med. Biol. Eng.* **39** 54–69

[13] Garcés Correa A, Orosco L L, Diez P and Laciar Leber E 2019 Adaptive filtering for epileptic event detection in the EEG *J. Med. Biol. Eng.* **39** 912–8

[14] Adrian E D and Matthews B H C 1934 The Berger rhythm: potential changes from the occipital lobes in man *Brain* **57** 355–85
[15] Barry R J and Blasio F M D 2017 EEG differences between eyes-closed and eyes-open resting remain in healthy ageing *Biol. Psychol.* **129** 293–304
[16] Boytsova Y A and Danko S G 2010 EEG differences between resting states with eyes open and closed in darkness *Hum. Physiol.* **36** 367–9
[17] Ahmadlou M and Adeli H 2011 Fuzzy synchronization likelihood with application to attention-deficit/hyperactivity disorder *Clin. EEG Neurosci.* **42** 6–13
[18] Mormann F, Lehnertz K, David P and Elger C E 2000 Mean phase coherence as a measure for phase synchronization and its application to the EEG of epilepsy patients *Physica* D **144** 358–69
[19] Dauwels J, Vialatte F, Musha T and Cichocki A 2010 A comparative study of synchrony measures for the early diagnosis of Alzheimer's disease based on EEG *NeuroImage* **49** 668–93
[20] Aviyente S and Mutlu A Y 2011 A time-frequency-based approach to phase and phase synchrony estimation *IEEE Trans. Signal Process.* **59** 3086–98
[21] Tass P *et al* 1998 Detection of *n*:*m* phase locking from noisy data: application to magnetoencephalography *Phys. Rev. Lett.* **81** 3291–4
[22] Rosenblum M G, Pikovsky A S and Kurths J 1996 Phase synchronization of chaotic oscillators *Phys. Rev. Lett.* **76** 1804–7
[23] Aviyente S, Bernat E M, Evans W S and Sponheim S R 2011 A phase synchrony measure for quantifying dynamic functional integration in the brain *Hum. Brain Mapp.* **32** 80–93
[24] Lachaux J P, Rodriguez E, Martinerie J and Varela F J 1999 Measuring phase synchrony in brain signals *Hum. Brain Mapp.* **8** 194–208
[25] Vinck M, Oostenveld R, van Wingerden M, Battaglia F and Pennartz C M A 2011 An improved index of phase-synchronization for electrophysiological data in the presence of volume-conduction, noise and sample-size bias *NeuroImage* **55** 1548–65
[26] Nolte G, Bai O, Wheaton L, Mari Z, Vorbach S and Hallett M 2004 Identifying true brain interaction from EEG data using the imaginary part of coherency *Clin. Neurophysiol.* **115** 2292–307
[27] Stam C J, Nolte G and Daffertshofer A 2007 Phase lag index: assessment of functional connectivity from multi channel EEG and MEG with diminished bias from common sources *Hum. Brain Mapp.* **28** 1178–93
[28] Bruna R, Maestu F and Pereda E 2018 Phase locking value revisited: teaching new tricks to an old dog *J. Neural Eng.* **15** 056011
[29] Ghanbari Z and Moradi M H 2020 FSIFT-PLV: an emerging phase synchrony index *Biomed. Signal Process. Control* **57** 101764
[30] Lowe D G 2004 Distinctive image features from scale-invariant keypoints *Int. J. Comput. Vis.* **60** 91–110
[31] Lowe D G 1999 Object recognition from local scale-invariant features *Proc. of the Seventh IEEE Int. Conf. on Computer Vision* vol. 2 1150–7
[32] Goldmann R E *et al* 2003 Late frontal brain potentials distinguish true and false recognition *Neuroreport* **14** 1717–20
[33] Poli R, Kennedy J and Blackwell T 2007 Particle swarm optimization *Swarm Intell.* **1** 33–57
[34] Lee S-H, Yoon S, Kim J-I, Jin S-H and Chung C K 2014 Functional connectivity of resting state EEG and symptom severity in patients with post-traumatic stress disorder *Progress Neuro-Psychopharmacol. Biol. Psychiatry* **51** 51–7

[35] Nessler D, Mecklinger A and Penney T B 2001 Event related brain potentials and illusory memories: the effects of differential encoding *Brain Res. Cogn. Brain Res.* **10** 283–301

[36] Pecora L M and Carroll T L 1998 Master stability functions for synchronized coupled systems *Phys. Rev. Lett.* **80** 2109–12

[37] Kemp A H *et al* 2010 Disorder specificity despite comorbidity: resting EEG alpha asymmetry in major depressive disorder and post-traumatic stress disorder *Biol. Psychol.* **85** 350–4

[38] Jalili M, Barzegaran E and Knyazeva M 2014 Synchronization of EEG: bivariate and multivariate measures *IEEE Trans. Neural Syst. Rehabil. Eng.* **22** 212–21

Chapter 3

Weighted complex network based framework for epilepsy detection from EEG signals

Supriya Supriya, Siuly Siuly, Hua Wang, Yanchun Zhang

Institute for Sustainable Industries & Liveable Cities, Victoria University, Australia

This chapter presents a weighted complex network based framework to identify one of the most challenging brain disorders, namely epilepsy. The pattern of action potentials fluctuates in epileptic disorders, and this can be best intelligible with the assistance of an electroencephalogram (EEG). The EEG is the core authorized biomarker that aids in enhancing the understanding of mental conditions and behaviors, to identify or diagnose any abnormal condition that occurs. The automated diagnosis of epileptic seizure activity using EEG signals is an area of profound ongoing attention in medical science as well as in research disciplines, as the traditional methods of diagnosis of epileptic disorders rely on monotonous visual inspection by highly expert clinicians of long-lasting EEG recordings. The study of complex networks has proved that the underlying dynamics of EEG signals can be best analysed by measuring the strength amongst the nodes of the network, as the topological invariants of the network are closely associated with the underlying dynamics of the EEG. Hence, this chapter introduces an innovative edge-weighted algorithm in the visibility graph for distinguishing epileptic EEG signals from healthy EEG recordings. This research aims to explore the efficacy of the proposed edge weights idea as well as the average weighted degree as efficient network features for identifying epileptic seizure activity using five prevalent machine learning classifiers.

3.1 Introduction

Epilepsy has been recognized as one of the most common chronic brain dysfunctions among the neurological diseases [1]. Epilepsy is basically a disorder of the central nervous system of the human brain, with a manifestation of abnormal brain activity

doi:10.1088/978-0-7503-3279-8ch3

that causes seizures. Around 50 million people at the global level suffer from such an epilepsy syndrome [2]. According to the WHO, the occurrence rate of epilepsy per year is 2.4 million, and it can affect any age group. The premature death rate for epileptic patients is two to three times higher than for a normal healthy person [3]. Epileptic patients have a greater risk of other complications, such as bleeding into the brain, brain tumors, cerebral palsy, Alzheimer's disease (in the later stage of life) and autism disorder, etc [4].

EEG is the predominant examination tool for the diagnosis, treatment and therapy planning related to epileptic disorders. EEG data are complex as well as high-dimensional in nature. Some clinics archive EEG data on paper. A half-hour EEG recording consumes around one cubic foot of paper [5]. The development of digital recording systems has overcome the paper recording issues of EEGs and has also provided ambulatory recording facilities. Lamentably, these digital and ambulatory systems have introduced new challenges for clinicians by introducing a large amount of data. Epilepsy detection comprises the visual inspection of long-term EEG recordings of several days by an expert neurologist. The non-stationary and complex nature of EEG signals make this task error-prone, time-consuming and expensive [6]. Visual EEG inspection is thus tiresome, fallible and not even cost-effective because the cost of studying the EEG is so high. An expert neurologist finding traces of seizure activity through visual examination of an EEG cannot be considered as a very reliable procedure. In addition, EEG analysis not only assists in diagnosis and the management of anti-disorder medication, but also plays a crucial role in all neurosurgery related to epilepsy. As a result, there is a need for automated seizure detection techniques that assist clinicians in the diagnosis of epilepsy through computer-based analysis of EEG and also reduce the high cost, fallibility and long duration of visual examination.

EEG data are available in the form of time series, therefore time series analysis methods are generally used for epilepsy detection. The existing epilepsy detection methods include a vast range of linear to non-linear techniques and mostly rely on traditional techniques [7–10]. Innumerable approaches [11–16] have been proposed for EEG analysis and classification for epilepsy detection. Mursalin *et al* [15] proposed a mixed approach for epilepsy detection by extracting parameters from the time and frequency domains. Mean, skewness, mode, standard deviation, median, minimum, kurtosis, maximum, and the first, third and interquartile ranges etc, were the parameters that were extracted in addition to the maximum, minimum, mean and standard deviation of the wavelet coefficient features. In order to select the most promising feature, a correlation-based feature selection technique was implemented, and then the selected parameters were passed to the random forest classifier. The experimental outcomes showed that the proposed study was a promising technique for classifying different test cases from the Bonn University epileptic EEG database. Rana *et al* [17] presented a method to detect epilepsy from multiple channels using the phase slope index (PSI). The PSI was used to distinguish epileptic and normal activity from an EEG database. The PSI quantified the link between two EEG channels and discovered an increase in the spatio-temporal interaction among the EEG channels to categorized different EEG signals. The research study was implemented on 258 h long EEG recordings of five patients with different categories of epileptic syndrome. The methodology was implemented on two different epochs of EEG signals with lengths of 20 s and 60 s. In the absence of any classifier, the proposed study achieved a 100% classification accuracy for 20 s epoch lengths. Siuly *et al* [18] developed a novel clustering approach for classifying epilepsy from EEGs using a least squares support vector machine (SVM) and reported a 94.18% classification accuracy. The importance of the sampling approach was demonstrated by Siuly *et al* [19] using a random sampling and optimal allocation sampling based approach for classifying epilepsy EEG signals. The study found that random sampling is a more promising approach for seizure classification compared to optimal allocation sampling. The proposed work used *k*-nearest neighbors (*k*-NN), multinomial logistic regression with a ridge estimator (MLR) and an SVM classifier. The *k*-NN classifier provided 100% classification results. Kabir *et al* [16] proposed the idea of a computer-aided analysis framework for the detection of epileptic disorder from EEGs. The *k*-means clustering approach was applied to determine various clusters of data by considering the similar and dissimilar behaviors between the patterns of EEG signals. The SVM, naïve Bayes and logistic regression were used to classifying normal EEG signals and those with epilepsy. One hundred per cent classification accuracy was reported for the SVM classifier.

Zainuddin *et al* [20] developed a real-time approach for epilepsy detection from EEG signals with the help of a wavelet transform. The feature sets, namely standard deviation, minimum and maximum measures of the wavelet coefficients, were extracted from each sub-band. The analysis included the study of the Gaussian wavelet, the Mexican hat wavelet and the Morlet wavelet by applying the wavelet neural network classification technique. They reported 98% classification performance results with the conclusion that the Morlet wavelet, with fourth-order Daubechies wavelets, provided better results compared to another wavelet. Pradhan *et al* [21] found that a linear prediction is a promising approach for better analysis and visualization of the EEG signal. In addition to this, the linear prediction method is an efficient technique to generate, store and transmit EEG signals. Altunay *et al* [22] observed that when the linear prediction method is used to detect epilepsy, the energy prediction error feature increases during seizure activity. Some other researchers also used the energy of the signal as a parameter for the detection of epileptic disease [23, 24]. Bajaj *et al* [25] used parameters based on images as well as empirical mode decomposition [26] and obtained high-performance results. Xie *et al* [27] attained 100% classification by proposing a new feature in principal component analysis (PCA) and also used the energy of the signal as a feature for epilepsy detection. Acharya *et al* [28] applied PCA with a Gaussian mixture model classifier to detect epilepsy and were able to acquire satisfactory clinical results. Nam *et al* [29] implemented independent component analysis (ICA) to find the ictal activity in temporal lobe epilepsy and observed that there is an increase in the lateralization from 75% to 96% during the seizure. Whitmer *et al* [30] found that ICA is able to distinguish different types of EEG signals with dissimilar sources. Arunkumar *et al* [31] found that ICA and the Hurst exponent perform better in seizure detection compared to PCA. Fathima *et al* [32] applied linear discriminant analysis by extracting the features variance, skewness and coefficient of variation, and achieved 96.9% classification accuracy with a linear classifier. Some researchers implemented the principal component, independent component and linear discriminant analysis approaches together for classifying epileptic seizures with the help of an SVM classifier and obtained promising results [33, 34]. Nigam *et al* [35] applied an artificial neural network based approach for epilepsy detection and reported 97.20% accuracy. Siuly *et al* [36] introduced a statistical system for classifying multi-category EEG signals using an optimum allocation approach for data representation based on definite time and variability within a class. By using a multiclass least squares SVM as a classifier, the study claimed high-performance results for epilepsy detection. Martis *et al* [37] applied empirical mode decomposition for the classification of normal, ictal and inter-ictal activity in epileptic EEG signals and achieved 95.3% classification accuracy. Chua *et al* [38] applied the higher order spectra (HOS) approach for classifying normal, background and higher order spectra, and analysed them by applying the Gaussian mixture model classifier and SVM classifier. Parameters were extracted from the power spectrum. It was reported that HOS based parameters are a more efficient approach to distinguish different EEG signals, achieving 93.11% accuracy. Acharya *et al* [39] proposed an epilepsy detection technique based on higher order spectra cumulants from wavelet packet decomposition coefficients and achieved 98.50% accuracy with an SVM classifier.

However, the existing techniques do not fill the gap between the non-linear EEGs and their underlying dynamical behavior. Moreover, the adoption rate and research and development in this field are still hindered by some fundamental problems inherent within the big EEG data paradigm. Network based EEG analysis is an alternative approach for visualizing the underlying as well as hidden patterns associated with time series [40]. The different properties of the network help in identifying the distinct nature (such as the chaotic or fractal behavior) of time series. The statistical parameters of a network play a significant role in obtaining valuable information associated with EEG time series data. In recent times, complex networks have become the most promising approaches in neuroscience for identifying brain abnormalities [41]. These network based techniques have introduced a new direction in the field of neuroscience for detecting epilepsy via exploring the changes that occur in the characteristic features of the network structure from EEG signals [42–45].

Among the several complex network approaches [46–48] for studying the time series and its underlying dynamical information, the visibility graph based network (VGBN) is considered one of the most promising approaches [49]. VGBN has the ability to characterize EEG signals in terms of network theory by inheriting the dynamical properties of the EEG data and representing them in the form of statistical parameters of the network. Therefore, VGBN can be utilized to obtain valuable or significant information about EEG signals. Moreover, Liu *et al* [50] also revealed that VGBN exhibits noise robustness and does not require any parameter selection (unlike the time series to complex network method [46] which requires a threshold value, and the recurrence plot-based network which also need to be determined by a threshold value [47]). The state-of-the-art visibility graph also reveals that the topological invariant associated with the VGBN plays a significant role in understanding the time series (EEG) data. The most important constraint of visibility graph based techniques relates to not considering the essential fact that the links between the nodes of the network have strength and different connections exhibit different values of strength. Moreover, the literature in the discipline of network theory has also revealed that a binary network exhibits information only about the existence of links, whereas the preservation of information regarding the weights among the links of the vertices of a network helps to determine more robust and reliable results. As a result, after considering the limitations of the VG-based methods for epilepsy detection, we introduce a graph theory-based framework for the analysis of the EEG data by considering the edge weight as essential.

The objectives of this chapter are as follows:

1. To explore the feasibility of weight in the graph approach for the identification of epileptic disorders from EEG signals.
2. To identify the complex network features that can effectively distinguish epileptic from non-epileptic sections of EEGs.
3. To introduce a framework based on a complex network approach which can provide valuable information for the identification of epilepsy.

The rest of the chapter is organized as follows: section 3.2 provides detailed information about the weighted complex network based framework, section 3.3 describes the analysed EEG data and the experimental results with necessary discussion, and the conclusion is presented in section 3.4.

3.2 Weighted complex network based framework

This chapter includes an innovative framework founded on graph theory for the automated analysis of EEG signals to extract valuable information for identifying epileptic seizure activity from diverse classes of EEG signals. Figure 3.1 illustrates a conceptual diagram of the proposed framework. The proposed framework consists of three sections. (i) The conversion of time series EEG signals into a weighted complex network (WCN) with the help of the VG theorem and by introducing an innovative weight formula. (ii) Statistical feature extraction from the WCN. (iii) Evaluation of the extracted feature set by employing two standard machine learning classifiers: the SVM classifier and discriminant analysis (DA) classifier. Then the performance of the anticipated framework is assessed via specific measurements. All three parts of the WCN based framework are described in detail in the following three subsections.

3.2.1 Conversion of EEG signals into the WCN

For the construction of the WCN, the first phase is to consider that each sample point of the EEG signal is the vertex of the network. If $G = (V, L)$ represents a network with $V = \{v_i\}$, $i = 1, 2, \ldots, n$ is the set of n number of vertices and $L = \{l_i\}$, $i = 1, 2, \ldots, m$ is the set of m number of links and $E = \{e_i\}$, $i = 1, 2, \ldots, n$ denotes an EEG time series signal with vertex v_i corresponding to sample point e_i. The links

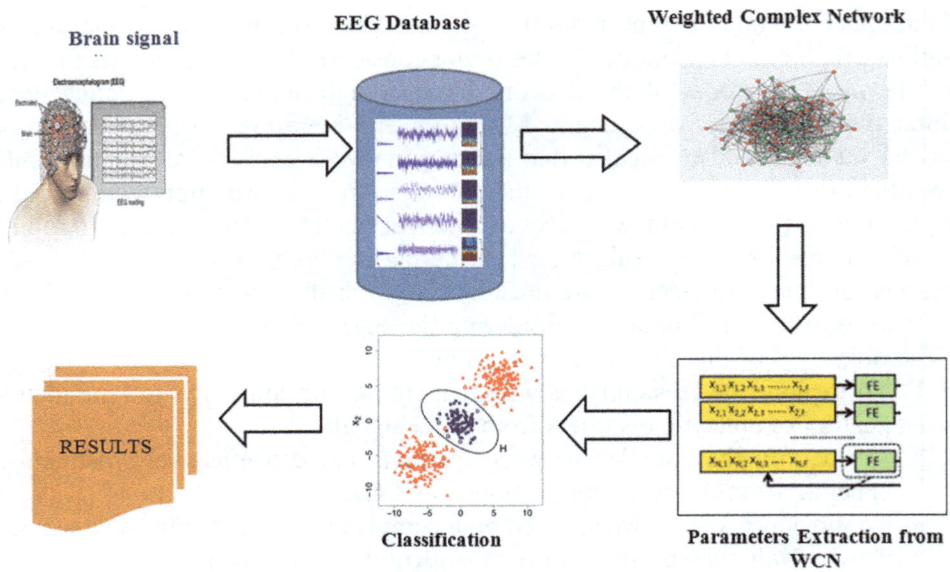

Figure 3.1. The block diagram of the WCN based framework.

among the different vertices of the network are only possible if they satisfy the following equation, which is based on the visibility graph method [49]:

$$v_b < v_a + (v_c - v_a)\frac{t_b - t_a}{t_c - t_a},\ c > b > a, \tag{3.1}$$

where, v_a, v_b and v_c are the vertices corresponding to the sample points e_a, e_b and e_c of the time series EEG signal E and t_a, t_b and t_c are their equivalent time events. The next phase is to evaluate the weight of the links among the vertices of the network. This study introduces the following weight equation:

$$w_{ab} = \frac{v_b - v_a}{t_b - t_a},\ b > a, \tag{3.2}$$

where w_{ab} represents the weight among the vertices v_a and v_b with their equivalent time events t_a and t_b. All the links are considered directional in nature, i.e. link l_{12} has direction from vertex v_1 to vertex v_2. In addition, the absolute value of the link's weight has been considered. After computing the link weights among all the vertices of the network, the next phase is the construction of the WCN of an EEG signal.

3.2.2 Statistical feature extraction from the WCN

Extracting the relevant statistical features of the network plays a crucial function during the classification of distinct EEG signals. In technical terms, a feature embodies a discernible dimension that can characterize the unique or distinguishable properties of a pattern or configuration. In the process of feature extraction, the vast EEG data are simplified into a feature vector based on the principle of the least possible loss of information [36]. Because an EEG signal is transformed into the

WCN, graphical parameters are mined in the feature extraction process and are used for the classification of different types of EEG signals. The measure of the network that is used as an extracted feature in this WCNF for epilepsy detection is called the average weighted degree (AWD).

$A_{l \times l} = \{a_{jk}\}$ denotes an adjacent matrix of WCN with l total nodes, where a_{jk} symbolize an edge from node j to k. The value of $a_{jk} = 1$ in the presence of a link from node j to k and $a_{jk} = 0$ if there is no link between them. According to Antoniou *et al* [51], the weighted degree of a node j is measured as the summation of all the edge weights of all the links joined to node j and is calculated as

$$wd_j = \sum\nolimits_{k \in N(j)} w_{jk}, \tag{3.3}$$

where wd_j represents the weighted degree of node j, $N(j)$ denotes the neighbors of j, and w_{jk} signifies the weight of the edge between nodes j and k. The AWD is measured as the average mean of the summation of all the edge weights incident on all the nodes of the network. During seizure activity, EEG signals present sudden fluctuations, which in turn affect the strengths by which all the nodes are connected to each other. As a result, the edge weight is changed and the result can be seen in the AWD feature, because the AWD is directly proportional to the weight of the edges. In this way, the AWD plays a crucial role in the analysis of diverse EEG signals.

3.2.3 Evaluation of the AWD using classifiers

In the procedure of classification, a mathematical function that called a classifier categorizes the EEG signals with distinct classes into their relevant category on the basis of feature sets. During classification, the complete EEG feature sets are divided into two parts: the training group set and the testing group set. The collection of unidentified observations (the testing group) is predicted or classified into the appropriate class by considering some criteria on the set of identified observations (the training group). This WCN based framework uses two prominent supervised classifiers of machine learning for evaluating the performance of the extracted feature AWD. The two classifiers are the SVM classifier and DA classifier.

3.2.3.1 Support vector machine (SVM) classifier

The SVM based on linear functions that make use of a hypothesis space with a feature space of high dimension, and are trained based on a learning process with an optimization principle for implementing a learning bias derivative based on a statistical learning model [52]. The mathematical learning model helps to generate a framework for analysing the problem of acquiring knowledge, to make predictions and to make decisions on the basis of the available dataset. In general, it helps in selecting the hyperplane space. Generalization is the capability of a hypothesis to perform accurate classification of data not present in the training group. The SVM is more proficient in terms of not performing over-generalization, whereas over-generalization is the major drawback of neural network classifiers. The linear classifiers execute a classification based on the principle of selecting appropriate hyperplanes and are affected by the shortcoming of choosing a hyperplane which is proficient at classifying one group of data and does not achieve good classification in other data sets. The SVM classifier overcomes this shortcoming by providing the concept of a hyperplane with a maximum margin and thus accomplishes efficient classification. The maximum margin is expressed as

$$\text{margin} = \arg\min_{l \in D} d(l) = \arg\min_{l \in D} \frac{l.\ w + m}{\sqrt{\sum_{j=1}^{d} w_j^2}}, \tag{3.4}$$

where w represents the weight vector and is orthogonal to the hyperplane, l denotes the input vector, m is the bias and d represents the margin of separation. Detailed information about equation (3.4) and all of the parameters are available in [53]. The hyperplane is defined as a subspace which has a dimension less than one in comparison to its ambient space. In general, a hyperplane partitions space into two parts. The standard equation of the hyperplane is

$$w \times l + m = 0. \tag{3.5}$$

The hyperplane based on w and m provides the following function for the correct classification of the training sample or data:

$$f(x) = \text{sign}(w \times l + m). \tag{3.6}$$

In the procedure of training the SVM, the following equation is used:

$$y_j(w \times l_j) \geqslant 1 \ \forall \quad 1 \leqslant j \leqslant n. \tag{3.7}$$

Equation (3.7) is solved to determine the hyperplane with a maximum margin for the value of w if y_j are the available data labels and l_j the feature vectors. The dual formulation of the SVM classifier is represented as

$$\begin{gathered} \min_{\alpha_j} \sum_{j=1}^{1} \alpha_j - \frac{1}{2} \sum_{j=1}^{1} \sum_{p=1}^{1} \alpha_j \alpha_p y_j y_j K(l_j, l_p) \\ 0 \leqslant \alpha_j \leqslant B, \ \forall j; \sum_j \alpha_j y_j = 0, \end{gathered} \tag{3.8}$$

where B denotes the cost penalty and α represents the Lagrange multipliers [54]. The SVM uses the concept of the kernel to map the non-linear input vector to the high dimensions. The function Φl functions as a transformational function for mapping the input vector into a high dimensions. The kernel function $K(l)$ is defined in terms of the dot product as

$$K(l_j, l_k) = \Phi(l_j) \times \Phi(l_k). \tag{3.9}$$

Different categories of the kernel have been developed for SVM. The following three types of kernels are the main ones used:

- Linear kernel function:

$$K(l_j, l_k) = l_j^T l_k. \tag{3.10}$$

- The radial basis kernel function has ∂ as a tunable parameter:

$$K(l_j, l_k) = \exp\left(-\frac{\|l_j - l_k\|^2}{2\partial^2}\right). \tag{3.11}$$

- The polynomial kernel function and z are tunable parameters showing variation from 1 to ~10:

$$K(l_j, l_k) = (l_j \times l_k + 1)^z. \tag{3.12}$$

The following are the applications of the SVM for binary classification:

- **Case 1.** The data have h training points and the data are linearly separable in nature.
  1. Construct matrix M, such that

$$M_{jk} = y_j y_k l_j \times l_k. \tag{3.13}$$

  2. Find the value of α for which

$$\sum_{j=1}^{h} \alpha_j - \frac{1}{2}\alpha^T M\alpha \tag{3.14}$$

  has a maximum value with respect to

$$\alpha_j \geqslant 0\ \forall_j \quad \text{and} \quad \sum_{j=1}^{h} \alpha_j y_j = 0 \tag{3.15}$$

  with the help of a quadratic programming solver.

  3. Compute

$$w = \sum_{j=1}^{h} \alpha_j y_j l_j. \tag{3.16}$$

  4. Find the S which is a set of support vectors via determining the indices where $\alpha_j > 0$.
  5. Calculate

$$m = \frac{1}{N_u}\sum_{u\in S}(y_u - \sum_{v\in S}\alpha_v y_v l_v \times l_u). \tag{3.17}$$

  6. Every new data point l' is classified via calculating

$$y' = \mathrm{sign}(w \times l' + m). \tag{3.18}$$

- **Case 2.** The data have h training points and the data are non-linearly separable in nature.
    1. Construct matrix M, such that

$$M_{jk} = y_j y_k l_j \times l_k. \tag{3.19}$$

    2. Select an appropriate value for the cost penalty parameter B to deal with a significant level of misclassification.
    3. Find that value of α for which

$$\sum_{j=1}^{h} \alpha_j - \frac{1}{2}\alpha^T M \alpha \tag{3.20}$$

    have the maximum value with respect to

$$0 \leqslant \alpha_j \leqslant B \ \forall_j \quad \text{and} \quad \sum_{j=1}^{h} \alpha_j y_j = 0, \tag{3.21}$$

    with the help of a quadratic programming solver.
    4. Compute

$$w = \sum_{j=1}^{h} \alpha_j y_j l_j. \tag{3.22}$$

    5. Find the S which is a set of support vectors via determining the indices where $0 < \alpha_{\mathrm{j}} < B$.
    6. Calculate

$$m = \frac{1}{N_u} \sum_{u \in S} (y_u - \sum_{v \in S} \alpha_v y_v l_v \times l_u). \tag{3.23}$$

    7. Every new data point l' is classified via calculating

$$y' = \mathrm{sign}(w \times l' + m). \tag{3.24}$$

- **Case 3.** The data have h training points and the data are non-linearly separable in nature; kernel function based classification.
    1. Select the appropriate kernel and map $l \mapsto \Phi(l)$.
    2. Construct matrix M, such that

$$M_{jk} = y_j y_k \Phi(l_j) \times \Phi(l_k). \tag{3.25}$$

    3. Select an appropriate value for the cost penalty parameter B to deal with the significant level of misclassification.
    4. Find that value of α in which

$$\sum_{j=1}^{h}\alpha_j - \frac{1}{2}\alpha^T M\alpha \tag{3.26}$$

have the maximum value with respect to

$$0 \leqslant \alpha_j \leqslant B\,\forall_j \quad \text{and} \quad \sum_{j=1}^{h}\alpha_j y_j = 0, \tag{3.27}$$

with the help of a quadratic programming solver.

5. Compute

$$w = \sum_{j=1}^{h}\alpha_j y_j \Phi(l_j). \tag{3.28}$$

6. Find the S which is a set of support vectors via determining the indices where $0 < \alpha_\mathrm{j} < B$.
7. Calculate

$$m = \frac{1}{N_u}\sum_{u\in S}\left(y_u - \sum_{v\in S}\alpha_v y_v \Phi(l_v) \times \Phi(l_u)\right). \tag{3.29}$$

8. Every new data point l' is classified via calculating

$$y' = \mathrm{sign}(w \times \Phi(l') + m). \tag{3.30}$$

3.2.3.2 Discriminant analysis (DA) classifier

Ronald Fisher is known as the developer of the discriminant analysis (DA) classifier. DA has played a critical role in classification by solving various problems [55]. The discriminant analysis method is used to categorize the entities into mutually exclusive and exhaustive sets on the basis of quantifiable parameters or features of the entities. The DA classifiers are categorized into two groups: the first is linear discriminant analysis (LDA) and the second is quadratic discriminant analysis (QDA). The significant difference between LDA and QDA is that the LDA classifier has a linear decision surface, whereas QDA has a non-linear decision boundary. Both DA classifiers are based on simple probabilistic models with a conditional distribution of the data $P(L|w = j)$ for each class j. Predictions are obtained with the help of Bayes' rule:

$$P(w = j|L) = \frac{P(L|w = j)P(w = j)}{P(L)}. \tag{3.31}$$

Generally, the class with maximum conditional probability is selected.

Construction of the DA classifier. The DA classifier is built on the basis of two steps. The first step is constructing the building model for the classifier. The second step is how an unknown data sample is classified by the DA classifier. The following presents the structure of the steps [56].

Step 1. Construction of the building model of the classifier.

1. Build a feature matrix L that has K samples, such as $[l_j]_{j=1}^{K}$, where each sample is signified as a column of p length and l_j denotes the jth sample.
2. Calculate each class's mean $\mu_j(p \times 1)$ as

$$\mu_j = \frac{1}{k_j}\sum_{j=1}^{k_j} l_j,\ l_j \in w_j,\ \forall_j = 1, 2, \dots, b, \tag{3.32}$$

where k_j denotes the number of data samples in the jth class and w_j are the sets of b classes.

3. For each class, compute the *a priori* probability, i.e. $P(w_j) = \frac{kj}{K}$.
4. For each class, calculate the covariance matrix using

$$\sum\nolimits_j = \frac{1}{k_j}\sum\nolimits_{j\in w_j}(l - \mu_j)(l - \mu_j)^T,\ \forall_j = 1, 2, \dots, b. \tag{3.33}$$

5. For all (class w_j, $j = 1, 2, \dots, b$) do.
6. Compute the discriminant function (f_j) as

$$f_j(l) = \ln P(w = w_j \mid l) = P(l \mid w = w_j) \tag{3.34}$$

$$P(w_j) = \ln\frac{1}{\sqrt{(2\pi)^p\left|\sum_i\right|}}\exp\left(-\frac{1}{2}(l - \mu_j)^T\sum\nolimits_j^{-1}(l - \mu_j)\right) + \ln(P(w_j)). \tag{3.35}$$

7. End for.

Step 2. How the unknown data sample (UNDS) is classified by the DA classifier.

1. The input is the unknown data samples ($U(p \times 1)$).
2. The output will be the class label (w_j).
3. For all (f_j,discriminant functions that are computed in step 1) do.
4. Put the value of the UNDS (U) in the discriminant function (f_j).
5. End for.
6. Class label (w_{max}) is assigned to the UNDS (U), while (w_{max}) denotes the class with the maximum value of the discriminant function.

3.2.4 Evaluation of performance

This study evaluates the performance of the WCN framework on the testing dataset using the following standard measuring parameters: accuracy, sensitivity and specificity. These criteria allow the behavior of the classifiers on the extracted feature data to be estimated [19].

$$\text{Sensitivity} = \frac{\text{True Positive (TP)}}{\text{TP} + \text{False Negative (FN)}}. \tag{3.36}$$

$$\text{Specificity} = \frac{\text{True Negative (TN)}}{\text{TN} + \text{False Positive (FP)}}. \tag{3.37}$$

$$\text{Accuracy} = \frac{\text{TP} + \text{TN}}{\text{TP} + \text{FN} + \text{TN} + \text{FP}}. \tag{3.38}$$

Here:

- True positive (TP): the number of epileptic seizure (S) samples classified correctly by the proposed model as S.
- False positive (FP): the number of non-seizure (NS) samples classified incorrectly by the proposed model as S.
- True negative (TN): the number of NS samples classified correctly by the proposed model as NS.
- False negative (FN): the number of S samples classified incorrectly by the proposed model as NS.

3.3 Experimental results and discussion

This section presents the experimental exploration of the WCN framework. All experiments are conducted on MATLAB R2015b for the simulation analysis. Each EEG channel is converted into a WCN and after extracting the AWD parameters from the WCN, the classification task is applied.

3.3.1 Experimental data

The WCN framework has been implemented on an epileptic EEG database that is available online, provided by Bonn University, Germany. The database is composed of five sets (labeled Z, O, F, N and S) of different categories of EEG signals. Each set includes 100 single-channel segments of EEG and the duration of each is 23.6 s with 4097 data sample values. Set Z contains EEG segments collected by recording the surface EEG of five healthy people with their eyes open. Set O contains recordings of the same healthy people with their eyes closed. Set F comprises the EEG recordings of five epileptic patients during the non-seizure interval from the epileptogenic zone of the brain. Set N includes the EEG recordings of the same five epileptic patients during the non-seizure interval in the hippocampal formation region of the opposite hemisphere area of the brain. Set S comprises the EEG recordings of the same five epileptic patients during epileptic seizure activity. The recording of all the EEG sets was performed at a 173.61 Hz sampling rate and using a 128 channel amplifier system. The filter called the band-pass filter was used with a value range of 0.53–40 Hz. More detailed information about how these data were collected is available in [57].

3.3.2 Results and discussion

To execute the classification task, the above-mentioned five sets of EEG signals are organized into the following four classification test-group problems:

- Group 1: Set Z and set S.
- Group 2: Set O and set S.
- Group 3: Set N and set S.
- Group 4: Set F and set S.

The cross-validation technique is used during the classification analysis. Cross-validation is a promising technique to assess the general performance as well as the potential of the classifier. In cross-validation, segmentation is performed on the independent dataset to generate the training subset and validation subsets. The subsets are rotated for evaluation. In the end, the average of the outcomes of multiple evaluations is considered for reducing the variation of the evaluation performance. In this research, *k*-fold cross-validation has been performed with $K = 10$ and is called ten-fold cross-validation. In ten-fold cross-validation, the whole feature set is partitioned randomly into ten subsets of equal size. During classification, one subset is used for testing purposes, and the remaining nine subsets are utilized for training of the classification model. This procedure is repeated ten times and, in the end, the average of the (ten) evaluation outcomes are considered. In this chapter, the SVM-linear, SVM-rbf, SVM-polynomial, LDA and QDA classifiers have been used for classifying the EEG signals. Tables 3.1–3.3 present the classification performances of the different classifiers, according to different test-group problems.

The experiments are executed to evaluate the different factors or parameters (such as sensitivity, specificity and accuracy) according to the requirements of the systems. If we consider the sensitivity parameter for group 1, then we find that all of the five classifiers present excellent performance. For group 2, LDA, SVM-linear and SVM-rbf provide 100% sensitivity results in contrast to QDA (with 97%) and SVM-poly (with 96%). Similarly, for group 3 the sensitivity values are 98% for LDA, 96% for QDA, SVM-linear and SVM-bf, and 94% for SVM-poly. For group 4, LDA and SVM-linear provide 100% sensitivity results, with 99% for SVM-rbf and SVM-poly,

Table 3.1. The classification performance in terms of sensitivity for different test groups.

	Classification test-group problems			
Classifiers	Group 1 (%)	Group 2 (%)	Group 3 (%)	Group 4 (%)
LDA	100	100	98	100
QDA	100	97	96	98
SVM-linear	100	100	96	100
SVM-rbf	100	100	96	99
SVM-poly	100	96	94	99

Table 3.2. The classification performance in terms of specificity for different test groups.

	Classification test-group problems			
Classifiers	Group 1 (%)	Group 2 (%)	Group 3 (%)	Group 4 (%)
LDA	76	74	75	76
QDA	100	93	82	99
SVM-linear	91	83	82	85
SVM-rbf	91	85	86	91
SVM-poly	100	94	95	98

Table 3.3. The classification performance in terms of accuracy for different test groups.

	Classification test-group problems			
Classifiers	Group 1 (%)	Group 2 (%)	Group 3 (%)	Group 4 (%)
LDA	88	87	86.5	88
QDA	100	95	89	98.5
SVM-linear	95.5	91.5	89	92.5
SVM-rbf	96.5	92.5	91	95
SVM-poly	100	95	94.5	98.5

and 98% for QDA. Table 3.1 shows that different classifiers provide diverse performances for all four test-group problems. Overall, the LDA classifier is efficient compared to the others in terms of sensitivity for the Bonn epileptic EEG database.

Similarly, the second experiment was executed to measure the specificity performance of the different test-group problems of the Bonn epileptic EEG database. Table 3.2 illustrates the classification outcomes in terms of specificity. As we can see from table 3.2, for group 1, QDA and SVM-poly provide 100% specificity, with 91% for SVM-linear and SVM-rbf, and 76% for LDA. If we consider group 2, the classification outcomes for SVM-poly (94%) and QDA (93%) are very close to each other and are better than those for the other classifiers (85% for SVM-rbf, 83% for SVM-linear and 74% for LDA). The specificity results for group 3 are 95% for SVM-poly, 86% for SVM-rbf, 82% for SVM-linear and QDA, and 75% for LDA. Similarly, for group 4, QDA has a 99% sensitivity, with 98% for SVM-poly, 91% for SVM-rbf, 85% for SVM-linear and 76% for LDA. It is clearly visible from table 3.2 that, overall, the SVM-poly classifier performed very well with better results compared to the other classifiers for all four different classification test-group problems.

The third experiment was performed to evaluate the classification accuracy performance measure for different test-group problems of the Bonn epileptic EEG database. Table 3.3 presents the accuracy outcomes for different classifiers. For group 1 the accuracy outcomes are 100% for SVM-poly and QDA, 96.5% for

SVM-rbf, 95.5% for SVM-linear and 88% for LDA. Similarly, for group 2, the accuracy was 95% for SVM-poly and QDA, 92.5% for SVM-rbf, 91.5% for SVM-linear and 87% for LDA. For group 3, SVM-poly provides an accuracy of 94.5%, with 91% for SVM-rbf, 89% for both QDA and SVM-linear, and 86.5% for LDA. For group 4, 98.5% accuracy was provided by SVM-poly and QDA, with 95% for SVM-rbf, 92.5% for SVM-linear, 98.5% for QDA and 88% for LDA. In short, if we analyse all three tables carefully, then we can say that the SVM classifier with the polynomial kernel function achieved the highest accuracy performance in comparison to the other classifiers for all four classification test-group problems.

Table 3.4 reports the comparative analysis of the WVNBF with different state-of-the-art techniques that have used similar Bonn EEG data in their experiments. Table 3.4 shows that for group 1 (i.e. set Z and set S), different researchers have used different numbers of extracted features for classification. Srinivasan *et al* [58] used the Elman network technique and extracted five different features: spike

Table 3.4. Comparison analysis of the WCN framework with different state-of-the-art techniques that also used similar Bonn EEG data for their experiments.

Classification test-group	Authors	Features	Accuracy (%)
Group 1	Srinivasan *et al* (2005) [58]	5	99.6
	Guo *et al* (2011) [59]	1	99.85
	Siuly *et al* (2011) [18]	9	99.9
	Nicolaou and Georgiou (2012) [60]	1	93.42
	Zhu *et al* (2014) [61]	2	99.0
	Husain and Rao (2014) [62]	–	99.8
	Ghayab *et al* (2016) [63]	9	99.90
	Martinez-del-Rincon *et al* (2017) [64]	1	99.85
	Siuly *et al* (2019) [68]	3	99.50
	Taran *et al* [69]	2	98
	WCN	1	100
Group 2	Siuly *et al* (2011) [18]	9	93.6
	Zhu *et al* (2014) [61]	2	97.0
	WVNBF	1	95
Group 3	Siuly *et al* (2011) [18]	9	96.20
	Xiang *et al* (2015) [65]	1	87.6
	WCN	1	94.5
Group 4	Siuly *et al* (2011) [18]	9	93.60
	Nicolaou and Georgiou (2012) [60]	1	83.13
	Kumar *et al* (2012) [66]	1	93
	Zhu *et al* (2014) [61]	2	93
	Riaz *et al* (2016) [67]	6	93
	Siuly *et al* (2019) [68]	3	97.5
	WCN	1	98.5

rhythmicity, relative spike amplitude, average power, spectral entropy and normalized spectral entropy. The combined set of these five features was used for classification analysis and 99.6% was the highest accuracy achieved with a single input parameter. Guo *et al* [59] used genetic programming for EEG signal analysis. A genetic programming based tree structure was used for feature extraction and after applying the *k*-NN classifier a 99.85% classification accuracy was obtained. Siuly *et al* [18] applied a clustering technique to extract nine parameters from EEGs and then implemented the least squares SVM classifier. A 99.9% accuracy output was obtained with the nine parameters, namely median, minimum, mode, interquartile range, mean, first quartile, maximum, standard deviation and third quartile. Nicolaou *et al* [60] invented the implementation of the permutation entropy parameter in the EEG signal for measuring epileptic activity, and obtained 93.42% classification accuracy using this feature with the SVM classifier. Zhu *et al* [61] introduced the idea of a fast weighted graph based on the horizontal visibility graph. Different types of features were extracted, but degree and strength were the two features that provided a 99% accuracy result. Husain *et al* [62] implemented a Hilbert transform on the intrinsic mode functions and the feed-forward back-propagation algorithm was applied for classification with a 99.8% accuracy. Ghayab *et al* [63] use a combination of two methods, namely random sampling and sequential feature selection. The selected nine features were analysed by the classifier (least squares SVM) and 99.90% accuracy was obtained. Martinez-del-Rincon *et al* [64] used a combination of two methods, the kernel trick and a bag-of-words model, for analysing EEG signals and achieved 99.85% results. Siuly *et al* [68] applied a Hermite transform and after extracting three different types of parameters, namely permutation entropy, statistical features and histogram-based features, the method achieved a 99.50% result in terms of accuracy. Taran *et al* [69] extracted a combination of two features, namely optimum allocation sampling and the Teager energy operator, for epilepsy detection. This method [69] achieved 98% results. Compared to all the above-mentioned techniques, our proposed framework achieved a 100% accuracy result, which is the highest of all.

Similarly, table 3.4 also demonstrates the classification results in terms of accuracy for group 2 (i.e. set O and set S) obtained by different researchers. Using the clustering technique, Siuly *et al* [18] achieved 93.6% results using nine features. Zhu *et al* [61] achieved 97% results with two features extracted from EEG signals. The proposed WCN achieved a 95% result for this group. For group 3 (i.e. set N and set S), table 3.4 presents the accuracy output of different researchers. Siuly *et al* [18] achieved 96.20% with a clustering technique which is close to the proposed WCN result of 94.5%. Xiang *et al* [65] applied fuzzy entropy and achieved 87.6% accuracy, which is much lower compared to our WCN. For group 4 (i.e. set F and set S), table 3.4 illustrates the comparison of the results from different researchers. Siuly *et al* [18] obtained 93.60% and Nicolaou *et al* [60] achieved 83.13% accuracy with one feature. Kumar *et al* [66] applied a discrete wavelet transform technique and used approximate entropy for parameter extraction and achieved a 93% result. Zhu *et al* [61], with a graph based technique, obtained 93% accuracy. Riaz *et al* [67] applied an empirical mode decomposition framework with temporal statistics and spectral statistics for epilepsy identification and achieved 93% accuracy. Siuly *et al* [68] achieved a 97.5% classification performance in terms of accuracy. In comparison to the above-mentioned methods, the proposed WCN achieved 98.5% accuracy, which is the highest of all. Therefore, on the basis of the comparative analysis presented in table 3.4, we can say that the proposed WCN framework is effective in classifying the diverse types of EEG signals and is a promising technique for epilepsy identification.

The above experimental analysis reveals that the AWG parameter of the WCN framework plays a significant role in characterizing the underlying dynamic behavior or nature of different EEG signals. This study also finds that the chaotic nature of EEG signals can be analysed with the help of a network based technique. The experimental results also demonstrate that among all five classifiers, the SVM-poly classifier is more efficient for providing higher performance results within the WCN framework. The information presented in this chapter can be helpful for neuroscientists, researchers and technicians for developing reliable and appropriate techniques for analysing epilepsy or an automated software system for epilepsy identification.

3.4 Conclusion

This chapter presents the ideas behind the importance of edge weight in the network for the detection of an epileptic seizure. Subsequently, a WCN framework based on weighted graph theory is proposed. An innovative edge weight computing method is used in the visibility graph. In addition, a statistical parameter called the average weighted degree is used as a graph feature for feature extraction. The WCN based framework is evaluated with five different classifiers: LDA, QDA, SVM-linear, SVM-rbf and SVM-poly. The experimental outcomes find that SVM with a polynomial kernel is more suitable and provides higher classification results compared to the other four classifiers and outperforms existing methods.

The experimental results of this chapter explain that the WCN based framework is superior and useful for distinguishing different categories of EEG signals. Also, this proposed method is proficient in distinguishing epileptic seizure activity from healthy subjects' EEG signals.

References

[1] Siuly L Y and Zhang Y 2016 EEG signal analysis and classification: techniques and applications *Health Information Science* (Berlin: Springer)

[2] Alçin Ö, Siuly S, Bajaj V, Guo Y, Şengür A and Zhang Y 2016 Multi-category EEG signal classification developing time-frequency texture features based Fisher vector encoding method *Neurocomputing* **218** 251–8

[3] Acharya U, Sree S, Alvin A and Suri J 2012 Use of principal component analysis for automatic classification of epileptic EEG activities in wavelet framework *Expert Syst. Appl.* **39** 9072–8

[4] Ghosh-Dastidar S, Adeli H and Dadmehr N 2008 Principal component analysis-enhanced cosine radial basis function neural network for robust epilepsy and seizure detection *IEEE Trans. Biomed. Eng.* **55** 512–8

[5] Alarcon G, Binnie C, Elwes R and Polkey C 1995 Power spectrum and intracranial EEG patterns at seizure onset in partial epilepsy *Electroencephalogr. Clin. Neurophysiol.* **94** 326–37
[6] Hassan A R, Siuly S and Zhang Y 2016 Epileptic seizure detection in EEG signals using tunable-Q factor wavelet transform and bootstrap aggregating *Comput. Methods Programs Biomed.* **137** 247–59
[7] Hsu M and Hsu D 2005 A piece-wise harmonic Langevin model of EEG dynamics: theory and application to EEG seizure detection *Neurocomputing* **65–66** 469–74
[8] Osterhage H and Lehnertz K 2007 Nonlinear time series analysis in epilepsy *Int. J. Bifurcat. Chaos* **17** 3305–23
[9] Vavadi H, Ayatollahi A and Mirzaei A 2010 A wavelet-approximate entropy method for epileptic activity detection from EEG and its sub-bands *J. Biomed. Sci. Eng.* **03** 1182–9
[10] Bellegdi S and Arafat S 2017 Automatic detection of epilepsy using EEG energy and frequency bands *Int. J. Appl. Math. Electron. Comput.* **1** 36–41
[11] Kabir E, Siuly, Cao J and Wang H 2018 A computer aided analysis scheme for detecting epileptic seizure from EEG data *Int. J. Comput. Intell. Syst.* **11** 663
[12] Zarei R, He J, Siuly and Zhang Y 2019 Exploring Douglas Peucker algorithm in the detection of epileptic seizure from multiclass EEG signals *BioMed Res. Int.* **2019** 1–19
[13] Ghayab H, Li Y, Siuly S and Abdulla S 2018 Epileptic seizures detection in EEGs blending frequency domain with information gain technique *Soft Comput.* **23** 227–39
[14] Wang X, Gong G and Li N 2019 Automated recognition of epileptic EEG states using a combination of Symlet wavelet processing, gradient boosting machine, and grid search optimizer *Sensors* **19** 219
[15] Mursalin M, Zhang Y, Chen Y and Chawla N 2017 Automated epileptic seizure detection using improved correlation-based feature selection with random forest classifier *Neurocomputing* **241** 204–14
[16] Hegde M, Santos-Sanchez C, Hess C, Kabir A and Garcia P 2012 Seizure exacerbation in two patients with focal epilepsy following marijuana cessation *Epilepsy Behav.* **25** 563–6
[17] Rana P, Lipor J, Lee H, van Drongelen W, Kohrman M and Van V B 2012 Seizure detection using the phase-slope index and multichannel ECoG *IEEE Trans. Bio-Med. Eng.* **59** 1125–34
[18] Siuly, Li Y and Wen P 2011 Clustering technique-based least square support vector machine for EEG signal classification *Comput. Methods Programs Biomed.* **104** 358–72
[19] Siuly S and Li Y 2015 Designing a robust feature extraction method based on optimum allocation and principal component analysis for epileptic EEG signal classification *Comput. Methods Programs Biomed.* **119** 29–42
[20] Zainuddin Z, Huong L and Pauline O 2012 On the use of wavelet neural networks in the task of epileptic seizure detection from electroencephalography signals *Procedia Comput. Sci.* **11** 149–59
[21] Pradhan N and Dutt D 1994 Data compression by linear prediction for storage and transmission of EEG signals *Int. J. Biomed. Comput.* **35** 207–17
[22] Altunay S, Telatar Z and Erogul O 2010 Epileptic EEG detection using the linear prediction error energy *Expert Syst. Appl.* **37** 5661–5
[23] Tsipouras M 2019 Spectral information of EEG signals with respect to epilepsy classification *EURASIP J. Adv. Signal Process.* **10** 1–17
[24] Yoo J, Yan L, El-Damak D, Altaf M, Shoeb A and Chandrakasan A 2013 An 8-channel scalable EEG acquisition SoC with patient-specific seizure classification and recording processor *IEEE J. Solid-State Circuits* **48** 214–28

[25] Bajaj V, Rai K, Kumar A and Sharma D 2017 Time-frequency image based features for classification of epileptic seizures from EEG signals *Biomed. Phys. Eng. Express* **3** 015012

[26] Bajaj V and Pachori R 2012 Classification of seizure and nonseizure EEG signals using empirical mode decomposition *IEEE Trans. Inf. Technol.* B **16** 1135–42

[27] Xie S, Jin F and Krishnan S 2011 Sparse approximation of long-term biomedical signals for classification via dynamic PCA *Conf. Proc. IEEE Engineering in Medicine and Biology Society* pp 7167–70

[28] Rajendra Acharya U, Vinitha Sree S, Alvin A and Suri J 2012 Use of principal component analysis for automatic classification of epileptic EEG activities in wavelet framework *Expert Syst. Appl.* **39** 9072–8

[29] Nam H, Yim T G, Han S K, Oh J B and Lee S K 2002 Independent component analysis of ictal EEG in medial temporal lobe epilepsy *Epilepsia* **43** 160–4

[30] Whitmer D, Worrell G, Stead M, Lee I and Makeig S 2010 Utility of independent component analysis for interpretation of intracranial EEG *Front. Hum. Neurosci.* **4** 1–13

[31] Arunkumar N, Balaji V S, Ramesh S and Natarajan S 2012 Automatic detection of epileptic seizures using independent component analysis algorithm *IEEE-Int. Conf. On Advances in Engineering, Science and Management (ICAESM-2012)* pp 542–4

[32] Fathima T, Khan Y U, Bedeeuzzaman M and Farooq O 2011 Discriminant analysis for epileptic seizure detection *IEEE Int. Conf. on Devices and Communications (ICDeCom, Feb 24)* pp 1–5

[33] Gursoy M I and Subast A 2008 A comparison of PCA, ICA, and LDA in EEG signal classification using SVM *IEEE 16th Signal Processing, Communication and Applications Conf. (SIU 2008)* pp 1–4

[34] Subasi A and Ismail Gursoy M 2010 EEG signal classification using PCA, ICA, LDA and support vector machines *Expert Syst. Appl.* **37** 8659–66

[35] Nigam V and Graupe D 2004 A neural-network-based detection of epilepsy *Neurol. Res.* **26** 55–60

[36] Siuly S and Li Y 2014 A novel statistical algorithm for multiclass EEG signal classification *Eng. Appl. Artif. Intell.* **34** 154–67

[37] Martis R *et al* 2012 Application of empirical mode decomposition (EMD) for automated detection of epilepsy using EEG signal *Int. J. Neural Syst.* **22** 1250027

[38] Chua K, Chandran V, Acharya U and Lim C 2010 Application of higher order spectra to identify epileptic EEG *J. Med. Syst.* **35** 1563–71

[39] Acharya U R, Sree S V and Suri J S 2011 Automatic detection of epileptic EEG signals using higher-order cumulant features *Int. J. Neural Syst.* **21** 1–12

[40] Zhang J and Small M 2006 Complex network from pseudoperiodic time series: topology versus dynamics *Phys. Rev. Lett.* **96** 238701

[41] Stam C and van-Straaten E 2012 The organization of physiological brain networks *Clin. Neurophysiol.* **123** 1067–87

[42] Bassett D and Bullmore E 2006 Small-world brain networks *Neuroscientist* **12** 512–23

[43] Wilke C, Worrell G and He B 2009 Analysis of epileptogenic network properties during ictal activity *2009 Annual Int. Conf. of the IEEE Engineering in Medicine and Biology Society*

[44] Ni Y, Wang Y, Yu T and Li X 2014 Analysis of epileptic seizures with complex network *Comput. Math. Methods Med.* **14** 1–6

[45] Petkov G, Goodfellow M, Richardson M and Terry J 2014 A critical role for network structure in seizure onset: a computational modeling approach *Front. Neurol.* **5** PMC4259126

[46] Wang F, Meng Q, Zhou W and Chen S 2013 The feature extraction method of EEG signals based on degree distribution of complex networks from nonlinear time series *Intelligent Computing Theories* (Lecture Notes in Computer Science) (Berlin: Springer) pp 354–61
[47] Thiel M, Romano M, Read P and Kurths J 2004 Estimation of dynamical invariants without embedding by recurrence plots *Chaos* **14** 234–43
[48] Niknazar M, Mousavi S, Vosoughi V B and Sayyah M 2013 A new framework based on recurrence quantification analysis for epileptic seizure detection *IEEE J. Biomed. Health Inform.* **17** 572–8
[49] Lacasa L, Luque B, Ballesteros F, Luque J and Nuño J 2008 From time series to complex networks: the visibility graph *Proc. Natl Acad. Sci.* **105** 4972–5
[50] Liu J, Liu H, Huang Z and Tang Q 2015 Differ multivariate time series from each other based on a simple multiplex visibility graphs technique *2015 Sixth Int. Conf. on Intelligent Control and Information Processing (ICICIP)*
[51] Antoniou I and Tsompa E 2008 Statistical analysis of weighted networks *Discrete Dyn. Nat. Soc.* **2008** 1–16
[52] Jakkula V 2006 *Tutorial on Support Vector Machine (SVM)* (School of EECS, Washington State University) p 37 https://course.ccs.neu.edu/cs5100f11/resources/jakkula.pdf
[53] Cristianini N and Shawe-Taylor J 2014 *An Introduction to Support Vector Machines and Other Kernel-based Learning Methods* (Cambridge: Cambridge University Press)
[54] Vapnik V 1998 *Statistical Learning Theory* (New York: Wiley)
[55] Sapatinas T 2005 Discriminant analysis and statistical pattern recognition *J. R. Stat. Soc.* A **168** 635–6
[56] Tharwat A 2016 Linear vs quadratic discriminant analysis classifier: a tutorial *Int. J. Appl. Pattern Recogn.* **3** 145
[57] Andrzejak R, Lehnertz K, Mormann F, Rieke C, David P and Elger C 2001 Indications of nonlinear deterministic and finite-dimensional structures in time series of brain electrical activity: dependence on recording region and brain state *Phys. Rev.* E **64** 061907
[58] Srinivasan V, Eswaran C and Sriraam A 2005 Artificial neural network based epileptic detection using time-domain and frequency-domain features *J. Med. Syst.* **29** 647–60
[59] Guo L, Rivero D, Dorado J, Munteanu C and Pazos A 2011 Automatic feature extraction using genetic programming: an application to epileptic EEG classification *Expert Syst. Appl.* **38** 10425–36
[60] Nicolaou N and Georgiou J 2012 Detection of epileptic electroencephalogram based on permutation entropy and support vector machines *Expert Syst. Appl.* **39** 202–9
[61] Zhu G, Li Y and Wen P 2014 Epileptic seizure detection in EEGs signals using a fast weighted horizontal visibility algorithm *Comput. Methods Programs Biomed.* **115** 64–75
[62] Husain S and Rao K S 2014 An artificial neural network model for classification of epileptic seizures using Huang–Hilbert transform *Int. J. Soft Comput.* **5** 23–33
[63] Ghayab H, Li Y, Abdulla S, Diykh M and Wan X 2016 Classification of epileptic EEG signals based on simple random sampling and sequential feature selection *Brain Inform.* **3** 85–91
[64] Martinez-del-Rincon J *et al* 2017 Non-linear classifiers applied to EEG analysis for epilepsy seizure detection *Expert Syst. Appl.* **86** 99–112
[65] Xiang J, Li C, Li H, Cao R, Wang B, Han X and Chen J 2015 The detection of epileptic seizure signals based on fuzzy entropy *J. Neurosci. Methods* **243** 18–25

[66] Kumar Y, Dewal M and Anand R 2012 Epileptic seizures detection in EEG using DWT-based ApEn and artificial neural network *Signal Image Video Process.* **8** 1323–34
[67] Riaz F, Hassan A, Rehman S, Niazi I and Dremstrup K 2016 EMD-based temporal and spectral features for the classification of EEG signals using supervised learning *IEEE Trans. Neural Syst. Rehabil. Eng.* **24** 28–35
[68] Siuly S, Alcin O, Bajaj V, Sengur A and Zhang Y 2019 Exploring Hermite transformation in brain signal analysis for the detection of epileptic seizure *IET Sci. Meas. Technol.* **13** 35–41
[69] Taran S, Bajaj V and Siuly S 2017 An optimum allocation sampling based feature extraction scheme for distinguishing seizure and seizure-free EEG signals *Health Inf. Sci. Syst.* **5** 2–7

IOP Publishing

Modelling and Analysis of Active Biopotential Signals in Healthcare, Volume 1

Varun Bajaj and G R Sinha

Chapter 4

Epileptic seizure prediction and onset zone localization using intracranial and scalp electroencephalographic and magnetoencephalographic signals

Hamid Reza Marateb[1,2], Carolina Migliorelli[2,3], Alejandro Bachiller[2], Tayebe Azimi[1], Farzad Ziaie Nezhad[1], Marjan Mansourian[2,4], Joan Francesc Alonso[2,3], Javier Aparicio[5], Maria Victoria San Antonio-Arce[5,6], Sergio Romero[2,3], Miguel Ángel Mañanas[2,3]

[1]*Biomedical Engineering Department, Engineering Faculty, University of Isfahan, Isfahan, Iran*

[2]*Biomedical Engineering Research Centre (CREB), Automatic Control Department (ESAII) Universitat Politécnica de Catalunya-Barcelona Tech (UPC), Barcelona, Spain*

[3]*Biomedical Research Networking Center in Bioengineering, Biomaterials, and Nanomedicine (CIBER-BBN), Spain*

[4]*Epidemiology and Biostatistics Department, Health School, Isfahan University of Medical Sciences, Isfahan, Iran*

[5]*Epilepsy Unit, Department of Neuropediatrics, Hospital Sant Joan de Déu, Barcelona, Catalunya, Spain*

[6]*Department of Neuropediatrics, University of Freiburg Hospital, Freiburg, Baden-Württemberg, Germany*

Epilepsy, a significant cause of disability and mortality and one of the most common neurological disorders, is associated with seemingly random incidences of recurrent seizures. It affects one percent of the world's population and individuals of all ages and has a worldwide distribution. About two-thirds of such patients could benefit from antiepileptic drugs. Resective surgery is a possible treatment for the other group (i.e. refractory epilepsy). In focal epilepsy, where a limited area in the brain is involved in abnormal neural activity, the identification of the epileptogenic zone (i.e. the seizure onset zone) is critical. In addition, preventing seizures using medication is facilitated by the prediction of epileptic seizures. In this chapter, we discuss how intracranial or scalp electroencephalographic and magnetoencephalographic recordings could be used for epileptic seizure prediction and onset zone localization. The signal processing methods and the challenges of the state-of-the-art are discussed.

4.1 Epileptic seizure prediction

Epilepsy is a chronic neurological pathology affecting the nervous system. It is one of the most prevalent neurological diseases and affects about one percent of the world population [1]. The disability-adjusted life year (DALY), calculated as the sum of the years of life lost (YLL) and the years lost due to disability (YLD) for people living, are shown in figure 4.1 from 1990 to 2017 in different World Health Organization (WHO) regions, showing an increasing trend in the African and American regions.

Epilepsy is characterized by recurrent paroxysmal events, called seizures, resulting in neurobiological, cognitive, psychological and social consequences [2]. A seizure is a brief, temporary, hyper-synchronized disturbance of the electrical activity of the brain. Epilepsy is often related to conditions of poverty and a loss of earnings due to disability, which affects people with epilepsy as well as people caring for children with epilepsy. This is essential because epilepsy is more likely to first develop during childhood. It is the most chronic disorder of the brain in childhood, affecting ~0.5%–1% of children [3].

Anti-convulsant drugs provide seizure-free outcomes in approximately 70% of cases [4], but according to the WHO approximately one-third of patients with epilepsy have refractory epilepsy or are resistant to medical therapy. Refractory epilepsy is a debilitating neurological condition, in which the uncertain nature of seizures adversely affects the patient's life quality. It affects their cognitive function and the maturation of the brain is thus impaired due to the impossibility of establishing regular cognitive networks [5]. The timely prediction of seizures could help patients or their caregivers to reduce seizure-related injuries.

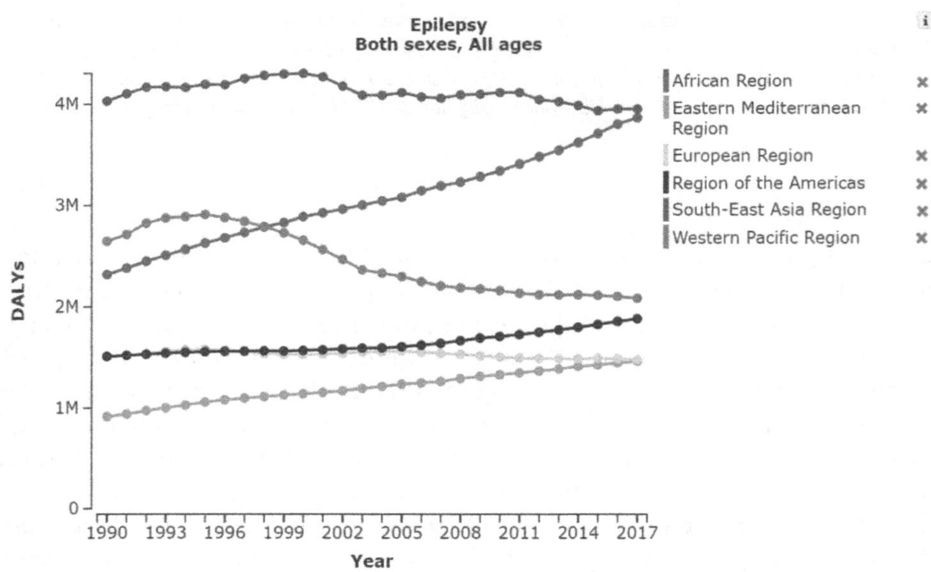

Figure 4.1. DALYs from 1990 to 2017 in different WHO regions. The graph was generated using the Global Burden of Diseases (GBD) visualization tool, available from https://vizhub.healthdata.org/gbd-compare/.

If the epilepsy is partial (focal epilepsy as opposed to generalized epilepsy), removing the epileptogenic zone (EZ) by surgery is the most promising treatment. The EZ is a theoretical zone defined as the minimum amount of cortex to resect to avoid seizure [6, 7]. Neurosurgeons have to delimit the EZ to obtain a successful surgical outcome. In fact, in the presurgical evaluation of refractory epilepsy, non-invasive and invasive electromagnetic recordings play a key role.

The functional (or eloquent) areas must be preserved while removing the EZ in epilepsy surgery [8]. In the majority of cases (75% to 50%), neuroimaging and non-invasive electrophysiology techniques provide proper EZ localization and proper eloquent area mapping. The presurgical assessment often involves an in-depth examination of structural and functional magnetic resonance imaging (MRI and fMRI), computerized tomography (CT), magnetoencephalography (MEG) and electroencephalography (EEG) recordings, positron emission tomography (PET) scans, etc. Invasive electrophysiology is required for patients whose EZ is not possible to localize with non-invasive technologies [9].

The two approaches that are currently being used are electrocorticography (ECoG), where a subdural grid is implanted directly in contact with the brain cortex, and stereo-electroencephalography (SEEG), where needle-shaped electrodes are implanted to map several regions of the brain.

Until the late 1990s, neuroscientists have thought that epileptic seizures begin suddenly, a few seconds before their clinical manifestation [10]. Viglione and colleagues predicted a seizure for the first time in 1970 [11]. The studies on seizure prediction in the 1970s and 1980s addressed the quantification of noticeable features in the EEG, such as sharp waves or spikes [10]. In the late 1980s and 1990s neuroscientists focused on predicting seizures based on the application of non-linear dynamics. They proposed that when seizures approached, the EEG became progressively less chaotic. In the early 2000s, researchers discovered some proof that seizures develop from minutes to hours prior to clinical onset, and they started to perform long-term EEG recordings [10, 12–14].

In 2007, a comprehensive review by Mormann *et al* did not draw a definite conclusion on whether seizures are predictable [13]. Nevertheless, several innovations have been proposed in the last decade, providing convincing evidence that the development of a device that predicts seizures is possible [14]. A better understanding of the physiological factors leading to the transition to epileptic seizures should now be taken into account when performing seizure prediction [14].

Seizure prediction algorithms mainly use scalp EEG or intracranial EEG (iEEG) recordings [15, 16]. Scalp EEG is mostly used by researchers since recording iEEG signals requires a lot of experience [16]. Other techniques, such as neuroimaging, could be used for the localization of epileptic lesions, but neuroimaging tools mainly depict a static measure of the brain. Thus EEG provides a useful measure of neural activity.

Seizure prediction algorithms typically follow similar schema. The raw data are continuously cleaned with an artifact rejection protocol, and various time–frequency analysis methods are applied for feature extraction. Then a classifier is usually applied. Finally, the advisor system sends a notification to the patient's caregivers (figure 4.2).

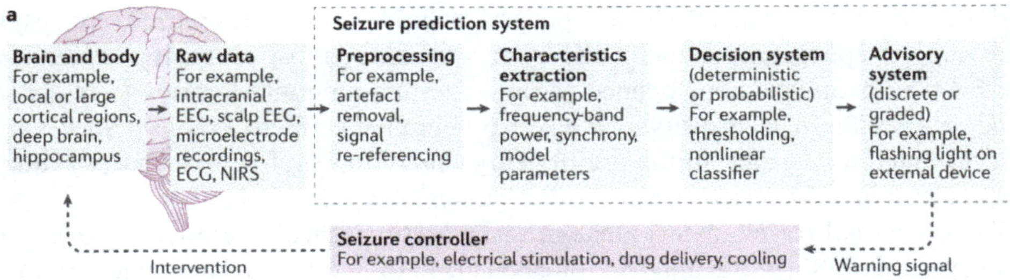

Figure 4.2. The structure of a seizure prediction system. After recording and filtering the invasive or non-invasive biosignals, the features are extracted. Then, such features are processed to detect changes indicating a seizure. (Reproduced with permission from [14].)

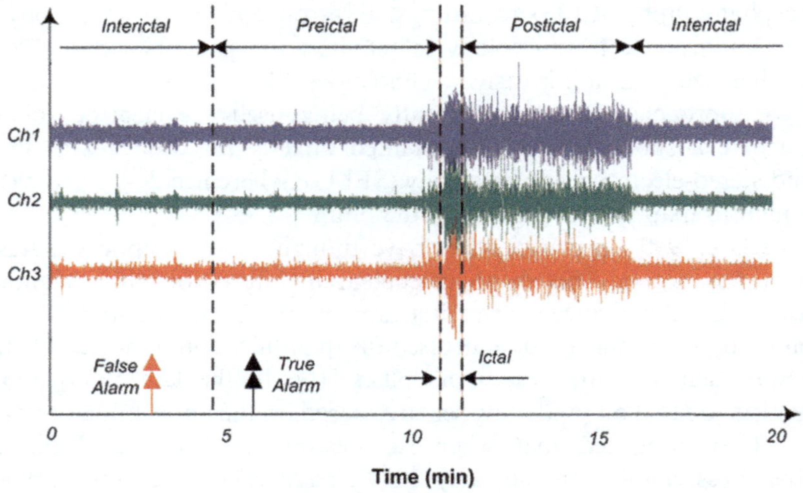

Figure 4.3. The interictal, preictal, ictal and postictal epileptic states are shown on multichannel electroencephalogram recordings. (Reproduced with permission from [17].)

Neurologists classify epileptogenic EEG recordings into four well-defined groups, interictal, preictal, ictal and postictal, having distinct patterns (figure 4.3). The preictal state precedes the seizure onset from several seconds up to several hours. Ictal is the interval where the seizure occurs, lasting from a few seconds to 2–3 min, and is initially marked by experts. The seizure prediction algorithms originally identify the preictal period–which precedes the seizure–as much as possible from the three other states [17]. The beginning interval right after one postictal seizure state prior to the next seizure's preictal state is called the interictal state.

A variety of algorithms were proposed in the literature for epileptic seizure prediction [18–20]. Such algorithms mostly follow the basic procedure shown in figure 4.4. Denoising, channel or feature selection, and classification are the basic steps of such predictions. Different signals such as EEG [20–33], iEEG [30, 33–43] and magnetoencephalography (MEG) [44] were used in the literature for such a prediction. A flowchart of a generic EEG/iEGG epileptic seizure prediction system is shown in figure 4.4.

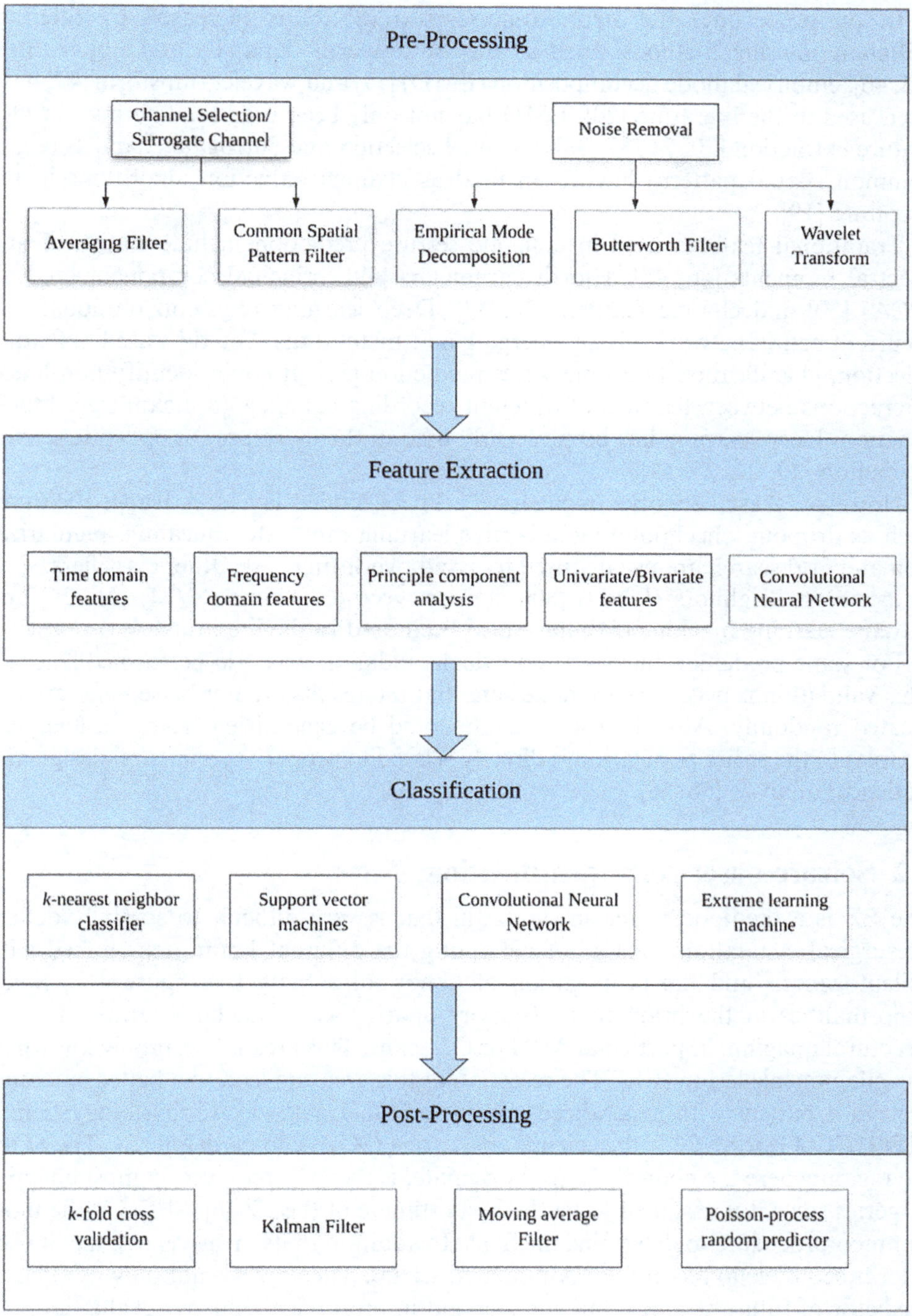

Figure 4.4. Epileptic seizure prediction systems, including pre-processing, feature extraction, classification and post-processing. (Reproduced from [20] with permission from Elsevier.)

In pre-processing, the signal-to-noise-ratio (SNR) is increased by filtering. Different filtering methods, such as the Butterworth bandpass and notch filters [45, 46], empirical mode decomposition (EMD) [17] and wavelet transform [47] have been used in the literature [20]. EMD has not only been used for filtering but also feature extraction [20, 21, 32, 38]. Channel selection and surrogate channels (e.g. a common spatial pattern) have been used as channel reduction algorithms in the literature [19].

Traditional features used in epileptic seizure prediction include statistical and spectral moments [46, 48], Hjorth parameters [49], principal component analysis (PCA) [50] and chaotic features [20, 37]. Deep learning (e.g. convolutional and recurrent neural networks) is an emerging new method that could be used in feature selection, classification and time-series prediction [51]. It could identify non-linear interactions between features of different recording channels to maximize a fitness function. Deep learning has been recently used in the literature for epileptic seizure prediction [20, 22, 25, 52].

However, these networks might have a lot of hidden layers. Advanced features such as dropout, checkpointing, adaptive learning rate, rate annealing, regularization and grid search are usually used to avoid overfitting [53]. Other classifiers, such as *k*-nearest neighbors (*k*-NN) [30], support vector machine (SVM) [21, 36] and extreme learning machine [54], have also been used in the literature.

For some prediction methods, statistical validation was also performed [21, 55]. This validation is necessary to make sure that the results are not biased and are not created randomly. Also, how the results could be generalized from the analyzed samples to the entire population is directly related to proper experimental design and statistical analysis [56–58].

4.2 Seizure onset zone identification

The EZ is a theoretical area in the brain that is very difficult to spatially define. Presurgical evaluation consists of delimiting the different brain areas linked with epileptogenesis and the propagation of abnormal activity (see figure 4.5). Some abnormalities in the brain tissue (tumors or dysplasia) can be determined using structural imaging, in particular MRI or CT scans. This area is commonly known as an epileptogenic lesion (EL). The resection of this area can lead to a better outcome, even in a patient with generalized epilepsies [59]. Using electrophysiology (EEG, MEG, ECoG or SEEG), the seizure onset zone (SOZ) can be delimited. The SOZ, the region where the clinical seizures originate, is the area removed in most epilepsy surgeries since it is assumed to be the best estimate of the EZ [6]. SEEG is the most common procedure to determine the SOZ. In addition to its invasiveness, the SOZ is determined by ictal recordings. Automated seizure detectors are not only useful as a predictor of future seizures, but can also aid in presurgical epilepsy evaluation.

Determining the SOZ using ictal invasive recordings faces several difficulties. First, clinical seizures are complicated to predict and therefore capture using SEEG. To capture seizures, it is necessary to record the signal for several days. Thus, taking into account the time, cost and risk, ictal recordings are not ideal [61]. Other useful transient

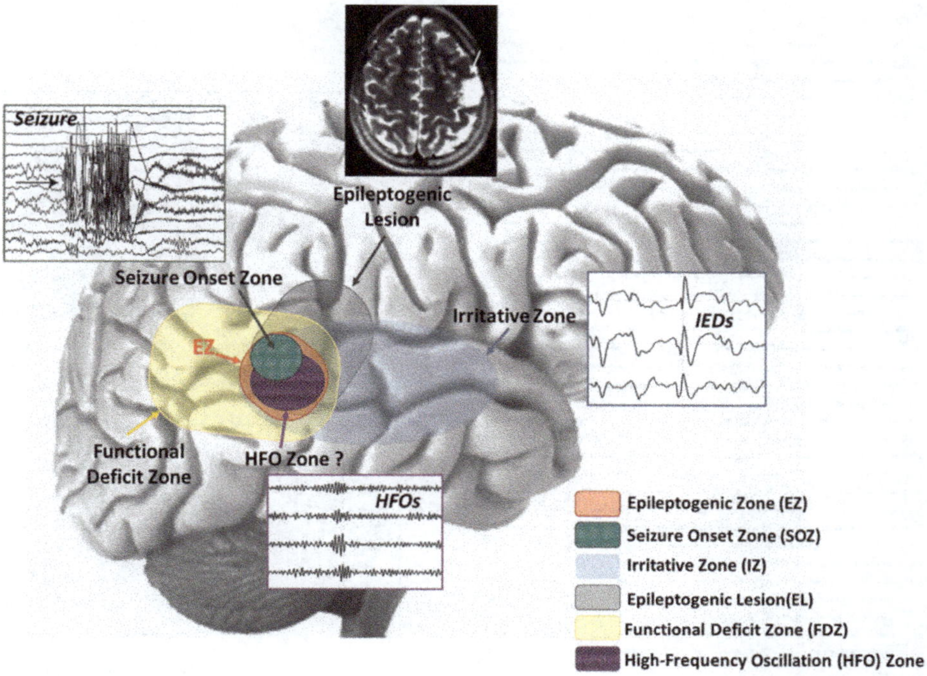

Figure 4.5. The overlapping epileptical cortical zones. Using a variety of imaging and electrophysiology techniques, these zones can be defined and used by the epileptologist for EZ dissection. (Reproduced with permission from [60].)

epilepsy-related perturbations can be recorded during the interictal period in rest. The most critical epileptiform abnormalities are the interictal epileptiform discharges (IEDs) that constitute the irritative zone (IZ) [62] and the novel biomarkers, called pathological high-frequency oscillations (HFOs), that comprise the HFO zone (HFOZ) [63–65].

The second difficulty in current medical practice is using invasive electrophysiology (ECoG and SEEG). These techniques provide a much higher SNR than non-invasive methods and an excellent spatial resolution for the area where they are implanted. However, in addition to their high invasiveness and high cost, invasive EEG sometimes does not provide a proper SOZ delimitation because it only provides a limited neurophysiological picture of the brain [67]. In this sense, there is increasing interest in non-invasive techniques to map the focal sources generating the epileptogenic activity (see figure 4.6). MEG and EEG can provide a whole-head picture of the epileptogenic processes and not only be useful to detect the focus but also to determine the principal and secondary epileptogenic networks [60]. Estimation of the location of the recorded activity is performed using source reconstruction methods, which are considered inverse problems. Several methods were tested to detect the SOZ in non-invasive signals: equivalent current dipole (ECD) [68], linear constrained minimum variance beamformer (LCMV-beamformer) [69], minimum norm estimation (MNE) [70] and low-resolution electromagnetic tomography (LORETA) [71], among others. These methods can be used to localize the SOZ with ictal and preictal recordings, as well as to localize it with

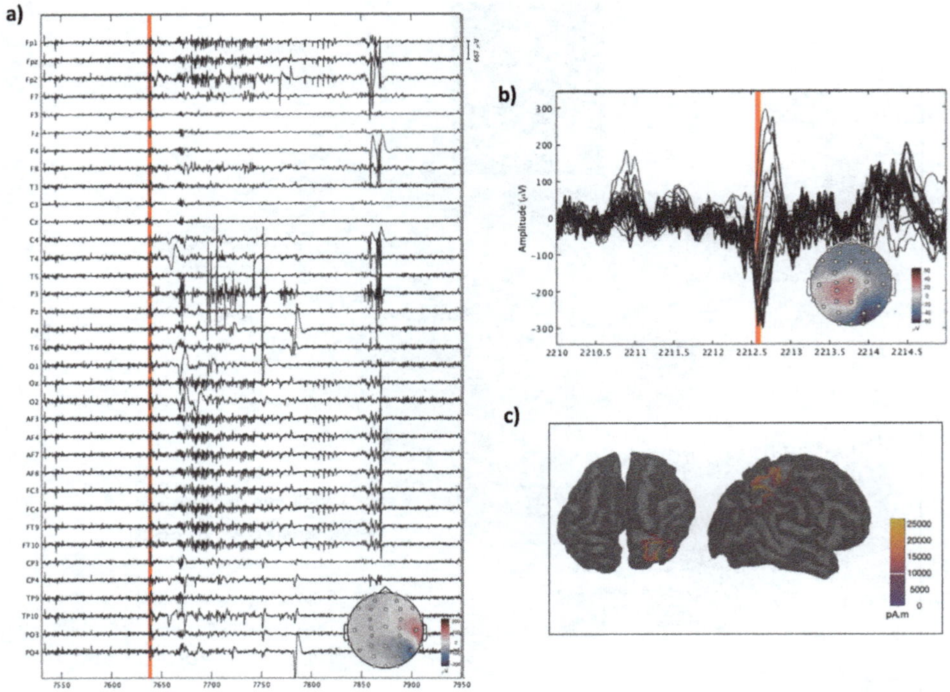

Figure 4.6. EEG visualizations of epileptic events recorded from an epileptic patient at the Hospital Sant Joan de Déu, Barcelona. (a) Representation of an epileptic seizure onset with a channel-by-channel visualization and the related topographic map of the SOZ. (b) Representation of an IED in butterfly mode and its related topographic map. Note that the localization of the IED does not precisely match the localization of the SOZ. (c) Source representation of the SOZ using MNE source analysis. Visualizations were obtained using an open-source program called Brainstorm [66], (http://neuroimage.usc.edu/brainstorm).

interictal biomarkers such as IEDs and novel HFOs. A variety of artifacts, including muscular, ocular and cardiac [72], and metallic interference [73], affect non-invasive EEG and MEG recordings. Thus, proper signal processing methods must be used to reduce such artifacts before any further processing. One of the frequently used methods is blind source separation (BSS) in the filtering step [73, 74]. Source analysis can also aid in the reduction of interference and improving the SNR, acting as a spatial filter [65, 67].

Detection of IEDs in non-invasive interictal signals using the equivalent current dipole (ECD) is a common practice in clinical settings [75]. In focal epilepsy, where the cortical activity is generated by a single dipolar source, it is possible to use this model. Several studies have validated its accuracy for epileptic focus localization [76–79]. However, IEDs may appear spread throughout the irritative area, which may include zones that are not associated with the epileptic discharge generation but the epileptiform propagation. On the other hand, in invasive recordings, specific biomarkers such as HFOs were evaluated for epileptogenicity [7]. These oscillations can be categorized as fast ripples (200–500 Hz) and ripples (80–200 Hz) and are usually above the baseline noise [80]. IEDs seem to be consistent with IEDs. However, their association is not entirely confirmed, and HFOs could be better

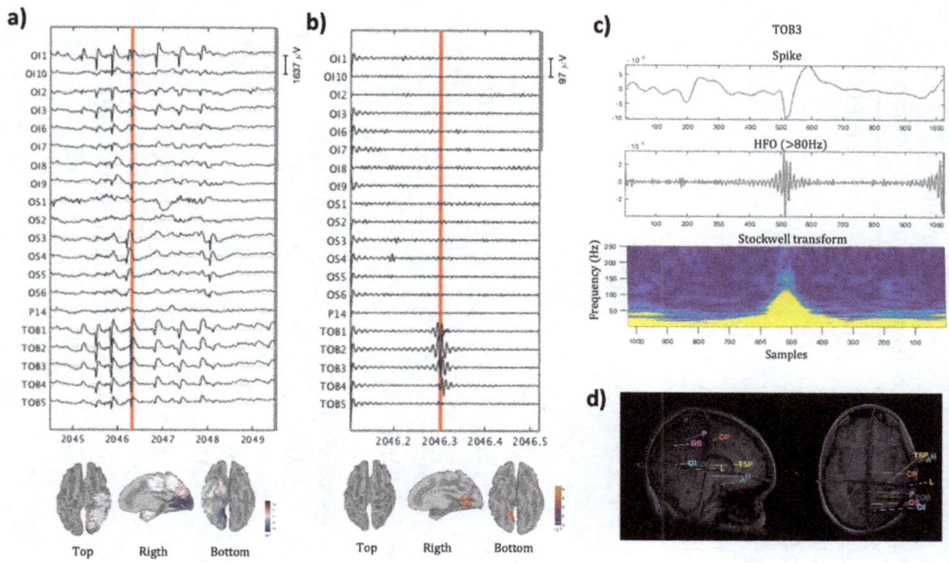

Figure 4.7. SEEG visualizations of epileptic interictal events recorded from an epileptic patient at the Hospital Sant Joan de Déu, Barcelona. (a) Visualization of the most representative electrode group of an interictal epileptiform discharge, and the topographical activation in the brain. (b) Visualization of the most representative electrode group of HFO and the topographical activation in the brain. Note that the area comprising the HFO is smaller than the area covered by the IEDs. (c) Representation of a spike + HFO event. By using the Stockwell transform, the HFO event at frequencies higher than 150 Hz can be observed and differentiated from the spike (whose energy reaches 100 Hz, approximately). (d) Placement of SEEG electrodes. Visualizations (a), (b) and (d) were obtained using an open-access program called Brainstorm [66] (http://neuroimage.usc.edu/brainstorm).

linked to the SOZ compared to IEDs as interictal biomarkers [7]. Visual detection is time-consuming and, in general, experts have to review the bandpass filtered and non-filtered signal to decide if an event is a fast oscillation, and agreement between experts should exist. For this reason, there is increasing interest in the automated detection of HFOs. Several automated algorithms have been proposed in the literature for HFO detection using invasive signals [65, 81–86]. Most advanced algorithms rely on time–frequency decompositions (see figure 4.7(c)). In the time–frequency map, events should be isolated from low-frequency activity to be considered as a real HFO and differentiated from noise or artifacts. HFOs require an excellent recording quality because they have a lower SNR and lower spatial distribution than IEDs. Accordingly, the detection using non-invasive signals is still an open discussion. Several algorithms have recently assessed its detectability based on non-invasive signals [60, 64, 87–91], but more efforts should be made in this direction.

Some of the methods and signal modalities proposed in the literature for SOZ detection are listed in table 4.1. For example, Elahian *et al* computed the phase-locking value (PLV) of ECoG recordings between the lower frequency rhythms (4–30 Hz) and high gamma activity (80–150 Hz). Such features were fed to the logistic regression classifier to identify whether an electrode is an SOZ or not, and an accuracy of 83.0% was reached [92].

Table 4.1. Some of the SOZ detection algorithms.

Ref.	Year	Method	# Electrodes	Performance measures	# Subjects	Sampling rate	Signal
[93]	2010	Local tangent space alignment method (LTSA) and using incremental learning	3	Sensitivity = 98.8% Average uninteresting false positives rate = 0.24 Average interesting false positives rate = 0.25 Average detection delay = 10.8	21	256	iEEG
[94]	2011	Continuous calculation of short-time energy within a particular frequency band and thresholding	40–128	Specificity = 96.4% Sensitivity = 83.8%	5	500	iEEGh
[95]	2012	Hemisphere-lobe-surface (HLS) model and hemisphere-lobe (HL) model	32 or 64	Sensitivity (ictal MEG source localization on HL level) = 95.8% Sensitivity (ictal MEG source localization on HLS level) = 70.6% Specificity (ictal MEG source localization on HL level) = 90.0% Specificity (ictal MEG source localization on HLS level) = 73.1%	47	—	MEG/iMEG/iEEG
[96]	2013	Delay permutation entropy (DPE) methodology and support vector machine (SVM) classification	—	Accuracy = 84.0%	10	512 or 1024	iEEG

[97]	2014	*k*-means algorithm with non-random centroid seeds	80–120 per patient	AUROC IF/time = 0.93 AUROC CF/time = 0.93	12	1000	ECOG
[98]	2015	Total ripple length (TRL), rippleness (R) and autoregressive model residual variation (ARR)	7–15 per patient	AUROC(TRL) = 0.79 AUROC(R) = 0.64 AUROC(ARR) = 0.82	9	1000	iEEG
[99]	2016	EEG source imaging (ESI) combined with functional connectivity analysis	32–204	Median LEconn (204 electrode) = 12 (mm) Median LEmin (204 electrode) = 6.93 (mm) Median LEpow (204 electrode) = 44.72 (mm)	5	250 or 1000	HD-EEG
[92]	2017	The logistic regression classifier based on phase-locking value	32–98	Accuracy = 83.0% Precision = 90.0% AUROC = 0.69	10	1000	ECOG
[100]	2019	Generalized linear model (GLM)	65	Accuracy = 89.0%	2	5000	EEG/fMRI
[101]	2020	Multiresolutional analysis and adaptive filtering (MRAF)	23	Accuracy = 86.7% Precision = 88.9%	23	256	EEG

4.3 Performance indices

In addition to the traditional performance indices used to assess two-class classification problems, i.e. the overall accuracy, sensitivity and specificity [18, 21], precision [102, 103], F1-score [31, 103], Matthews correlation coefficient (MCC) [103] and area under the ROC curve (AUC) [52, 104], the performance of an epileptic seizure prediction algorithm is further assessed using the following indices:

- False alarm per patient, defined as the number of false predictions during patient monitoring or in the model evaluation process per sample case [105].
- Average prediction time (APT), which is the average of the maximum time that the predictor foresees the seizure occurrence, measured in minutes [31].
- Time in warning (TIW), defined as the interictal period predicted as preictal, over the interictal period [106].
- Seizure occurrence period (SOP), which is the expected time interval of seizure occurrence [31].
- False alarm rate per hour (FPR), which is the number of times that the model prediction has false positives (also known as false alarms) per hour [31, 107].
- Seizure prediction horizon (SPH), defined as the time from seizure prediction to seizure occurrence [52].

For SOZ identification and related denoising methods, in addition to sensitivity, specificity, accuracy, AUC, precision, F1-score [65, 87], Youden index [108], false discovery rate (FDR) [109], Pearson correlation coefficient [110], measurement of agreement (Kappa) [111], and the Mann–Whitney [112] and McNemar tests [113], the following performance indices have also been used in the literature:

- Concordance with the area delimited by experts and post-surgical outcomes [114].
- Distances between the source positions of different approaches [115].
- Average HFO rate [116].
- LEmin, which is the distance between the source closest to the RZ/driving patch [99].
- Dispersion, as the distance between each equivalent current dipole analysis (ECD) and the central position [79].
- LEconn, defined as the distance between the source with the highest out-degree and the RZ/driving patch [99].
- LEpow, as the distance between the source with the highest power and the RZ/driving patch [99].

4.4 Conclusion and future scope

Epileptic seizure prediction and SOZ identification have received much attention recently as epilepsy accounts for more than 0.5% of the global burden of disease. In refractory epilepsy, the localization of the SOZ is needed for successful resection surgery. Although different methods have been used for SOZ localization, recent progress suggests the promising role of high-frequency activity in delineating the SOZ accurately. However, HFO acquisition typically requires invasive EEG or expensive MEG recordings. Additional research on non-invasive EEG outcomes is needed. On the other hand, such a prediction is necessary for a patient to create proper alarms to avoid accidents and falls. However, traditional EEG electrodes are not comfortable. Accordingly, the application of EEG dry electrodes has become necessary in practice.

Moreover, the fabrication of portable prediction devices requires the possibility of using a limited number of recording channels, reliable and real-time signal processing, and classification methods implemented on programmable systems on a chip (PSOCs) [117]. This is one of the main topics currently being investigated by start-ups (e.g. https://www.medicalstartups.org/top/epilepsy/ and https://www.eu-startups.com/tag/epilepsy/).

As far as SOZ identification goes, the methods have improved from brief intraoperative EcOG (the EZ as a lesion) to SEEG (the EZ as a network). However, surgical outcomes remain the same [118]. The EZ can remain in the brain in some patients even after dissecting the SOZ and adjacent tissue, indicating that the current methods are not good enough to delineate the EZ. Now is the time to upgrade the SOZ identification algorithms and to fabricate comfortable devices to assist the millions of people with epilepsy, not only in clinical practice but also in home-care settings.

Acknowledgments

This work was supported by the Ministry of Economy and Competitiveness (MINECO), Spain, under contract DPI2017-83989-R and the Ministry of Science and Innovation (MICINN), Spain, under contract PRE2018-085387. CIBER-BBN is an initiative of the Instituto de Salud Carlos III, Spain. J F Alonso and A Bachiller are Serra Hunter Fellows. The research leading to these results has also received funding from the European Union's Horizon 2020 research and innovation program under the Marie Skłodowska-Curie grant agreement No 712949 (TECNIOspring PLUS) and from the Agency for Business Competitiveness of the Government of Catalonia, University of Isfahan and Isfahan University of Medical Sciences.

References

[1] Ramey W L, Martirosyan N L, Lieu C M, Hasham H A, Lemole G M Jr and Weinand M E 2013 Current management and surgical outcomes of medically intractable epilepsy *Clin. Neurol. Neurosurg.* **115** 2411–8

[2] World Health Organization 2019 *Epilepsy: A Public Health Imperative: Summary* (Geneva: World Health Organization)

[3] Jallon P, Loiseau P and Loiseau J 2001 Newly diagnosed unprovoked epileptic seizures: presentation at diagnosis in CAROLE study *Epilepsia* **42** 464–75

[4] Wiebe S and Jette N 2012 Pharmacoresistance and the role of surgery in difficult to treat epilepsy *Nat. Rev. Neurol.* **8** 669–77

[5] Berg A T 2011 Epilepsy, cognition, and behavior: the clinical picture *Epilepsia* **52** 7–12

[6] Lüders H O, Najm I, Nair D, Widdess-Walsh P and Bingman W 2006 The epileptogenic zone: general principles *Epileptic Disord.* **8** S1–9

[7] Jacobs J *et al* 2012 High-frequency oscillations (HFOs) in clinical epilepsy *Prog. Neurobiol.* **98** 302–15

[8] Rosenow F and Lüders H 2001 Presurgical evaluation of epilepsy *Brain* **124** 1683–700

[9] Pittau F, Grouiller F, Spinelli L, Seeck M, Michel C M and Vulliemoz S 2014 The role of functional neuroimaging in pre-surgical epilepsy evaluation *Front. Neurol.* **5** 31

[10] Litt B and Echauz J 2002 Prediction of epileptic seizures *Lancet Neurol.* **1** 22–30

[11] Viglione S S and Walsh G O 1975 Proceedings: epileptic seizure prediction *Electroencephalogr. Clin. Neurophysiol.* 39 435–6

[12] Litt B *et al* 2001 Epileptic seizures may begin hours in advance of clinical onset: a report of five patients *Neuron* **30** 51–64

[13] Mormann F, Andrzejak R G, Elger C E and Lehnertz K 2007 Seizure prediction: the long and winding road *Brain* **130** 314–33

[14] Kuhlmann L, Lehnertz K, Richardson M P, Schelter B and Zaveri H P 2018 Seizure prediction—ready for a new era *Nat. Rev. Neurol.* **14** 618–30

[15] Noachtar S and Rémi J 2009 The role of EEG in epilepsy: a critical review *Epilepsy Behav.* **15** 22–33

[16] Paul Y 2018 Various epileptic seizure detection techniques using biomedical signals: a review *Brain Inform.* **5** 6

[17] Rasekhi J, Mollaei M R K, Bandarabadi M, Teixeira C A and Dourado A 2015 Epileptic seizure prediction based on ratio and differential linear univariate features *J. Med. Signals Sens.* **5** 1–11

[18] Carney P R, Myers S and Geyer J D 2011 Seizure prediction: methods *Epilepsy Behav.* **22** S94–101

[19] Usman S M, Usman M and Fong S 2017 Epileptic seizures prediction using machine learning methods *Comput. Math. Methods Med.* **2017** 9074759

[20] Usman S M, Khalid S, Akhtar R, Bortolotto Z, Bashir Z and Qiu H 2019 Using scalp EEG and intracranial EEG signals for predicting epileptic seizures: review of available methodologies *Seizure* **71** 258–69

[21] Bajaj V and Pachori R B 2012 Classification of seizure and nonseizure EEG signals using empirical mode decomposition *IEEE Trans. Inf. Technol. Biomed.* **16** 1135–42

[22] Ullah I, Hussain M, Qazi E-U-H and Aboalsamh H 2018 An automated system for epilepsy detection using EEG brain signals based on deep learning approach *Expert Syst. Appl.* **107** 61–71

[23] Wang L, Xue W, Li Y, Luo M, Huang J, Cui W and Huang C 2017 Automatic epileptic seizure detection in EEG signals using multi-domain feature extraction and nonlinear analysis *Entropy* **19** 222

[24] Thara D K, PremSudha B G and Xiong F 2019 Auto-detection of epileptic seizure events using deep neural network with different feature scaling techniques *Pattern Recognit. Lett.* **128** 544–50

[25] Wei X, Zhou L, Zhang Z, Chen Z and Zhou Y 2019 Early prediction of epileptic seizures using a long-term recurrent convolutional network *J. Neurosci. Methods* **327** 108395

[26] Thara D K, PremSudha B G and Xiong F 2019 Epileptic seizure detection and prediction using stacked bidirectional long short term memory *Pattern Recognit. Lett.* **128** 529–35

[27] Yang Y, Zhou M, Niu Y, Li C, Cao R, Wang B, Yan P, Ma Y and Xiang J 2018 Epileptic seizure prediction based on permutation entropy *Front. Comput. Neurosci.* **12** 55

[28] Tjepkema-Cloostermans M C, de Carvalho R C V and van Putten M J A M 2018 Deep learning for detection of focal epileptiform discharges from scalp EEG recordings *Clin. Neurophysiol.* **129** 2191–6
[29] Shafique A, Sayeed M and Tsakalis K 2018 Nonlinear dynamical systems with chaos and big data: a case study of epileptic seizure prediction and control *Guide to Big Data Applications* ed S Srinivasan (Cham: Springer) pp 329–69
[30] Alickovic E, Kevric J and Subasi A 2018 Performance evaluation of empirical mode decomposition, discrete wavelet transform, and wavelet packed decomposition for automated epileptic seizure detection and prediction *Biomed. Signal Process. Control* **39** 94–102
[31] Chu H, Chung C K, Jeong W and Cho K-H 2017 Predicting epileptic seizures from scalp EEG based on attractor state analysis *Comput. Methods Programs Biomed.* **143** 75–87
[32] Cho D, Min B, Kim J and Lee B 2017 EEG-based prediction of epileptic seizures using phase synchronization elicited from noise-assisted multivariate empirical mode decomposition *IEEE Trans. Neural Syst. Rehabil. Eng.* **25** 1309–18
[33] Zhang Z and Parhi K K 2016 Low-complexity seizure prediction from iEEG/sEEG using spectral power and ratios of spectral power *IEEE Trans. Biomed. Circuits Syst.* **10** 693–706
[34] Schelter B, Feldwisch-Drentrup H, Timmer J, Gotman J and Schulze-Bonhage A 2010 A common strategy and database to compare the performance of seizure prediction algorithms *Epilepsy Behav.* **17** 154–6
[35] Kiral-Kornek I *et al* 2018 Epileptic seizure prediction using big data and deep learning: toward a mobile system *EBioMedicine* **27** 103–11
[36] Shiao H-T, Cherkassky V, Lee J, Veber B, Patterson E E, Brinkmann B H and Worrell G A 2017 SVM-based system for prediction of epileptic seizures from iEEG signal *IEEE Trans. Biomed. Eng.* **64** 1011–22
[37] Aarabi A and He B 2017 Seizure prediction in patients with focal hippocampal epilepsy *Clin. Neurophysiol.* **128** 1299–307
[38] Zheng Y, Wang G, Li K, Bao G and Wang J 2014 Epileptic seizure prediction using phase synchronization based on bivariate empirical mode decomposition *Clin. Neurophysiol.* **125** 1104–11
[39] Eftekhar A, Juffali W, El-Imad J, Constandinou T G and Toumazou C 2014 Ngram-derived pattern recognition for the detection and prediction of epileptic seizures *PLoS One* **9** e96235
[40] Aarabi A and He B 2014 Seizure prediction in hippocampal and neocortical epilepsy using a model-based approach *Clin. Neurophysiol.* **125** 930–40
[41] Li S, Zhou W, Yuan Q and Liu Y 2013 Seizure prediction using spike rate of intracranial EEG *IEEE Trans. Neural Syst. Rehabil. Eng.* **21** 880–6
[42] Cook M J *et al* 2013 Prediction of seizure likelihood with a long-term, implanted seizure advisory system in patients with drug-resistant epilepsy: a first-in-man study *Lancet Neurol.* **12** 563–71
[43] Park Y, Luo L, Parhi K K and Netoff T 2011 Seizure prediction with spectral power of EEG using cost-sensitive support vector machines *Epilepsia* **52** 1761–70
[44] Soriano M C, Niso G, Clements J, Ortín S, Carrasco S, Gudín M, Mirasso C R and Pereda E 2017 Automated detection of epileptic biomarkers in resting-state interictal MEG data *Front. Neuroinform.* **11** 43

[45] Rasekhi J, Mollaei M R K, Bandarabadi M, Teixeira C A and Dourado A 2013 Preprocessing effects of 22 linear univariate features on the performance of seizure prediction methods *J. Neurosci. Methods* **217** 9–16
[46] Bandarabadi M, Teixeira C A, Rasekhi J and Dourado A 2015 Epileptic seizure prediction using relative spectral power features *Clin. Neurophysiol.* **126** 237–48
[47] Ocak H 2009 Automatic detection of epileptic seizures in EEG using discrete wavelet transform and approximate entropy *Expert Syst. Appl.* **36** 2027–36
[48] Andrzejak R G, Mormann F, Widman G, Kreuz T, Elger C E and Lehnertz K 2006 Improved spatial characterization of the epileptic brain by focusing on nonlinearity *Epilepsy Res.* **69** 30–44
[49] Alexandre Teixeira C, Direito B, Bandarabadi M, Le Van Quyen M, Valderrama M, Schelter B, Schulze-Bonhage A, Navarro V, Sales F and Dourado A 2014 Epileptic seizure predictors based on computational intelligence techniques: a comparative study with 278 patients *Comput. Methods Programs Biomed.* **114** 324–36
[50] Williamson J R, Bliss D W, Browne D W and Narayanan J T 2012 Seizure prediction using EEG spatiotemporal correlation structure *Epilepsy Behav.* **25** 230–8
[51] Goodfellow I, Bengio Y and Courville A 2016 *Deep Learning* (Cambridge, MA: MIT Press)
[52] Truong N D, Nguyen A D, Kuhlmann L, Bonyadi M R, Yang J, Ippolito S and Kavehei O 2018 Convolutional neural networks for seizure prediction using intracranial and scalp electroencephalogram *Neural Netw.* **105** 104–11
[53] Belharbi S 2018 *Neural Networks Regularization Through Representation Learning* arXiv:1807.05292
[54] Cui S, Duan L, Qiao Y and Xiao Y 2018 Learning EEG synchronization patterns for epileptic seizure prediction using bag-of-wave features *J. Ambient Intell. Human. Comput.* https://doi.org/10.1007/s12652-018-1000-3
[55] Kuhlmann L, Freestone D, Lai A, Burkitt A N, Fuller K, Grayden D B, Seiderer L, Vogrin S, Mareels I M Y and Cook M J 2010 Patient-specific bivariate-synchrony-based seizure prediction for short prediction horizons *Epilepsy Res.* **91** 214–31
[56] Bossuyt P M, Reitsma J B, Bruns D E, Gatsonis C A, Glasziou P P, Irwig L, Lijmer J G, Moher D, Rennie D and De Vet H C 2015 STARD 2015: an updated list of essential items for reporting diagnostic accuracy studies *Clin. Chem.* **61** 1446–52
[57] Colquhoun D 2014 An investigation of the false discovery rate and the misinterpretation of *p*-values *R. Soc. Open Sci.* **1** 140216
[58] Moons K G M, Altman D G, Reitsma J B, Ioannidis J P A, Macaskill P, Steyerberg E W, Vickers A J, Ransohoff D F and Collins G S 2015 Transparent reporting of a multivariable prediction model for individual prognosis or diagnosis (TRIPOD): explanation and elaboration *Ann. Intern. Med.* **162** W1–73
[59] Wyllie E, Lachhwani D K, Gupta A, Chirla A, Cosmo G, Worley S, Kotagal P, Ruggieri P and Bingaman W E 2007 Successful surgery for epilepsy due to early brain lesions despite generalized EEG findings *Neurology* **69** 389–97
[60] Tamilia E, Madsen J R, Grant P E, Pearl P L and Papadelis C 2017 Current and emerging potential of magnetoencephalography in the detection and localization of high-frequency oscillations in epilepsy *Front. Neurol.* **8** 14
[61] Staba R J, Stead M and Worrell G A 2014 Electrophysiological biomarkers of epilepsy *Neurotherapeutics* **11** 334–46

[62] Chang W-D, Cha H-S, Lee C, Kang H-C and Im C-H 2016 Automatic identification of interictal epileptiform discharges in secondary generalized epilepsy *Comput. Math. Methods Med.* **2016** 8701973

[63] Cimbalnik J, Brinkmann B, Kremen V, Jurak P, Berry B, Van Gompel J, Stead M and Worrell G 2018 Physiological and pathological high frequency oscillations in focal epilepsy *Ann. Clin. Transl. Neurol.* **5** 1062–76

[64] Thomschewski A, Hincapié A-S and Frauscher B 2019 Localization of the epileptogenic zone using high frequency oscillations *Front. Neurol.* **10** 94

[65] Migliorelli C, Bachiller A, Alonso J F, Romero S, Aparicio J, Jacobs J, Mananas M A and San Antonio-Arce V 2020 SGM: a novel time-frequency algorithm based on unsupervised learning improves high-frequency oscillation detection in epilepsy *J. Neural Eng.* **17** 026032

[66] Tadel F, Baillet S, Mosher J C, Pantazis D and Leahy R M 2011 Brainstorm: a user-friendly application for MEG/EEG analysis *Comput. Intell. Neurosci.* **2011** 879716

[67] Muthukumaraswamy S D 2013 High-frequency brain activity and muscle artifacts in MEG/EEG: a review and recommendations *Front. Hum. Neurosci.* **7** 138

[68] Tanaka N and Stufflebeam S M 2014 Clinical application of spatiotemporal distributed source analysis in presurgical evaluation of epilepsy *Front. Hum. Neurosci.* **8** 62

[69] Samadzadehaghdam N, Makkiabadi B, Masjoodi S, Mohammadi M and Mohagheghian F 2019 A new linearly constrained minimum variance beamformer for reconstructing EEG sparse sources *Int. J. Imaging Syst. Technol.* **29** 686–700

[70] Kanamori Y *et al* 2013 Minimum norm estimates in MEG can delineate the onset of interictal epileptic discharges: a comparison with ECoG findings *Neuroimage Clin.* **2** 663–9

[71] Worrell G A, Lagerlund T D, Sharbrough F W, Brinkmann B H, Busacker N E, Cicora K M and O'Brien T J 2000 Localization of the epileptic focus by low-resolution electromagnetic tomography in patients with a lesion demonstrated by MRI *Brain Topogr.* **12** 273–82

[72] Stufflebeam S M, Tanaka N and Ahlfors S P 2009 Clinical applications of magnetoencephalography *Hum. Brain Mapp.* **30** 1813–23

[73] Migliorelli C, Alonso J F, Romero S, Mañanas M A, Nowak R and Russi A 2015 Automatic BSS-based filtering of metallic interference in MEG recordings: definition and validation using simulated signals *J. Neural Eng.* **12** 046001

[74] Romero S, Mañanas M A and Barbanoj M J 2009 Ocular reduction in EEG signals based on adaptive filtering, regression and blind source separation *Ann. Biomed. Eng.* **37** 176–91

[75] Anderson C T, Carlson C E, Li Z and Raghavan M 2014 Magnetoencephalography in the preoperative evaluation for epilepsy surgery *Curr. Neurol. Neurosci. Rep.* **14** 446

[76] Stefan H, Hummel C, Scheler G, Genow A, Druschky K, Tilz C, Kaltenhäuser M, Hopfengärtner R, Buchfelder M and Romstöck J 2003 Magnetic brain source imaging of focal epileptic activity: a synopsis of 455 cases *Brain* **126** 2396–405

[77] Fischer M J M, Scheler G and Stefan H 2005 Utilization of magnetoencephalography results to obtain favourable outcomes in epilepsy surgery *Brain* **128** 153–7

[78] Oishi M, Kameyama S, Masuda H, Tohyama J, Kanazawa O, Sasagawa M and Otsubo H 2006 Single and multiple clusters of magnetoencephalographic dipoles in neocortical epilepsy: significance in characterizing the epileptogenic zone *Epilepsia* **47** 355–64

[79] Migliorelli C, Alonso J F, Romero S, Mañanas M A, Nowak R and Russi A 2016 Influence of metallic artifact filtering on MEG signals for source localization during interictal epileptiform activity *J. Neural Eng.* **13** 026029

[80] Worrell G A, Jerbi K, Kobayashi K, Lina J M, Zelmann R and Le Van Quyen M 2012 Recording and analysis techniques for high-frequency oscillations *Prog. Neurobiol.* **98** 265–78
[81] Staba R J, Wilson C L, Bragin A, Fried I and Engel J Jr 2002 Quantitative analysis of high-frequency oscillations (80–500 Hz) recorded in human epileptic hippocampus and entorhinal cortex *J. Neurophysiol.* **88** 1743–52
[82] Gardner A B, Worrell G A, Marsh E, Dlugos D and Litt B 2007 Human and automated detection of high-frequency oscillations in clinical intracranial EEG recordings *Clin. Neurophysiol.* **118** 1134–43
[83] Crépon B, Navarro V, Hasboun D, Clemenceau S, Martinerie J, Baulac M, Adam C and Le Van Quyen M 2010 Mapping interictal oscillations greater than 200 Hz recorded with intracranial macroelectrodes in human epilepsy *Brain* **133** 33–45
[84] Burnos S, Hilfiker P, Sürücü O, Scholkmann F, Krayenbühl N, Grunwald T and Sarnthein J 2014 Human intracranial high frequency oscillations (HFOs) detected by automatic time-frequency analysis *PLoS One* **9** e94381
[85] Dümpelmann M, Jacobs J and Schulze-Bonhage A 2015 Temporal and spatial characteristics of high frequency oscillations as a new biomarker in epilepsy *Epilepsia* **56** 197–206
[86] Roehri N, Pizzo F, Bartolomei F, Wendling F and Bénar C-G 2017 What are the assets and weaknesses of HFO detectors? A benchmark framework based on realistic simulations *PLoS One* **12** e0174702
[87] Migliorelli C, Alonso J F, Romero S, Nowak R, Russi A and Mañanas M A 2017 Automated detection of epileptic ripples in MEG using beamformer-based virtual sensors *J. Neural Eng.* **14** 046013
[88] von Ellenrieder N, Pellegrino G, Hedrich T, Gotman J, Lina J-M, Grova C and Kobayashi E 2016 Detection and magnetic source imaging of fast oscillations (40–160 Hz) recorded with magnetoencephalography in focal epilepsy patients *Brain Topogr.* **29** 218–31
[89] Lu Y, Worrell G A, Zhang H C, Yang L, Brinkmann B, Nelson C and He B 2014 Noninvasive imaging of the high frequency brain activity in focal epilepsy patients *IEEE Trans. Biomed. Eng.* **61** 1660–7
[90] Amiri M, Lina J-M, Pizzo F and Gotman J 2016 High frequency oscillations and spikes: separating real HFOs from false oscillations *Clin. Neurophysiol.* **127** 187–96
[91] Kramer M A *et al* 2019 Scalp recorded spike ripples predict seizure risk in childhood epilepsy better than spikes *Brain* **142** 1296–309
[92] Elahian B, Yeasin M, Mudigoudar B, Wheless J W and Babajani-Feremi A 2017 Identifying seizure onset zone from electrocorticographic recordings: a machine learning approach based on phase locking value *Seizure* **51** 35–42
[93] Zhang Y, Xu G, Wang J and Liang L 2010 An automatic patient-specific seizure onset detection method in intracranial EEG based on incremental nonlinear dimensionality reduction *Comput. Biol. Med.* **40** 889–99
[94] Medvedev A V, Murro A M and Meador K J 2011 Abnormal interictal gamma activity may manifest a seizure onset zone in temporal lobe epilepsy *Int. J. Neural Syst.* **21** 103–14
[95] Medvedovsky M *et al* 2012 Sensitivity and specificity of seizure-onset zone estimation by ictal magnetoencephalography *Epilepsia* **53** 1649–57
[96] Zhu G, Li Y, Wen P P, Wang S and Xi M 2013 Epileptogenic focus detection in intracranial EEG based on delay permutation entropy *AIP Conf. Proc., American Institute of Physics* https://aip.scitation.org/doi/abs/10.1063/1.4824993

[97] Burns S P, Santaniello S, Yaffe R B, Jouny C C, Crone N E, Bergey G K, Anderson W S and Sarma S V 2014 Network dynamics of the brain and influence of the epileptic seizure onset zone *Proc. Natl Acad. Sci. USA* **111** E5321–30

[98] Geertsema E E, Visser G H, Velis D N, Claus S P, Zijlmans M and Kalitzin S N 2015 Automated seizure onset zone approximation based on nonharmonic high-frequency oscillations in human interictal intracranial EEGs *Int. J. Neural Syst.* **25** 1550015

[99] Staljanssens W, Strobbe G, Van Holen R, Birot G, Gschwind M, Seeck M, Vandenberghe S, Vulliémoz S and van Mierlo P 2017 Seizure onset zone localization from ictal high-density EEG in refractory focal epilepsy *Brain Topogr.* **30** 257–71

[100] Ebrahimzadeh E, Nikravan M, Nikravan M, Manuchehri M S, Amoozegar S, Dolatabad M R, Bagheri M and Soroush M Z 2019 Simultaneous EEG-fMRI: a novel approach to localize the seizure onset zone *Int. J. Biol. Med.* 130–9

[101] Yedurkar D P and Metkar S P 2020 Multiresolution approach for artifacts removal and localization of seizure onset zone in epileptic EEG signal *Biomed. Signal Process. Control* **57** 101794

[102] Rodríguez Aldana Y, Marañón Reyes E J, Macias F S, Rodríguez V R, Chacón L M, Van Huffel S and Hunyadi B 2019 Nonconvulsive epileptic seizure monitoring with incremental learning *Comput. Biol. Med.* **114** 103434

[103] Rosas-Romero R, Guevara E, Peng K, Nguyen D K, Lesage F, Pouliot P and Lima-Saad W-E 2019 Prediction of epileptic seizures with convolutional neural networks and functional near-infrared spectroscopy signals *Comput. Biol. Med.* **111** 103355

[104] Truong N D, Zhou L and Kavehei O 2019 Semi-supervised seizure prediction with generative adversarial networks *Conf. Proc. IEEE Eng. Med. Biol. Soc.* **2019** 2369–72

[105] Parvez M Z and Paul M 2016 Epileptic seizure prediction by exploiting spatiotemporal relationship of EEG signals using phase correlation *IEEE Trans. Neural Syst. Rehabil. Eng.* **24** 158–68

[106] Bou Assi E, Nguyen D K, Rihana S and Sawan M 2018 A functional-genetic scheme for seizure forecasting in canine epilepsy *IEEE Trans. Biomed. Eng.* **65** 1339–48

[107] Rukhsar S, Khan Y U, Farooq O, Sarfraz M and Khan A T 2019 Patient-specific epileptic seizure prediction in long-term scalp EEG signal using multivariate statistical process control *IRBM* **40** 320–31

[108] Fluss R, Faraggi D and Reiser B 2005 Estimation of the Youden index and its associated cutoff point *Biom. J.* **47** 458–72

[109] Rouam S 2013 False discovery rate (FDR) *Encyclopedia of Systems Biology* ed W Dubitzky, O Wolkenhauer, K-H Cho and H Yokota (New York: Springer) 731–2

[110] Benesty J, Chen J, Huang Y and Cohen I 2009 Pearson correlation coefficient *Noise Reduction in Speech Processing* ed I Cohen, Y Huang, J Chen and J Benesty (Berlin: Springer) pp 1–4

[111] Viera A J and Garrett J M 2005 Understanding interobserver agreement: the kappa statistic *Fam. Med.* **37** 360–3

[112] Nachar N *et al* 2008 The Mann–Whitney *U*: a test for assessing whether two independent samples come from the same distribution *Tutor. Quant. Methods Psychol.* **4** 13–20

[113] Eliasziw M and Donner A 1991 Application of the McNemar test to non-independent matched pair data *Stat. Med.* **10** 1981–91

[114] Tanaka H, Gotman J, Khoo H M, Olivier A, Hall J and Dubeau F 2019 Neurophysiological seizure-onset predictors of epilepsy surgery outcome: a multivariable analysis *J. Neurosurg.* 1–10

[115] Gritsch G, Hartmann M M, Perko H, Fürbass F, Ossenblok P and Kluge T 2011 Automatic detection of the seizure onset zone based on ictal EEG *Conf. Proc. IEEE Eng. Med. Biol. Soc.* **2011** 3901–4

[116] Malinowska U, Bergey G K, Harezlak J and Jouny C C 2015 Identification of seizure onset zone and preictal state based on characteristics of high frequency oscillations *Clin. Neurophysiol.* **126** 1505–13

[117] Stacey W C 2018 Seizure prediction is possible—now let's make it practical *EBioMedicine* **27** 3–4

[118] Jehi L 2018 The epileptogenic zone: concept and definition *Epilepsy Curr.* **18** 12–6

Chapter 5

Automatic drowsiness detection based on variational non-linear chirp mode decomposition using electroencephalogram signals

Smith K Khare[1], Varun Bajaj[1], G R Sinha[2]

[1]*PDPM-Indian Institute of Information Technology, Design and Manufacturing, Jabalpur, India*

[2]*Myanmar Institute of Information Technology, Mandalay, Myanmar*

The timely detection of drowsiness could save millions of lives. The detection of a driver's drowsiness involves various behavioral and physiological elements. Electroencephalogram (EEG) signals provide information regarding these behavioral and physiological changes during drowsiness. Extracting the information directly from EEG signals is difficult due to its complex and non-stationary nature. To extract the vital information from signals it is necessary to split them into multiple components. In this chapter, variational non-linear chirp mode decomposition (VNCMD) based drowsiness detection is proposed. VNCMD decomposes the signal into subbands of instantaneous amplitude and frequency. Various linear and non-linear features are extracted from these instantaneous amplitudes and instantaneous frequencies of EEG signals. The dominant features are selected by using the Kruskal–Wallis test. Highly discriminant features are classified using different classifiers. Multiple performance parameters are evaluated to test the system's performance. Finally, the effectiveness of the proposed method is evaluated by comparing it to other existing methods using the same dataset.

5.1 Introduction

Drowsy driving is one of the key reasons for road accidents worldwide. Every decade in the USA, around 328 000 road accidents are reported. Thirty percent of

doi:10.1088/978-0-7503-3279-8ch5

these accidents cause severe injuries and sometimes result in death. The number of road accidents is increasing gradually each year [1]. As per the statistics of the national highway traffic safety administration, the main cause of road accidents is a lack of driver concentration during drowsiness [2]. Many automobile companies are aiming to develop an automatic drowsiness detection system to provide timely warnings to the driver regarding drowsiness. Also, the industry is spending billions on the development of an intelligent driverless system [3]. Many methods have been proposed. The identification of drowsiness and control has been carried out using skin reflex [4]. Drowsiness has been detected by heart rate variability using electrocardiography (ECG) [5–7]. Electrooculograms (EOGs) have been used to identify drowsiness with the aid of wearable sensors and convolutional neural networks [8, 9]. However, the models based on skin reflex, ECG and EOG have the possibility of false predictions. To overcome the possibility of false predictions, electroencephalogram (EEG) signals are used to measure brain electrical activity. EEG signals provide an accurate solution for the detection of drowsiness. The electrical activities of the brain cannot be changed deliberately and provide factual information on the variation in brain activity during drowsiness. Moreover, a solution provided by EEG signals is non-invasive, making it a viable choice for the detection of drowsiness.

Researchers have proposed many methods of appraising EEG signals to detect drowsiness. The extraction of the alpha band spectrum has been used for separating drowsy signals from alert signals [10]. Filters have also been used for the detection of drowsiness. The separation of various rhythms using suitable filters was demonstrated in [11]. A filtering technique in conjunction with a support vector machine (SVM) have been used in the identification of drowsiness [12, 13]. Filters, isolation circuits, modulators and demodulators are used in [14]. The alpha bands of EEG signals have been analysed using filtering for the detection of drowsiness [15]. Single source EEG signals and body movement have been used to detect drowsy signals using filters [16]. A sigmoid wake probability model based on the filtering of rhythms is used in [17]. The detection of drowsiness using filters to analyse physiological behavior was achieved in [18, 19]. Higher-order-statistics (HOS) along with filtering and bispectrum analysis were used for the detection of drowsiness [20]. A bandpass filter (BPF) technique is used in [21] for the detection of drowsy signals. BPF and fast Fourier transform with the Hanning window are used in [22]. BPF and time–frequency domain feature extraction was classified using an SVM in [23].

The fast Fourier transform (FFT) along with artificial neural networks (ANNs) and an SVM have been explored for the detection of drowsiness [24–27]. FFT and power spectral density (PSD) based features have been classified using a multilayer perceptron model (MLP) for automatic drowsiness detection [28]. The FFT and wavelet transform (WT) have been used for the detection of driver drowsiness using an SVM [29]. Speech analysis and the FFT, Mel filtering and discrete cosine transform (DCT) are used in [30]. Welch's FFT and principal component analysis (PCA) are used in [31]. The FFT using the Hanning window and SVM has been used to evaluate and classify the PSD based features [32]. The FFT and Welch methods are used to evaluate the spectral densities in [33]. FFT and Bayesian-copula discriminant classifier based drowsiness detection is used in [34]. FFT and filtering techniques have been used to separate the drowsy signal using an SVM [35]. The FFT and wavelet packet decomposition (WPD) have been employed for the extraction of features then classified using a subtractive fuzzy inference classifier [36]. Filtering, the FFT and ANNs are used in [37] for the classification of alert and drowsy signals. The FFT, PSD and SVM have been used for feature extraction and the classification of drowsy signals [38]. FFT, spectral analysis and power spectral analysis features have been classified using *k*-nearest neighbors (*k*-NN), linear discriminant analysis (LDA) and decision trees [39]. The FFT and K-singular value decomposition (K-SVD) have been used in [40] for the identification of drowsiness. The separation of rhythms [41, 42] and hierarchical cluster analysis [43] have also been used. The independent component analysis method [44] and Higuchi's fractal dimension are used in [45] for drowsiness detection. Iterative filtering based on multiresolution analysis and deep neural networks (DNNs) has been used for feature extraction and drowsiness detection [46].

The Hilbert–Huang transform (HHT) has been used for the detection of drowsiness [47, 48]. Higuchi and Petrosian's fractal dimensions are used in [49]. The HHT and empirical mode decomposition (EMD) have been used for feature extraction, which were then classified using a bagged tree classifier [50]. EMD is used for the detection of drowsiness in [51]. Adaptive EMD and a least squares support vector machine (LS-SVM) are used in [52]. The discriminant analysis method is used in [53]. Entropy and frequency domain features are used to detect drowsiness in [54, 55]. The utility of ANNs has been explored for automatic drowsiness segregation [56, 57]. An ANN-based on the Widrow–Hoff algorithm, Levenberg–Marquardt rule and learning vector quantization has been used for the recognition of alert and drowsy states [58]. An autoregressive model and SVM are used in [59]. A hidden Markov model with global and local contextual effects is used in [60]. A discrete wavelet transform is explored for the classification of drowsiness [61, 62]. The power analysis of delta, theta, alpha, sigma, beta and gamma has been used to differentiate the drowsy state [63]. PSD [64] and the WT [65] have been used for the detection of drowsiness. Spectral correlation of an EEG log power spectrum and coherence are used in [66–68]. Three-dimensional deep neural networks and semi-supervised gradient boosting machines have been used for the detection of driver drowsiness [69]. Drowsiness classification though *k*-means clustering and LDA is used in [70]. Filtering, spectral analysis using the FFT and correlation, PCA and a fuzzy neural network have been used for the detection of drowsiness [71]. Mean comparison tests and the short-time Fourier transform (STFT) are used in [72, 73]. PSD, autoregressive (AR) modeling and multiscale entropy features have been classified using step-wise linear discriminant analysis [74]. The WT and *k*-means clustering have been used for the detection of drowsiness [75, 76]. The WT, energy coefficient distribution and a neural network have also been used for the detection of drowsiness [77]. Wavelet packet decomposition (WPD) and fuzzy WPD are used in [78–81]. Automatic detection and analysis of drowsiness have been achieved using multimodal analysis [82, 83]. Spectral analysis based on the Mardia test and Mahalanobis distance [84] and a paired *t*-test along with an SVM [85] have been used for drowsiness detection.

Rhythm-based Tsallis entropy [86] and closed-loop lapse detection [87] have been used. PCA and multivariate linear regression have been used for classification [88]. The STFT in combination with k-NN and MCT is used in [89, 90]. Power spectral features, approximate entropy, fractal dimension and Lempel–Ziv complexity [91] and BPF, WPD and SVM [92] have been used for drowsiness detection. The Mahalanobis distance [93] and adaptive Hermite decomposition with an extreme learning machine [94] have been used for the identification of drowsiness signals. A hybrid model based on the tunable Q wavelet transform (TQWT), long short term memory and a convolutional neural network have been used for drowsiness detection [95]. Filtering, the FFT and SVM have been used for the identification of drowsiness from EEG signals [96]. The logarithmic powers of rhythms have been classified using LDA, SVM and k-NN [97]. Spectral power has been used for the separation of drowsy signals using fuzzy LDA [98]. An information fusion based decision algorithm has been used for drowsiness detection [99]. Phase space reconstruction and recurrence quantification analysis have been used for drowsiness detection [100]. The TQWT and an extreme learning machine are proposed in [101] for the separation of drowsy and alert signals.

The FFT suffers from the issue of localization of time–frequency. Accurate selection of the type and length of the window is required in the STFT; moreover, it assumes the signal to be stationary. A filter-based method requires the selection of filter coefficients. The wavelet requires the selection of boundaries that can affect system performance. EMD is purely experimental and lacks mathematical proof. To overcome these shortcomings, variational non-linear chirp mode decomposition (VNCMD) based drowsiness detection is proposed in this chapter. VNCMD decomposes the signal into narrowband instantaneous amplitudes and frequencies. The rest of this chapter is organized as follows: section 5.2 covers the methodology, section 5.3 presents detailed results and discussion of the method, and section 5.4 presents the conclusion of this chapter.

5.2 Methodology

The methodology used in this chapter consists of a dataset, variational non-linear chirp mode decomposition, feature extraction, classification and performance parameters. The steps involved in the proposed methodology are shown in figure 5.1.

5.2.1 Dataset

A polysomnographic MIT/BIH dataset containing the electrophysiology signals of 16 male subjects is used in this chapter. A detailed description of the dataset can be found in [102, 103]. Single-channel EEG recording and a total of 18 EEG recordings are used here. Channel C3–O1 and channel C4–A1 were used to capture 11 and five of the recordings, respectively, while two recordings of channel O2–A1 are used. The weight of the subjects varied from 89 to 152 kg and they had a mean age of 43 years. The EEG recordings used a sampling frequency of 250 Hz. A 30 s epoch is used to annotate different sleep stages. The annotation is carried out maintaining the Rechtschaffen and Kales criteria [104]. Power-line noise, and artifacts due to eye

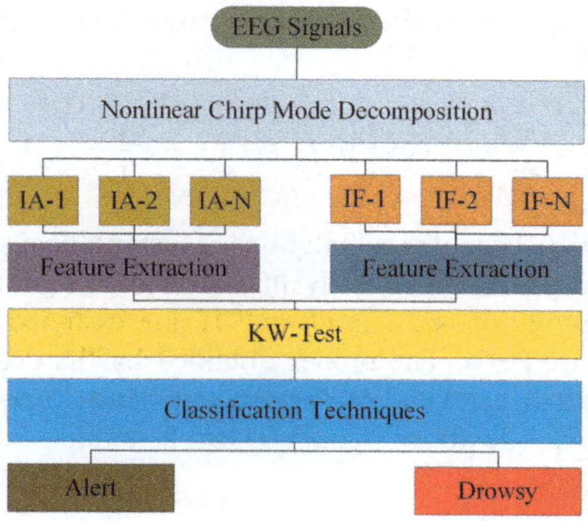

Figure 5.1. Steps used in automatic drowsiness detection using VNCMD.

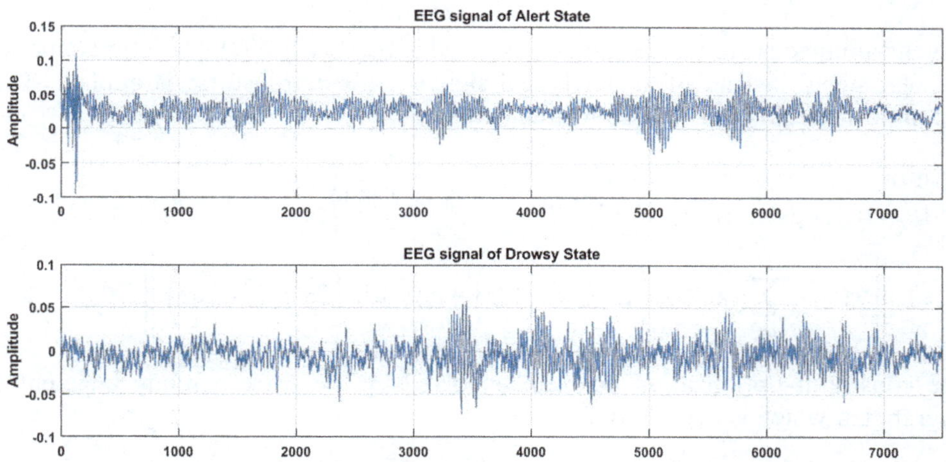

Figure 5.2. EEG recordings of alert and drowsy states.

movements and muscle movements are identified. These artifacts are removed using an electrooculogram and electromyogram [105]. Two states, i.e. the awake state denoted as 'W' and the drowsy state denoted as '1' are used. The alert state consists of 667 signals and the drowsy state consists of 488 signals. Each signal in both states contains 7500 samples. EEG recordings of alert and drowsy states are shown in figure 5.2.

5.2.2 Variational non-linear chirp mode decomposition (VNCMD)

The main idea behind VNCMD is to split a non-stationary chirp signal into its constituent signal modes. The signal modes which are obtained by VNCMD are called non-linear chirp modes. The non-linear chirp modes are essentially amplitude modulated and frequency modulated functions [106]. These AM–FM components are expressed as

$$v(t) = \sum_{n=1}^{N} a_n(t) \cos\left(2\pi \int_0^t f(s)\mathrm{d}s + \varphi\right) + n(t), \tag{5.1}$$

where $a(t) > 0$ and $f(t) > 0$ denote the functions of instantaneous amplitude (IA) and instantaneous frequency (IF), respectively, $n(t)$ is the noise signal, N is the number of modes and ϕ is the initial phase. The IA and IF are assumed to be varying more slowly than their phase [107]. The modes obtained by the decomposed VNCMD may not always be narrowband and can have a wideband property as IF can vary in a wide range [108, 109]. The above signal $v(t)$ can be modeled into two forms given by

$$v(t) = \sum_{n=1}^{N} u_n(t)\cos\left(2\pi \int_0^t \tilde{f}_n(s)\mathrm{d}s\right) + w_n(t)\sin\left(2\pi \int_0^t \tilde{f}_n(s)\mathrm{d}s\right) + n(t). \tag{5.2}$$

The instantaneous amplitude can be recovered as $a_n(t) = \sqrt{u_n^2(t) + w_n^2(t)}$. To model the real signal instead of an analytical signal, Hilbert transform is applied. The decomposition problem can be formulated as

$$\begin{aligned} &\min_{\{u_n(t)\},\, \{w_n(t)\},\, \{\tilde{f}_n(t)\}} \left\{ \sum_{n=1}^{N} \left(\| u_n(t)\|_2^2 + \| w_n(t)\|_2^2 \right) \right\} \\ &\text{s.t.} \left\| v(t) - \sum_{n=1}^{N} u_n(t)\cos\left(2\pi \int_0^t \tilde{f}_n(s)\mathrm{d}s\right) + w_n(t)\sin\left(2\pi \int_0^t \tilde{f}_n(s)\mathrm{d}s\right) \right\|_2 \leqslant \varepsilon. \end{aligned} \tag{5.3}$$

The constraint optimization problem is solved by the corresponding augmented Lagrangian which is expressed as

$$\begin{aligned} L_a\left(\{u_n(t)\},\, \{w_n(t)\},\, \{\tilde{f}_n(t)\},\, \omega,\, \lambda\right) = I_{c_\varepsilon}(\omega) &+ \sum (\| \Omega u_n \|_2^2 + \| \Omega w_n \|_2^2) \\ &+ \lambda^T\left(\omega + \sum (A_n u_n + B_n w_n) - v\right) \\ &+ \frac{\alpha}{2} \left\| \left(\omega + \sum (A_n u_n + B_n w_n) - v\right) \right\|_2^2. \end{aligned} \tag{5.4}$$

Equivalently, it can be written as

$$\begin{aligned} L_a(\{u_n(t)\},\, \{w_n(t)\},\, \{\tilde{f}_n(t)\},\, \omega,\, \lambda) = I_{c_\varepsilon}(\omega) &+ \sum \left(\| \Omega u_n \|_2^2 + \| \Omega w_n \|_2^2\right) \\ &+ \frac{\alpha}{2} \left\| \left(\omega + \sum (A_n u_n + B_n w_n) - v - \frac{1}{\alpha}\lambda\right) \right\|_2^2 \\ &- \frac{1}{2\alpha} \| \lambda \|_2^2, \end{aligned} \tag{5.5}$$

where α and λ are the penalty factor and Lagrangian multiplier, respectively. The optimization of the above problem is carried out by the alternating direction method of multipliers (ADMM) [110]. The optimization of the problem by ADMM is explained in algorithm 5.1.

Algorithm 5.1. VNCMD.

1. Initialize $v,\ \alpha,\ \mu,\ f_1^1,\ A_n^1 \leftarrow \mathrm{diag}\left[\cos\left(2\pi\int f_1^1(t)\mathrm{d}t\right)\right],\ B_n^1 \leftarrow \mathrm{diag}\left[\sin\left(2\pi\int f_1^1(t)\mathrm{d}t\right)\right],$
   $u_n^1 \leftarrow \left(\frac{2}{\alpha}\Omega^T\Omega + (A_n^1)^T A_n^1\right)(A_n^1)^T v,\ w_n^1 \leftarrow \left(\frac{2}{\alpha}\Omega^T\Omega + (B_n^1)^T B_n^1\right)(B_n^1)^T v,$ for $n = 1\ldots, N;\ \lambda \leftarrow 0$
2. repeat
3. $k \leftarrow k + 1$
4. $\omega^{k+1} \leftarrow P_{ce}\left(v - \sum_n (A_n^k u_n^k + B_n^k w_n^k) - \frac{1}{\alpha}\lambda^k\right)$
5. for $n \leftarrow 1{:}N$
6. $u_n^{k+1} \leftarrow \underset{u_n}{\arg\min}\left\{L_a\left(\left\{u_{m<n}^{k+1}, u_{m\geqslant n}^k\right\}, \left\{w_{m<n}^{k+1}, w_{m\geqslant n}^k\right\}, \left\{f_{m<n}^{k+1}, f_{m\geqslant n}^k\right\}, \omega^{k+1}, \lambda^k\right)\right\}$
7. $u_n^{k+1} \leftarrow \underset{u_n}{\arg\min}\left\{L_a\left(\left\{u_{m<n}^{k+1}, u_{m\geqslant n}^k\right\}, \left\{w_{m<n}^{k+1}, w_{m\geqslant n}^k\right\}, \left\{f_{m<n}^{k+1}, f_{m\geqslant n}^k\right\}, \omega^{k+1}, \lambda^k\right)\right\}$
8. $\Delta f_n^{k+1}(t) \leftarrow \dfrac{w_n^{k+1}(t)\left(u_n^{k+1}(t)\right)' - u_n^{k+1}(t)\left(w_n^{k+1}(t)\right)'}{2\pi\left(\left(u_n^{k+1}(t)\right)^2 + \left(w_n^{k+1}(t)\right)^2\right)}$
9. $f_n^{k+1} \leftarrow f_n^k + \frac{1}{2}\left(\frac{2}{\mu}\Omega^T\Omega + I\right)^{-1}\Delta f_n^{k+1}$
10. $A_n^{k+1} \leftarrow \mathrm{diag}\left[\cos\left(2\pi\int f_n^{k+1}(t)\mathrm{d}t\right)\right]$
11. $B_n^{k+1} \leftarrow \mathrm{diag}\left[\sin\left(2\pi\int f_n^{k+1}(t)\mathrm{d}t\right)\right]$
12. $v_n^{k+1} \leftarrow (A_n^{k+1}u_n^{k+1} + B_n^{k+1}w_n^{k+1})$
13. end for
14. $\lambda^{k+1} \leftarrow \lambda^k + \alpha\left(\omega^{k+1} + \sum_n v_n^{k+1} - v\right)$
15. If $\left\|\omega^{k+1} + \sum_n v_n^{k+1} - v\right\|_2^2 > \|v\|_2^2$ then
16. $\lambda^{k+1} \leftarrow 0$
17. for $n \leftarrow 1{:}N$
18. $u_n^{k+1} \leftarrow \left(\frac{2}{\alpha}\Omega^T\Omega + (A_n^{k+1})^T A_n^{k+1}\right)(A_n^{k+1})^T v,$

19. $w_n^{k+1} \leftarrow \left(\frac{2}{\alpha}\Omega^T\Omega + (B_n^{k+1})^T B_n^{k+1}\right)(B_n^{k+1})^T v$
20. end for
21. end if
22. until $\sum_n \| v_n^{k+1} - v_n^k \|_2^2 / \| v_n^k \|_2^2$ converges.

5.2.3 Feature extraction

To extract dominant information from the modes obtained from VNCMD various statistical parameters are evaluated. Multiple linear and non-linear features are extracted from the IA and IF. Six features from IA, namely interquartile range (IQR), covariance, Fano factor, kurtosis, root mean square (RMS) and mean, are selected using the Kruskal–Wallis (KW) test. Three features from IF are selected using the KW test, namely mean, maxima and variance.

5.2.3.1 Features of instantaneous amplitude

i. *Interquartile range.*

The IQR is the measure of statistical variation. It is equal to the difference between the higher quartile and the lower quartile and is mathematically expressed as

$$\begin{aligned} \text{IQR} &= Q3 - Q1 \\ Q1 &= \frac{1}{4}(M + 1) \\ Q3 &= \frac{3}{4}(M + 1). \end{aligned} \tag{5.6}$$

ii. *Root mean square.*

RMS is the measure of the magnitude of a signal containing a set of numbers. It is the measure of variation in the data points and is denoted as

$$\text{RMS} = \sqrt{\frac{1}{M}\sum_{m=1}^{M} u(m)^2}. \tag{5.7}$$

iii. *Kurtosis.*

Kurtosis is the measure of the deviation of tails of a distribution from the tails of a normal distribution. It is denoted as

$$\text{Kurtosis} = \frac{\sum_{m=1}^{M}(u(m) - \mu)^4 / M}{\sigma^4}. \tag{5.8}$$

iv. *Variance.*

Variance is the measure of quality of the signal being divergent or inconsistent and this is represented by

$$\text{Variance} = \frac{\sum_{m=1}^{M}(u(m) - \mu)^2}{M - 1}. \tag{5.9}$$

v. *Mean.*

The mean is the measure of the statistical property of the signal that measures the central tendency of the signal and is given as

$$\mu = \frac{\sum_{m=1}^{M} u(m)}{M}. \tag{5.10}$$

vi. *Fano factor.*

The Fano factor measures the dispersion in the probability distribution. It is defined as the ratio of variance to the mean. It is represented by

$$\text{Fano factor} = \frac{\sigma^2}{\mu}. \tag{5.11}$$

5.2.3.2 *Features of instantaneous frequency*

i. *Mean*

$$\mu_f = \frac{\sum_{m=1}^{M} \Omega(m)}{M}. \tag{5.12}$$

ii. *Maxima*

$$\text{Max}_f = \max(\Omega(m)). \tag{5.13}$$

iii. *Variance*

$$\sigma_f^2 = \frac{\sum_{m=1}^{M}(\Omega(m) - \mu_f)^2}{M - 1}, \tag{5.14}$$

where M is the number of samples in the signal. $u(m)$ and $\Omega(m)$ are the modes of the IA and IF, respectively, μ and μ_f are the means of the IA and IF, respectively, and σ and σ_f are the standard deviation of the modes of the IA and IF, respectively.

5.2.4 Classifiers

Classification techniques are employed to classify binary or multiclass signals. In this chapter, multiple classification techniques are employed to classify the features of the IA and IF. In this chapter, the complex tree, linear and quadratic discriminant analysis, support vector machine, *k*-nearest neighbors, logistic regression and bagged tree classifiers are employed. Tree-based classifiers are based on the principle of splitting the data into a tree structure. The SVM draws a hyperplane, *k*-NN uses a number of neighbors to assign a class, and bagged tree uses a tree structure and a number of learners. Detailed discussions of these classifiers are provided in [111–115].

5.3 Results and discussion

Extracting the information from EEG signals is very difficult and therefore the signals are decomposed into subbands to discover important information. VNCMD is one such technique that splits the signal into multiple components. Decomposition of a signal using VNCMD results in the generation of the instantaneous amplitude (IA) and instantaneous frequency (IF) bands. In this chapter, five bands are chosen to split the signal into modes. The number of modes is selected to be five. The frequency range is kept at {0–4}, {4–8}, {8–12}, {12–30} and {30–60} for the five modes, respectively. The tolerance is set to 10–5. The decomposition of signals results in five IA and IF modes. Figures 5.3 and 5.4 show the bands of the IA for drowsiness and alertness states.

As can be seen from figures 5.3 and 5.4, the IAs of the alert state are dominant over the drowsy state and are easily distinguishable even by visual inspection. The modes obtained by VNCMD are used to extract various features. Multiple features from the IA and IF are evaluated. To select the highly differentiable features, the KW test is applied. The KW test is a non-parameterized analysis of variance method. It provides the value probability of χ. The values of probability less than

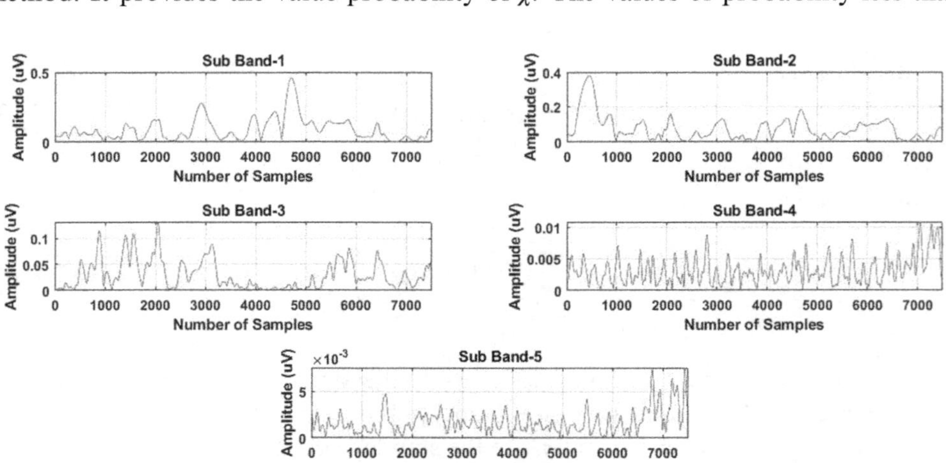

Figure 5.3. Modes of instantaneous amplitude for alertness.

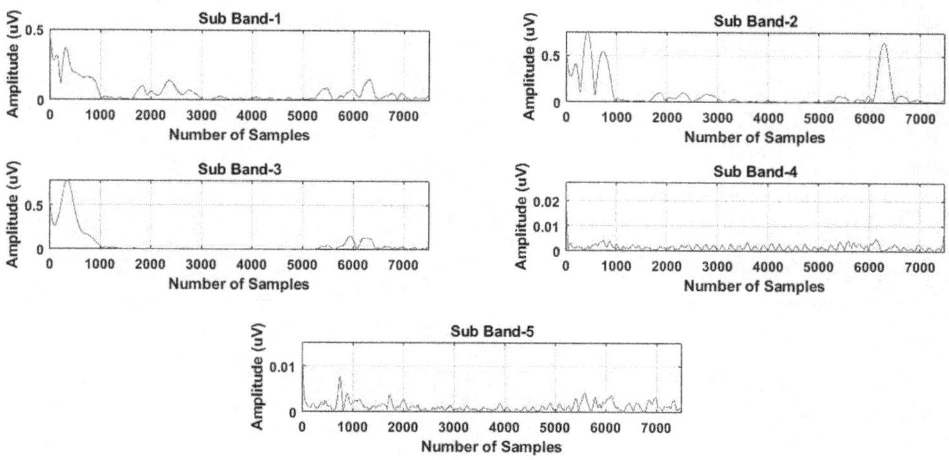

Figure 5.4. Modes of instantaneous amplitude for drowsiness.

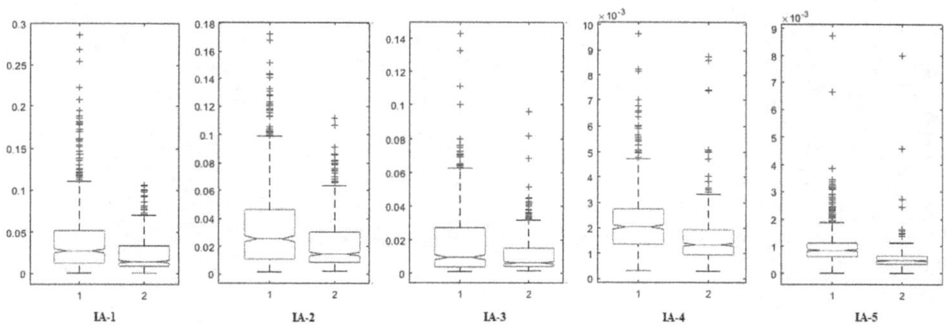

Figure 5.5. KW results for the interquartile range of the IA.

0.05 are considered to be significant for classification. Based on the KW test, six features of the IA and three features of the IF are selected. The IQR, Fano factor, mean, variance, kurtosis and RMS are selected for the IA. The variance, mean and maxima are selected for the IF. Figures 5.5–5.10 show the KW test results for the IQR, Fano factor, mean, variance, kurtosis and RMS, respectively, for the IA.

The KW results for the mean, max and variance of the IF are shown in figures 5.11–5.13, respectively.

The results of tables 5.1 and 5.2 and figures 5.5–5.13 show that the selected features are highly discriminable and can be used for the classification of drowsy and alert signals. The highly discriminable features are given as the input to different classifiers. In this chapter, the tree, discriminant analysis, *k*-nearest neighbors, support vector machine, logistic regression (LR) and bagged tree (BT) are employed. The number of splits for a tree is set to four, the number of neighbors is set to ten, the Gaussian kernel is selected for the SVM and the number of learners is set to 30 for BT. The classification accuracies obtained by the different classifiers using the IA are shown in table 5.3.

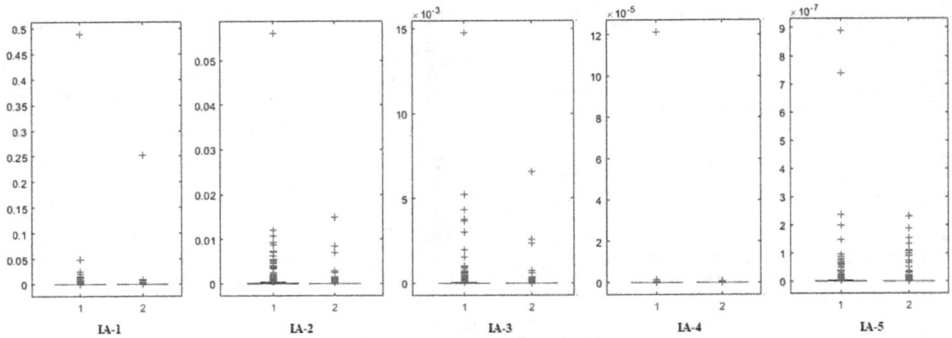

Figure 5.6. KW results for the Fano factor of the IA.

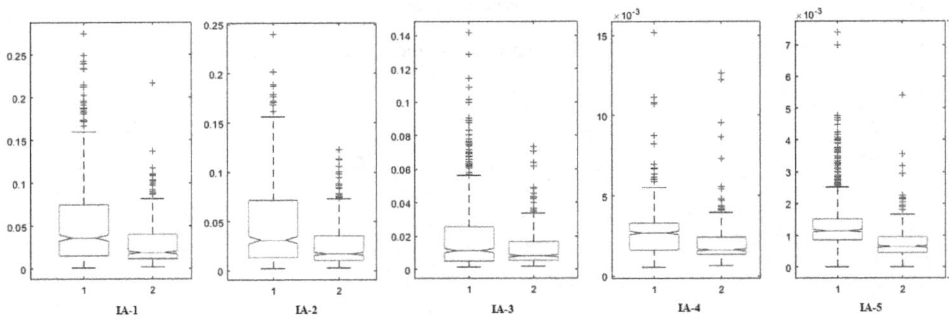

Figure 5.7. KW results for the mean of the IA.

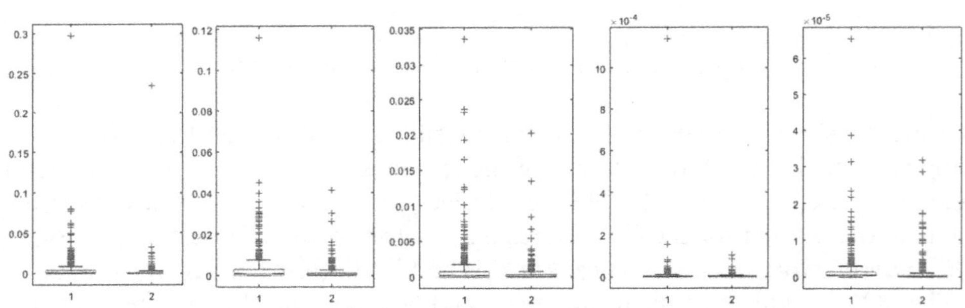

Figure 5.8. KW results for variance of the IA.

As can be seen from table 5.3, the classification accuracies obtained using the complex tree are 66.8%, 84.6%, 65.3%, 62.9% and 72.6% for IA-1, IA-2, IA-3, IA-4 and IA-5, respectively. The accuracies obtained for IA-1, IA-2, IA-3, IA-4 and IA-5 are, respectively, 67.3%, 73.2%, 68.5%, 68% and 74.5% for cubic *k*-NN, 67%, 74.9%, 72.4%, 69.1% and 74.6% for cosine *k*-NN, and 66.2%, 75%, 75.8%, 64.8% and 73.2% for weighted *k*-NN. LDA produced accuracies of 57.8%, 51.4%, 59.8% 45.8% and 56.5% and for QDA the accuracies are 61.9%, 61.8%, 73.6%, 65.4% and 71.7% for IA-1, IA-2, IA-3, IA-4 and IA-5, respectively. LR provides accuracies of 62.8%,

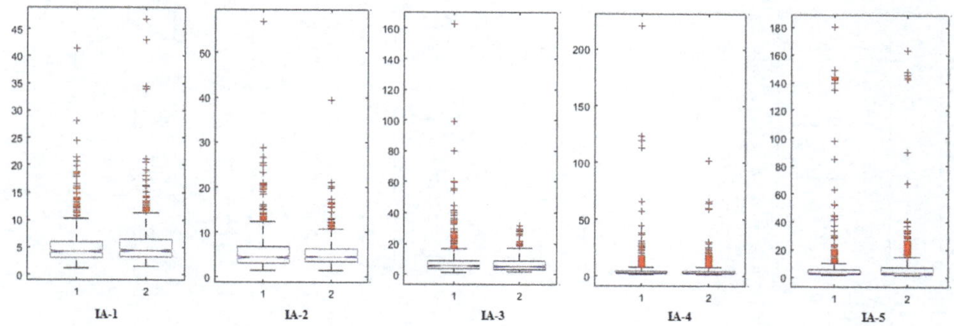

Figure 5.9. KW results for kurtosis of the IA.

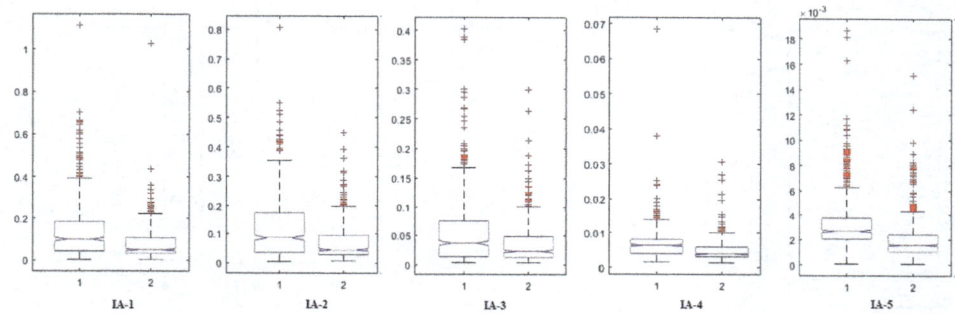

Figure 5.10. KW results for RMS of the IA.

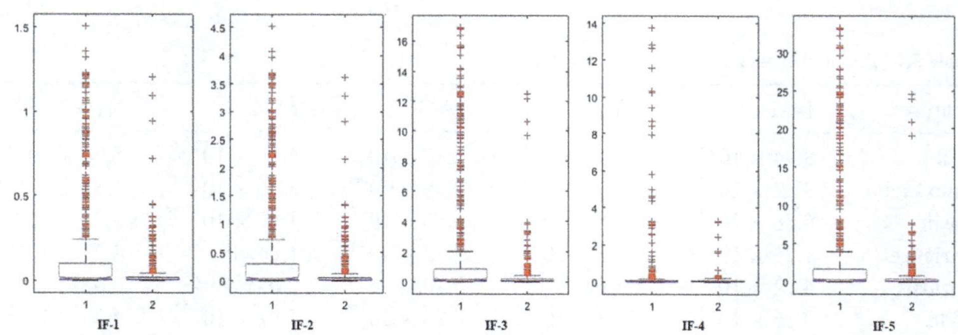

Figure 5.11. KW results for the mean of the IF.

69.6%, 62.5%, 66.9% and 74.8%, for SVM the accuracies are 63%, 75.8%, 74.1%, 66.8% and 75.2%, and for BT they are 73.2%, 92.4%, 63.7%, 64.2% and 73.2%, for IA-1, IA-2, IA-3, IA-4 and IA-5, respectively.

The classification accuracies for alert and drowsy signals for the IF are shown in table 5.4. IF-3 shows the highest classification accuracy for all the classifiers, while IF-5 provides the worst performance. The classification accuracies for IF-3 are 76.6%, 81%, 79.6%, 81.3%, 79.4%, 78.4%, 82.3%, 81.1% and 79.4% for the complex tree, cubic *k*-NN, cosine *k*-NN, weighted *k*-NN, QDA, LDA, SVM, BT and LR,

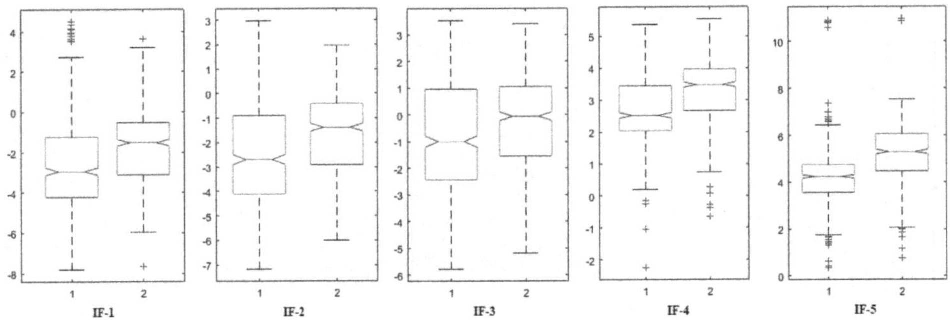

Figure 5.12. KW results for the maxima of the IF.

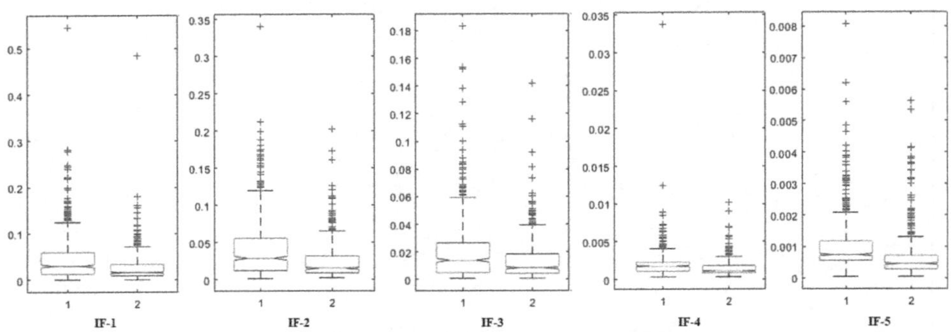

Figure 5.13. KW result for the variance of the IF.

Table 5.1. Probability of χ for features of the IA.

Features	IA-1	IA-2	IA-3	IA-4	IA-5
IQR	8.94×10^{-15}	1.54×10^{-14}	1.17×10^{-4}	5.95×10^{-31}	5.04×10^{-73}
Fano factor	3.22×10^{-13}	1.62×10^{-15}	3.98×10^{-8}	8.86×10^{-23}	1.51×10^{-34}
Mean	6.38×10^{-18}	3.44×10^{-19}	5.76×10^{-5}	1.87×10^{-34}	2.07×10^{-64}
Variance	1.26×10^{-14}	8.27×10^{-17}	7.73×10^{-8}	1.66×10^{-25}	8.23×10^{-40}
Kurtosis	4.12×10^{-3}	0.0171	0.105	0.021	0.022
RMS	3.36×10^{-17}	3.53×10^{-19}	8.94×10^{-7}	1.91×10^{-31}	4.30×10^{-51}

respectively. IF-5 provides the worst separation of drowsy and alert signals with accuracies of 65.8%, 56.6%, 57.7%, 52.1%, 51.6%, 52.3%, 51.8%, 50.2% and 50.6% for the complex tree, cubic *k*-NN, cosine *k*-NN, weighted *k*-NN, QDA, LDA, SVM, BT and LR classifiers, respectively.

Bar-graph representations of the classification accuracies for the IA and IF using different classification techniques are shown in figures 5.14 and 5.15. From the results in figures 5.14 and 5.15 and tables 5.4 and 5.5, it is evident that the performance of the IA is better than that of the IF. Moreover, the performance of

Table 5.2. Probability of χ for features of the IF.

Features	IF-1	IF-2	IF-3	IF-4	IF-5
Mean	4.40×10^{-4}	2.15×10^{-4}	1.55×10^{-3}	3.15×10^{-5}	1.24×10^{-3}
Maxima	6.25×10^{-3}	5.87×10^{-4}	9.88×10^{-3}	7.12×10^{-4}	4.19×10^{-3}
Variance	1.17×10^{-3}	2.65×10^{-2}	4.15×10^{-2}	1.54×10^{-3}	1.47×10^{-3}

Table 5.3. Classification accuracy of the IA (%).

Classifiers	IA-1	IA-2	IA-3	IA-4	IA-5
Complex tree	66.8	84.6	65.3	62.9	72.6
Cubic *k*-NN	67.3	73.2	68.5	68	74.5
Cosine *k*-NN	67	74.9	72.4	69.1	74.6
Weighted *k*-NN	66.2	75	75.8	64.8	73.2
QDA	57.8	51.4	59.8	45.8	56.5
LDA	61.9	61.8	73.6	65.4	71.7
SVM	63	75.8	74.1	66.8	75.2
BT	73.2	92.4	63.7	64.2	73.2
LR	62.8	69.6	62.5	66.9	74.8

Table 5.4. Classification accuracy of IF (%).

Classifiers	IF-1	IF-2	IF-3	IF-4	IF-5
Complex tree	68.9	73.5	76.6	68.9	65.8
Cubic*k*-NN	66.3	75.3	81	58.6	56.6
Cosine*k*-NN	66.7	75.4	79.6	54.9	57.7
Weighted*k*-NN	66.8	75.1	81.3	56.8	52.1
QDA	62.1	70.4	79.4	57.9	51.6
LDA	67	74.1	78.4	54.4	52.3
SVM	70.4	77.8	82.3	62.2	51.8
BT	71.7	77.4	81.1	63.5	50.2
LR	68.9	74	79.4	65.4	50.6

bagged tree is superior to the other classification techniques, hence the performance parameters of IA are evaluated for BT. In this chapter four performance parameters are evaluated, namely accuracy (ACR), sensitivity (SENS), specificity (SPEC) and precision (PREC). Table 5.4 shows the performance parameters obtained for the instantaneous amplitude function using the bagged tree classifier.

For IA-1, the ACC, SENS, SPEC and PREC are 73.2%, 76.69%, 68.42% and 77.27%, for IA-2 they are 90.8%, 90.28%, 91.57% and 94.24%, for IA-3 they are

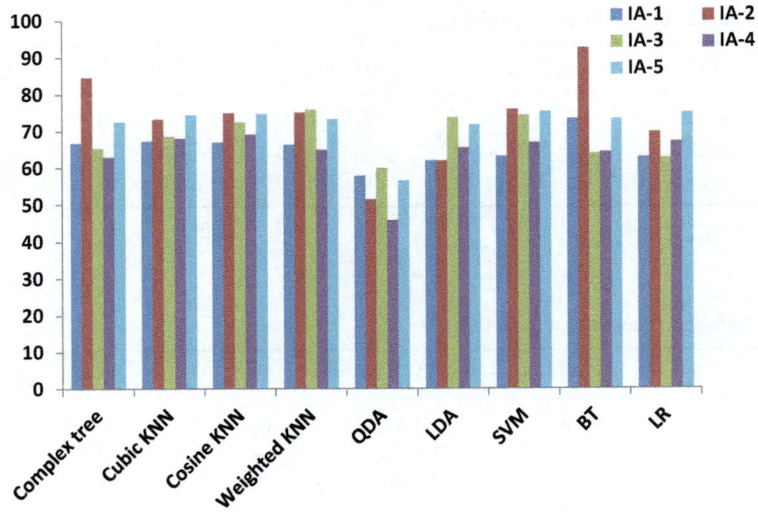

Figure 5.14. Classification accuracy of the IA with different classifiers.

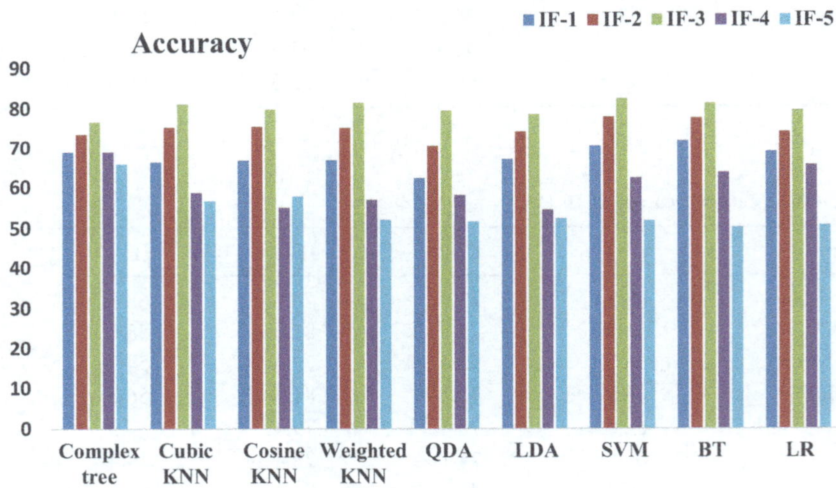

Figure 5.15. Classification accuracy of the IF with different classifiers.

Table 5.5. Performance parameters for the IA function.

Parameter	Accuracy %	Sensitivity %	Specificity %	Precision %
IA-1	73.2	76.69	68.42	77.27
IA-2	90.8	90.28	91.57	94.24
IA-3	63.7	66.95	56.78	71.52
IA-4	64.2	69.21	57.44	68.79
IA-5	73.2	76.85	64.86	72.42

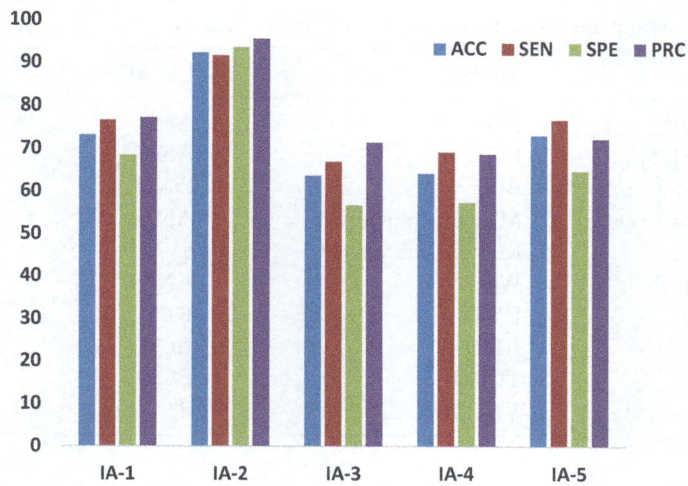

Figure 5.16. Performance parameters for BT.

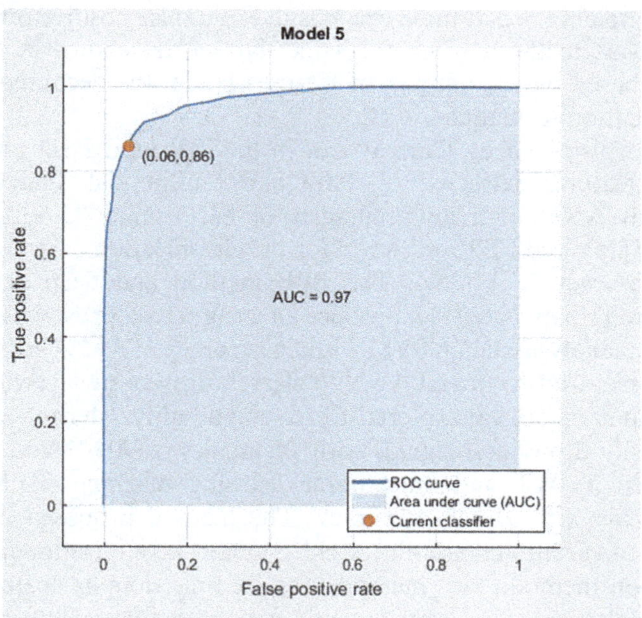

Figure 5.17. The ROC and AUC for IA-2.

63.7%, 66.95%, 56.78% and 71.52%, for IA-4 they are 64.2%, 69.21%, 57.44% and 68.79%, and for IA-5 they are 73.2%, 76.85%, 64.86% and 72.42%. As can be seen from the table, IA-2 performs best while IA-3 performs worst for ACR, SENS and SPEC. A bar-graph representation of the performance parameters is shown in figure 5.16.

The area under the curve (AUC) and receiver operator characteristic (ROC) are shown in figure 5.17. As the best performing instantaneous amplitude function is IA-2, the ROC and AUC are plotted for this IA.

Table 5.6. Comparison to the results for existing state-of-the-art methods.

Authors	Method	Classifier	Accuracy %
Correa *et al* [65]	PSD	ANN	84.1
Belakhdar *et al* [25]	FFT	ANN	84.75
Tripathy *et al* [46]	BPF	DNN	85.51
Correa *et al* [83]	Multimodal analysis	ANN	85.66
Anitha *et al* [82]	Multimodal analysis	SVM	87.2
Boonnak *et al* [77]	WT	ANN	90.7
Taran *et al* [50]	EMD	BT	90.6
Taran *et al* [94]	AHD	ELM	92.28
Bajaj *et al* [101]	TQWT	ELM	91.84
Proposed	VNCMD	BT	92.4

As can be seen from the figure, the AUC is 97% while the ROC is obtained at a false positive rate of 0.06 and a true positive rate of 0.86. The current classifier characteristics are obtained at these true positive and false positive rates. The results for state-of-the-art methods using the same dataset are shown in table 5.6. The comparison is based on the number of features taken, the decomposition method and the classification technique.

The methods proposed by Correa *et al* in [65, 83] used PSD and multimodal analysis. The features extracted by PSD and multimodal analysis have been classified by an ANN with an accuracies of 84.1% and 85.66%, respectively. Belakhdar *et al* [25] used FFT and ANN for the classification of drowsiness signals, claiming an accuracy of 84.75%. The BPF method and deep neural networks (DNNs) used by Tripathy *et al* [46] produce an accuracy of 85.51%. Anitha *et al* [82] used multimodal analysis and an SVM with an accuracy of 87.2% and Boonnak *et al* [77] used a wavelet transform and ANN to classify drowsy signals with an accuracy of 90.7%. Taran *et al* [50, 94] explored EMD and adaptive Hermite decomposition (AHD) to classify drowsiness signals with accuracies of 90.6% and 92.28%. Bajaj *et al* [101] used TQWT and an extreme learning machine (ELM) to classify drowsiness signals with 91.84% accuracy. The method proposed in this chapter uses the utility of variational non-linear chirp mode decomposition and the bagged tree classification method. This method uses six time domain features to classify drowsiness signals from alert signals using a bagged tree classifier. The method achieves an accuracy of 92.4% using instantaneous amplitude functions and 82.3% using instantaneous frequency functions with a support vector machine. The proposed method proved to be superior to other state-of-the-art methods using the same dataset.

5.4 Conclusion

Accurate detection of drowsiness can help to provide early warnings to drivers to avoid road accidents. The decomposition of EEG signals helps in extracting the hidden information. In this chapter, variational non-linear chirp mode decomposition is used to split the signals into multiple instantaneous amplitudes and frequency modes. Extracted features from instantaneous amplitudes and frequency are classified using various classification techniques. The instantaneous amplitude is helpful in the identification of drowsiness signals compared to the instantaneous frequency. The modes of instantaneous amplitude provide 92.4% accurate classification whereas 82.3% of the signals of drowsiness and alertness are classified accurately using instantaneous frequency signals. The method proposed in this chapter can be utilized for the modeling of real-time drowsiness detection devices.

References

[1] Tefft B C 2014 *Prevalence of Motor Vehicle Crashes Involving Drowsy Drivers, USA, 2009–13* (Washington, DC: AAA Foundation for Traffic Safety)

[2] Royal D 2002 *National Survey of Distracted and Drowsy Driving Attitudes and Behavior: 2002 (DOT HS 809 566)* (Washington, DC: National Highway Traffic Safety Administration)

[3] BMW Assistance/intelligent-driving (Accessed: 20 Feb 2020) https://bmw.co.uk/bmw-ownership/connecteddrive/driver

[4] Yamamoto Y and Isshiki H 1992 Instrument for controlling drowsiness using galvanic skin reflex *Med. Biol. Eng. Comput.* **30** 562–4

[5] Fujiwara K *et al* 2019 Heart rate variability-based driver drowsiness detection and its validation with EEG *IEEE Trans. Biomed. Eng.* **66** 1769–78

[6] Vicente J, Laguna P, Bartra A and Bailón R 2011 Detection of driver's drowsiness by means of HRV analysis *2011 Computing in Cardiology Conf. (Hangzhou)* pp 89–92

[7] Warwick B, Symons N, Chen X and Xiong K 2015 Detecting driver drowsiness using wireless wearables *2015 IEEE 12th Int. Conf. on Mobile Ad Hoc and Sensor Systems (Dallas, TX)* pp 585–8

[8] Thum C C, Mustafa M M, Hussain A, Hendi S F and Majlis B Y 2005 Development of vehicle driver drowsiness detection system using electrooculogram (EOG) *2005 1st Int. Conf. on Computers, Communications, and Signal Processing with Special Track on Biomedical Engineering (Kuala Lumpur, Malaysia)* pp 165–8

[9] Zhu X, Zheng W, Lu B, Chen X, Chen S and Wang C 2014 EOG-based drowsiness detection using convolutional neural networks *2014 Int. Joint Conf. on Neural Networks (IJCNN) (Beijing)* pp 128–34.

[10] Lin C, Chang C, Lin B, Hung S, Chao C and Wang I 2010 A real-time wireless brain–computer interface system for drowsiness detection *IEEE Trans. Biomed. Circuits Syst.* **4** 214–22

[11] AlZu'bi H S, Al-Nuaimy W and Al-Zubi N S 2013 EEG-based driver fatigue detection *2013 Sixth Int. Conf. on Developments in eSystems Engineering (Abu Dhabi)* pp 111–4

[12] Hwang T, Kim M, Hong S and Park K S 2016 Driver drowsiness detection using the in-ear EEG *2016 38th Annual Int. Conf. of the IEEE Engineering in Medicine and Biology Society (EMBC) (Orlando, FL)* pp 4646–9

[13] Dey G *et al* 2017 Automatic detection of drowsiness in EEG records based on time analysis *2017 Innovations in Power and Advanced Computing Technologies (i-PACT) (Vellore)* pp 1–5

[14] Dhavala D 2011 Classification of EEG signals for drowsiness detection in brain and computer interface *Comput. Sci. Telecommun.* **3** 1–12
[15] Ben Dkhil M, Chawech N, Wali A and Alimi A M 2017 Towards an automatic drowsiness detection system by evaluating the alpha band of EEG signals *2017 IEEE 15th Int. Symp. on Applied Machine Intelligence and Informatics (SAMI) (Herl'any)* pp 000371–6
[16] Pritchett S, Zilberg E, Xu Z M, Karrar M, Burton D and Lal S 2011 Comparing accuracy of two algorithms for detecting driver drowsiness—single source (EEG) and hybrid (EEG and body movement) *7th Int. Conf. on Broadband Communications and Biomedical Applications (Melbourne, VIC)* pp 179–84
[17] Hassan A R, Kabir M, Keshavarz B, Taati B and Yadollahi A 2019 Sigmoid wake probability model for high-resolution detection of drowsiness using electroencephalogram *2019 41st Annual Int. Conf. of the IEEE Engineering in Medicine and Biology Society (EMBC) (Berlin, Germany)* pp 7080–3
[18] Li P, Meziane R, Otis M J, Ezzaidi H and Cardou P 2014 A smart safety helmet using IMU and EEG sensors for worker fatigue detection *2014 IEEE Int. Symp. on Robotic and Sensors Environments (ROSE) Proc. (Timisoara)* pp 55–60
[19] Pritchett S, Zilberg E, Xu Z, Karrar M M, Lal S and Burton D 2011 Strengthening association between driver drowsiness and its physiological predictors by combining EEG with measures of body movement *7th Int. Conf. on Broadband Communications and Biomedical Applications (Melbourne)* pp 103–7
[20] Swarnkar V, Abeyratne U and Hukins C 2010 Objective measure of sleepiness and sleep latency via bispectrum analysis of EEG *Med. Biol. Eng. Comput.* **48** 1203–13
[21] Deepa T P and Reddy V 2013 EEG based drowsiness detection using mobile device for intelligent vehicular system *Int. J. Eng. Trends Technol.* **6** 21–4
[22] Ko L-W *et al* 2015 Single channel wireless EEG device for real-time fatigue level detection *2015 Int. Joint Conf. on Neural Networks (IJCNN) (Killarney)* pp 1–5
[23] Ribeiro D, Teixeira C and Cardoso A 2017 EEG-based drowsiness detection platform to compare different methodologies *2017 4th Experiment@Int. Conf. (Faro)* pp 318–22
[24] Ben Dkhil M, Wali A and Alimi M 2015 Drowsy driver detection by EEG analysis using fast Fourier transform *2015 15th Int. Conf. on Intelligent Systems Design and Applications (ISDA) (Marrakech)* pp 313–8
[25] Belakhdar I, Kaaniche W, Djmel R and Ouni B 2016 A comparison between ANN and SVM classifier for drowsiness detection based on single EEG channel *2016 2nd Int. Conf. on Advanced Technologies for Signal and Image Processing (ATSIP) (Monastir)* pp 443–6
[26] Awais M, Badruddin N and Drieberg M 2014 A simulator based study to evaluate driver drowsiness using electroencephalogram *2014 5th Int. Conf. on Intelligent and Advanced Systems (ICIAS) (Kuala Lumpur)* pp 1–5
[27] Leng H, Lin Y and Mourant R 2008 An experimental study using EEG to detect driver drowsiness *SAE Technical Papers* 2008-01-2693
[28] Hashemi A, Saba V and Resalat S N 2014 Real time driver's drowsiness detection by processing the EEG signals stimulated with external flickering light *Basic Clin. Neurosci.* **5** 22–7
[29] Li G and Chung W 2013 Detection of driver drowsiness using wavelet analysis of heart rate variability and a support vector machine classifier *Sensors* **13** 16494–511
[30] Dhupati L S, Kar S, Rajaguru A and Routray A 2010 A novel drowsiness detection scheme based on speech analysis with validation using simultaneous EEG recordings *2010 IEEE Int. Conf. on Automation Science and Engineering (Toronto, ON)* pp 917–21

[31] Wei C, Lin Y, Wang Y, Jung T, Bigdely-Shamlo N and Lin C 2015 Selective transfer learning for EEG-based drowsiness detection *2015 IEEE Int. Conf. on Systems, Man, and Cybernetics (Kowloon)* pp 3229–32

[32] Yu S *et al* 2013 Support vector machine based detection of drowsiness using minimum EEG features *2013 Int. Conf. on Social Computing (Alexandria, VA)* pp 827–35

[33] Awais M, Badruddin N and Drieberg M 2014 Driver drowsiness detection using EEG power spectrum analysis *2014 IEEE Region 10 Symp. (Kuala Lumpur)* pp 244–7

[34] Qian D *et al* 2017 Drowsiness detection by Bayesian-copula discriminant classifier based on EEG signals during daytime short nap *IEEE Trans. Biomed. Eng.* **64** 743–54

[35] Li G and Chung W Y 2015 A context-aware EEG headset system for early detection of driver drowsiness *Sensors* 15 20873–93

[36] Murugappan M, Wali M K, Ahmmad R B and Murugappan S 2013 Subtractive fuzzy classifier based driver drowsiness levels classification using EEG *2013 Int. Conf. on Communication and Signal Processing (Melmaruvathur)* pp 159–64

[37] King L M, Nguyen H T and Lal S K L 2006 Early driver fatigue detection from electroencephalography signals using artificial neural networks *2006 Int. Conf. of the IEEE Engineering in Medicine and Biology Society (New York, NY)* pp 2187–90

[38] Zhang X *et al* 2017 Design of a fatigue detection system for high-speed trains based on driver vigilance using a wireless wearable EEG *Sensors* **17** 486

[39] Mehreen A, Anwar S M, Haseeb M, Majid M and Ullah M O 2019 A hybrid scheme for drowsiness detection using wearable sensors *IEEE Sens. J.* **19** 5119–26

[40] Zhang Z, Luo D, Rasim Y, Li Y, Meng G, Xu J and Wang C 2016 A vehicle active safety model: vehicle speed control based on driver vigilance detection using wearable EEG and sparse representation *Sensors* **16** 242

[41] Park J, Xu L, Sridhar V, Chi M and Cauwenberghs G 2011 Wireless dry EEG for drowsiness detection *2011 Annual Int. Conf. of the IEEE Engineering in Medicine and Biology Society (Boston, MA)* pp 3298–301

[42] Van Hal B, Rhodes S, Dunne B and Bossemeyer R 2014 Low-cost EEG-based sleep detection *2014 36th Annual Int. Conf. of the IEEE Engineering in Medicine and Biology Society (Chicago, IL)* pp 4571–4

[43] Wei C, Lin Y, Wang Y, Lin C and Jung T 2018 A subject-transfer framework for obviating inter- and intra-subject variability in EEG-based drowsiness detection *NeuroImage* **174** 407–19

[44] Chin-Teng L, Ruei-Cheng W, Sheng-Fu L, Wen-Hung C, Yu-Jie C and Tzyy-Ping J 2005 EEG-based drowsiness estimation for safety driving using independent component analysis *IEEE Trans. Circuits Syst.* I **52** 2726–38

[45] Noori S M and Mikaeili M 2016 Driving drowsiness detection using fusion of electroencephalography, electrooculography, and driving quality signals *J. Med. Signals Sens.* **6** 39–46

[46] Tripathy R K and Acharya U R 2018 Use of features from RR-time series and EEG signals for automated classification of sleep stages in deep neural network framework **38** 890–902

[47] Sharabaty H, Jammes B and Esteve D 2008 EEG analysis using HHT: one step toward automatic drowsiness scoring *22nd Int. Conf. on Advanced Information Networking and Applications—Workshops (AINA Workshops 2008) (Okinawa)* pp 826–31

[48] Wang R, Wang Y and Luo C 2015 EEG-based real-time drowsiness detection using Hilbert–Huang transform *2015 7th Int. Conf. on Intelligent Human-Machine Systems and Cybernetics (Hangzhou)* pp 195–8

[49] Mardi Z, Ashtiani S N and Mikaili M 2011 EEG-based drowsiness detection for safe driving using chaotic features and statistical tests *J. Med. Signals Sens.* **1** 130–7
[50] Taran S and Bajaj V 2018 Drowsiness detection using instantaneous frequency based rhythms separation for EEG signals *2018 Conf. on Information and Communication Technology (CICT) (Jabalpur, India)* pp 1–6
[51] Kaur R and Singh K 2013 Drowsiness detection based on EEG signal analysis using EMD and trained neural network. *Int. J. Sci. Res.* **2** 2 157–61
[52] Taran S, Bajaj V and Sharma D 2017 TEO separated AM–FM components for identification of apnea EEG signals *2017 IEEE 2nd Int. Conf. on Signal and Image Processing (ICSIP) (Singapore)* pp 391–5
[53] Skipper J H and Wierwille W W 1986 Drowsy driver detection using discriminant analysis *Hum. Factors* **28** 527–40
[54] Awais M, Badruddin N and Drieberg M 2017 A hybrid approach to detect driver drowsiness utilizing physiological signals to improve system performance and wearability *Sensors* **17** 1991
[55] Mu Z, Hu J and Min J 2017 Driver fatigue detection system using electroencephalography signals based on combined entropy features *Appl. Sci.* **7** 150
[56] Sayed R and Eskandarian A 2001 Unobtrusive drowsiness detection by neural network learning of driver steering *Proc. Inst. Mech. Eng.* D **215** 969–75
[57] Vuckovic A, Popovic D and Radivojevic V 2002 Artificial neural network for detecting drowsiness from EEG recordings *6th Seminar on Neural Network Applications in Electrical Engineering (Belgrade, Yugoslavia)* pp 155–8
[58] Vučković A, Radivojevic V, Chen A C and Popovic D B 2002 Automatic recognition of alertness and drowsiness from EEG by an artificial neural network *Med. Eng. Phys.* **24** 349–60
[59] Lawhern V, Hairston W D, McDowell K, Westerfield M and Robbins K 2012 Detection and classification of subject-generated artifacts in EEG signals using autoregressive models *J. Neurosci. Methods* **208** 181–9
[60] Huang R S, Kuo C J, Tsai L-L and Chen O T C 1996 EEG pattern recognition-arousal states detection and classification *Proc. of Int. Conf. on Neural Networks (ICNN'96) (Washington, DC)* vol 2 pp 641–6
[61] Silveira T, Kozakevicius A J and Rodrigues C R 2015 Drowsiness detection for single channel EEG by DWT best m-term approximation *Res. Biomed. Eng.* **31** 107–15
[62] Han H and Song K Y 2015 Improvement of EEG-based drowsiness detection system using discrete wavelet transform *J. Korean Inst. Commun. Inf. Sci.* **40** 1731–3
[63] Patrick K C A, Imtiaz S A, Bowyer S and Rodriguez E V 2016 An algorithm for automatic detection of drowsiness for use in wearable EEG systems *2016 38th Annual Int. Conf. of the IEEE Engineering in Medicine and Biology Society (EMBC) (Orlando, FL)* pp 3523–6
[64] Lim C A, Wai C C and Siew W C 2014 A mobile driver safety system: analysis of single-channel EEG on drowsiness detection *2014 Int. Conf. on Computational Science and Technology (ICCST) (Kota Kinabalu)* pp 1–5
[65] Correa A G and Laciar L E 2010 An automatic detector of drowsiness based on spectral analysis and wavelet decomposition of EEG records *2010 Annual Int. Conf. of the IEEE Engineering in Medicine and Biology (Buenos Aires)* pp 1405–8
[66] Jung T, Makeig S, Stensmo M and Sejnowski T J 1997 Estimating alertness from the EEG power spectrum *IEEE Trans. Biomed. Eng.* **44** 60–9

[67] Makeig S and Inlow M 1993 Lapses in alertness: coherence of fluctuations in performance and EEG spectrum *Electroencephalogr. Clin. Neurophysiol.* **86** 23–35

[68] Awais M, Badruddin N and Drieberg M 2017 EEG brain connectivity analysis to detect driver drowsiness using coherence *2017 Int. Conf. on Frontiers of Information Technology (FIT) (Islamabad)* pp 110–4

[69] Huynh X P, Park S M and Kim Y G 2017 Detection of driver drowsiness using 3D deep neural network and semi-supervised gradient boosting machine *Computer Vision—ACCV 2016 Workshops. ACCV 2016. Lecture Notes in Computer Science* vol 10118 ed C S Chen, J Lu and K K Ma (Cham: Springer)

[70] Yin Y, Zhu Y, Xiong S and Zhang J 2011 Drowsiness detection from EEG spectrum analysis *Informatics in Control, Automation and Robotics. Lecture Notes in Electrical Engineering* vol 133 ed D Yang (Berlin: Springer)

[71] Ruei-Cheng W, Chin-Teng L, Sheng-Fu L, Teng-Yi H and Tzyy-Ping J 2004 EEG-based fuzzy neural network estimator for driving performance *2004 IEEE Int. Conf. on Systems, Man and Cybernetics (IEEE Cat. No. 04CH37583) (The Hague)* vol 4 pp 4034–40

[72] Picot A, Charbonnier S and Caplier A 2008 On-line automatic detection of driver drowsiness using a single electroencephalographic channel *2008 30th Annual Int. Conf. of the IEEE Engineering in Medicine and Biology Society (Vancouver, BC)* pp 3864–7

[73] Picot A, Charbonnier S and Caplier A 2012 On-line detection of drowsiness using brain and visual information *IEEE Trans. Syst. Man. Cybern.* A **42** 764–75

[74] Ogino M and Mitsukura Y 2018 Portable drowsiness detection through use of a prefrontal single-channel electroencephalogram *Sensors* **18** 4477

[75] Gurudath N and Riley B H 2014 Drowsy driving detection by EEG analysis using wavelet transform and *k*-means clustering *Procedia Comput. Sci.* **34** 400–9

[76] Tsai P, Hu W, Kuo T B and Shyu L 2009 A portable device for real time drowsiness detection using novel active dry electrode system *2009 Annual Int. Conf. of the IEEE Engineering in Medicine and Biology Society* pp 3775–8

[77] Boonnak N, Kamonsantiroj S and Pipanmaekaporn L Wavelet transform enhancement for drowsiness classification in EEG records using energy coefficient distribution and neural network *Int. J. Machine Learning Comput.* **5** 288–93

[78] Silveira T, Kozakevicius A J and Rodrigues C R 2016 Automated drowsiness detection through wavelet packet analysis of a single EEG channel *Expert Syst. Appl.* **55** 559–65

[79] Zhang A and Chen Y 2012 EEG feature extraction and analysis under drowsy state based on energy and sample entropy *2012 5th Int. Conf. on BioMedical Engineering and Informatics (Chongqing)* pp 501–5

[80] Chen L, Zhao Y, Zhang J and Zou J 2015 Automatic detection of alertness/drowsiness from physiological signals using wavelet-based nonlinear features and machine learning *Expert Syst. Appl.* **42** 7344–55

[81] Khushaba R N, Kodagoda S, Lal S and Dissanayake G 2011 Driver drowsiness classification using fuzzy wavelet-packet-based feature-extraction algorithm *IEEE Trans. Biomed. Eng.* **58** 121–31

[82] Anitha C 2019 Detection and analysis of drowsiness in human beings using multimodal signals *Digital Business. Lecture Notes on Data Engineering and Communications Technologies* vol 21 ed S Patnaik, X S Yang, M Tavana, F Popentiu-Vlădicescu and F Qiao (Cham: Springer)

[83] Correa A G, Orosco L and Laciar E 2014 Automatic detection of drowsiness in EEG records based on multimodal analysis *Med. Eng. Phys.* **36** 244–9

[84] Pal N R *et al* 2008 EEG-based subject- and session-independent drowsiness detection: an unsupervised approach *EURASIP J. Adv. Signal Process.* **2008** 519480

[85] Hu S and Zheng G 2009 Driver drowsiness detection with eyelid related parameters by support vector machine *Expert Syst. Appl.* **36** 7651–8

[86] Zhang A, Bi J and Sun S 2013 A method for drowsiness detection based on Tsallis entropy of EEG *IFMBE Proc., World Congress on Medical Physics and Biomedical Engineering (Beijing, China, 26–31 May, 2012) vol 39M Long (Berlin: Springer)*

[87] Wang Y, Huang K, Wei C, Huang T, Ko L, Lin C, Cheng C and Jung T 2014 Developing an EEG-based on-line closed-loop lapse detection and mitigation system *Front. Neurosci.* **8** 321

[88] Chang C-W, Ko L-W, Lin F-C, Su T-P, Jung T-P, Lin C-T and Chiou J C 2010 Drowsiness monitoring with EEG-based MEMS biosensing technologies *GeroPsych: J. Gerontopsychol. Geriatr. Psychiatry* **23** 107–13

[89] Sauvet F, Bougard C, Coroenne M, Lely L, Beers P V, Elbaz M, Guillard M, Leger D and Chennaoui M 2014 In-flight automatic detection of vigilance states using a single EEG channel *IEEE Trans. Biomed. Eng.* **61** 2840–7

[90] Jalilifard A and Pizzolato E B 2016 An efficient *k*-NN approach for automatic drowsiness detection using single-channel EEG recording *2016 38th Annual Int. Conf. of the IEEE Engineering in Medicine and Biology Society (EMBC) (Orlando, FL)* pp 820–4

[91] Peiris M, Davidson P, Bones P J and Jones R 2011 Detection of lapses in responsiveness from the EEG *J. Neural Eng.* **8** 016003

[92] Lee B-G, Lee B-L and Chung W-Y 2014 Mobile healthcare for automatic driving sleep-onset detection using wavelet-based EEG and respiration signals *Sensors* **14** 17915–36

[93] Deshmukh P, Somani M S, Mishra M S and Soni M D EEG based drowsiness estimation using Mahalanobis distance *Int. J. Adv. Res. Comput. Sci. Electron. Eng.* http://ijarcsee.org/index.php/IJARCSEE/article/view/146

[94] Taran S and Bajaj V 2018 Drowsiness detection using adaptive Hermite decomposition and extreme learning machine for electroencephalogram signals *IEEE Sens. J.* **18** 8855–62

[95] Budak U, Bajaj V, Akbulut Y, Atila O and Sengur A 2019 An effective hybrid model for EEG-based drowsiness detection *IEEE Sens. J.* **19** 7624–31

[96] Li G, Lee B and Chung W 2015 Smartwatch-based wearable EEG system for driver drowsiness detection *IEEE Sens. J.* **15** 7169–80

[97] Wei C, Wang Y, Lin C and Jung T 2018 Toward drowsiness detection using non-hair-bearing EEG-based brain–computer interfaces *IEEE Trans. Neural Syst. Rehabil. Eng.* **26** 400–6

[98] Nguyen T, Ahn S, Jun S and Kim J 2017 Utilization of a combined EEG/NIRS system to predict driver drowsiness *Sci. Rep.* **7** 43933

[99] Wei L, Mukhopadhyay S C, Jidin R and Chen C 2013 Multi-source information fusion for drowsy driving detection based on wireless sensor networks *2013 Seventh Int. Conf. on Sensing Technology (ICST) (Wellington)* pp 850–7

[100] Shabani H, Mikaili M and Noori S M R 2016 Assessment of recurrence quantification analysis (RQA) of EEG for development of a novel drowsiness detection system *Biomed. Eng. Lett.* **6** 196–204

[101] Bajaj V, Taran S, Khare S and Sengur A 2020 Feature extraction method for classification of alertness and drowsiness states EEG signals *Appl. Acoust.* **163** 107224

[102] Ichimaru Y and Moody G B 1999 Development of the polysomnographic database on CD-ROM *Psychiatry Clin. Neurosci.* **53** 175–7

[103] Goldberger A L *et al* 2000 PhysioBank, PhysioToolkit, and PhysioNet: components of a new research resource for complex physiologic signals *Circulation* **101** e215–20

[104] Rechtschaffen A and Kales A 1968 *A Manual Of Standardized Terminology, Techniques, And Scoring System For Sleep Stages Of Human Subjects* (Washington, DC: US Government Printing Office)

[105] Dornhege G, Blankertz B, Curio G and Muller K 2004 Boosting bit rates in noninvasive EEG single-trial classifications by feature combination and multiclass paradigms *IEEE Trans. Biomed. Eng.* **51** 993–1002

[106] Chen S, Peng Z K, Zhang W and Meng G 2017 Nonlinear chirp mode decomposition: a variational method *IEEE Trans. Signal Process.* **62** 6024–37

[107] Kowalski M, Meynard A and Wu H-T 2016 Convex optimization approach to signals with fast varying instantaneous frequency *Appl. Comput. Harmon. Anal.* **44** 89–122

[108] Gilles J 2013 Empirical wavelet transform *IEEE Trans. Signal Process.* **61** 3999–4010

[109] Dragomiretskiy K and Zosso D 2014 Variational mode decomposition *IEEE Trans. Signal Process.* **62** 531–44

[110] Boyd S, Parikh N, Chu E, Peleato B and Eckstein J 2011 Distributed optimization and statistical learning via the alternating direction method of multipliers *Found. Trends Mach. Learn.* **3** 1–122

[111] Song Y Y and Lu Y 2015 Decision tree methods: applications for classification and prediction *Shanghai Arch. Psychiatry* **27** 130–5

[112] Awad M and Khanna R 2015 Support vector machines for classification *Efficient Learning Machines* (Berkeley, CA: Apress)

[113] Adeniy D A, Wei Z and Yong-quan Y 2016 Automated web usage data mining and recommendation system using *k*-nearest neighbor (KNN) classification method **12** 10

[114] Bühlmann P 2012 Bagging, boosting and ensemble methods *Handbook of Computational Statistics* (Berlin: Springer)

[115] Tharwat A 2016 Linear vs quadratic discriminant analysis classifier: a tutorial *Int. J. Appl. Pattern Recogn.* **3** 145

IOP Publishing

Modelling and Analysis of Active Biopotential Signals in Healthcare, Volume 1

Varun Bajaj and G R Sinha

Chapter 6

Noise removal and classification of EEG signals using the Fourier decomposition method

Virender Kumar Mehla[1], Ashish Kumar[1], Amit Singhal[1], Pushpendra Singh[2]

[1]*Department of Electronics and Communication Engineering, Bennett University, Greater Noida, India*

[2]*Department of Electronics and Communication Engineering, National Institute of Technology, Hamirpur, India*

Through rapid innovation in the health care sector, several types of biomedical signals, including the electrocardiogram (ECG), electroencephalogram (EEG), positron emission tomography (PET), electromyogram (EMG) and electrogastrogram (EGG), now play an important role in the accurate analysis and detection of many health problems, such as heart related problems, sleep disorders, brain abnormalities etc. In the current chapter, an efficient approach based on the Fourier decomposition method (FDM) is proposed for noise removal and classification of epileptic and healthy EEG signals. This work is composed of three main steps. In the first, preprocessing of the EEG signal is performed to remove noise or artifacts present in EEG recordings. In the second step, the FDM is utilized to fragment non-stationary data into a set of orthogonal basis functions known as Fourier intrinsic band functions (FIBFs). In the third step, FIBF based features are extracted and fed to various models for the accurate classification of EEG signals. The simulation results reveal that the current study has achieved superior performance compared to the existing techniques discussed in the literature.

6.1 Introduction

The primary use of the electroencephalogram (EEG) technique is to record the electrical activity produced by the excitation of neurons by placing electrodes on the scalp or the brain [1]. During EEG recording, electrodes are mostly placed on the scalp of the patient, primarily to detect the electrical activity of the neurons in the cerebral cortex [2]. EEG mainly records signals from the small areas of the brain

doi:10.1088/978-0-7503-3279-8ch6

surrounding each electrode. It primarily measures the post-synaptic potentials or changes in membrane potential that are elicited by neuro-transmitters binding to receptors on the post-synaptic membrane. The electrical activity generated in the human brain is represented in the form of brain waves with different frequencies, amplitudes and shapes. The typical electrical activity that occurs in association with an event is sometimes referred to as the event-related potential.

EEG is a useful process that is utilized for the detailed examination, interpretation and visualization of various brain disorders such as epilepsy, stroke, cancers, dementia, sleep disorders and identification of alcoholism [2]. In general, trained professionals and neurologists analyse the EEG recordings manually, which is a challenging process which may produce inaccurate results. Also, the presence of noise during EEG recording may corrupt the quality of the signal which makes visual inspection unreliable. To solve this issue, various algorithms are employed for automatic identification and recognition of brain disorders using EEG signals. These methods are mainly composed of four stages: preprocessing, decomposition, feature extraction and classification.

Preprocessing is a technique used to remove the noise and artifacts introduced during recording of EEG signals [3]. Various types of noise and artifacts are as follows:

i. *Power-line interference.* This is basically 50 or 60 Hz noise which may enter the EEG recording when it is operated using a power source. Power-line interference can be eliminated using a notch filter [4].
ii. *Baseline interference.* This is a low frequency noise, with a frequency range of 0.05–0.7 Hz. This noise is generated due to the breathing of the patient or variation in the skin–electrode impedance [4].
iii. *Electrode contact noise.* This noise occurs either due to a loose contact point or due to poor conductivity between the electrodes and skin.
iv. *Muscle artifacts.* These artifacts are generated by muscle activities near the head which are outlined by high frequency activity (>20 Hz). Low pass filters are commonly employed for removal of muscle artifacts.
v. *Eye blink artifacts.* These may occur due to eye movement or eye blinks. This artifact is the most elevated compared to other artifacts and may produce ten to a few hundred times the amplitude compared to the normal potential in the EEG. The blind source separation and regression techniques are commonly used methods for removal of these artifacts. To remove these artifacts, thresholding and wavelet transformation are also often employed.

An EEG recording is mainly divided into different brain waves and transients. The brain waves are divided into different frequency bands which follow specific nomenclature. Table 6.1 shows the frequency bands of human EEG waves along with their characteristics.

6.2 Related work

Different approaches have been discussed by many researchers in the last few decades for the automated detection of many health problems such as epilepsy,

Table 6.1. Five different frequency bands in EEG signals.

Rhythm band	Frequency (Hz)	Amplitude (μV)	Brain function
δ	Less than 4	20–200	Sleeping
θ	4–7	<20	Dreaming/meditation
α	8–13	30–50	Relaxed, eyes closed
β	13–36	5–10	Active thinking, high alert
γ	>36	5–10	Conscious perception

stroke, dementia, Alzheimer's, alcoholism, sleep disorders etc. For automatic detection of epileptic seizures, various techniques have presented, using time domain or frequency domain methods, time–frequency analysis and some non-linear methods. Both the time and frequency domain representations are known as classical representations. In [5], the fast Fourier transform is utilized to extract the features from the EEG signals. Further, the features obtained from the Fourier coefficients are used for the training and testing of the artificial neural network (ANN) to discriminate EEG signals. The disadvantage of the Fourier transform is that it only provides relevant information for stationary signals, not for non-stationary signals. To reduce this problem, the short time Fourier transform was presented [6, 7]. A new approach based on the discrete short time Fourier transform (DSTFT) is discussed in [6] for epileptic seizure detection using EEG signals. A multi-layer perceptron classifier is used to differentiate between healthy and seizure EEG signals by applying the coefficients of the rational DSTFT. The STFT has a fixed window size, which results either in poor time resolution or frequency resolution depending on the size of the window. To achieve better time–frequency localizations, the wavelet transform is used for processing the EEG signal [8, 9]. Thereafter, an ANN is utilized for the discrimination of healthy and abnormal EEG signals. However, a major concern with the wavelet-based methods is optimal selection of mother wavelets. Various qualitative measures, such as the shape of the wavelet transform, number of vanishing moments, compact support, orthogonality and symmetry, are used in the literature for the selection of the mother wavelet. Based on these measures, many researchers have employed different wavelet transforms in their respective works [10–14], however, it is very difficult to match the shape of the wavelet transform with the shape of the given signal.

In recent years, researchers have presented several signal decomposition approaches, including empirical mode decomposition (EMD) [15, 46–48, 52], variational mode decomposition (VMD) [16], eigen-value decomposition (EVD) [17] and the synchro-squeezed wavelet transform (SSWT) [18] to analyse non-stationary physiological signals. Bajaj *et al* [15] presented an EMD based process to discriminate between seizure and seizure-free EEG signals. This method decomposes non-stationary EEG data into stationary and orthogonal basis functions, referred to as the intrinsic mode functions (IMFs). Two features, the AM and FM bandwidths, extracted from various IMFs, are fed to the machine learning model to categorize two classes of EEG signals.

Priya *et al* in [19] used EMD to decompose EEG signals into various IMFs using a sifting process. Kurtosis, skewness, mean, entropy and negentropy were obtained from various IMFs and fed to a supervised learning classifier for discrimination of two groups of EEG signals, i.e. normal and alcoholic EEG signals. Pachori *et al* demonstrated EMD and Fourier–Bessel (FB) expansion as a significant useful tool for the proper diagnosis of ictal and seizure-free EEG signals [20–22]. The EEG signal is segmented into band-limited IMFs and FB expansion is employed to measure the mean frequency from each IMF. The mean frequency, as a feature vector, is used to test and train the classifier which provides better simulation results in terms of classification accuracy. Hassan *et al* [22] presented an enhanced version of the EMD procedure, referred to as complete ensemble EMD along with adaptive noise (CEEMDAN), for the analysis of physiological signals. Three statistical attributes, namely kurtosis, variance and mean, are extracted from the portions of EEG signals segmented by CEEMDAN. The ANN utilizes these features for training and testing purposes, which provides accurate detection of seizures. A novel approach based on multi-rate filter-bank structure is presented in [23] to categorize EEG signals as seizure or seizure-free. These signals are decomposed to obtain five distinct frequency bands. These brain rhythms are modeled using a random Gaussian process. Two parameters, the Hurst exponent and auto-regressive moving average, are employed to train and test the binary classifier for classification of EEG signals. In [24], a fractional linear prediction approach is presented to distinguish EEG signals as ictal or seizure-free. Two features, the energy of signal and the error, are utilized as inputs to train and test a support vector machine (SVM) classifier that obtains a classification accuracy of 95.33%. Various adaptive signal decomposition techniques such as EMD, VMD, SSWT, EVD etc, have been presented in the literature, presuming that the Fourier method is not appropriate for the investigation of non-linear or non-stationary data. Although EMD is the most popular technique for analysis of such data, it has some drawbacks: mode mixing problems, end-effect artifacts and sensitivity to noise. It is based on the sifting process and there is no mathematical background for this method.

To overcome these problems, Singh *et al* [25] proposed a new method, based on Fourier theory, known as the Fourier decomposition method (FDM) for the analysis of non-linear and non-stationary signals. The FDM has been employed in many applications such as gravitational wave analysis [25, 26, 43, 44], noise removal from physiological signals [4], speech signal analysis and seismic signal analysis [29, 44]. In the current study, the FDM technique is employed for accurate detection of various classes of EEG signals by applying different machine learning models and the simulation results are compared with the existing procedures explored in the literature.

6.3 Proposed work

6.3.1 Dataset

This chapter employs a publicly accessible dataset developed by the Epileptology Department, Bonn University, to capture the significant features for the detection of seizures [27]. Figure 6.1 presents samples of EEG signals which include five subsets, namely, Z, O, N, F and S. The description of these subsets is provided in table 6.2.

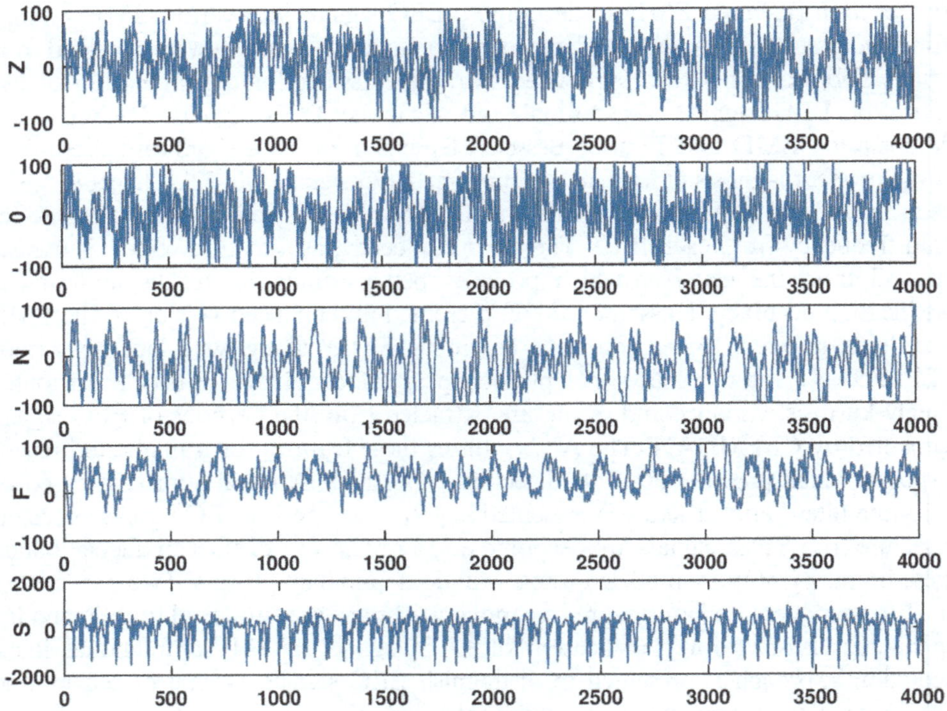

Figure 6.1. Samples of the five EEG datasets.

Subsets Z and O denote the healthy class, N and F comprise the seizure-free (SF) class, while subset S denotes the seizure class. Subsets N and F contain the recorded EEG signals from epileptic subjects before the occurrence seizure attacks in the epileptogenic zone. Each subset holds 100 segmented single-channel EEG signals. The sampling frequency of the EEG signals is 173.61 Hz. In the present work, a total of $173.61 \times 23.6 = 4097$ samples are used. All subjects have age groups of 19–60 years. The combination of the four subsets Z, O, N and F is represented as the non-seizure (NS) class. The raw dataset is utilized to compute the performance metrics for the proposed technique.

6.3.2 The Fourier decomposition method

To construct a discrete-time signal, a continuous-time signal is generally discretized using an analog to digital converter (ADC) before being used by a computing device. In this chapter, a Fourier based method, known as the Fourier decomposition method (FDM), proposed in [25], is applied for analysing non-linear, non-stationary and discrete-time signals. This method represents an adaptive signal decomposition procedure and it separates the EEG signal into a set of orthogonal basis functions, referred to as the Fourier intrinsic band functions (FIBFs), which are complete, local, adaptive and orthogonal or linearly independent non-orthogonal yet energy

Table 6.2. Details of the EEG data.

Set name	Condition	Patient's state	Duration (seconds)	Number of channels	Sampling frequency (Hz)
Z	Healthy	Eye opened		100	173.61
O	Healthy	Eye closed		100	
N	Seizure-free	Inter-ictal	23.6	100	
F	Seizure-free	Inter-ictal		100	
S	Seizure	Ictal		100	

preserving (LINOEP). Let $p[n]$ denote a discrete signal with length M. By applying the discrete Fourier transform (DFT), $p[n]$ can be expressed as

$$p[n] = \sum_{k=0}^{M-1} P[k]\exp\left(\frac{j2\pi kn}{M}\right), \tag{6.1}$$

where $P[k] = \frac{1}{M}\sum_{n=0}^{M-1} p[n]\exp(\frac{-j2\pi kn}{M})$ is known as the DFT for signal $p[n]$. Let M refer to an even number (follow in the same manner if M is odd), then $P[0]$ and $P[M/2]$ are the real numbers, therefore $p[n]$ is expressed as

$$\begin{aligned} p[n] = p[0] &+ \sum_{k=1}^{\frac{M}{2}-1} P[k]\exp\left(\frac{j2\pi kn}{M}\right) + P\left[\frac{M}{2}\right]\exp(j\pi n) \\ &+ \sum_{k=\frac{M}{2}+1}^{M-1} P[k]\exp\left(\frac{j2\pi kn}{M}\right). \end{aligned} \tag{6.2}$$

Since $p[n]$ is real, therefore,

$$z_1[n] \triangleq \sum_{k=1}^{\frac{M}{2}-1} P(k)\exp\left(\frac{j2\pi kn}{M}\right) \tag{6.3}$$

is a complex conjugate of

$$z_2[n] \triangleq \sum_{k=\frac{M}{2}+1}^{M-1} P(k)\exp\left(\frac{j2\pi kn}{M}\right) \tag{6.4}$$

and equation (6.2) can be written as

$$p[n] = P(0) + 2\,\mathrm{Re}\{z_1[n]\} + P\left[\frac{M}{2}\right](-1)^n, \tag{6.5}$$

where $\mathrm{Re}\{z_1[n]\}$ represents the real part of the complex $z_1[n]$. The analytic signal $z_1[n]$ can be written as

$$\sum_{k=1}^{\frac{M}{2}-1} P[k]\exp\left(\frac{j2\pi kn}{M}\right) = \sum_{i=1}^{S} a_i[n]\exp(j\phi_i[n]), \tag{6.6}$$

where, in a forward search of analytic Fourier intrinsic band functions (AFIBFs), one can obtain

$$a_1[n]\exp(j\phi_1[n]) = \sum_{k=1}^{M_1} P[k]\exp\left(\frac{j2\pi kn}{M}\right) \tag{6.6a}$$

$$a_2[n]\exp(j\phi_2[n]) = \sum_{k=(M_1+1)}^{M_2} P[k]\exp\left(\frac{j2\pi kn}{M}\right) \tag{6.6b}$$

$$\vdots$$

$$a_S[n]\exp(j\phi_S[n]) = \sum_{k=(N_{S-1}+1)}^{\frac{M}{2}-1} P[k]\exp\left(\frac{j2\pi kn}{M}\right) \tag{6.6c}$$

and finally, in general,

$$a_i[n]\exp(j\phi_i[n]) = \sum_{k=M_{i-1}+1}^{M_i} P[k]\exp\left(\frac{j2\pi kn}{M}\right) \tag{6.6d}$$

with $M_0 = 0$ and $M_S = (M/2 - 1)$. To achieve minimum possible FIBFs in a low to high frequency scan, for each i we proceed from $(M_{i-1} + 1)$ to $(M/2 - 1)$ to obtain a maximum value of M_i such that $(M_{i-1} + 1) \leqslant M_i \leqslant (M/2 - 1)$ and phase $\Phi_i[n]$ remains a monotonically increasing function, i.e. an estimate of the discrete instantaneous frequency (IF) [49]

$$\omega_i[n] = (\phi_i[n + 1] - \phi_i[n]) \geqslant 0, \quad \forall\, n \tag{6.7}$$

or

$$\omega_i[n] = \left(\frac{\phi_i[n + 1] - \phi_i[n - 1]}{2}\right) \geqslant 0, \quad \forall\, n \tag{6.8}$$

and amplitude $a_i[n] \geqslant 0$, $\forall n$, for $i = 1, 2, \ldots, M$.

The block diagram of the FDM realized from the DFT based zero-phase filter-bank (ZPF) is shown in figure 6.2, where a signal, $y[n]$, is segmented into a set of adaptive and orthogonal basis functions $\{y_1[n], y_2[n], y_3[n], \ldots y_M[n]\}$ as follows:

$$y[n] = b_0 + \sum_{i=1}^{M} y_i[n] = b_0 + \sum_{i=1}^{M} c_i[n], \tag{6.9}$$

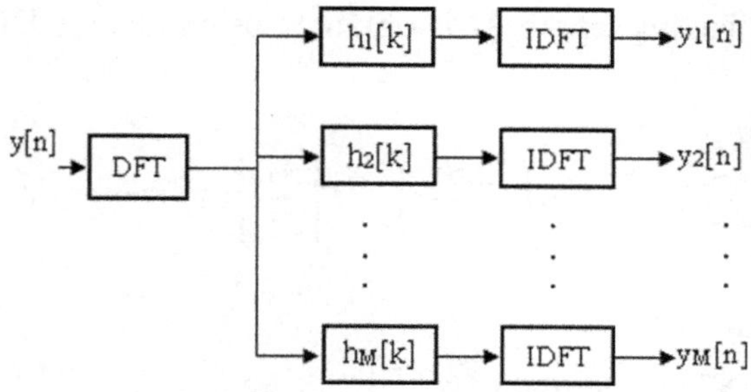

Figure 6.2. Block diagram for FDM realization from the DFT filter-banks.

where b_0 represents the average of the signal $y[n]$, and $y_i[n]$ and $c_i[n]$ represent M orthogonal and LINOEP FIBFs, respectively, for $i = 1, 2, \ldots, M$.

The FDM method is used to fragment non-linear, non-stationary signals into a set of FIBFs which satisfy the following conditions [25, 28–31]:

1. The FIBFs, $y_i(t)$, $1 \leqslant i \leqslant M$, have a zero mean, i.e. $\int_u^v y_i(t)\mathrm{d}t = 0$.
2. The FIBFs should satisfy the orthogonality condition, i.e. $\int_u^v y_i(t)y_k(t)\mathrm{d}t = 0, \forall\ i \neq k$.
3. The FIBFs provide analytical FIBFs representation with instantaneous frequency (IF), $\omega_i(t) = \frac{\mathrm{d}}{\mathrm{d}t}\phi_i(t) \geqslant 0$ and $a_i(t) \geqslant 0, \forall\ t$, where $\phi_i(t)$ is the instantaneous phase and $a_i(t)$ is the instantaneous amplitude.

The FDM method discussed in this study, based on the DFT with a Gabor analytic signal (GAS) representation, is made computationally efficient by implementing it using the fast Fourier transform (FFT) algorithm. The GAS representation is based on a complex-valued function $z[n]$, that contains only positive frequency components in its Fourier spectrum, and is given by

$$z[n] = p[n] + j\hat{p}[n], \tag{6.10}$$

where the real part of the complex $z[n]$ represents the original signal while the imaginary part represents the Hilbert transform (HT) of this signal. Also, both the real and imaginary parts exhibit orthogonality to each other. The HT of the signal is obtained as

$$\hat{p}[n] = H\{p[n]\} = p[n]*h[n] = \sum_{m=-\infty}^{\infty} p[m]\,h[n-m],\ h[n] = \frac{1-\cos(\pi n)}{\pi n}, \tag{6.11}$$

where (*) denotes the convolution operator, H represents the Hilbert operator and the impulse response $h[n]$ is known as the Hilbert kernel. As the impulse response of the Hilbert transform is unstable, non-causal and also has infinite time-support, it cannot be implemented in real applications. The GAS representation obtained from the real signal $p[n]$ having length M is obtained using inverse DFT (IDFT) and is defined as

$$z[n] = P[0] + \sum_{K=1}^{\frac{M}{2}-1} 2P[K]\exp\left(\frac{j2\pi kn}{M}\right) + P\left[\frac{M}{2}\right]\exp(j\pi n), \quad \text{if } M \text{ is even}$$

and (6.12)

$$z[n] = P[0] + \sum_{K=1}^{\frac{(M-1)}{2}} 2P[K]\exp\left(\frac{j2\pi kn}{M}\right), \quad \text{if } M \text{ is odd,}$$

where $P[K] = \frac{1}{M}\sum_{n=1}^{M-1} p[n]\exp(-\frac{j2\pi kn}{M})$ signifies the DFT of $p[n]$, $0 \leqslant n, k \leqslant M-1$, $\exp(j\pi n) = (-1)^n$ and $P[M/2]$ represents the highest frequency signal present in the Fourier spectrum. This approach, as discussed above, for obtaining the GAS representation confirms to the following properties:

i. The Fourier spectrum contains only positive frequency components.
ii. The imaginary part is obtained by applying the Hilbert transform to the real part, i.e. $\hat{p}[n] = H\{p[n]\}$.
iii. Both these parts, real and imaginary, satisfy the orthogonality condition, i.e. $\langle p[n], \hat{p}[n]\rangle = 0$.
iv. The real part can be obtained as the Hilbert transform of the imaginary part with a negative sign, i.e.

$$p[n] = -H\{\hat{p}[n]\} = -H^2\{p[n]\}$$

or

$$p[n] = H^4\{p[n]\}.$$

The polar form for the GAS (6.10) is given by

$$z[n] = p[n] + j\hat{p}[n] = a[n]\exp(j\phi[n]), \tag{6.13a}$$

$$a[n] = \sqrt{p^2[n] + \hat{p}^2[n]} \tag{6.13b}$$

$$\phi[n] = \arctan\left(\frac{\hat{p}[n]}{p[n]}\right) \tag{6.13c}$$

and

$$f[n] = \frac{\omega[n]}{2\pi} = \frac{\phi_d[n]}{2\pi}, \tag{6.13d}$$

where $a[n]$, $\Phi[n]$ and $f[n]$ represent the instantaneous amplitude (IA), the instantaneous phase (IP) and the instantaneous frequency (IF), respectively.

The phase in (6.13c) provides a result in the range $(-\pi, \pi]$ and it also eliminates the problems of division of zero. It is important to note from (6.13d) that the IF is only valid for mono-component signals as the defined instantaneous frequency can become negative for some time intervals for multi-component signals, thereby lacking any physical importance. To resolve this issue and to make the IF positive for all time, we introduce the concept of the phase unwrapping fact and multi-valued function using an inverse tangent function, that is $\tan(\phi[n]) = \tan(\phi[n] + kn\pi)$, $\forall\ k,\ \ n \in Z$, then the IF $\omega[n]$ is defined as

$$w[n] = \begin{cases} \phi_d[n], & \text{if } \phi_d[n] \geqslant 0 \\ \phi_d[n] + \pi, & \text{otherwise} \end{cases}. \tag{6.14}$$

The equation in (6.14) ensures that the IF is positive, i.e. $0 \leqslant \omega[n] \leqslant \pi$ in units of radians/sample, corresponding to $[0,\ Fs/2]$ in hertz at all time instants n. This equation is applicable for mono-component as well as multi-component signals.

In the proposed method, the ZPF is used, in which the frequency response of the filter is one (i.e. $H_i[k] = 1$) in the desired frequency band, while it is zero for the other bands:

$$\left.\begin{aligned} H_i[k] &= 1, \qquad (K_{i-1}+1) \leqslant k \leqslant K_i \text{ and} \\ &\qquad (M - K_i) \leqslant k \leqslant (M - K_{i-1} - 1), \\ &= 0, \qquad \text{otherwise} \end{aligned}\right\}. \tag{6.15}$$

Algorithm 6.1 illustrates the usage of the FDM algorithm to decompose a signal, p, into a set of LINOEP vectors such that $p = g_0 + \sum_{i=1}^{N} g_i$ and $g_i \perp \sum_{l=i+1}^{N} g_l$.

Algorithm 6.1. FDM algorithm to obtain the LINOEP vectors.

%PART A

$g_0 = \text{mean}(p);$

$p_1 = p - g_0;$

for $i = 1$ to $N - 1$ do

$\quad y_i = \text{ZPHP } F_i(p_i, f_{gi});$

$\quad r_i = p_i - y_i;$

$\quad \alpha_i = \frac{\langle y_i, r_i \rangle}{\langle r_i, r_i \rangle};$

$\quad g_i = y_i - \alpha_i r_i;$

$\quad \tilde{g}_{i+1} = (1 + \alpha_i) r_i;$

$\quad p_{i+1} = \tilde{p}_{i+1}$

$g_M = \tilde{g}_M$

%PART B

$g_0 = \text{mean}(p);$

$p_1 = p - g_0;$

for $i = 1$ to $N - 1$ do

$\quad y_i = \text{ZPLP } F_i(p_i, f_{gi});$

$\quad r_i = p_i - y_i;$

$\quad \alpha_i = \frac{\langle r_i, y_i \rangle}{\langle y_i, y_i \rangle};$

$\quad g_i = (1 + \alpha_i) y_i;$

$\quad \tilde{g}_{i+1} = r_i - \alpha_i y_i;$

$\quad p_{i+1} = \tilde{g}_{i+1}$

$g_M = \tilde{g}_M;$

Part A of the algorithm is used to obtain the LINOEP vectors from the highest frequency to lowest frequency components, while part B of the algorithm is used to obtain the lowest to highest frequency AM–FM components.

A Gaussian filter can be used for zero-phase filtering in the segmentation of a non-stationary signal. The FDM technique advocates applying ZPF as it conserves some important features, including the locations of peaks in the filtered waveform in the same position as they occupied in the unfiltered waveform. It has also been observed that conventional filtering techniques transfer these features in the filtered waveform and also cannot be employed to obtain a meaningful time–frequency–energy (TFE) representation. Many transformations such as the DFT, discrete sine transform (DST), discrete cosine transform (DCT) and other decomposition methods such as the STFT, discrete wavelet transform (DWT), EWT, VMD, EVD, EMD etc, have been used to obtain the zero-phase filtering of the signal. Depending on the type of application, anyone can employ three perspectives to split the complete bandwidth of the signal into a set of finite numbers of (i) equal frequency bands, (ii) dyadic frequency bands, and (iii) any other desired frequency bands. The proposed methodology is illustrated in figure 6.3. The complete process of the proposed scheme is as follows. Initially, the EEG signal is divided into a set of band-limited FIBFs using the FDM method. After obtaining the FIBFs, the next step is to extract their relevant features. Three features, i.e. the mean, variance and kurtosis, are obtained from the various FIBFs. Finally, these features are used as inputs to train and test numerous machine learning models to differentiate seizure and healthy EEG signals. The features obtained from FIBFs are defined as follows.

Mean. This is most common statistic used to determine the 'center of gravity' of a given dataset. It is determined by adding all the data points in a signal and dividing

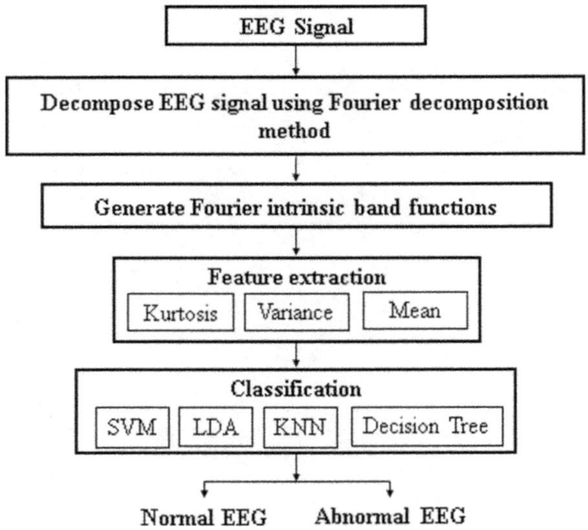

Figure 6.3. Flow chart of proposed methodology.

by the number of points. It has less complexity and better estimate of data values as compared to higher order moments. The mean of the data is defined as

$$\bar{g} = \frac{1}{m}\sum_{i=1}^{m} y_i, \tag{6.16}$$

where m denotes the number of terms and y_i is the list of numbers being averaged.

Variance. Variance, a statistical measure, shows how the data are spread about the mean or expected value. A larger spreading of data points indicates larger variance. Variance is defined as

$$\sigma^2 = \frac{\sum_{i=1}^{m}(g_i - \bar{g})}{m}, \tag{6.17}$$

where g_i is the ith data point and $\bar{g}$ is the mean.

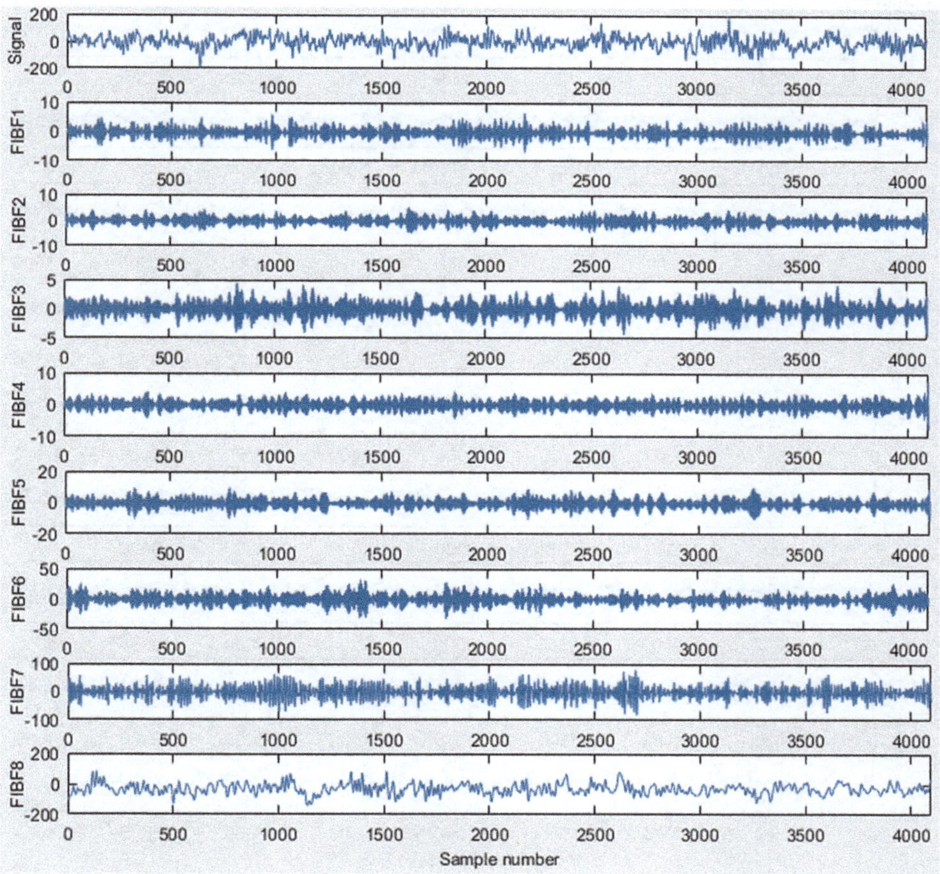

Figure 6.4. Fourier decomposition of a healthy signal of 23.6 s duration into eight FIBFs using equal frequency bands.

Kurtosis. This is defined as the extent to which a frequency distribution is peaked or flat. Large peaked values of the frequency distribution lead to high values of kurtosis and vice versa. Kurtosis is defined as

$$\text{Kurt}[S] = \frac{\mu_4}{\sigma^4}, \tag{6.18}$$

where μ_4 is the fourth central moment and σ is the standard deviation.

6.4 Classification

i. The linear discriminant analysis (LDA) technique is used to reduce the number of variables or dimensions in the given dataset by retaining as much information as possible [32]. This technique is identical to principal component analysis (PCA). LDA, proposed by R A Fisher, is employed for classification problems. In LDA, a linear combination of variables is

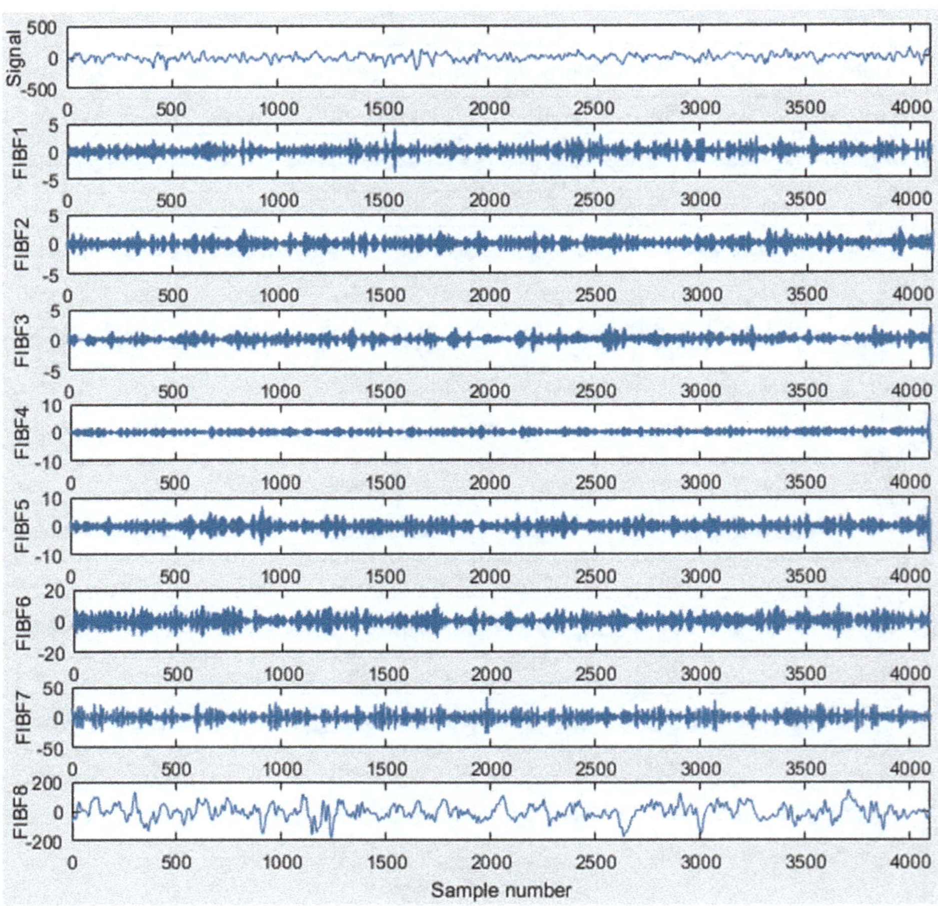

Figure 6.5. Fourier decomposition of a seizure-free signal of 23.6 s duration into eight FIBFs using equal frequency bands.

employed which plays a significant role in the best separation of two classes (targets).

ii. The *k*-nearest neighbors (*k*-NN) classifier is used for regression as well as classification problems [33]. This classifier is the simplest algorithm for classification problems. This algorithm is useful for non-linear data as there is no assumption regarding the data in the algorithm. The *k*-NN algorithm is applied in many areas, such as handwriting detection, and speech, image and video recognition. This classifier employs feature similarity to predict new data points where new data points are assigned a value based on how closely they match the data points in the given training set.

iii. The decision tree (DT) classifier is a kind of supervised machine learning (ML) algorithm where the dataset is continuously broken into smaller and smaller subsets according to a certain parameter [34]. This classifier is employed in many applications, including financial analysis, biomedical applications, astronomy, system control etc. It can be applied to solve classification and regression problems. The main advantages of this classifier

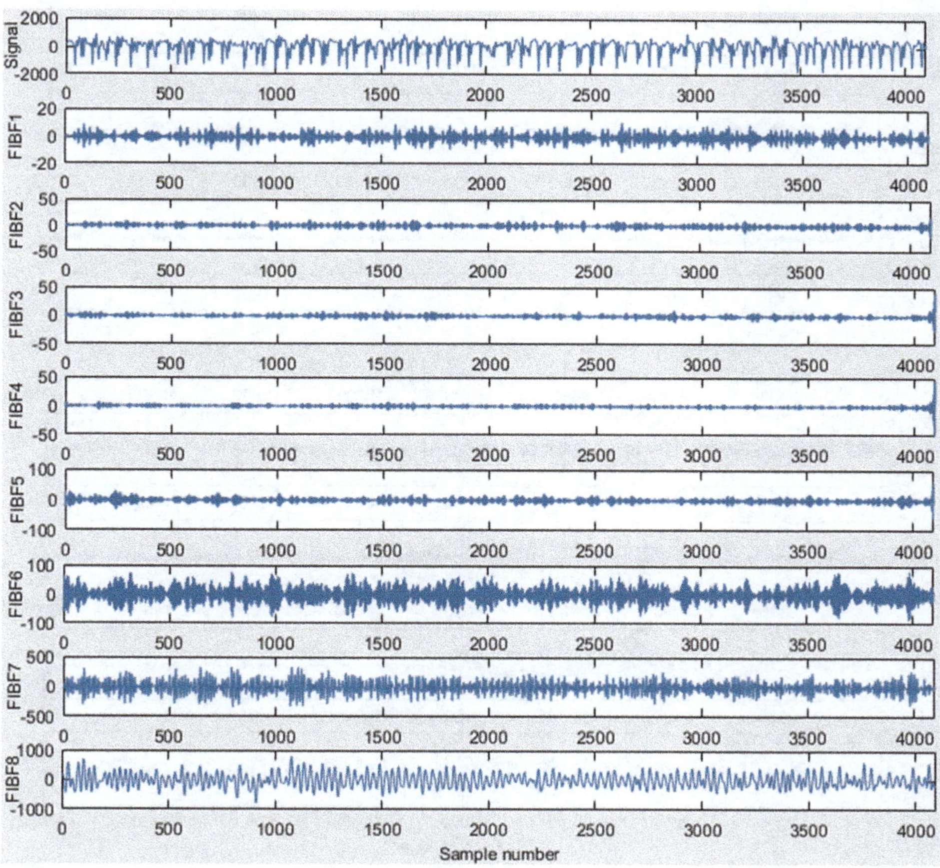

Figure 6.6. Fourier decomposition of a seizure signal of 23.6 s duration into eight FIBFs using equal frequency bands.

are that it is easy to assess and interpret, require less effort to prepare data, and non-linear specifications do not degrade its performance.

iv. The SVM classifier, a supervised machine learning model, is widely used to solve classification and regression problems [35]. In comparison to other existing algorithms, the SVM has a solid theoretical background. By transforming low-dimensional input space into higher-dimensional space, the SVM employs a technique known as the kernel trick. In other words, kernel functions convert non-separable problems into separable problems. The commonly used kernel basis functions in SVM are linear, polynomial and RBF.

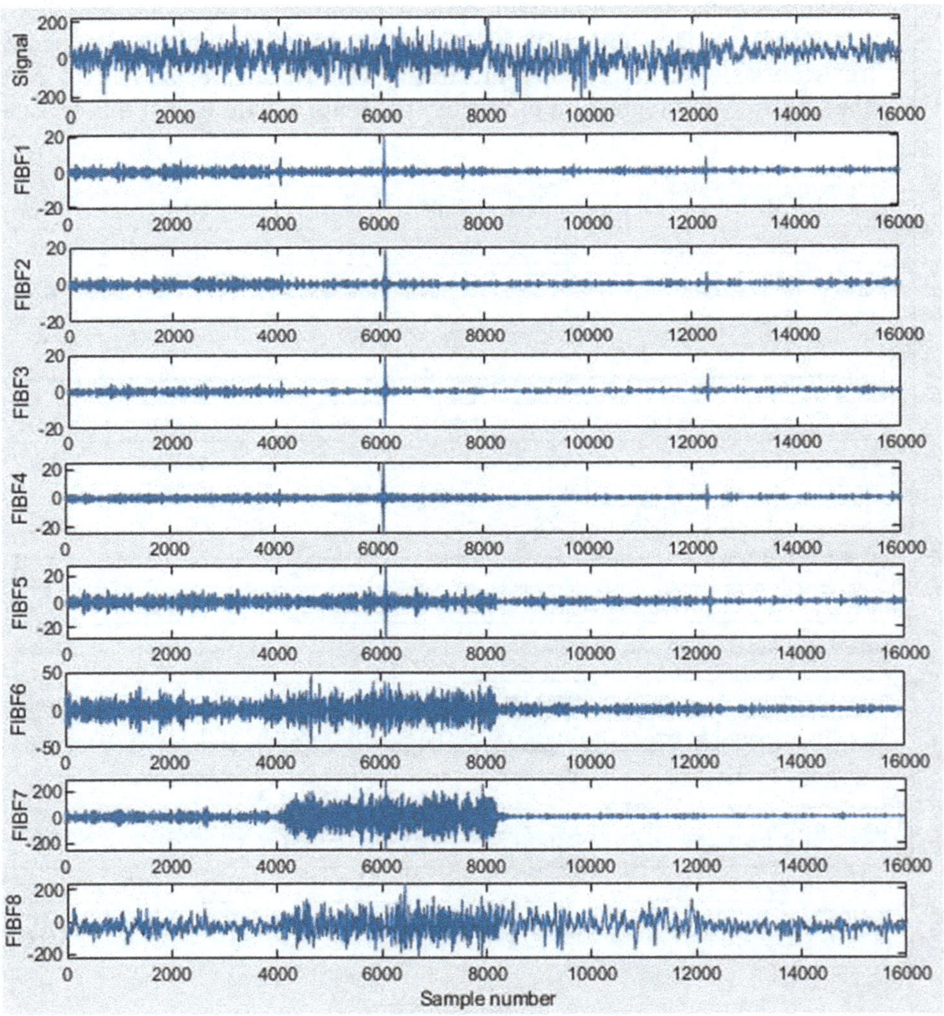

Figure 6.7. Fourier decomposition process of a non-seizure signal of 23.6 s duration into eight FIBFs using equal frequency bands.

6.5 Experimental results and discussion

In order to estimate the performance of the current study, three main evaluation metrics, namely sensitivity, specificity and accuracy, are employed for binary classification problems. These metrics are measured as

$$\text{Sensitivity} = \frac{\text{TP}}{\text{TP} + \text{FN}} \times 100 \tag{6.19}$$

$$\text{Specificity} = \frac{\text{TN}}{\text{TN} + \text{FP}} \times 100 \tag{6.20}$$

$$\text{Accuracy} = \frac{\text{TP} + \text{TN}}{\text{TN} + \text{TP} + \text{FP} + \text{FN}} \times 100, \tag{6.21}$$

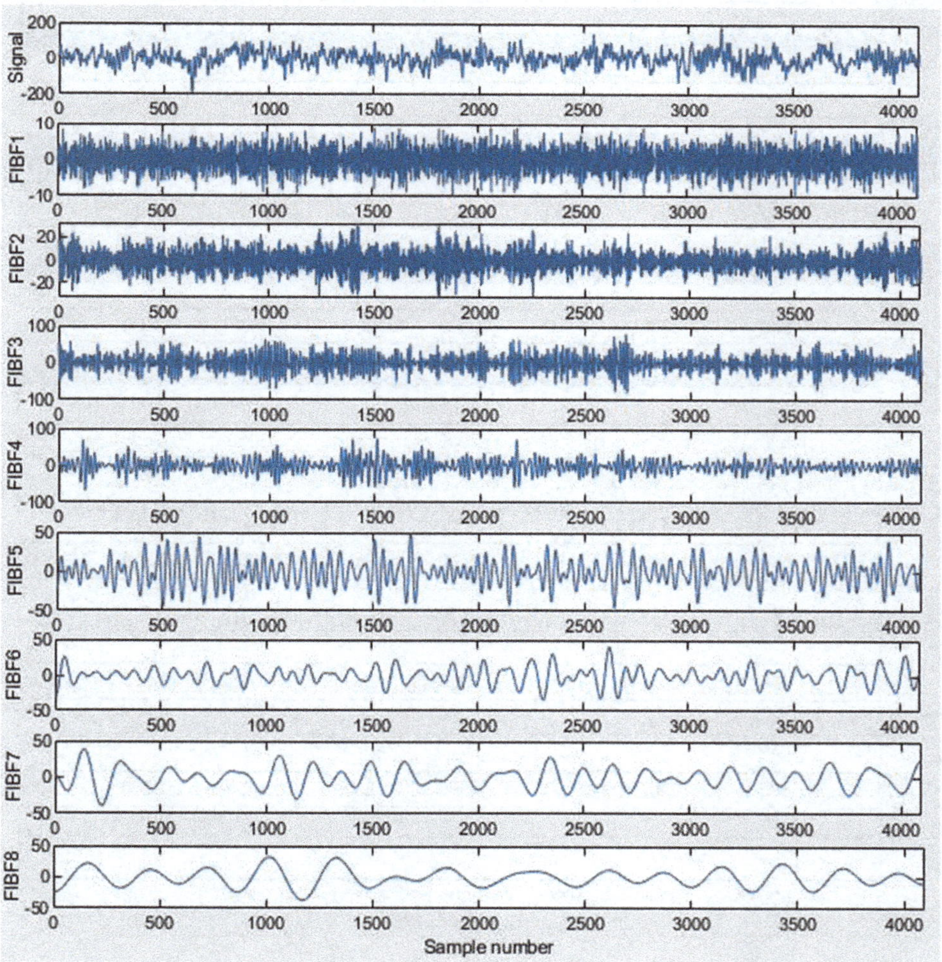

Figure 6.8. Fourier decomposition of a healthy signal using dyadic frequency bands into a set of eight FIBFs.

where TP, TN, FP and FN denote the true positive events (representing positive cases correctly recognized as positive), true negative events (representing negative cases correctly recognized as negative), false positive events (representing negative cases predicted incorrectly as positive) and false negative events (representing positive cases predicted incorrectly as negative). The current work has three main objectives: (i) to differentiate between seizure and healthy EEG signals, (ii) to differentiate seizure-free versus seizure signals, (iii) to differentiate seizure versus non-seizure EEG signals. The publicly available dataset used in this chapter incorporates five subsets, namely, Z, O, N, F and S, respectively. Fragmenting the

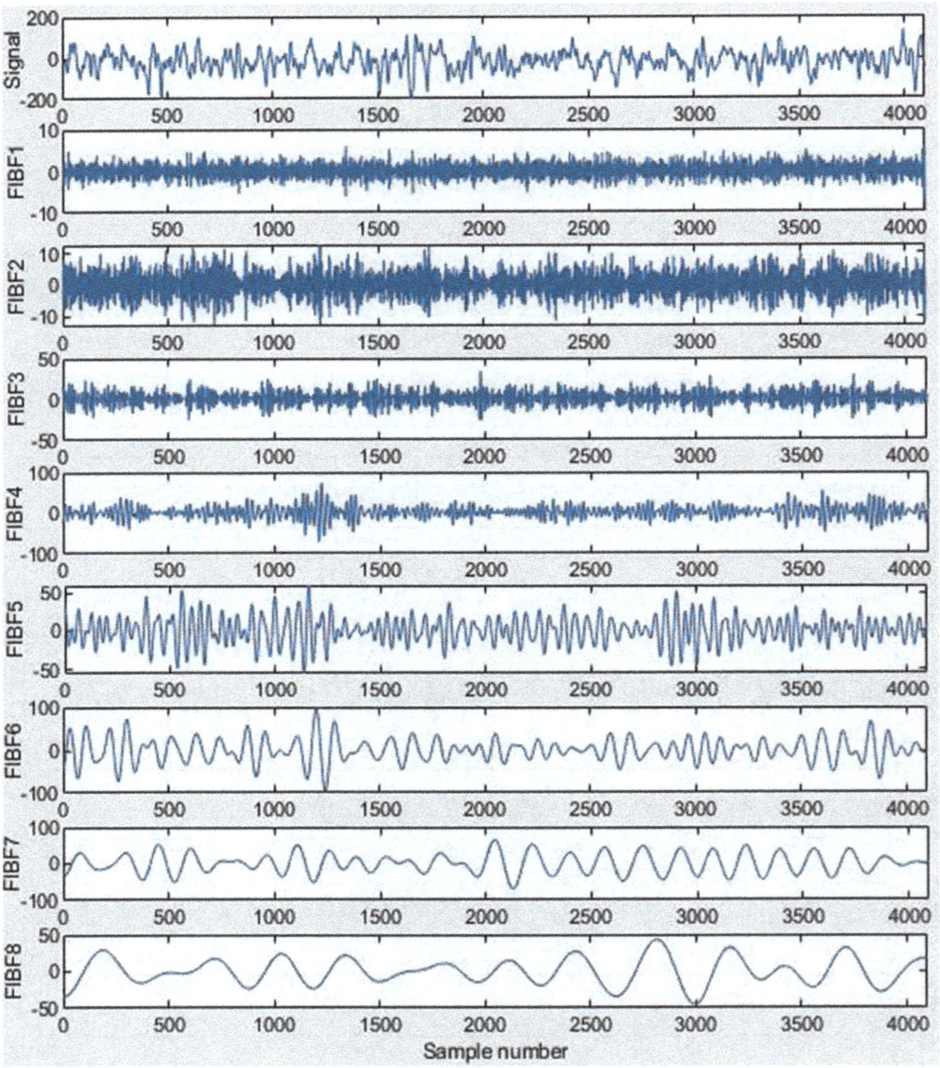

Figure 6.9. Fourier decomposition of a seizure-free signal using dyadic frequency bands into a set of eight FIBFs.

whole bandwidth of a signal into an equal number of frequency bands, the decomposition of healthy, seizure-free, seizure and non-seizure EEG signals using FDM is depicted in figures 6.4–6.7, respectively. Using dyadic frequency bands, the decomposition of the three cases as discussed above using FDM are shown in figures 6.8–6.11, respectively.

The current study has employed a ten-fold cross validation (CV) approach to compute the performance of the machine learning model. The entire data sample is divided into ten groups where each group is selected as a training set at some point. During the first iteration, the first group is employed to test the model and the

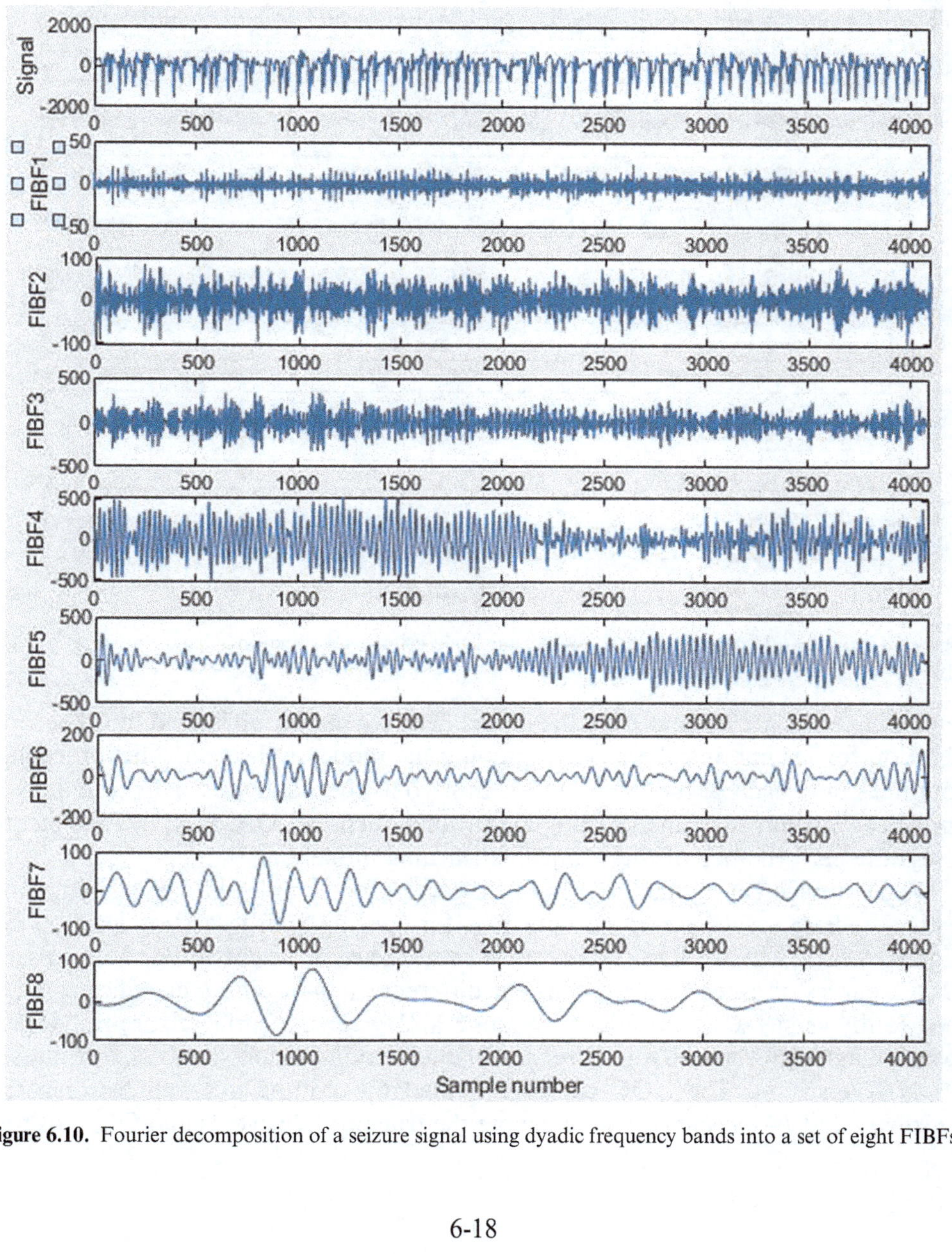

Figure 6.10. Fourier decomposition of a seizure signal using dyadic frequency bands into a set of eight FIBFs.

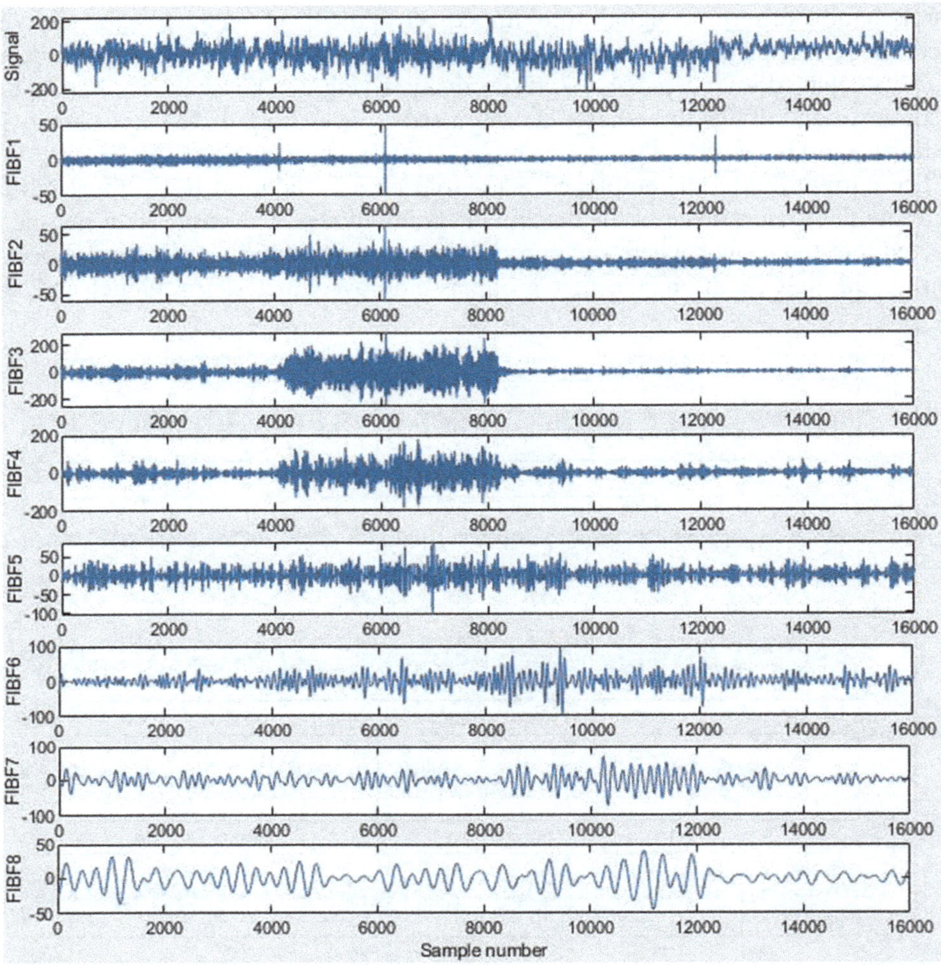

Figure 6.11. Fourier decomposition of a non-seizure signal using dyadic frequency bands into a set of eight FIBFs.

remaining parts are used to train the model. The second group is used in the second iteration for testing and the rest for training the model and so on. This procedure continues until each of the the groups is utilized as the testing set. This technique is applied to various machine learning algorithms, such as LDA, *k*-NN, SVM etc, to assess the performance of these machine learning models.

While implementing this method, all groups are employed for training and testing, where each group is utilized for testing purposes at least once. Figures 6.12(a), (b), 6.13(a), (b), and 6.14 show the receiver operating characteristic (ROC) curves of *k*-nearest neighbors for different cases of binary classification, i.e. the healthy versus seizure classes (Z versus S and O versus S), the seizure-free versus seizure classes (N versus S and F versus S) and the non-seizure versus seizure classes (ZONF versus S). The ROC curve is created by plotting a graph between the *sensitivity* and *1−specificity*, which denote the true positive events and false positive

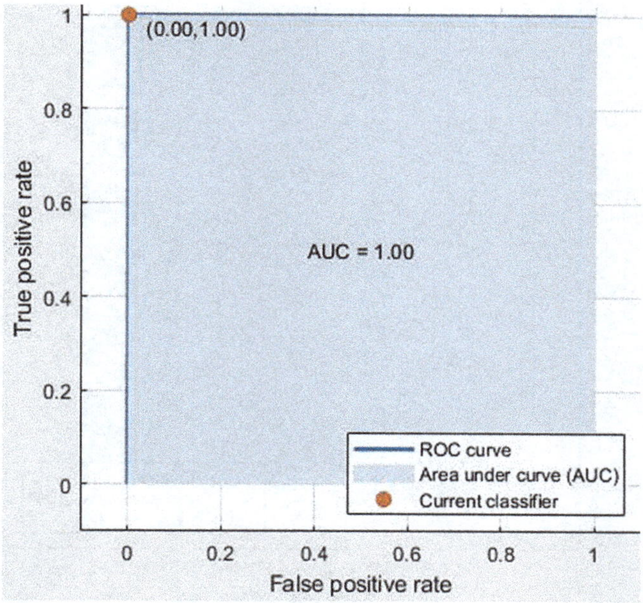

(a)

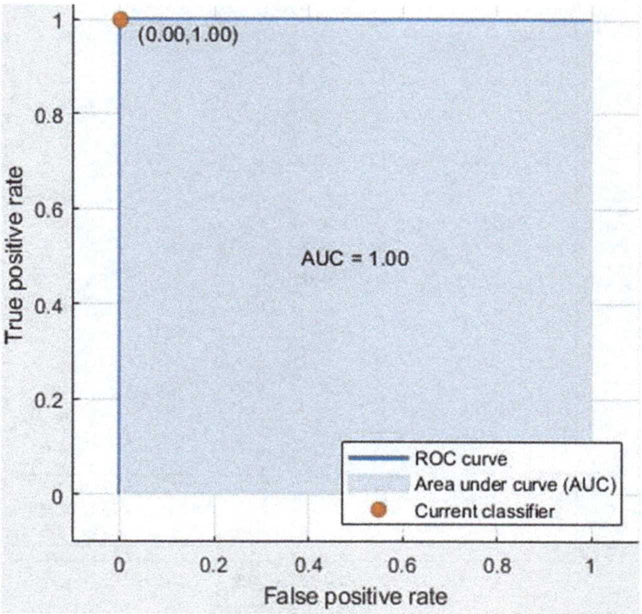

(b)

Figure 6.12. ROC curve for the *k*-NN classifier: (a) class Z versus class S and (b) class O versus class S.

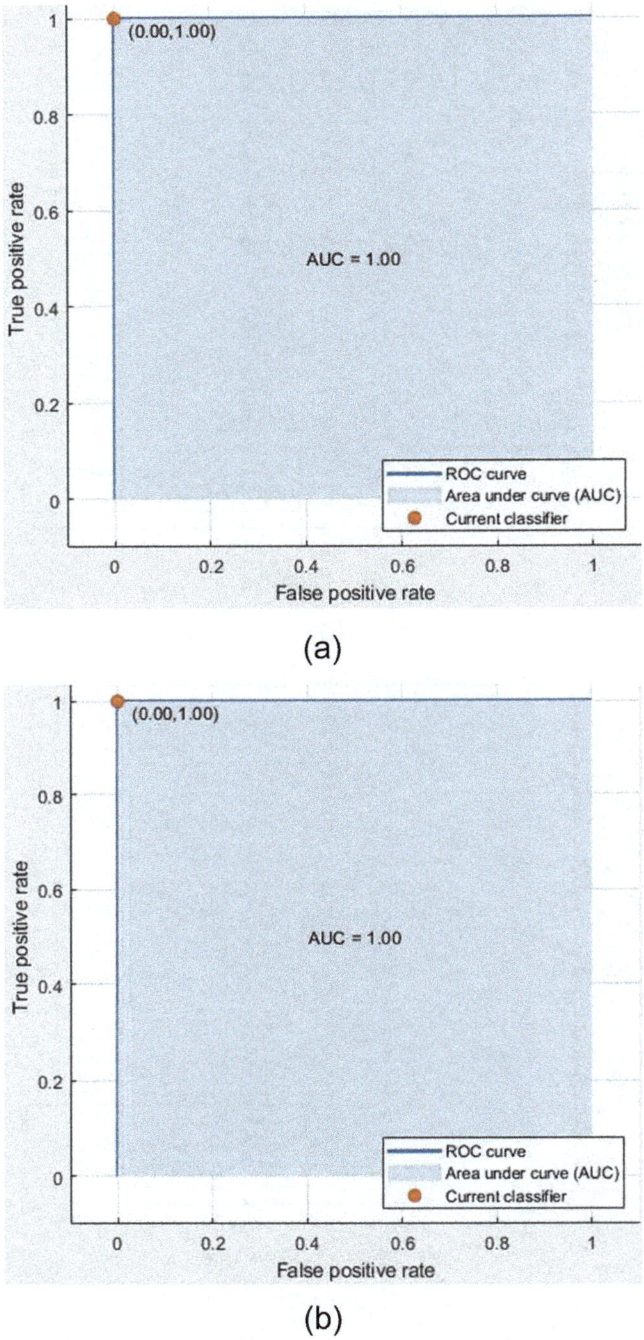

Figure 6.13. ROC curve for the *k*-NN classifier: (a) class N versus class S and (b) class F versus class S.

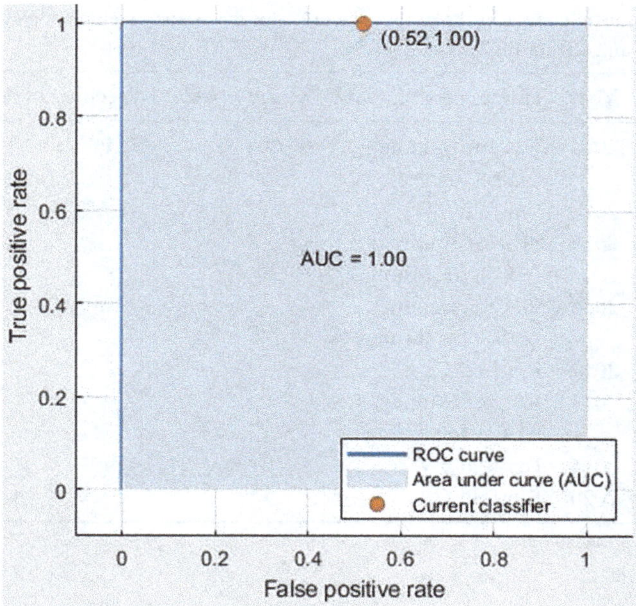

Figure 6.14. ROC curve for the *k*-NN classifier (class ZONF versus class S).

Table 6.3. Performance analysis of the current work and the existing methodologies for discriminating between seizure and healthy EEG signals.

Authors	Year	Training/testing data	Sensitivity (%)	Specificity (%)	Accuracy (%)
Guo *et al* [11]	2010	50% for training, 50% for testing	100	99.20	99.85
Subasi *et al* [36]	2005	60% for training, 40% for testing	93.6	92.4	93
Sharma *et al* [21]	2018	Ten-fold CV	99.50	99.50	99.50
Kannathal *et al* [37]	2005	50% for training, 50% for testing	–	–	90
Vidyaratne *et al* [40]	2017	Five-fold CV	99.00	100.0	99.8
Kaya *et al* [41]	2014	Ten-fold CV	–	–	99.5
Fu *et al* [49]	2015	Ten-fold CV	–	–	99.85
Peker *et al* [50]	2016	60% for training, 40% for testing	–	–	100
Tiwari *et al* [38]	2017	Ten-fold CV	–	–	99.45
Proposed work	2020	Ten-fold CV	–	–	100

events, respectively. The performance analysis of the present study with the existing methodologies is presented in tables 6.3–6.5. Tables 6.3 and 6.4 reported the classification performance of seizure versus healthy EEG signals and seizure-free versus seizure EEG signals in relation to the existing methods, while the

Table 6.4. A comparative analysis of the proposed work with the existing methodologies for discrimination between seizure-free and seizure EEG signals.

Authors	Year	Training/testing data	Sensitivity (%)	Specificity (%)	Accuracy (%)
Joshi *et al* [24]	2014	50% for training, 50% for testing	96.00	95.00	95.33
Kaya *et al* [41]	2014	Ten-fold CV	–	–	97.00
Samiee *et al* [6]	2015	50% for training, 50% for testing	–	–	99.8
Kumar *et al* [42]	2015	50% for training, 50% for testing	–	–	99.33
Bajaj *et al* [15]	2012	Ten-fold CV	–	–	99.5–100
Sriraam *et al* [51]	2017	30% for training, 70% for testing	–	–	84.58
Pachori *et al* [52]	2014	Ten-fold CV	–	–	95.75
Proposed work	2020	Ten-fold CV	–	–	100

Table 6.5. A comparative analysis of proposed work with the existing methodologies for discrimination between seizure and non-seizure EEG signals.

Authors	Year	Training/testing data	Sensitivity (%)	Specificity (%)	Accuracy (%)
Guo *et al* [11]	2010	50% for training, 50% for testing	99.00	95.50	98.27
Fu *et al* [49]	2015	Ten-fold CV	–	–	98.80
Peker *et al* [50]	2016	60% for training, 40% for testing	100	98.01	99.33
Hassan *et al* [22]	2015	40% for training, 60% for testing	–	–	98.87
Tzallas *et al* [39]	2007	50% for training, 50% for testing	–	–	97.73
Kaleem *et al* [53]	2013	Ten-fold CV	–	–	98.20
Jaiswal *et al* [54]	2017	Ten-fold CV	98.30	98.82	98.72
Wang *et al* [55]	2017	Ten-fold CV	–	–	99.25
Proposed work	2020	Ten-fold CV	–	–	99.6

performance analysis of various methods for non-seizure versus seizure EEG signals is shown in table 6.5. It is concluded from tables 6.3–6.5 that the proposed approach achieved better simulation results in terms of classification accuracy compared to those of existing state-of-the-art methods.

6.6 Conclusion and proposed future scope

This study discussed an effective method based on the FDM for the decomposition of EEG signals to enable automatic diagnosis of seizure activity in epileptic patients. A set of orthogonal signal components called FIBFs were obtained through decomposition using the FDM. Three features were extracted from the FIBFs and fed to various machine learning algorithms to compute the classifier performance. It can be noted from tables 6.3 and 6.4 that the present method attained a classification accuracy of 100%, when comparing healthy versus seizure EEG signals and seizure-free versus seizure EEG signals using a k-NN classifier, which is better than for the existing methods presented in the literature. Further, it is noted from table 6.5 that the current study has also attained 99.6% accuracy when discriminating between seizure-free and seizure data.

After adequate trials on various large datasets, the present work could be validated for the real-time interpretation of various diseases in the health sector. In the future, the FDM could be adopted in various areas such as climate research, speech and audio processing, bio-engineering and molecular science. Moreover, the efficacy of the FDM can be investigated for forecasting data obtained from financial time-series, earthquake data analysis, water image identification, and audio and video signal analysis.

References

[1] Homan R W, Herman J and Purdy P 1987 Cerebral location of international 10–20 system electrode placement *Electroencephalogr. Clin. Neurophysiol.* **66** 376–82

[2] Electroencephalography *Science Direct* https://www.sciencedirect.com/topics/psychology/electroencephalography

[3] Jiang X, Bian G B and Tian Z 2019 Removal of artifacts from EEG signals: a review *Sensors* **19** 987

[4] Singhal A, Singh P, Fatimah B and Pachori R B 2020 An efficient removal of power-line interference and baseline wander from ECG signals by employing Fourier decomposition technique *Biomed. Signal Process. Control* **57** 101741

[5] Bhople A D and Tijare P A 2012 Fast Fourier transform based classification of epileptic seizure using artificial neural network *Int. J. Adv. Res. Comput. Sci. Softw. Eng.* **2** 228–31

[6] Samiee K, Kovacs P and Gabbouj M 2015 Epileptic seizure classification of EEG time-series using rational discrete short-time Fourier transform *IEEE Trans. Biomed. Eng.* **62** 541–52

[7] Fadzal C C W, Mansor W, Khuan L Y and Zabidi A 2012 Short-time Fourier transform analysis of EEG signal from writing *IEEE Proc. 8th Int. Coll. on Signal Processing and its Applications* pp 525–7

[8] Hazarika N, Chen J Z, Tsoi A C and Sergejew A 1997 Classification of EEG signals using the wavelet transform *Signal Process.* **59** 61–72

[9] Bhattacharyya A, Singh L and Pachori R B 2018 Fourier–Bessel series expansion based empirical wavelet transform for analysis of non-stationary signals *Digital Signal Process.* **78** 185–96

[10] Gilles J 2013 Empirical wavelet transform *IEEE Trans. Signal Process.* **61** 3999–4010

[11] Guo L, Rivero D and Pazos A 2010 Epileptic seizure detection using multiwavelet transform based approximate entropy and artificial neural networks *J. Neurosci. Methods* **193** 156–63

[12] Kumar A, Komaragiri R and Kumar M 2018 Heart rate monitoring and therapeutic devices: a wavelet transform based approach for the modeling and classification of congestive heart failure *ISA Trans.* **79** 239–50
[13] Kumar A, Berwal D and Kumar Y 2018 Design of high-performance ECG detector for implantable cardiac pacemaker systems using biorthogonal wavelet transform *Circuits Syst. Signal Process.* **37** 3995–4014
[14] Berwal D, Kumar A and Kumar Y 2018 Design of high performance QRS complex detector for wearable healthcare devices using biorthogonal spline wavelet transform *ISA Trans.* **81** 222–30
[15] Bajaj V and Pachori R B 2011 Classification of seizure and non-seizure EEG signals using empirical mode decomposition *IEEE Trans. Inf. Technol. Biomed.* **16** 1135–42
[16] Zhang T, Chen W and Li M 2017 AR based quadratic feature extraction in the VMD domain for the automated seizure detection of EEG using random forest classifier *Biomed. Signal Process. Control* **31** 550–9
[17] Sharma R R and Pachori R B 2017 Time–frequency representation using IEVDHM–HT with application to classification of epileptic EEG signals *IET Sci. Meas. Technol.* **12** 72–82
[18] Daubechies I, Lu J and Wu H T 2011 Synchrosqueezed wavelet transforms: an empirical mode decomposition-like tool *Appl. Comput. Harmon. Anal.* **30** 243–61
[19] Priya A, Yadav P, Jain S and Bajaj V 2018 Efficient method for classification of alcoholic and normal EEG signals using EMD *J. Eng.* **2018** 166–72
[20] Pachori R B 2008 Discrimination between ictal and seizure-free EEG signals using empirical mode decomposition *Res. Lett. Signal Process.* **2008** 293056
[21] Sharma R R, Varshney P, Pachori R B and Vishvakarma S K 2018 Automated system for epileptic EEG detection using iterative filtering *IEEE Sens. Lett.* **2** 1–4
[22] Hassan A R and Haque M A 2015 Epilepsy and seizure detection using statistical features in the complete ensemble empirical mode decomposition domain *Proc. TENCON-2015 IEEE* pp 1–6
[23] Gupta A, Singh P and Karlekar M 2018 A novel signal modeling approach for classification of seizure and seizure-free EEG signals *IEEE Trans. Neural Syst. Rehabil. Eng.* **26** 925–35
[24] Joshi V, Pachori R B and Vijesh A 2014 Classification of ictal and seizure-free EEG signals using fractional linear prediction *Biomed. Signal Process. Control* **9** 1–5
[25] Singh P, Joshi S D, Patney R K and Saha K 2017 The Fourier decomposition method for nonlinear and non-stationary time series analysis *Proc. R. Soc.* A **473** 20160871
[26] Singh P, Singhal A and Joshi S D 2018 Time-frequency analysis of gravitational waves *Proc. 2018 Int. Conf. on Signal Processing and Communications (SPCOM) IEEE* pp 197–201
[27] Andrzejak R G, Lehnertz K, Mormann F, Rieke C, David P and Elger C E 2001 Indications of nonlinear deterministic and finite-dimensional structures in time series of brain electrical activity: dependence on recording region and brain state *Phys. Rev.* E **64** 061907
[28] Singh P and Joshi S D 2019 Some studies on multidimensional Fourier theory for Hilbert transform, analytic signal and AM–FM representation *Circuits Syst. Signal Process.* **38** 5623–50
[29] Singh P 2018 Novel Fourier quadrature transforms and analytic signal representations for nonlinear and non-stationary time-series analysis *R. Soc. Open Sci.* **5** 181131
[30] Singh P, Joshi S D, Patney R K and Saha K 2016 Fourier-based feature extraction for classification of EEG signals using EEG rhythms *Circuits Syst. Signal Process.* **35** 3700–15

[31] Singh P 2016 Some studies on a generalized Fourier expansion for nonlinear and nonstationary time series analysis *Doctoral Dissertation* IIT Delhi
[32] Brownlee J 2016 *Linear Discriminant Analysis for Machine Learning* vol 6 (Melbourne: Machine Learning Mastery)
[33] Harrison O 2018 *Machine Learning Basics with the K-Nearest Neighbors Algorithm* (Toronto: Towards Data Science)
[34] Chakure A 2019 *An Introduction to Decision Tree Classifier* (Toronto: Towards Data Science)
[35] Vapnik V 2013 *The Nature of Statistical Learning Theory* (Berlin: Springer)
[36] Subasi A, Alkan A, Koklukaya E and Kiymik M K 2005 Wavelet neural network classification of EEG signals by using AR model with MLE preprocessing *Neural Netw.* **18** 985–97
[37] Kannathal N, Choo M L, Acharya U R and Sadasivan P K 2005 Entropies for detection of epilepsy in EEG *Comput. Methods Programs Biomed.* **80** 187–94
[38] Tiwari A K, Pachori R B, Kanhangad V and Panigrahi B K 2016 Automated diagnosis of epilepsy using key-point-based local binary pattern of EEG signals *IEEE J. Biomed. Health Inform.* **21** 888–96
[39] Tzallas A T, Tsipouras M G and Fotiadis D I 2007 Automatic seizure detection based on time-frequency analysis and artificial neural networks *Comput. Intell. Neurosci.* **2007** 1–31
[40] Vidyaratne L S and Iftekharuddin K M 2017 Real-time epileptic seizure detection using EEG *IEEE Trans. Neural Syst. Rehabil. Eng.* **25** 2146–56
[41] Kaya Y, Uyar M, Tekin R and Yıldırım S 2014 1D-local binary pattern-based feature extraction for classification of epileptic EEG signals *Appl. Math. Comput.* **243** 209–19
[42] Kumar T S, Kanhangad V and Pachori R B 2015 Classification of seizure and seizure-free EEG signals using local binary patterns *Biomed. Signal Process. Control* **15** 33–40
[43] Gupta A, Joshi S D and Singh P 2018 On the approximate discrete KLT of fractional Brownian motion and applications *J. Franklin Inst.* **355** 8989–9016
[44] Singh P 2018 Breaking the limits: redefining the instantaneous frequency *Circuits Syst. Signal Process.* **37** 3515–36
[45] Singh P and Pachori R B 2017 Classification of focal and non-focal EEG signals using features derived from Fourier-based rhythms *J. Mech. Med. Biol.* **17** 1740002
[46] Huang N E, Shen Z, Long S R, Wu M C, Shih H H, Zheng Q, Yen N C, Tung C C and Liu H H 1998 The empirical mode decomposition and the Hilbert spectrum for nonlinear and non-stationary time series analysis *Proc. R. Soc. Lond.* A **454** 903–95
[47] Singh P, Joshi S D, Patney R K and Saha K 2014 Some studies on nonpolynomial interpolation and error analysis *Appl. Math. Comput.* **244** 809–21
[48] Boashash B 1992 Estimating and interpreting the instantaneous frequency of a signal. I. fundamentals *Proc. IEEE* **80** 520–38
[49] Fu K, Qu J, Chai Y and Zou T 2015 Hilbert marginal spectrum analysis for automatic seizure detection in EEG signals *Biomed. Signal Process. Control* **18** 179–85
[50] Peker M, Sen B and Delen D 2015 A novel method for automated diagnosis of epilepsy using complex-valued classifiers *IEEE J. Biomed. Health Inform.* **20** 108–18
[51] Raghu S and Sriraam N 2017 Optimal configuration of multilayer perceptron neural network classifier for recognition of intracranial epileptic seizures *Expert Syst. Appl.* **89** 205–21

[52] Pachori R B and Patidar S 2014 Epileptic seizure classification in EEG signals using second-order difference plot of intrinsic mode functions *Comput. Methods Programs Biomed.* **113** 494–502

[53] Kaleem M, Guergachi A and Krishnan S 2013 EEG seizure detection and epilepsy diagnosis using a novel variation of empirical mode decomposition *Proc. 35th Annual Int. Conf. of the IEEE Engineering in Medicine and Biology Society (EMBC)* pp 4314–7

[54] Jaiswal A K and Banka H 2017 Local pattern transformation-based feature extraction techniques for classification of epileptic EEG signals *Biomed. Signal Process. Control* **34** 81–92

[55] Wang L, Xue W, Li Y, Luo M, Huang J, Cui W and Huang C 2017 Automatic epileptic seizure detection in EEG signals using multi-domain feature extraction and nonlinear analysis *Entropy* **19** 222

IOP Publishing

Modelling and Analysis of Active Biopotential Signals in Healthcare, Volume 1

Varun Bajaj and G R Sinha

Chapter 7

Reliable and accurate information extraction from surface electromyographic signals

Hamid Reza Marateb[1,2], Mislav Jordanic[2], Monica Rojas-Martínez[3], Joan Francesc Alonso[2,4], Leidy Yanet Serna[2], Mehdi Shirzadi[2], Marjan Nosouhi[1], Miguel Ángel Mañanas[2,4], Kevin C McGill[5]

[1]*Biomedical Engineering Department, Engineering Faculty, University of Isfahan, Isfahan, Iran*

[2]*Biomedical Engineering Research Centre (CREB), Automatic Control Department (ESAII) Universitat Politècnica de Catalunya-Barcelona Tech (UPC), Barcelona, Spain*

[3]*Department of Bioengineering, Universidad El Bosque, Bogotá, Colombia*

[4]*Biomedical Research Networking Center in Bioengineering, Biomaterials, and Nanomedicine (CIBER-BBN), Spain*

[5]*Veterans Affairs, Palo Alto Health Care System, Palo Alto, CA, USA [retired]*

The electrical activity generated in contracting muscles is measured using electromyographic (EMG) signals. By placing an array of electrodes on the surface of the skin, surface EMG (sEMG) signals are recorded non-invasively. The sEMG signal is a stochastic signal whose amplitude is generally between 1 and 10 mV, with the most dominant spectral power between 50 and 150 Hz. The sEMG signal is often corrupted by various types of noise, such as movement artifacts, power-line interference and activity from the other muscles. Moreover, the electrode placement could affect the recorded signals. They make analysing and classifying sEMG signals difficult. sEMG has applications in rehabilitation, sport science, kinesiology, ergonomics, muscle architecture identification, neurological disease diagnosis, prosthesis control and human–machine interfaces. In this chapter, the sEMG signal processing methods used in such applications are discussed, and critical issues are considered. Finally, a basic prosthesis control example is provided for interested readers.

7.1 Surface electromyography

Skeletal muscles are under the voluntary control of the somatic nervous system. Muscle fibers are activated by the terminal branches of spinal motoneurons. The

doi:10.1088/978-0-7503-3279-8ch7

neuromuscular junctions are clustered in a narrow area of a muscle called the innervation zone (IZ) or motor point. The location of the IZ depends on the type of muscle (non-pennate, pennate, bipennate, or multipennate), it can shift under the skin during muscular contraction (isometric or dynamic) and in some muscles, such as those of the forearm, there can be several IZs. The location of the IZ can also change as a result of neuromuscular disease and new ones can appear during rehabilitation after neurological injuries because of neuroplasticity [1–3].

Each motoneuron innervates multiple muscle fibers. The motoneuron and all the muscle fibers that it activates make up a motor unit (MU), the smallest physiological muscle unit that the nervous system can control individually. When a nerve impulse reaches the neuromuscular junction, it excites muscle fiber action potentials (APs) that travel from the IZ to the fiber-endings at a conduction velocity (CV) that ranges between 3 and 6 m s^{-1} [4].

The net potential produced by the simultaneous discharge of all the fibers in a motor unit is called the MUAP and constitutes the basic unit of the EMG signal. The MUAP duration is inversely proportional to the CV of the muscle fiber APs (figure 7.1).

Surface electromyography involves recording muscular electrical activity by placing electrodes on the skin overlying the muscles. The resulting signal is known as the surface EMG (sEMG) signal. During a muscular contraction, usually several motor units are activated repeatedly and the recording electrodes detect the superimposed sequences of MUAPs (figure 7.2). In contrast to intramuscular electromyography (iEMG), sEMG is non-invasive. It does not involve piercing the skin and is not painful. Thus, it is highly recommended for motor rehabilitation.

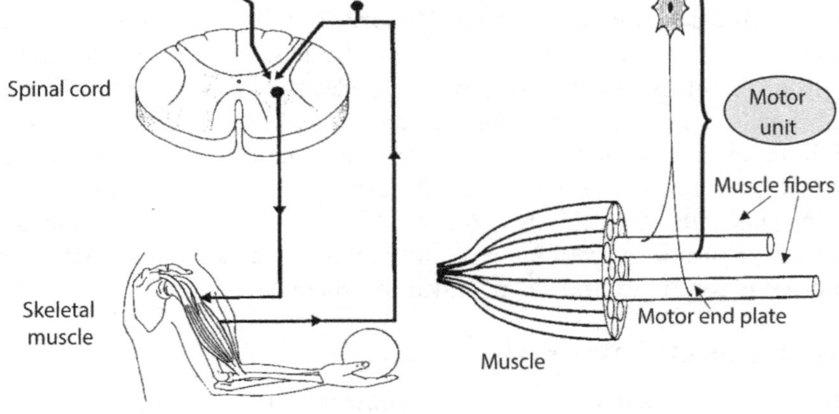

Figure 7.1. Motor control procedures. (Reproduced with permission from [5].)

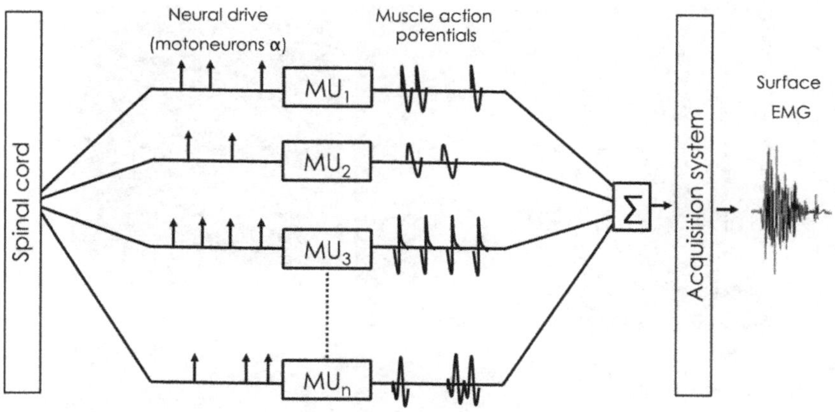

Figure 7.2. The generation of the sEMG signal. (Reproduced from [6] with permission from Springer.)

The sEMG signal can be recorded successfully from many skeletal muscles, and many researchers and groups have focused their efforts on the standardization of electrode placement, best practice and reporting EMG data [2, 7–9].

The sEMG can be recorded using a reference electrode that is placed on an electrically inactive area of the body and one or two electrodes placed over the muscle. In monopolar recording, the sEMG signal is recorded as the potential difference between an active sEMG electrode and the reference electrode, while in bipolar recording, the signal is the potential difference between two sEMG electrodes.

More detailed information can be obtained using arrays of electrodes (figure 7.3). Multiple electrodes may be placed in a single line, forming a linear electrode array, or in a quadrature grid, forming a 2D electrode array which is called high-density sEMG (HD-sEMG) when the number of electrodes is high (commonly more than 64), providing information about electrical activity in both space and time [10, 11].

When the root-mean-square (RMS) value of the signal is calculated over a 100 ms time window, for example, a 2D myoelectric activation map of the recorded area can be created, providing insight into the spatial activation of muscles (see the example in figure 7.4). This is one of the applications of HD-sEMG, where the signals are converted into image frames.

7.2 Surface EMG applications

Surface EMG is becoming more and more widely used in areas that require non-invasive assessment of muscular function [12] and has a variety of applications in neurophysiology (e.g. the study of movement strategies, reflex modulation, age, training and neuromuscular disease effects) [13–16], ergonomics (e.g. prevention of injuries, posture and kinematic studies) [17–20], motion analysis (e.g. gait analysis and support for clinical decisions, adjustment of musculoskeletal rehabilitation) [21–24], sports science (e.g. motor control assessment, coordination, fatigue) [25–28], human–machine interfaces (e.g. gesture recognition, powered prostheses, orthoses or exoskeletons, with applications in neurorehabilitation, substitution, or augmented normal limb function) [29–34] and non-invasive neural decoding [35–38]. Spasticity treatment

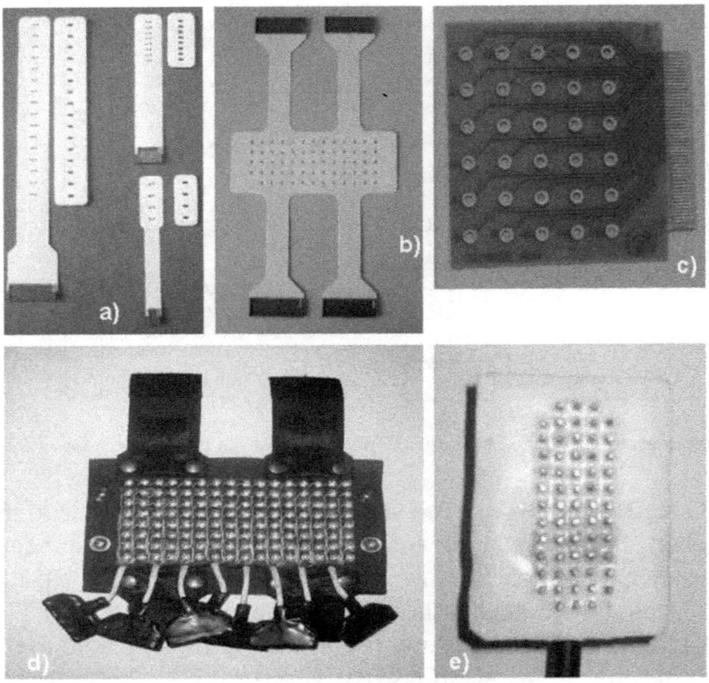

Figure 7.3. (a) Different linear electrode arrays. (b) A 2D 64-channel flexible electrode array (LISiN-Spes Medica, Italy). (c) A number of 65 electrodes in the form of a printed circuit board. (d) Silver-coated eyelets on a textile support. (e) A 61 silver pin-electrode array. (Reproduced from [11] with permission from Elsevier.)

using botulinum toxin injections and identifying the type of episiotomy during childbirth are other sEMG applications [39–41].

7.3 Challenges in sEMG recording

Similar to the recording of other biopotentials, several sources of noise and interference can affect the recording of sEMG [42]. These are primarily power-line interference, motion artifacts and crosstalk (i.e. activity from other muscles). Coupling of the power-line to the recording causes power-line interference. It appears at a power-line frequency (50 Hz or 60 Hz) and its harmonics acts as a common-mode voltage (VCM) that is transferred to a differential-mode voltage due to the potential driving effect (PDE) [43, 44].

Motion artifacts are low-frequency components in the high-amplitude caused by movement of the electrodes over the skin [44]. These components are commonly localized in the time domain and have a frequency spectrum of up to 20 Hz [45]. Similarly, there are artifacts associated with the movement of the recording cables, in particular if they are not fixed. These artifacts have a frequency spectrum of up to 50 Hz. Crosstalk is an unwanted sEMG signal generated by a close muscle and superimposed on the myoelectric activity of the recorded muscle. It is difficult to identify the crosstalk in the sEMG recording and thus some methods have been proposed in the literature for its suppression [5].

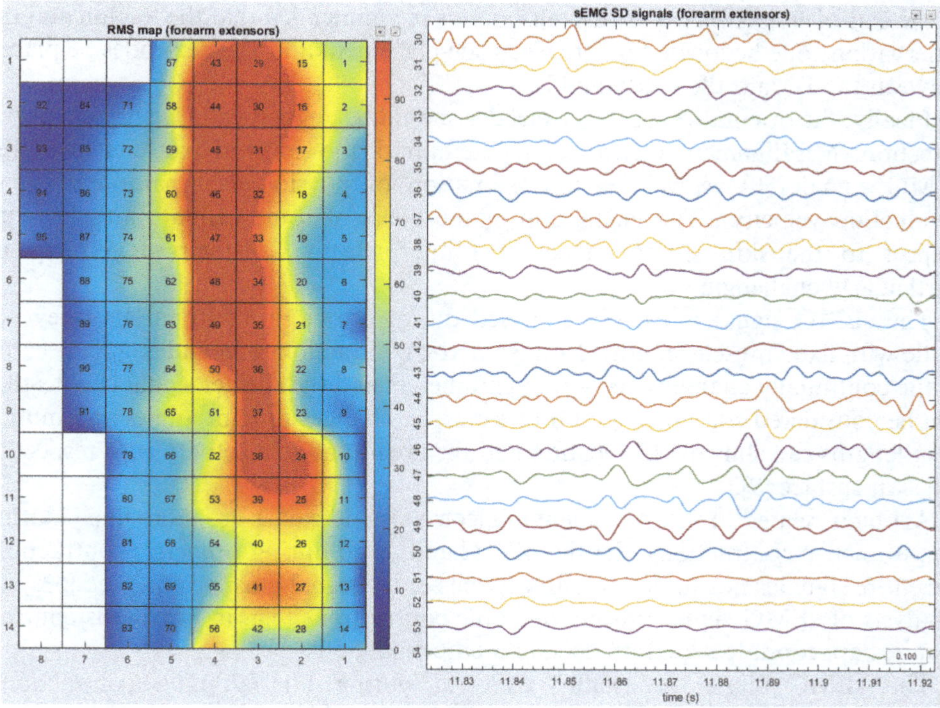

Figure 7.4. A 2D myoelectric activation map (left) displaying the RMS values of the sEMG signal (right) from the forearm during the activation of the extensor digitorum communis. Propagated MUAPs along the muscle fibers can be observed mainly between channels 44 and 49 of the sEMG signals.

There are also multiple factors associated with electrode placement that affect the recorded sEMG signal. Even a seemingly insignificant change in the placement of the electrode on the same subject can result in an sEMG signal with different parameters. In a bipolar recording, one of these factors is the inter-electrode distance (IED). The IED affects the frequency spectrum of the signal. A bipolar recording acts as a spatial filter and the zeros in its transfer function produce dips in the frequency spectrum of the recorded signal [9].

Also, IED determines the detection volume of the recording. A small IED provides higher spatial selectivity by recording a small region of muscle fibers. Increasing the IED increases the detection volume. This can allow recording of the deeper muscle fibers, but it also increases the effect of crosstalk by including a greater volume of neighboring muscles [46]. Too large an IED could also introduce spatial aliasing [47, 48]. Common values for IED are between 0.5 and 2 cm. Note that both spatial and temporal aliasing must be taken into account in sEMG [9].

Another factor affecting the sEMG signal is the location of the electrodes concerning the muscle IZ. Since action potentials propagate from the IZ along the muscle fibers in both directions, the bipolar signal recorded between electrodes placed on opposite sides of the IZ will have a misleading amplitude because of amplitude cancellation [49]. The extinction of the action potential at the junction between the muscle fiber and the tendon creates a non-propagating potential known as the end-of-fiber effect. For this reason, it is recommended that the tendon and the innervation zone be located before electrode placement and the electrodes be placed between the IZ and the tendon [40].

Ideally, the orientation of the electrodes should be aligned with the muscle fiber's direction. Misalignment can affect the temporal and spectral parameters of the sEMG signal [50] as well as the estimation of physiological parameters (e.g. conduction velocity). The inclination of the fibers of some pennate muscles with respect to the skin surface makes the interpretation of their sEMG signals particularly challenging [51].

The sEMG signal is usually analysed during isometric contractions (they are static with fixed muscle length). Non-isometric contractions are challenging because of the continually changing recording parameters and the shorter signal epochs that can be considered stationary [52]. For example, changes in joint angles and muscle fiber lengths can shift the IZ location and the area of muscle activity with respect to the skin surface [9].

Artifacts related to the movement of electrodes and cables are more likely during non-isometric contractions. On the other hand, even during isometric contraction, the entire signal is not stationary, and epochs of the signal must be analysed [53, 54]. Analysis of sEMG signals during dynamic contractions (e.g. in prosthesis control) requires appropriate signal processing methods [55, 56].

The sEMG signal is usually analysed during non-fatiguing contractions. Myoelectric fatigue is manifested in the sEMG as a shift in the power spectrum to lower frequencies and an increase in signal amplitude [57]. Also, the MUAP duration is shortened as a result of the decrease in the CV. The spatial distribution of myoelectric activity over the muscle can also change due to fatigue, and also due to other effects such as the duration of movement, the level of contraction and the position of the joint [58–61].

Such issues must also be taken into account when analysing sEMG signals [62–64]. Other factors related to the time-dependent change of the sEMG can be related to the electrode–skin impedance. During sustained recording, perspiration can influence the impedance of the electrode–skin contact. The material, surface preparation and position of the electrode or array of electrodes can affect the perspiration rate and, thus, the quality of the sEMG signal.

Moreover, the type of electrodes themselves can affect the recording. Conductive gel significantly decreases the electrode–skin impedance and, therefore, increases the signal quality. Disposable pre-gelled adhesive electrodes are commonly used. However, gel dehydrates over time, increasing the electrode–skin impedance and compromising the stability of recording conditions. For multichannel sEMG electrode arrays, the conductive gel is applied manually, usually using a gel dispenser or a syringe. When recording a huge number of channels, e.g. hundreds of channels in HD-sEMG, the manual application of the conductive gel can be time-consuming and require the assistance of trained personnel. It may also be uncomfortable for the subject. This is one of the issues which makes current HD-sEMG technology impractical for autonomous and wearable use. Emerging technology may develop dry electrodes that are better suited for practical applications [65–68].

The challenges in sEMG recording can be summarized as follows:

- Power-line interference, motion artifacts, crosstalk, the type of muscle (parallel-fibered or pinnate), and the placement of the recording electrodes in terms of the muscle IZ, location of the tendons and orientation of the muscle fibers can all affect the recorded sEMG signal.
- During sEMG recording, both spatial and temporal aliasing must be considered.
- Signal processing methods differ for isometric and dynamic muscle contractions, as well as for fatiguing and non-fatiguing contractions.
- The type of electrodes in terms of dry or wet electrodes can affect the sEMG signal quality and its practical applications.

7.4 Detection of atypical signals in HD-sEMG

The recording of the propagation of MUAPs using HD-sEMG has gained much interest since it was first introduced by Roeleveld *et al* in 1997 [69]. Contrary to traditional bipolar surface recordings, HD-sEMG can sense the activity of MUAPs over a wide area of the muscle, allowing the analysis of physiological properties that cannot be observed otherwise. For example, it can capture spatial inhomogeneities that originated in central nervous system activation due to regions with different properties within the muscle [61, 70, 71]. It also offers redundancy due to a large number of recording channels [72]. In a typical HD-sEMG arrangement, at least 64 channels are simultaneously recorded, and current commercial systems are capable of recording more than 100 channels.

Similar to other non-invasive recording methods over the skin, HD-sEMG signals are susceptible to different sources of external and internal contamination. Common contaminants include environmental noise such as power-line interference and high-frequency electronic noise, movement artifacts, amplifier saturation, electrode detachments, shifts and conductivity changes, and, for some muscles, cardiac interference [73].

In figure 7.5, examples of common contaminants in both the temporal and frequency domains are provided. These kinds of contamination can cause what are called *atypical signals*: low-quality signals that could result in misleading conclusions regarding muscular activation or errors in the control of interfaces. Given the increasing interest in HD-sEMG for diagnosis and evaluation of neuromuscular diseases [74], sports, ergonomics, telemedicine [75], control of human–machine interfaces [30, 76] and biometrics [77] among others, there is a need for checking the quality of HD-sEMG signals using either online or offline methods before any further processing.

When dealing with bipolar recordings, the quality of the signals can be checked online by measuring the electrode–skin impedance. However, this approach becomes impractical with an increasing number of channels [75]. Also, the quality of the contact between the recording electrode and the skin can change over time due to factors such as transpiration, drying gel and electrode detachment. Therefore, impedance measures are not stable and reliable. To address this issue, research on

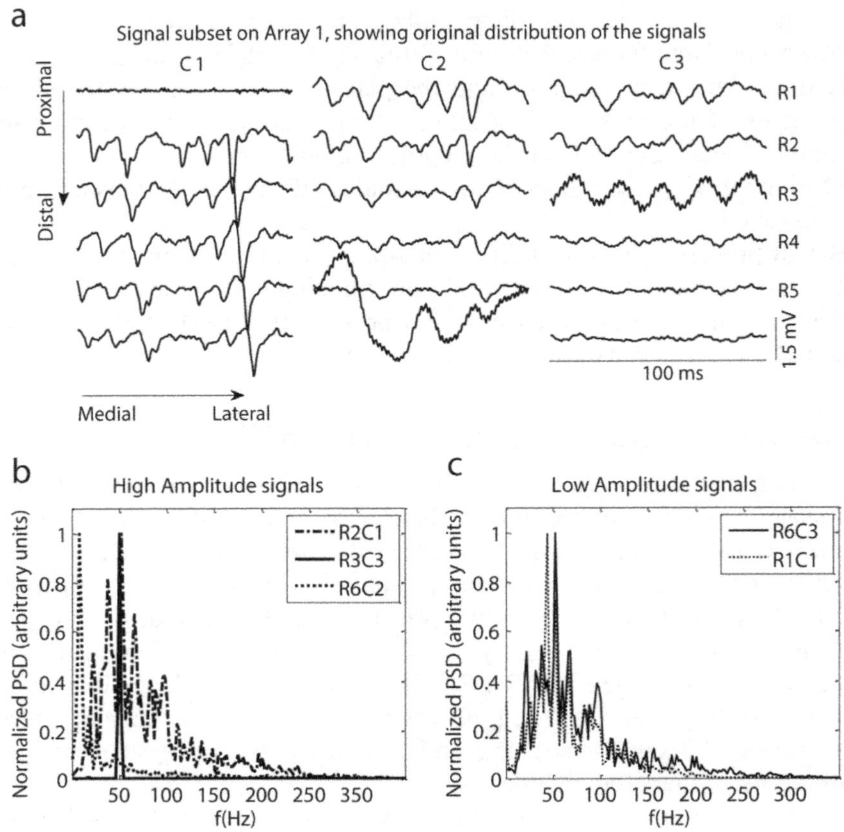

Figure 7.5. Subset of HD-sEMG recordings. (a) Signals in the temporal domain. Atypical signals can be observed in C1R1, C2R6 and C3R3. (b) Normalized power spectral density of high-amplitude signals in channels R2C1 (good-quality signal) and R2C1 and R6C2 (atypical signals). (c) Normalized power spectral density of low amplitude signals for an atypical signal (R1C1) and a good-quality signal (R6C3). (Reproduced with permission from [70].)

automated methods for the selection and identification of atypical EMG signals has increased over the last few years. Examples can be found in [70, 73, 75, 78–81].

In general, a method for detecting low-quality or atypical signals contains the following steps: (i) segmentation of the signal in small windows to address changes during the recording; (ii) feature extraction; and (iii) application of a detection method. Afterward, the outlier signals can be discarded or cleaned for further analysis.

If one wants to clean the signals, then it is necessary to correctly identify the contaminant type and then mitigate the interference with a proper method. An intermediate step can also include a quantification to provide an estimate of the contamination severity [82]. In some applications, e.g. muscle activity map extraction, it is possible to reconstruct outlier channels from nearby channels using image processing and surface reconstruction methods [83].

7.4.1 Feature extraction

Different features were defined to detect atypical signals in the literature. These features are designed to capture either the spectral or temporal properties of signals that differ from good-quality signals. When dealing with HD-sEMG, such features can also take into account the properties of other signals in the proximal neighborhood.

The most common features for atypical signal detection include the relative power in the frequency bands in which different types of noise (power-line interference, high-frequency interference, motion artifacts, or ECG) might be expected, power spectrum deformation and maximum drop in power density [70, 75, 81]. The signal to interference ratio, where interference can be understood as motion, ECG, power-line interference, or simply added noise, is also estimated [82].

The RMS value, standard deviation (SD) and mean absolute value (MAV) are common choices for detecting abnormal changes in signal amplitude (see the examples in figure 7.5) [70, 75, 80, 84]. Features related to the fractal dimension, autoregressive models and wavelet coefficients have also been used [73, 75, 84]. Regarding HD-sEMG, normalized mutual information and cross-correlation have been used to exploit and characterize the expected similarity between adjacent channels [79, 81].

Finally, given the dimensionality of some feature sets and the large number of possibilities for selecting features, some studies have investigated dimensionality reduction, employing principal or independent component analysis, pairwise correlations, or backward feature elimination [73, 75, 81, 85].

7.4.2 Detection methods

In the same way that numerous features have been proposed for the characterization of atypical signals, a variety of methods have been proposed for their detection. One approach consists of considering the atypical signals as outliers. For this purpose, Gronlund *et al* proposed a bivariate extension of the boxplot [80] with the disadvantage that a threshold had to be tuned for the detection. A similar thresholding method for detecting outliers was presented in [79]. Marateb *et al* proposed a method using a local distance-based outlier detection that did not require tuning [81]. Finally, Ijaz and Choi applied an anomaly detection method in which features were projected into a lower-dimensional space that best separated between outliers and good-quality signals [73].

Recent approaches also apply machine learning methods for the detection of atypical signals. For example, in [75] a one-class support vector machine (SVM) was proposed. In this case, only good-quality EMG signals (i.e. not contaminated) are used in the training set, and the optimization goal is to find a hyperplane maximizing the separation between the training data and the origin in a multi-dimensional space. SVMs with two classes have also been used to differentiate between atypical (in this case, a mixture of signals contaminated with various types of artifacts) and good-quality signals [84] and multiclass SVM was employed to identify the type of contaminant of EMG signals [82] .

Regarding the detection methods, one last remark can be made: even when EMG signals can look clean, different contaminants may be present that cannot be visually detected by a human operator. This is why recent approaches simulate and add contaminants over 'noise-free signals' and check how well the method performs. However, in general, these contaminants are added to real recorded signals that can only be clean to some extent or to oversimplistic artificial EMG signals that may not capture the complexity of the EMG [82]. Therefore, further research employing simulated signals with more complex models in a multichannel configuration is still needed.

7.5 Myoelectric prosthesis control, a hot topic

Among the applications of sEMG in rehabilitation technology, the myoelectric control of active upper limb prostheses has relevant clinical impact. For this purpose, sEMG provides information on the neural control of the motor intention of the users [86]. Although current robotic technology has reached a satisfactory level, EMG-based controlling schemes are still coarse. One paradigm for myoelectric control is pattern recognition. Such systems are trained to identify patterns from the sEMG and to select the corresponding function to execute.

When sEMG features could be repeated for some movements and are different from others, this control method could be used [87]. The typical structure of myoelectric control based on pattern recognition is shown in figure 7.6. Briefly, signal conditioning and segmentation are performed on the raw sEMG signal. Feature extraction is usually performed on stationary epochs, and then classifiers are trained to predict the movement class [88].

Now, we provide a simple tutorial on movement identification using the Ninapro V.2 dataset (also known as a hand prosthesis control benchmark on intact subjects). The second version of the Ninapro database was used for reporting results [89–91]. The database consists of recordings with twelve sEMG wireless double-differential (DD) electrodes (DelsysTringo Wireless System; OttoBock MyoBock 13E200 sEMG electrodes) at a sampling rate of 2 ksps. Eight DD electrodes were located at the biceps brachii, the triceps brachii, the flexor and extensor digitorum superficialis, and the forearm. The main activity spots were identified by palpation [90].

The dataset is freely available from http://ninaweb.hevs.ch/. The Matlab® v. 2019b code 'extract_emg_movement_1.m' uses the file 'S1_E1_A1.mat', which is part of 'DB2_s1.zip' that can be downloaded from the Ninapro homepage.

In this function, the first eight sEMG channels are used to classify the first 12 movements (i.e. basic movement class). The signals are filtered using the sixth order band-pass Butterworth filter with cut-off frequencies of 20 and 500 Hz. The middle 100 ms epoch of each movement is used, and the RMS values of the eight channels are extracted. Alternatively, it is possible to use basic Hudgins features [92] in the code.

The first three repetitions were used for training, and the other three ones were used to test the linear discriminant analysis (LDA) classifier. The overall accuracy was then shown on the test set. When running the code on the first subject data, an

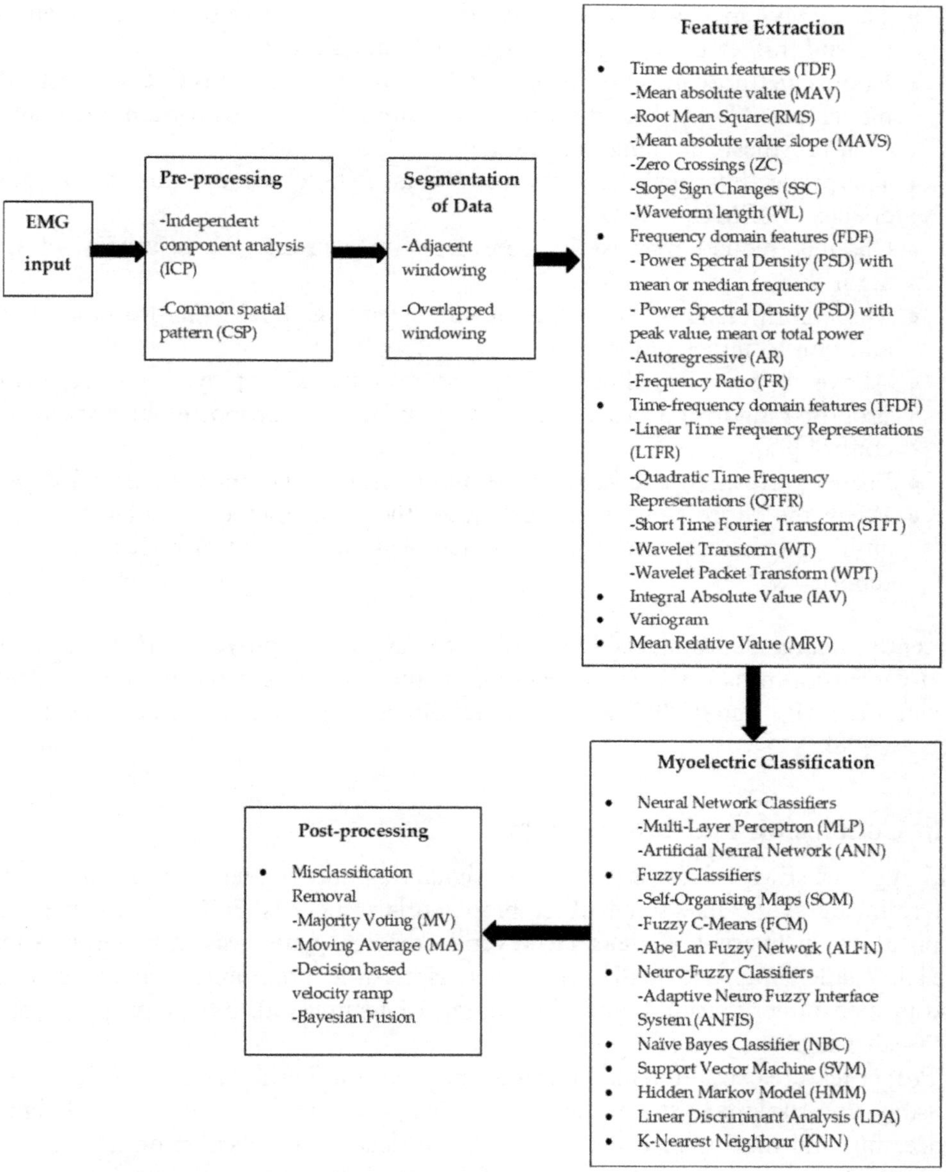

Figure 7.6. General pattern recognition schemes for myoelectric control. (Reproduced with permission from [88].)

overall accuracy of 78% was achieved. Increasing the epoch width to 400 ms increased the overall accuracy to 89%.

To simulate realistic conditions, the following issues must be considered:

- The window length must be 100 ms at maximum, and the processing time for each epoch must be less than 50 ms for proper real-time applications [93].

- The sEMG epochs must be analysed from the beginning of the movement to the end, rather than only using the stable middle part.
- Robust features must be proposed that are not too sensitive to electrode movements [94]. Otherwise, extensive training is required for amputees after each prosthesis donning and doffing.
- The proper filtering procedure must be taken into account due to the short epochs of sEMG signals.
- Channel reduction must be considered to increase the efficiency of the algorithm.
- The algorithm must be tested on amputees (e.g. DB3 Ninapro dataset) in addition to intact subjects.
- Above all, proper methods and procedures must be employed to upgrade the proposed algorithm in real-time simultaneous and proportional myoelectric control [95].
- Proper performance indices for the multiclass classifier must be provided [96].
- When the entire subjects are analysed, the total number of subjects whose entire movements are correctly identified, as a clinical index, must be reported.

Recently, neural decoding methods [93] were used to improve the control of the next-generation prosthesis. Such decoding methods are based on real-time decomposing the EMG signals [97], which enables direct assessment of the neural drive to muscles [98].

7.6 Conclusion and future scope

Although the sEMG signal is a noise-like signal that does not have a straightforward deterministic shape like some other biopotentials such as ECG, it has essential applications in a variety of areas. However, its recording, analysis and interpretation are still challenging. The quality of the signals must be monitored during recording, and proper outlier detection and filtering must be employed to have proper signals for further analysis.

Proper feature extraction methods must be used considering the sEMG signal as a possibly non-stationary stochastic signal. Then, the application of classical signal processing and machine learning methods can lead to practical programmes and devices. The reader is invited to read the following books which discuss the frontiers of sEMG and its applications for further information in detail [5, 12].

Acknowledgments

This work was supported by the Ministry of Economy and Competitiveness (MINECO), Spain, under contract DPI2017-83989-R and Ministry of Science and Innovation (MICINN), Spain, under contract PRE2018-085387. CIBER-BBN is an initiative of the Instituto de Salud Carlos III, Spain. JFA is a Serra Hunter Fellow. This work was also supported in part by the University of Isfahan.

References

[1] Nishihara K, Kawai H, Chiba Y, Kanemura N and Gomi T 2013 Investigation of innervation zone shift with continuous dynamic muscle contraction *Comput. Math. Methods Med.* **2013** 174342

[2] Beretta Piccoli M, Rainoldi A, Heitz C, Wüthrich M, Boccia G, Tomasoni E, Spirolazzi C, Egloff M and Barbero M 2014 Innervation zone locations in 43 superficial muscles: toward a standardization of electrode positioning *Muscle Nerve* **49** 413–21

[3] Jahanmiri-Nezhad F, Barkhaus P E, Rymer W Z and Zhou P 2015 Innervation zones of fasciculating motor units: observations by a linear electrode array *Front. Hum. Neurosci.* **9** 239

[4] Moritani T, Stegeman D and Merletti R 2004 Basic physiology and biophysics of EMG signal generation *Electromyography* (New York: Wiley) pp 1–25

[5] Merletti R and Parker P J 2004 *Electromyography: Physiology, Engineering, and Non-Invasive Applications* (New York: Wiley)

[6] Hug F and Tucker K 2017 Surface electromyography to study muscle coordination *Handbook of Human Motion* ed B Müller, S I Wolf, G-P Brueggemann, Z Deng, A McIntosh, F Miller and W S Selbie (Cham: Springer) pp 1–21

[7] Stegeman D and Hermens H 2008 Standards for surface electromyography: the European project (SENIAM) *Surface Electromyography Application Areas and Parameters* ed H J Hermens, G Rau, C Disselhorst-Klug and B Freriks *Proc. of the Third General SENIAM Workshop on surface electromyography, Aachen, Germany* pp 108–12

[8] Merletti R 2014 Standards for reporting EMG data *J. Electromyogr. Kinesiol.* **24** I–II

[9] Merletti R and Muceli S 2019 Surface EMG detection in space and time: best practices *J. Electromyogr. Kinesiol.* **49** 102363

[10] Urbanek H and van der Smagt P 2016 iEMG: imaging electromyography *J. Electromyogr. Kinesiol.* **27** 1–9

[11] Merletti R, Botter A, Troiano A, Merlo E and Minetto M A 2009 Technology and instrumentation for detection and conditioning of the surface electromyographic signal: state of the art *Clin. Biomech.* **24** 122–34

[12] Merletti R and Farina D 2016 *Surface Electromyography: Physiology, Engineering, and Applications* (New York: Wiley)

[13] Enoka R, Baudry S, Rudroff T, Farina D, Klass M and Duchateau J 2010 Unraveling the neurophysiology of muscle fatigue *J. Electromyogr. Kinesiol.* **21** 208–19

[14] Baudry S, Penzer F and Duchateau J 2014 Input–output characteristics of soleus homonymous Ia afferents and corticospinal pathways during upright standing differ between young and elderly adults *Acta Physiol.* **210** 667–77

[15] Negro F, Muceli S, Castronovo M, Holobar A and Farina D 2016 Multi-channel intramuscular and surface EMG decomposition by convolutive blind source separation *J. Neural Eng.* **13** 026027

[16] Stålberg E *et al* 2019 Standards for quantification of EMG and neurography *Clin. Neurophysiol.* **130** 1688–729

[17] Januario L B, Moreira R F C, Cid M M, Samani A, Madeleine P and Oliveira A B 2016 Effects of active pause pattern of surface electromyographic activity among subjects performing monotonous tasks: a systematic review *J. Electromyogr. Kinesiol.* **30** 196–208

[18] Neblett R 2016 Surface electromyographic (SEMG) biofeedback for chronic low back pain *Healthcare* **4** 27

[19] Martín-Fuentes I, Oliva-Lozano J M and Muyor J M 2020 Electromyographic activity in deadlift exercise and its variants. A systematic review *PLoS One* **15** e0229507

[20] Tang D *et al* 2020 Strain-insensitive elastic surface electromyographic (sEMG) electrode for efficient recognition of exercise intensities *Micromachines* **11** 239

[21] Simonsen E B 2014 Contributions to the understanding of gait control *Dan. Med. J.* **61** B4823

[22] Parel I, Jaspers E, De Baets L, Amoresano A and Cutti A G 2016 Motion analysis of the shoulder in adults: kinematics and electromyography for the clinical practice *Eur. J. Phys. Rehabil. Med.* **52** 575–82

[23] Serrao M, Ranavolo A and Casali C 2018 Neurophysiology of gait *Handbook of Clinical Neurology* vol 154 ed M Manto and T A G M Huisman (Amsterdam: Elsevier) ch 18, pp 299–303

[24] Roche N, Bonnyaud C, Reynaud V, Bensmail D, Pradon D and Esquenazi A 2019 Motion analysis for the evaluation of muscle overactivity: a point of view *Ann. Phys. Rehabil. Med.* **62** 442–52

[25] Stastny P, Gołaś A, Blazek D, Maszczyk A, Wilk M, Pietraszewski P, Petr M, Uhlir P and Zając A 2017 A systematic review of surface electromyography analyses of the bench press movement task *PLoS One* **12** e0171632

[26] Howard R M, Conway R and Harrison A J 2018 Muscle activity in sprinting: a review *Sports Biomech.* **17** 1–17

[27] McCrary J M, Ackermann B J and Halaki M 2018 EMG amplitude, fatigue threshold, and time to task failure: a meta-analysis *J. Sci. Med. Sport* **21** 736–41

[28] Vigotsky A D, Halperin I, Lehman G J, Trajano G S and Vieira T M 2018 Interpreting signal amplitudes in surface electromyography studies in sport and rehabilitation sciences *Front. Physiol.* **8** 985

[29] Micera S, Carpaneto J and Raspopovic S 2010 Control of hand prostheses using peripheral information *IEEE Rev. Biomed. Eng.* **3** 48–68

[30] Jordanić M, Rojas-Martínez M, Mañanas A M, Alonso F J and Marateb R H 2017 A novel spatial feature for the identification of motor tasks using high-density electromyography *Sensors* **17** E1597

[31] Li X, Samuel O W, Zhang X, Wang H, Fang P and Li G 2017 A motion-classification strategy based on sEMG-EEG signal combination for upper-limb amputees *J. Neuroeng. Rehabil.* **14** 2

[32] Rivela D, Scannella A, Pavan E E, Frigo C A, Belluco P and Gini G 2018 Analysis and comparison of features and algorithms to classify shoulder movements from sEMG signals *IEEE Sens. J.* **18** 3714–21

[33] Wei W, Wong Y, Du Y, Hu Y, Kankanhalli M and Geng W 2019 A multi-stream convolutional neural network for sEMG-based gesture recognition in muscle–computer interface *Pattern Recognit. Lett.* **119** 131–8

[34] Kapelner T, Sartori M, Negro F and Farina D 2020 Neuro-musculoskeletal mapping for man–machine interfacing *Sci. Rep.* **10** 5834

[35] Mohebian M R, Marateb H R, Karimimehr S, Mañanas M A, Kranjec J and Holobar A 2019 Non-invasive decoding of the motoneurons: a guided source separation method based on convolution kernel compensation with clustered initial points *Front. Comput. Neurosci.* **13** 14

[36] Farina D, Merletti R and Enoka R M 2014 The extraction of neural strategies from the surface EMG: an update *J. Appl. Physiol.* **117** 1215–30

[37] Nawab S H, Chang S-S and De Luca C J 2010 High-yield decomposition of surface EMG signals *Clin. Neurophysiol.* **121** 1602–15

[38] Chen M and Zhou P 2016 A novel framework based on FastICA for high density surface EMG decomposition *IEEE Trans. Neural Syst. Rehabil. Eng.* **24** 117–27

[39] Marateb H R, Farahi M, Rojas M, Mañanas M A and Farina D 2016 Detection of multiple innervation zones from multi-channel surface EMG recordings with low signal-to-noise ratio using graph-cut segmentation *PLoS One* **11** e0167954

[40] Barbero M, Merletti R and Rainoldi A 2012 *Atlas of Muscle Innervation Zones: Understanding Surface Electromyography and Its Applications* (Berlin: Springer)

[41] Soares F A, Carvalho J L A, Miosso C J, de Andrade M M and da Rocha A F 2015 Motor unit action potential conduction velocity estimated from surface electromyographic signals using image processing techniques *Biomed. Eng. Online* **14** 84

[42] Clancy E A, Morin E L and Merletti R 2002 Sampling, noise-reduction and amplitude estimation issues in surface electromyography *J. Electromyogr. Kinesiol.* **12** 1–16

[43] Merletti R and Parker P J 2004 *Electromyography: Physiology, Engineering, and Non-Invasive Applications* (Hoboken, NJ: Wiley-IEEE)

[44] Webster J G 2009 *Medical Instrumentation: Application and Design* (New York: Wiley)

[45] De Luca C J, Gilmore L D, Kuznetsov M and Roy S H 2010 Filtering the surface EMG signal: movement artifact and baseline noise contamination *J. Biomech.* **43** 1573–9

[46] Vieira T M, Botter A, Muceli S and Farina D 2017 Specificity of surface EMG recordings for gastrocnemius during upright standing *Sci. Rep.* **7** 13300

[47] Afsharipour B, Ullah K and Merletti R 2015 Amplitude indicators and spatial aliasing in high density surface electromyography recordings *Biomed. Signal Process. Control* **22** 170–9

[48] Afsharipour B, Ullah K and Merletti R 2014 Spatial aliasing and EMG amplitude in time and space: simulated action potential maps *XIII Mediterranean Conf. on Medical and Biological Engineering and Computing 2013* (Berlin: Springer) pp 293–6

[49] Merletti R, Aventaggiato M, Botter A, Holobar A, Marateb H and Vieira T M M 2010 Advances in surface EMG: recent progress in detection and processing techniques *Crit. Rev. Biomed. Eng.* **38** 305–45

[50] Merletti R, Rainoldi A and Farina D 2001 Surface electromyography for noninvasive characterization of muscle *Exerc. Sport Sci. Rev.* **29** 20–5

[51] Marateb H R, McGill K C, Holobar A, Lateva Z C, Mansourian M and Merletti R 2011 Accuracy assessment of CKC high-density surface EMG decomposition in biceps femoris muscle *J. Neural Eng.* **8** 066002

[52] Nazmi N, Abdul Rahman M A, Yamamoto S-I, Ahmad S A, Malarvili M B, Mazlan S A and Zamzuri H 2017 Assessment on stationarity of EMG signals with different windows size during isotonic contractions *NATO Adv. Sci. Inst. Ser. E Appl. Sci.* **7** 1050

[53] Farina D, Jiang N, Rehbaum H, Holobar A, Graimann B, Dietl H and Aszmann O C 2014 The extraction of neural information from the surface EMG for the control of upper-limb prostheses: emerging avenues and challenges *IEEE Trans. Neural Syst. Rehabil. Eng.* **22** 797–809

[54] Bilodeau M, Cincera M, Arsenault A B and Gravel D 1997 Normality and stationarity of EMG signals of elbow flexor muscles during ramp and step isometric contractions *J. Electromyogr. Kinesiol.* **7** 87–96

[55] Nazmi N, Abdul Rahman M A, Yamamoto S-I, Ahmad S A, Zamzuri H and Mazlan S A 2016 A review of classification techniques of EMG signals during isotonic and isometric contractions *Sensors* **16** 1304
[56] Farina D, Pozzo M, Merlo E, Bottin A and Merletti R 2004 Assessment of average muscle fiber conduction velocity from surface EMG signals during fatiguing dynamic contractions *IEEE Trans. Biomed. Eng.* **51** 1383–93
[57] De Luca C J 1984 Myoelectrical manifestations of localized muscular fatigue in humans *Crit. Rev. Biomed. Eng.* **11** 251–79
[58] Tucker K, Falla D, Graven-Nielsen T and Farina D 2009 Electromyographic mapping of the erector spinae muscle with varying load and during sustained contraction *J. Electromyogr. Kinesiol.* **19** 373–9
[59] Staudenmann D, van Dieën J H, Stegeman D F and Enoka R M 2014 Increase in heterogeneity of biceps brachii activation during isometric submaximal fatiguing contractions: a multichannel surface EMG study *J. Neurophysiol.* **111** 984–90
[60] Vieira T M M, Merletti R and Mesin L 2010 Automatic segmentation of surface EMG images: improving the estimation of neuromuscular activity *J. Biomech.* **43** 2149–58
[61] Holtermann A, Roeleveld K and Karlsson J S 2005 Inhomogeneities in muscle activation reveal motor unit recruitment *J. Electromyogr. Kinesiol.* **15** 131–7
[62] Georgakis A, Stergioulas L K and Giakas G 2003 Fatigue analysis of the surface EMG signal in isometric constant force contractions using the averaged instantaneous frequency *IEEE Trans. Biomed. Eng.* **50** 262–5
[63] Wang L, Wang Y, Ma A, Ma G, Ye Y, Li R and Lu T 2018 A comparative study of EMG indices in muscle fatigue evaluation based on grey relational analysis during all-out cycling exercise *Biomed. Res. Int.* **2018** 9341215
[64] Karthick P A, Venugopal G and Ramakrishnan S 2015 Analysis of surface EMG signals under fatigue and non-fatigue conditions using B-distribution based quadratic time frequency distribution *J. Mech. Med. Biol.* **15** 1540028
[65] Laferriere P, Lemaire E D and Chan A D C 2011 Surface electromyographic signals using dry electrodes *IEEE Trans. Instrum. Meas.* **60** 3259–68
[66] Fayyaz Shahandashti P, Pourkheyrollah H, Jahanshahi A and Ghafoorifard H 2019 Highly conformable stretchable dry electrodes based on inexpensive flex substrate for long-term biopotential (EMG/ECG) monitoring *Sens. Actuators* A **295** 678–86
[67] Di Flumeri G, Aricò P, Borghini G, Sciaraffa N, Di Florio A and Babiloni F 2019 The dry revolution: evaluation of three different EEG dry electrode types in terms of signal spectral features, mental states classification and usability *Sensors* **19** 1365
[68] Yamagami M, Peters K M, Milovanovic I, Kuang I, Yang Z, Lu N and Steele K M 2018 Assessment of dry epidermal electrodes for long-term electromyography measurements *Sensors* **18** E1269
[69] Roeleveld K, Stegeman D F, Vingerhoets H M and Van Oosterom A 1997 The motor unit potential distribution over the skin surface and its use in estimating the motor unit location *Acta Physiol. Scand.* **161** 465–72
[70] Rojas-Martínez M, Mañanas M A and Alonso J F 2012 High-density surface EMG maps from upper-arm and forearm muscles *J. Neuroeng. Rehabil.* **9** 85
[71] Rojas-Martínez M, Mañanas M A, Alonso J F and Merletti R 2013 Identification of isometric contractions based on high density EMG maps *J. Electromyogr. Kinesiol.* **23** 33–42

[72] Staudenmann D, Stegeman D F and van Dieën J H 2013 Redundancy or heterogeneity in the electric activity of the biceps brachii muscle? Added value of PCA-processed multi-channel EMG muscle activation estimates in a parallel-fibered muscle *J. Electromyogr. Kinesiol.* **23** 892–8

[73] Ijaz A and Choi J 2018 Anomaly detection of electromyographic signals *IEEE Trans. Neural Syst. Rehabil. Eng.* **26** 770–9

[74] Rasool G, Afsharipour B, Suresh N L, Hu X and Rymer W Z 2015 Spatial analysis of muscular activations in stroke survivors *Conf. Proc. IEEE Eng. Med. Biol. Soc.* **2015** 6058–61

[75] Fraser G D, Chan A D C, Green J R and MacIsaac D T 2014 Automated biosignal quality analysis for electromyography using a one-class support vector machine *IEEE Trans. Instrum. Meas.* **63** 2919–30

[76] Zhang X and Zhou P 2012 High-density myoelectric pattern recognition toward improved stroke rehabilitation *IEEE Trans. Biomed. Eng.* **59** 1649–57

[77] Al Harrach M, Boudaoud S and Nait-ali A 2020 Towards high density sEMG (HD-sEMG) acquisition approach for biometrics applications *Hidden Biometrics: When Biometric Security Meets Biomedical Engineering* ed A Nait-ali (Singapore: Springer) pp 101–12

[78] Agostini V and Knaflitz M 2012 An algorithm for the estimation of the signal-to-noise ratio in surface myoelectric signals generated during cyclic movements *IEEE Trans. Biomed. Eng.* **59** 219–25

[79] Bingham A, Jelfs B, Arjunan S P and Kumar D K 2018 Identifying noisy electrodes in high density surface electromyography recordings through analysis of spatial similarities *Conf. Proc. IEEE Eng. Med. Biol. Soc.* **2018** 2325–8

[80] Grönlund C, Roeleveld K, Holtermann A and Karlsson J S 2005 On-line signal quality estimation of multichannel surface electromyograms *Med. Biol. Eng. Comput.* **43** 357–64

[81] Marateb H R, Rojas-Martínez M, Mansourian M, Merletti R and Villanueva M A M 2012 Outlier detection in high-density surface electromyographic signals *Med. Biol. Eng. Comput.* **50** 79–89

[82] McCool P, Fraser G D, Chan A D C, Petropoulakis L and Soraghan J J 2014 Identification of contaminant type in surface electromyography (EMG) signals *IEEE Trans. Neural Syst. Rehabil. Eng.* **22** 774–83

[83] Ghaderi P and Marateb H R 2017 Muscle activity map reconstruction from high density surface EMG signals with missing channels using image inpainting and surface reconstruction methods *IEEE Trans. Biomed. Eng.* **64** 1513–23

[84] Moura K O A, Ruschel R S and Balbinot A 2018 Fault-tolerant sensor detection of sEMG signals: quality analysis using a two-class support vector machine *Conf. Proc. IEEE Eng. Med. Biol. Soc.* **2018** 5644–7

[85] Wand M, Himmelsbach A, Heistermann T, Janke M and Schultz T 2013 Artifact removal algorithm for an EMG-based silent speech interface *Conf. Proc. IEEE Eng. Med. Biol. Soc.* **2013** 5750–3

[86] Jordanić M, Rojas-Martínez M, Mañanas M A and Alonso J F 2016 Prediction of isometric motor tasks and effort levels based on high-density EMG in patients with incomplete spinal cord injury *J. Neural Eng.* **13** 046002

[87] Ameri A, Scheme E J, Kamavuako E N, Englehart K B and Parker P A 2014 Real-time, simultaneous myoelectric control using force and position-based training paradigms *IEEE Trans. Biomed. Eng.* **61** 279–87

[88] Parajuli N *et al* 2019 Real-time EMG based pattern recognition control for hand prostheses: a review on existing methods, challenges and future implementation *Sensors* **19** 4596
[89] Atzori M, Gijsberts A, Kuzborskij I, Elsig S, Hager A-G M, Deriaz O, Castellini C, Muller H and Caputo B 2015 Characterization of a benchmark database for myoelectric movement classification *IEEE Trans. Neural Syst. Rehabil. Eng.* **23** 73–83
[90] Gijsberts A, Atzori M, Castellini C, Muller H and Caputo B 2014 Movement error rate for evaluation of machine learning methods for sEMG-based hand movement classification *IEEE Trans. Neural Syst. Rehabil. Eng.* **22** 735–44
[91] Atzori M, Gijsberts A, Castellini C, Caputo B, Hager A-G M, Elsig S, Giatsidis G, Bassetto F and Müller H 2014 Electromyography data for non-invasive naturally-controlled robotic hand prostheses *Sci. Data* **1** 140053
[92] Hudgins B, Parker P and Scott R N 1993 A new strategy for multifunction myoelectric control *IEEE Trans. Biomed. Eng.* **40** 82–94
[93] Karimimehr S, Marateb H R, Muceli S, Mansourian M, Mañanas M A and Farina D 2017 A real-time method for decoding the neural drive to muscles using single-channel intra-muscular EMG recordings *Int. J. Neural Syst.* **27** 1750025
[94] Muceli S, Jiang N and Farina D 2014 Extracting signals robust to electrode number and shift for online simultaneous and proportional myoelectric control by factorization algorithms *IEEE Trans. Neural Syst. Rehabil. Eng.* **22** 623–33
[95] Jiang N, Rehbaum H, Vujaklija I, Graimann B and Farina D 2014 Intuitive, online, simultaneous, and proportional myoelectric control over two degrees-of-freedom in upper limb amputees *IEEE Trans. Neural Syst. Rehabil. Eng.* **22** 501–10
[96] Sokolova M and Lapalme G 2009 A systematic analysis of performance measures for classification tasks *Inf. Process. Manag.* **45** 427–37
[97] Webster J G 1999 Electromyographic (EMG) decomposition *Wiley Encyclopedia of Electrical and Electronics Engineering* vol 8 (Hoboken, NJ: Wiley) pp 1–15
[98] Farina D and Holobar A 2015 Human–machine interfacing by decoding the surface electromyogram *IEEE Signal Process. Mag.* **32** 115–20

IOP Publishing

Modelling and Analysis of Active Biopotential Signals in Healthcare, Volume 1

Varun Bajaj and G R Sinha

Chapter 8

Classification of physical actions from surface EMG signals using the wavelet packet transform and local binary patterns

Ömer F Alçin[1], Ümit Budak[2], Muzaffer Aslan[3], Yaman Akbulut[4], Zafer Cömert[5], Muhammed H Akpınar[6], Abdulkadir Şengür[6]

[1]*Malatya Turgut Ozal University, Faculty of Engineering and Natural Sciences, Department of Electrical Engineering, Malatya, Turkey*

[2]*Bitlis Eren University, Engineering Faculty, Electrical and Electronics Engineering Department, Bitlis, Turkey*

[3]*Bingol University, Engineering Faculty, Electrical-Electronics Engineering Department, Bingol, Turkey*

[4]*Firat University, Informatics Department, Elazig, Turkey*

[5]*Samsun University, Engineering Faculty, Software Engineering Department, Samsun, Turkey*

[6]*Firat University, Technology Faculty, Electrical and Electronics Engineering Department, Elazig, Turkey*

Physical action recognition is a hot topic in human–machine interactions. It has potential uses in helping disabled people and in various robotic applications. Electromyography (EMG) signals measure the electrical activity of the muscular systems involved in physical actions. In this chapter, an efficient approach is developed for physical action recognition in humans based on EMG signals. The proposed method is composed of signal decomposition, feature extraction and feature classification. The signal decomposition is carried out using the wavelet packet transform (WPT). The WPT successively decomposes an input signal into its approximation and detail coefficients, which offers a more productive signal analysis. A one-dimensional local binary pattern (LBP) is used to code the approximation and detail coefficients of the decomposed EMG signals. The histogram of the LBP is used as the feature vectors of the EMG physical action classes. The support vector machine (SVM), decision tree, linear discriminant, *k*-nearest neighbors (*k*-NN), boosted and bagged tree ensemble classifiers are used in the classification stage.

doi:10.1088/978-0-7503-3279-8ch8

A dataset taken from the UCI machine learning repository is used in the experiments. The Delsys EMG apparatus, which has eight electrodes, is used to record the surface EMG signals. Each class of the dataset contains three male subjects and one female subject. The dataset contains ten physical actions, namely hugging, jumping, bowing, clapping, handshaking, running, sitting, standing, walking and waving. The experiments are carried out on each electrode with a ten-fold cross-validation test, and the average accuracy score is calculated. The experimental results show that the proposed method is quite efficient in EMG signal classification. The calculated average accuracy is 100% for each electrode.

8.1 Introduction

Recent technologies have enabled the elderly and disabled people to interact with computers and machines. Interactions with computers and robots help the elderly and disabled people handle their daily tasks independently [1]. In addition, human–machine interactions may help the elderly and disabled people perform the physical actions of daily life. Surface electromyography (sEMG) is a clinical, non-invasive technique which enables the identification the muscular physical activities.

In the last two decades, there have been various attempts to monitor the daily actions of humans based on sEMG signals and machine learning techniques [2–15]. These techniques have generally been based on feature extraction from sEMG signals and feature classification. Xi *et al* [2] attempted to determine the best features and the best classifier for classification of daily human activity based on sEMG signals. They used a total of fifteen features that were extracted from either the time, frequency or time–frequency domains. These features were generally based on entropy, and five machine learning techniques were used in the classification stage of the work. Xi *et al* used a dataset that contains seven daily activities of humans. The Gaussian-kernel SVM classifier with permutation entropy achieved the best classification accuracy, where the accuracy score was 97.35%. Simão *et al* [3] presented an extensive review of machine intelligence and EMG signals. In their review, they discussed the collection of EMG signals, EMG signal filtering methods, feature extraction from EMG signals and feature dimension reduction methods. The reviewed feature extraction methods were generally based on signal processing techniques. New classification approaches for classification of non-trained actions were also reviewed. Further comparative analysis and discussion of advantages and drawbacks were also provided. Demir *et al* [4] proposed deep learning based approaches for the classification of physical actions based on sEMG signals. The authors initially converted the sEMG signals into sEMG images using the short time Fourier transform (STFT) approach. Fine-tuning and deep feature extraction approaches were employed using pre-trained convolutional neural network (CNN) models such as AlexNet and VGGNet. For deep features, the SVM classifier was considered. The UCI sEMG dataset was used and a 99.04% accuracy score was reported. Wang *et al* [5] proposed a novel approach for sEMG based hand gesture recognition. The proposed approach was mainly based on a wavelet transform and variable predictive model based class discrimination classifier. After denoising of the sEMG signals, the maximum values of the wavelet coefficients were used as feature vectors. These features were then used in the classification of various hand gestures based on the developed classifier. The obtained results suggest that the proposed method was quite effective in the classification of different hand gestures. Bajaj *et al* [6] used intrinsic mode functions to extract features from sEMG signals. The extracted features were mean frequency and singular value computation, respectively. The least squares SVM classification was adopted in the feature classification stage of the work. More specifically, the radial basis kernel was considered in the SVM classifier. The authors reported that 99.03% classification accuracy was produced by the proposed method. Sahinbegovic *et al* [7] used artificial neural networks (ANNs) for the classification of human behavior based on EMG recordings. Bi-spectrum signal analysis was used for feature extraction and a feed-forward ANN architecture was adopted for ten different human actions. The best classification accuracy score, 86.25%, was obtained by the ANN architecture where the hidden layer contained 17 neurons. Jana *et al* [8] used the adaptive neuro-fuzzy inference system for the classification of EMG signals. Two main human behaviors, namely normal and aggressive, were classified based on the wavelet features. The db4 wavelet function and five levels of decomposition produced the best accuracy score. The authors reported a 98% accuracy score using their proposed method. Abdullah *et al* [9] used sEMG signals for the classification of various myoelectric signals. Wavelet packet decomposition was used for feature extraction. The extracted features were then selected using a set of feature selection approaches. Various ensemble tree classifiers were used in the classification of the selected features. The best accuracy score, 92.1%, was produced by the random forest classifier. Bai *et al* [10] proposed an approach for shoulder motion classification. The proposed approach was based on maximizing mutual information from multiclass CSP selected spatial feature channels. The WPT features were extracted and were used in the classification stage of the study. The proposed method produced a 92% accuracy score. Tuncer *et al* [11] presented an sEMG signal recognition method using sEMG signals for the control of the prosthetic hands of amputees by means of a new triple pattern and discrete wavelet transform (TPDWT) with an extraction feature network. This study used the TPDWT network for sEMG signal decomposition, extracted features with two-level feature selection methods and made use of a *k*-NN classifier for classification. Four cases were studied in order to comprehensively evaluate the TPDWT based hand movement method with sEMG signals, using a sEMG dataset with three strength levels (low, medium and high) collected from amputee participants in their study. Using the TPDWT with sEMG classification using the *k*-NN classifier, they obtained accuracies of 97.78%, 93.33% and 92.96% for the low, medium and high power levels, respectively, and achieved a classification accuracy of 99.14% for all force levels. Kumari *et al* [12] attempted to classify the time domain and time scale characteristics of sEMG signals with the support vector machine to normalize the lives of partially disabled people. In their study, they extracted a time scale feature with a wavelet packet transform (WPT). The authors used 303 sEMG data points from eight participants between the ages of 18 and 30 to evaluate the performance of their proposed method. Of these data points, 114 were used for lengthening and 189 for flexion. The extension data were classified with 92.1% accuracy and the flexion data were classified with 90.5% accuracy. They also achieved 91.1% classification accuracy for elbow movement, which can be used to control any bidirectional robotic vehicle. Amma *et al* [13] used sEMG data to classify 27 different finger movements. The authors extracted features with high intensity sEMG (HD-sEMG) signals to determine muscle strength activity, recorded with 192 electrodes arranged in 8 × 24 arrays with root mean square (RMS). These features were classified using a naïve Bayes classifier with an accuracy of 90.4%. Wei *et al* [14] proposed a multi-flow convolutional neural network to recognize movements from deep learning, sEMG signals at the muscle–computer interface (MCI). The system was trained for each flow, with the features obtained from equal sized patches split from the original sEMG images at the time of multi-flow separation. A fusion network was implemented, combining features learned from all previous streams. Among the studies with different patch sizes, 99.7% recognition accuracy was achieved for the final patch sizes (8 × 2). Chaya *et al* [15] tried to identify the movement of the arm through SEMG signals received from the elbow and wrist areas of the arm using a real-time application. They extracted features from the averaged EMG signals such as average absolute value (MAV), variance (VAR), standard deviation (SD), RMS, power spectral entropy (PSE) and average frequency (MF). They achieved 93.3% success using an SVM in the classification of both the elbow and wrist positions after performing a size reduction for the optimum feature set. In Turlapaty *et al*'s study [16], EMG signals were first segmented for the classification of physical activities with multi-channel EMG data. Then feature extraction was performed using methods such as time statistics (TS), inter-channel statistics (ICS), spectral band power (SBP) and local binary patterns (LBP). Using probabilistic neural networks and *k*-NN, classification accuracies of 92.7% and 93.9%, respectively, were achieved. Ouyang *et al* [17] extracted the hidden dynamic features of EMG data through test recurrence quantification analysis (RQA) using nonlinear measures based on the recurrence graph during four different hand movements to recognize human hand grip movements from sEMG. By applying these features in an adaptive neuro-fuzzy inference system, they achieved 93.4% recognition in only linear measurements and 99.1% in combinations of linear and nonlinear measurements. Sapsanis [18] presented a machine learning method for classification. Eight features were extracted from each raw EMG dataset by decomposing them from intrinsic mode functions (IMFs) using the empirical mode decomposition (EMD). The sEMG dataset was obtained by means of two electrodes attached to two specific muscles of the hand. An average accuracy rate of 89.21% was achieved in the study. Sravani *et al* [19] performed feature extraction through decomposition into eight sub-bands as negative entropy using the flexible analytical wavelet transform (FAWT), average absolute value (MAV), variance (VAR), modified average absolute value type 1 (MAV1), waveform length, simple square integral (SSI), Tsallis entropy and integrated EMG (IEMG) in order to classify the physical actions. Then, these features were classified through application of the ELM classifier with sigmoid activation. The highest accuracy was 99.36% in all sub-bands. Qi *et al* [20] used linear discriminant analysis (LDA) and an extreme-learning machine (ELM) to reduce unnecessary information in sEMG signals in the recognition of hand movements and to increase the recognition efficiency and accuracy. In the study, the features were obtained primarily from EMG signals using the RMS, waveform length and median amplitude spectrum (MAS) methods. Then, these high-dimensional features were reduced using LDA and applied to the ELM input. Using a genetic algorithm to increase the performance of the ELM, they achieved 79.32% accuracy by optimizing the initial conditions and thresholds of the network.

In this chapter, we propose an approach to develop an efficient physical action recognition system. This developed system is mainly based on sEMG signals. The signal decomposition, feature extraction and feature classification stages are at the heart of the developed system. A flow chart of the proposed method is shown in figure 8.1. Signal decomposition is achieved using the WPT. The WPT successively decomposes the sEMG signals into approximation and detail coefficients. Then, the one-dimensional LBP approach is considered in order to code the approximation and detail coefficients. Finally, the histograms of the coded coefficients are used as features. The SVM, decision tree, linear discriminant, k-NN, boosted and bagged tree ensemble classifiers are used in the classification stage. A dataset taken from the

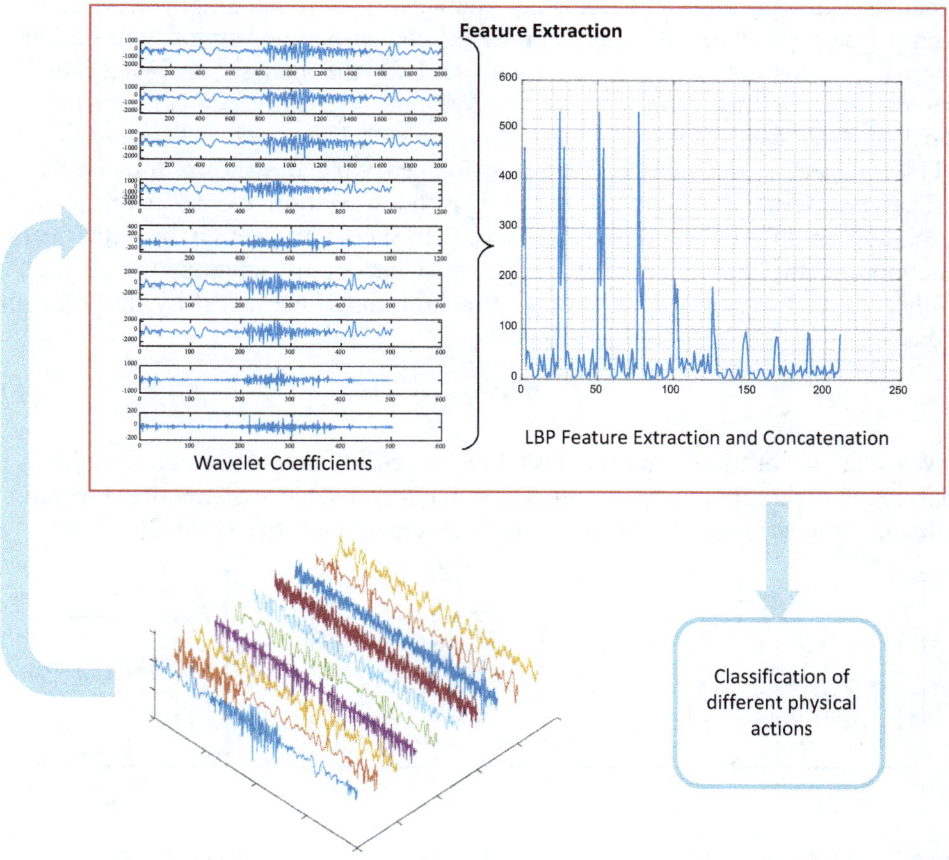

Figure 8.1. Flow diagram of the proposed method.

UCI machine learning repository is used in the experiments. Each class of the dataset contains three male subjects and one female subject. The dataset contains ten physical actions, namely bowing, clapping, handshaking, hugging, jumping, running, standing, sitting, walking and waving. The experiments are carried out on each electrode with a ten-fold cross-validation test and the average accuracy score is calculated. The experimental results show that the proposed method is quite efficient in EMG signal classification.

The remainder of this chapter is organized as follows. In the next section, the theoretical background is described in order to provide the reader with perspective about the details of the methods. The dataset, experimental methods and obtained results are presented in section 8.3. This chapter is concluded in section 8.4.

8.2 Materials and methods

8.2.1 The wavelet transform

The wavelet transform (WT) is a multi-resolution analysis method developed by Mallat [21]. While other classical methods, such as the Fourier transform (FT), analyse a signal only in the frequency domain, the WT has superior aspects that explain the signal in both the time and frequency domains. Electromyography (EMG) is an electrical sign produced by skeletal muscles and also time-varying non-stationary recordings. Thus the WT is an important tool for the analysis of EMG. The WT presents superior time resolution for the low and high-frequency components of a signal, and so it can handle a fixed time–frequency problem, such as the short time FT (STFT). A general framework of the time–frequency plane is shown in figure 8.2.

The most crucial feature of the WT is the variation of the window width depending on different decomposition levels. In such a wavelet analysis, the scaling function and the mother wavelet are the most important functions to consider. A mother wavelet $\Psi(t)$ is a zero average function and is expressed by the following equation:

$$\int_{-\infty}^{\infty} \Psi(t)\mathrm{d}t = 0. \tag{8.1}$$

Wavelet packet transformation (WPT) is a useful tool with special capabilities to analyse the higher frequency domains of a signal and achieve higher decomposition. Thus this tool is more suitable for signal processing than the WT [22].

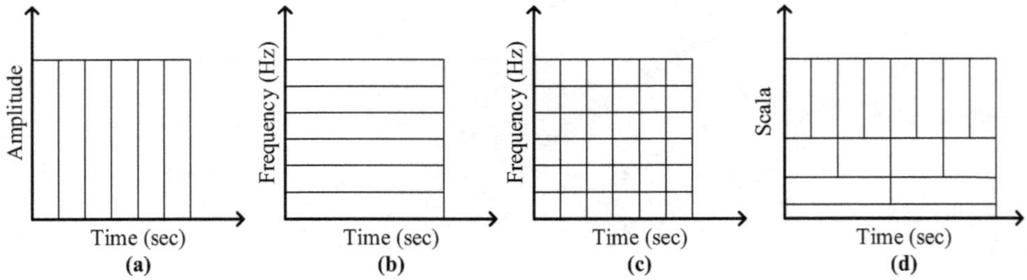

Figure 8.2. (a) Windowed time–amplitude plane, (b) windowed Fourier transform frequency–amplitude spectrum, (c) frequency–time plane with STFT and (d) windowed technique with the WT.

One needs to consider the tree of subspaces to understand WPT. In the tree, $\Omega_{j,k}$ notation is used to name the nodes [13, 14]. Here, the scale is denoted by j and the sub-band based on the current scale is denoted by k. $\Omega_{0,0}$ is the root node and represents the signal. In figure 8.3, an example WPT tree is presented. First, WPT decomposes from the root node to the approximation space ($\Omega_{j,k} \longrightarrow \Omega_{j+1,\,2k}$) and the detail space ($\Omega_{j,k} \longrightarrow \Omega_{j+1,\,2k+1}$). This process is implemented by separating $\{\phi_j(t - 2^j k)\}_{k\in Z}$ of $\Omega_{j,k}$ into new orthogonal bases $\{\phi_{j+1}(t - 2^{j+1}k)\}_{k\in Z}$ of $\Omega_{j+1,\,2k}$ and $\{\psi_{j+1}(t - 2^{j+1}k)\}_{k\in Z}$ of $\Omega_{j+1,\,2k}$. Here, $\phi_{j,k}(t)$ are the scale functions and $\psi_{j,k}(t)$ represent the wavelet functions [13, 14] given by

$$\phi_{j,k}(t) = \frac{1}{\sqrt{|2^j|}}\phi\left(\frac{t - 2^j k}{2^j}\right) \tag{8.2}$$

$$\psi_{j,k}(t) = \frac{1}{\sqrt{|2^j|}}\psi\left(\frac{t - 2^j k}{2^j}\right). \tag{8.3}$$

In equations (8.2) and (8.3), 2^j is the dilation parameter, which measures the degree of scaling, and $2^j k$ is the translation parameter which determines the time location of the wavelet [23, 24].

In this work, the scale j and sub-band parameter k are chosen empirically to be 3, taking into consideration computational complexity.

8.2.2 The one-dimensional LBP

Ojala *et al* propose the LBP as a powerful method to extract information on local texture in image processing applications [15]. Briefly, the LBP operator produces a code related to the local neighborhood. In the basic version of the LBP, there is a relation with eight neighbors of a pixel called the center pixel, here considered to be a 3 × 3 pixel. The relation is summarized by [25–28].

$$P_c = \sum_{i=0}^{P-1} f(x)2^i, \tag{8.4}$$

where P_c is the value of the center pixel, P shows the number of neighbors and $f(x)$ is the sign function given by

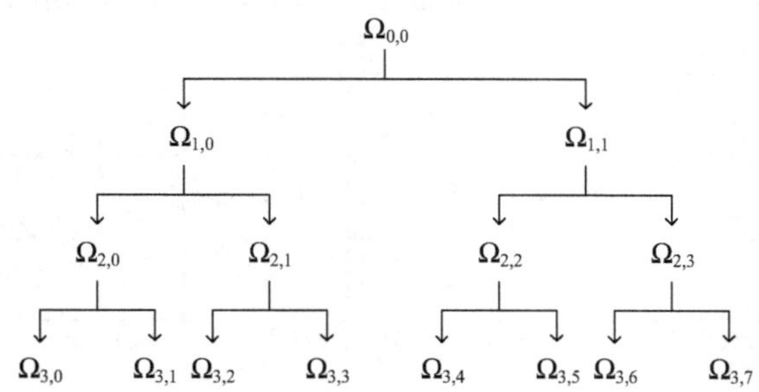

Figure 8.3. Representation of the WPT tree.

$$f(x) = \begin{cases} 0, & P_i \geqslant P_c \\ 1, & P_i < P_c \end{cases}, \tag{8.5}$$

where P_i is the ith neighbor pixel. The LBP operator is visualized in figure 8.4.

The center pixel is compared to its eight neighbors using equation (8.5); the resulting neighbors are encoded as 0 or 1, as shown figure 8.4(b). Then an 8 bit binary code is calculated according to the neighbors' values. The 8 bit binary code is converted to decimal and is assigned to the center pixel. After gaining popularity in image processing, the LBP has also been applied to the fields of speech and signal processing, as the one-dimensional LBP (1D-LBP) [27, 28]. In the 1D-LBP, a sliding window applies along with the signal. The window consists of nine samples instead of 3×3 pixels as in the original LBP. In the window, the fifth sample is an assumed center sample, and others are neighbor samples. The 1D-LBP scheme is visualized for a given sample x ($x = 17, 9, 40, 16, 27, 9, 31, 14, 33$) in figure 8.5.

In figures 8.5(a) and (b), the sample signal x is shown. The binary code value is depicted for the x signal in figure 8.5(c). The new value of the center sample, which is shown in figure 8.5(d), is calculated as in equation (8.4). The 1D-LBP process is applied along with a signal by using a sliding window [28]. In this way, the signal is quantized between 0–255. Then a histogram of the quantized signal is used as a feature [27, 28].

8.2.3 The support vector machine classifier

The support vector machine (SVM), which was developed by Vapnik [29], is a classifier with superior generalization capability based on the structural risk reduction principle, and is categorized as a supervised statistical learning theorem. The main idea of the SVM is to determine the optimum hyper-plane between two classes of data, called positive and negative [30].

Suppose that the two-class data are $(x_1, y_1), (x_2, y_2)\ldots, (x_i, y_i)$ and $i = 1, 2, \ldots, M$. Here, while x is the length of the input feature belonging to N-dimensional classes, y denotes class labels equal to $+1$ or -1. Thus, the linear separation of the two classes can be calculated with the following equation [31]:

$$f(x) = w^T x + b = 0. \tag{8.6}$$

Here w is an N-dimensional weight vector and b is the bias parameter used to determine the position of the hyper-plane. Two situations can be encountered in

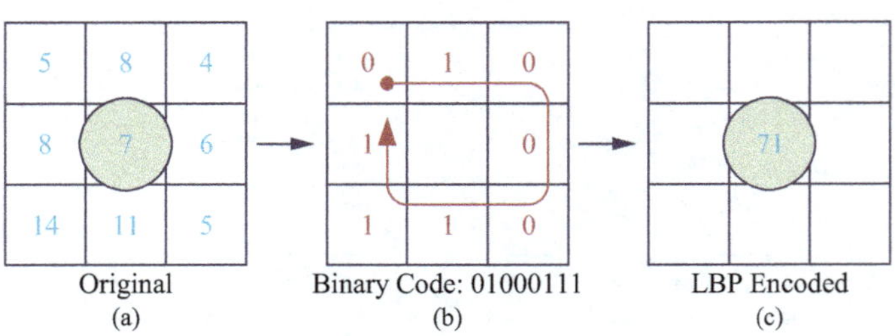

Figure 8.4. The procedure of the LBP.

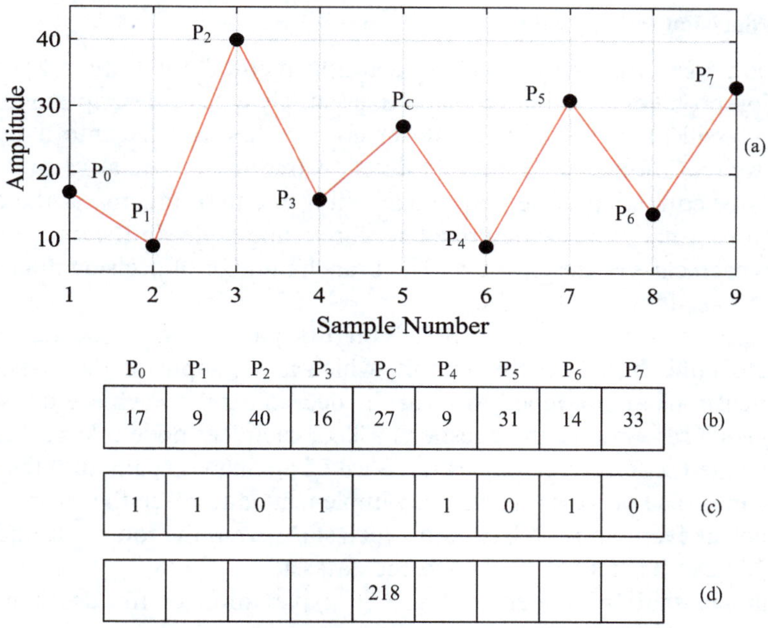

Figure 8.5. An example of the 1D-LBP.

SVMs depending on whether the data are in a structure that can be separated linearly or not [32]. In such a situation, it may be possible to overcome this problem by using different kernel functions such as the polynomial function, radial basis function (RBF), and so on. However, it is known that many planes separate the two classes linearly, but there is only one hyper-plane that maximizes the margin between the two classes. This optimum hyper-plane can be determined by the following equation [33]:

$$\begin{cases} \min \dfrac{\|w\|^2}{2} \\ y_i(w^T x_i + b) \geqslant 1 \quad i = 1, 2, \ldots, M \end{cases} . \tag{8.7}$$

In the case of classifications that cannot be separated linearly, the equation is revised by adding a C regularization parameter and a positive artificial ξ_i variable as follows [34, 35]:

$$\min f(w, \xi) = \frac{1}{2} \|w\|^2 + C\sum_{i=1}^{M} \xi_i. \tag{8.8}$$

SVMs are inherently designed as two-class classifiers. One of the best ways to do multiclass classification with SVMs is to use combinations of two-class based classification techniques, named one-versus-rest (mostly referred to as one-versus-all) and one-versus-one. One-versus-one two-class classifiers identify one class from another one, and one-versus-all classifiers separate one class from all other classes [35, 36].

8.2.4 The decision tree classifier

The decision tree is a useful machine learning method for both regression and classification problems [37–44]. Decision trees supply a basic description in order to carry out classification tasks. The dataset given is first divided into a small-scale subset as well as forming a correlated decision tree in a cumulative manner. The result is a tree consisting of decision nodes. In a tree that anticipates the optimum predictor, a root node can be described as a crowning node. In the tree, a leaf node symbolizes each class of the classifier. The branch number of decision nodes is equal to two or higher than two [37, 38].

The questions are constructed based on the parameters associated with the specified attributes based on the dataset, which is the input of the classifier. The prepared questions are designed to form the decision tree to achieve classification. Following each response to the questions asked, each leaf node takes a label based on the calculated score, the next question is asked again and again until the stopping criteria are met. The process is recursively implemented to reach the reasonable right of the branch at every node [37]. A path which starts from the root node and stops at the leaf node can be created to classify the dataset.

One can say that the understanding and explanation of the decision tree are straightforward. Also, the preparation of the dataset can be an easy task compared to most other classification methods. The decision tree is very suitable for binary classification. However, the decision tree may have a tendency of unstable performance for large input datasets. Also, the performance of the decision tree may be good only in the case of a large input dataset. As the dataset borderline has little variation, the performance of the decision tree may be adversely affected. The widely used decision tree algorithms are the iterative dichotomizer (ID3), C4.5, the classification and regression tree (CART), and the quick, unbiased and efficient statistical tree (QUEST) [37–44]. We briefly describe the C4.5 approach. Take a set S, $S = \{S_1, S_2, \ldots, S_m\}$ with m classes. A with n distinct values represents an attribute belonging to S. We separate S into k subsets, such that $S' = \{a_1, a_2, \ldots, a_m\}$. The entropy $E(S)$, which measures the information from S, is calculated from the given S dataset as

$$E(S) = -\sum_i^m p(S_i)\log_2 p(S_i), \tag{8.9}$$

where $p(S_i)$ shows the probability of the ith class. The information gain rate $G(S')$ is computed for the A attribute by

$$G(S') = \frac{H(S) - H_A(S')}{SI(S')}. \tag{8.10}$$

Here SI symbolizes the split information and is expressed as

$$SI(S') = -\sum_i^k p(a_i)\log_2 p(a_i). \tag{8.11}$$

By using the information gain rate, the dataset divides into k subsets to ensure maximum gain in each node. This process is repeated until each leaf contains observations belonging to one class. As a result of this process, a large and complex decision tree may be produced. The parts that do not affect the result are removed by pruning the tree. Thus a pure decision tree is obtained [40–44].

8.2.5 The ensemble bagging classifier

Ensemble paradigms involve using two or more individual algorithms by weighting them to obtain a single superior model [45–49]. There are two approaches in the ensemble paradigm: classifier selection and classifier fusion [46]. In this chapter, the ensemble bagging tree and ensemble boosting tree are chosen because of their widespread use.

Ensemble bagging (bootstrap aggregating) is intended to produce a superior composite bagged classifier I^* by assembling various trained classifier outputs. The ensemble bagging classifier algorithm is given in algorithm 8.1 [45].

Algorithm 8.1. The ensemble bagging algorithm.

Input: $S = \{x_m, y_m\}$ dataset, μ subset size, I inducer and T maximum iteration number
Ensure: C base classifier algorithm
Initialize: $t = 1$
Loop: $t < T$
$S_t = \text{bootstrap}(S)$ // Generate a bootstrap S_t sized μ from S
$h_t = C(S_t)$ // Train base classifier h_t by using I and S_t
$t = t + 1$
End
Output: $H(x) = \text{argmax}\sum_{t=1}^{T} f(x); f(x) = \begin{cases} 1, & y_m = h_t(x_m) \\ 0, & \text{otherwise} \end{cases}$

The composite bagged classifier gives the class most often predicted as a result. The result is better compared to a single classifier trained on the undivided original dataset. Breiman [48] predicted this situation correctly, particularly in the presence of unstable inducers, as ensemble bagging can address the instability of inducers and increase the accuracy of the classifier [45, 46, 48]. In ensemble bagging, the base classifier is a decision tree with C4.5.

8.2.6 The ensemble boosting classifier

The ensemble boosting classifier also builds composite classification models by resampling the dataset, as in the ensemble bagging classifier. The resampling procedure is used to obtain training data with useful information for each sequential classifier. The ensemble boosting algorithm is given in algorithm 8.2 [46, 49].

Algorithm 8.2. The ensemble boosting algorithm.

Input: $S = \{x_m, y_m\}$ dataset, T maximum iteration number
Ensure: C base classifier algorithm
Initialize: $t = 1$, $S_1(i) = 1/m$ // The weight distribution
For $t = 1, \ldots, T$
$h_t = C(S, S_t)$ // Train base classifier h_t from S using the S_t distribution
$e_t = Pr_{i \sim D_i}[h_t(x_i \neq y_i)]$ // The error of h_t
$\alpha_t = \frac{1}{2}\ln\frac{1-e_t}{e_t}$ // Determine the weight of h_t
$D_{t+1}(i) = \frac{D_t(i)\exp(-\alpha_t y_i h_t(x_i))}{Z_t}$ // Update the distribution, here Z_t represents to normalization
End
Output: $H(x) = \text{sign}(\sum_{t=1}^{T}\alpha_t h_t(x))$

In ensemble boosting, the base classifier is AdaBoosting.

8.2.7 The *k*-nearest neighbor classifier

The k-nearest neighbor (k-NN) algorithm is one of the simplest, most effective and most widely used methods for classification and regression in machine learning [50, 51]. It is generally used in pattern recognition and statistical predictions [51]. k-NN is not a learning phase or nonparametric, but instead memorizes the training set [3]. Basically, as seen in figure 8.6, it classifies all existing data and new data according to the measure of similarity [52, 53].

This also shows that the training phase is high-speed. The operation of the algorithm depends on the k value [54]. The number of neighbors is called the decision factor and is often chosen as an odd number. When a new value comes in, the distance between the incoming values is calculated by taking k elements [55]. However, the distance on the k-NN performance input area is of great importance [56]. While the Euclid function is mostly used in distance calculations, the Manhattan and Minkowski functions can also be used [57]. These distances are calculated as

$$\text{Euclid} = \sqrt{\sum_{i=1}^{k}(x_i - y_i)^2} \tag{8.12}$$

$$\text{Manhattan} = \sum_{i=1}^{k}|x_i - y_i|$$

$$\text{Minkowski} = \left(\sum_{i=1}^{k}(|x_i - y_i|)^q\right)^{1/q}$$

8.2.8 The linear discriminant classifier

The linear discriminant classifier (LDC) is a well-known classification approach that is based on a distance function. In the LDC concept, the test samples are classified

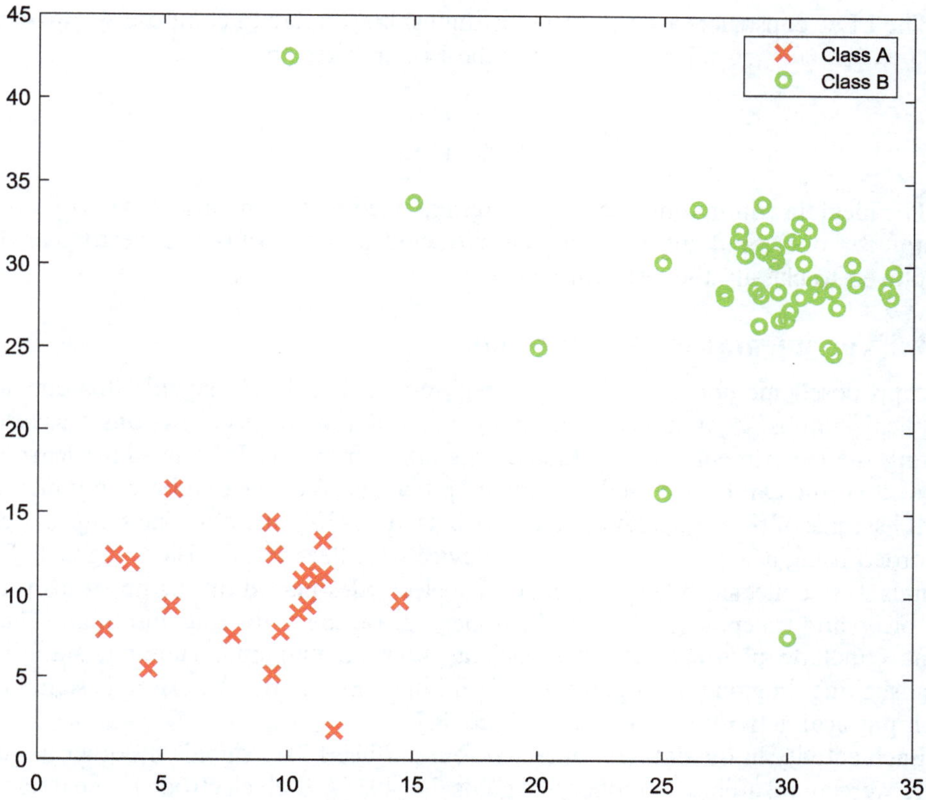

Figure 8.6. *k*-NN illustration.

based on the training samples, where training samples are used to produce a series of equations. The LDC employs two potential objectives in the classification of the test samples. In the first objective, a predictive equation is searched for classifying a test sample. In the second, the predictive equation is constructed for understanding the relationships that may exist among the training sample features [58–60].

Let X_j^i denote the jth training sample of class i. In addition, let X^i and X specify the average value of the training samples of class i and the average value of all training samples, respectively. Thus, the scatter matrices of the between-class and within-class M_b and M_w are calculated as

$$M_b = \frac{1}{N}\sum_{i=1}^{m} N_i(X_i - X)(X_i - X)^T \tag{8.13}$$

$$M_w = \frac{1}{N}\sum_{i=1}^{m}\sum_{j=1}^{N_i}\left(X_j^i - X^i\right)\left(X_j^i - X^i\right)^T, \tag{8.14}$$

where m indicates the number of classes, and N and N_i show the total number of training samples and the number of samples in class i, respectively.

The LDC constructs a set of ideal distinguishing vectors to compose a transform of $\Phi_d = [\varphi_1, \varphi_2, \ldots, \varphi_d]$ by maximizing the Fisher criterion as

$$J(\Phi) = \frac{\mathrm{tr}(\Phi_d^T M_b \Phi_d)}{\mathrm{tr}(\Phi_d^T M_w \Phi_d)}. \tag{8.15}$$

The ideal discriminating vectors are strengthened by benchmark maximization, a vector for one input instance can be extracted and the obtained vector can be employed to classify the test sample [58, 59].

8.3 Experimental work and results

The proposed method is aimed at classifying surface EMG signals that include physical actions. A database containing ten different physical actions was used during the experiments. The database was taken from the UCI machine learning repository and can be accessed through [61]. Each class of the database contains the EMG signals of three men and one woman, aged 25–30 years old. The samples were recorded using a wireless Delsys EMG device to collect surface EMG signals. The signals were collected by eight skin surface electrodes placed on the upper arms on the bicep and tricep, as well as on the upper legs on the hamstring and thigh. These signals include physical activities such as bowing, clapping, running, standing, handshaking, hugging, sitting, jumping, walking and waving. An example signal for each physical activity is provided in figure 8.7.

Each activity in the dataset contained four surface EMG signals. In other words, there were an insufficient number of signals, i.e. 40, for each electrode in the dataset. Therefore, we took into account a strategy to artificially augment the number of signals. To do this, we took random 1 × 2000 size samples from each raw signal and created 1000 samples for each class in the dataset. In this way we augmented the total number of samples in the dataset to 10 000 and each signal was directly fed into the input of the WPT. The decomposition level 3, 'db3' family, and Shannon entropy were preferred for the WPT parameters. Then a 1D-LBP approach was used to code the coefficients of the decomposed EMG signals. For the LBP operator, a total of eight neighbor signals were used four samples on the left and four samples on the right of the reference point. Detailed information about the LBP operator is provided in the previous section. The histogram of each LBP output was obtained using five bins in order to achieve the best classification results. Normalization was then achieved by dividing each of the one-dimensional histograms in the bins by the total number of histograms. A feature vector was obtained for each histogram, each being 1 × 5 in size. Consequently, the total size of the concatenated features was recorded as 1 × 45.

All of the experiments were conducted on a workstation equipped with an Intel (R) Core (TM) i7-3630QM 8GB RAM 2.4 GHz CPU with eight cores. In addition, the SVM classifier was first used to evaluate the performance of the proposed method. When evaluating the performance of the method, a ten-fold cross-validation approach was implemented. The parameters of the SVM were chosen intuitively during the experiments. The kernel function, polynomial order, kernel

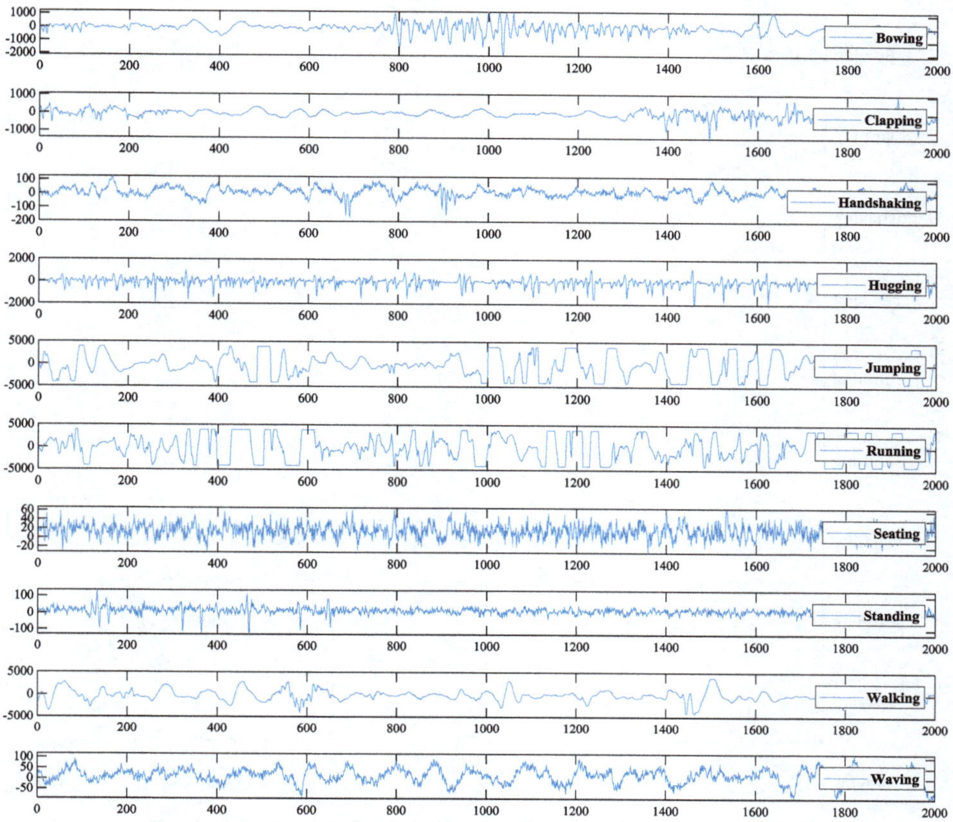

Figure 8.7. Example physical activity signals.

scale and C-regularization parameters were selected as polynomial, 2, auto and 1, respectively. The cumulative confusion matrix of the obtained results for all electrodes is shown in figure 8.8.

As seen in figure 8.8, all actions were classified with 100% classification accuracy. These scores showed that the proposed method was quite efficient for sEMG based action classification. The experiments were also run using various classifiers, namely the decision tree, linear discriminant, *k*-nearest neighbors (k-NN), boosted ensemble and bagged ensemble classifiers [63, 64]. The MATLAB classification learner tool was used to implement the classifiers. The MATLAB classification learner tool enables the user to employ a series of classifiers without the need for any coding. It can be considered as a useful interface for data mining applications. Decision trees have some advantages, such as easy implementation, fast classification and low memory usage. However, decision trees have low predictive accuracies in some high-dimensional applications. The maximum number of splits of the decision trees was assigned as 10. The linear discriminant is also another fast, accurate and effective classifier. The linear discriminant classifier creates linear boundaries between predicted classes. *k*-NN classifiers generally have high predictive accuracy in low dimensions. This can be seen as an advantage of the *k*-NN classifiers, but their high

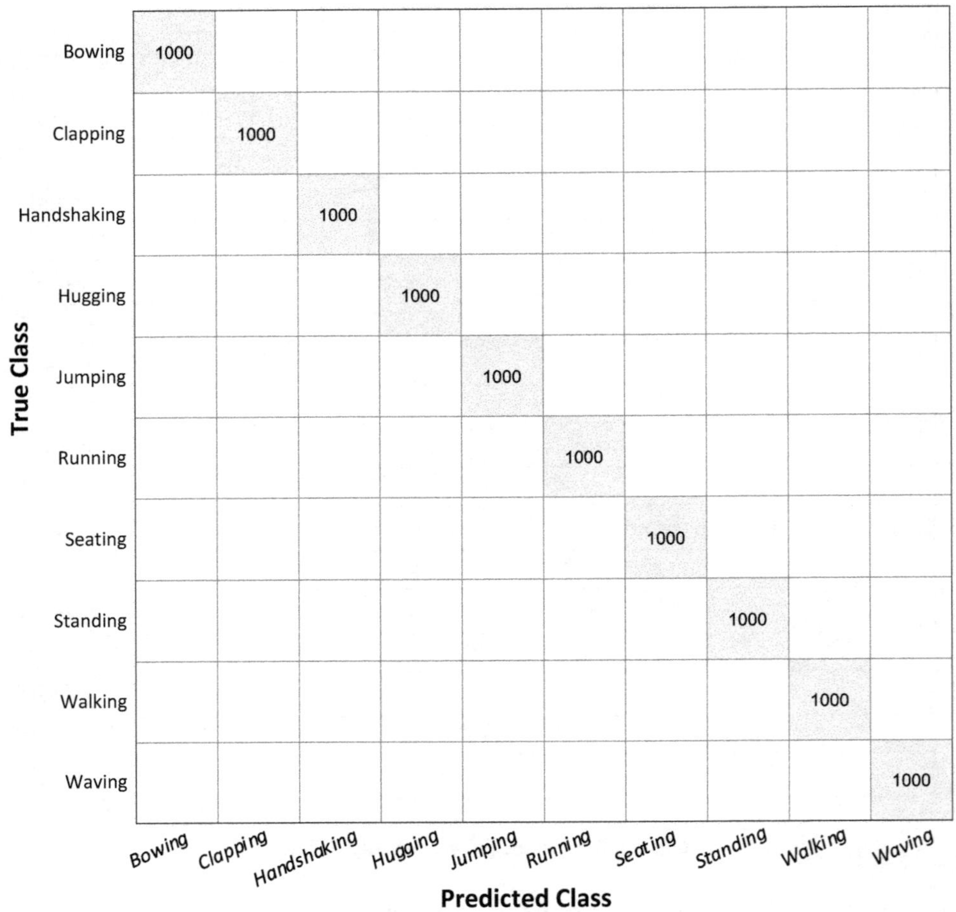

Figure 8.8. Confusion matrix for all electrodes.

memory usage and difficult to interpret properties are disadvantages. The number of neighbors k is chosen as 3. As ensemble learners encompass dozens of classifiers, they are generally regarded as slower approaches. Two different ensemble learner techniques are used in classification learner applications: boosted trees and bagged trees. In the boosted tree technique, the AdaBoost ensemble method is used while for bagged trees the random forest approach is adopted. Table 8.1 shows the obtained results for the decision tree, linear discriminant, k-NN, boosted ensemble and bagged ensemble classifiers, respectively. The rows of table 8.1 show the electrodes and the columns show the classifiers achievements.

From table 8.1, it can be observed that the classifiers produced quite close accuracy scores for the electrodes. For electrode 1, decision trees produced an 84.7% accuracy score. The linear discriminant, k-NN, boosted trees ensemble and bagged trees ensemble approaches produced 89.5%, 100.0%, 73.3% and 99.9% accuracy scores, respectively. Thus k-NN achieved the best performance for electrode 1. In fact, k-NN achieved the best performance for all electrodes. The bagged tree

Table 8.1. The obtained accuracy scores for EMG electrodes using each classifier.

	Decision tree	Linear discriminant	*k*-NN	Boosted tree ensemble	Bagged tree ensemble	Average accuracy
Electrode 1	84.7%	89.5%	100.0%	73.3%	99.9%	89.48%
Electrode 2	81.3%	90.4%	100.0%	76.7%	99.9%	89.66%
Electrode 3	85.4%	90.3%	100.0%	76.9%	99.9%	90.50%
Electrode 4	80.5%	87.9%	100.0%	72.6%	99.8%	88.16%
Electrode 5	80.0%	90.5%	100.0%	70.8%	99.9%	88.24%
Electrode 6	86.3%	90.7%	100.0%	76.6%	99.7%	90.66%
Electrode 7	85.1%	87.6%	100.0%	75.3%	99.8%	89.56%
Electrode 8	85.7%	91.3%	100.0%	76.0%	99.7%	90.54%
Average accuracy	83.63%	89.78%	100.0%	74.78%	99.83%	

ensemble method produced the second best accuracy score of 99.9% for electrode 1. For electrode 2, 81.3%, 90.4%, 100.0%, 76.7% and 99.9% accuracy scores were obtained by the decision trees, linear discriminant, *k*-NN, boosted ensemble and bagged ensemble classifiers, respectively. The best and the second best accuracy scores were obtained by the *k*-NN and bagged tree ensemble techniques. The worst accuracy scores for all the electrodes were produced by the boosted tree ensemble method. For electrode 3, the decision tree, linear discriminant, *k*-NN, boosted ensemble and bagged ensemble classifiers produced 85.4%, 90.3%, 100.0%, 76.9% and 99.9% accuracy scores, respectively. As can be observed in table 8.1, for electrodes 1, 2 and 3, the *k*-NN and bagged tree ensemble techniques produced identical accuracy scores. For electrode 4, the bagged tree ensemble method produced a 99.8% accuracy score. In addition, 80.5%, 87.9%, 100.0% and 72.6% accuracy scores were yielded by the decision tree, linear discriminant, *k*-NN and boosted ensemble methods, respectively. For electrode 5, 80.0%, 90.5%, 100%, 70.8% and 99.9% accuracy scores were obtained by the decision tree, linear discriminant, *k*-NN, boosted ensemble and bagged ensemble classifiers. For electrode 6, the decision tree, linear discriminant, *k*-NN, boosted ensemble and bagged ensemble classifiers produced 86.3%, 90.7%, 100.0%, 76.6% and 99.7% accuracy scores, respectively. For electrode 7, the decision tree, linear discriminant, *k*-NN, boosted ensemble and bagged ensemble classifiers produced 85.1%, 87.6%, 100.0%, 75.3% and 99.7% accuracy scores, respectively. For electrode 8, 85.7%, 91.3%, 100.0%, 76.0% and 99.7% accuracy scores were obtained, respectively.

In table 8.1, the average achievements of the electrodes and classifiers are also given. While the last row shows the average accuracy scores for the classifiers, the last column of table 8.1 tabulates the average achievements for each electrode, respectively. As mentioned earlier, the *k*-NN classifier produced the highest average accuracy score for EMG based action classification. The bagged tree ensemble method produced the second highest accuracy score, where the calculated average accuracy score was 99.83%. The worst average accuracy score, 74.78%, was obtained using the boosted tree ensemble method. The decision tree and linear discriminant methods produced 83.63% and 89.78% average accuracy scores, respectively. Moreover, when the average accuracy scores of the electrodes are examined, it is seen that the highest average accuracy score 90.66% was produced by electrode 6. Electrodes 8 and 3 produced 90.54% and 90.50% accuracy scores, which were the second and third best accuracy scores, respectively. The worst average accuracy score, 88.16%, was produced by electrode 4.

As electrode 6 produced the highest average accuracy score, the receiver operating characteristic (ROC) curves and the related area under curve (AUC) values for electrode 6 are given for the decision tree, linear discriminant, *k*-NN, boosted ensemble and bagged ensemble classifiers in figure 8.9.

As can be seen in figure 8.9, the *k*-NN and bagged tree ensemble methods produced identical ROC curves, which were better than the other ROC curves. The ROC curves for the decision tree, linear discriminant and boosted tree ensemble classifiers were lower than the ROC curves of the *k*-NN and bagged tree ensemble classifiers, which indicates a worse performance than for the latter two classifiers. Moreover, the AUC values for the linear discriminant, *k*-NN and bagged tree ensemble methods were 1.0, and the AUC scores for the decision tree and boosted tree ensemble methods were 0.98 and 0.99, respectively.

A performance comparison with the achievements of state-of-the-art methods is also given in table 8.2. In [9], Abdullah *et al* used the MSPCA and WPD coefficients as features, and a random forest classifier was used to recognize the sEMG based actions. The authors reported an 99.10% accuracy score. Sumukar *et al* [62] used VMD based statistical features and MC-LSSVM for sEMG based action classification achieving a 98.17% performance score. Sahinbegovic *et al* [7] used neural networks for sEMG signal classification and produced an 86.25% accuracy score. Demir *et al* [4] used deep features and the SVM classifier for sEMG based action classification achieving a 99.04% accuracy score. Sravani *et al* [19] used FAWT based statistical features and ELM classification for efficient sEMG signal recognition, obtaining a 99.36% accuracy score. As seen in table 8.2, until now, the best accuracy score, 99.36%, was reported in [19], and 99.10% and 99.04% accuracy scores were reported by Abdullah *et al* [9] and Demir *et al* [4], respectively. To the best of our knowledge, a 100% accuracy score has only been obtained in our work.

8.4 Conclusion

In this chapter, a simple but efficient approach is proposed for physical action recognition based on EMG signals. The proposed method is composed of signal decomposition, feature extraction and feature classification. Signal decomposition is carried out using the WPT. The WPT successively decomposes a signal into its approximation and detail coefficients. One-dimensional LBP and its histogram are used to characterize the EMG signals. The support vector machine (SVM) classifier is used in the classification stage. Various experiments were conducted and satisfactory results were obtained. The experiments show that 100% correct

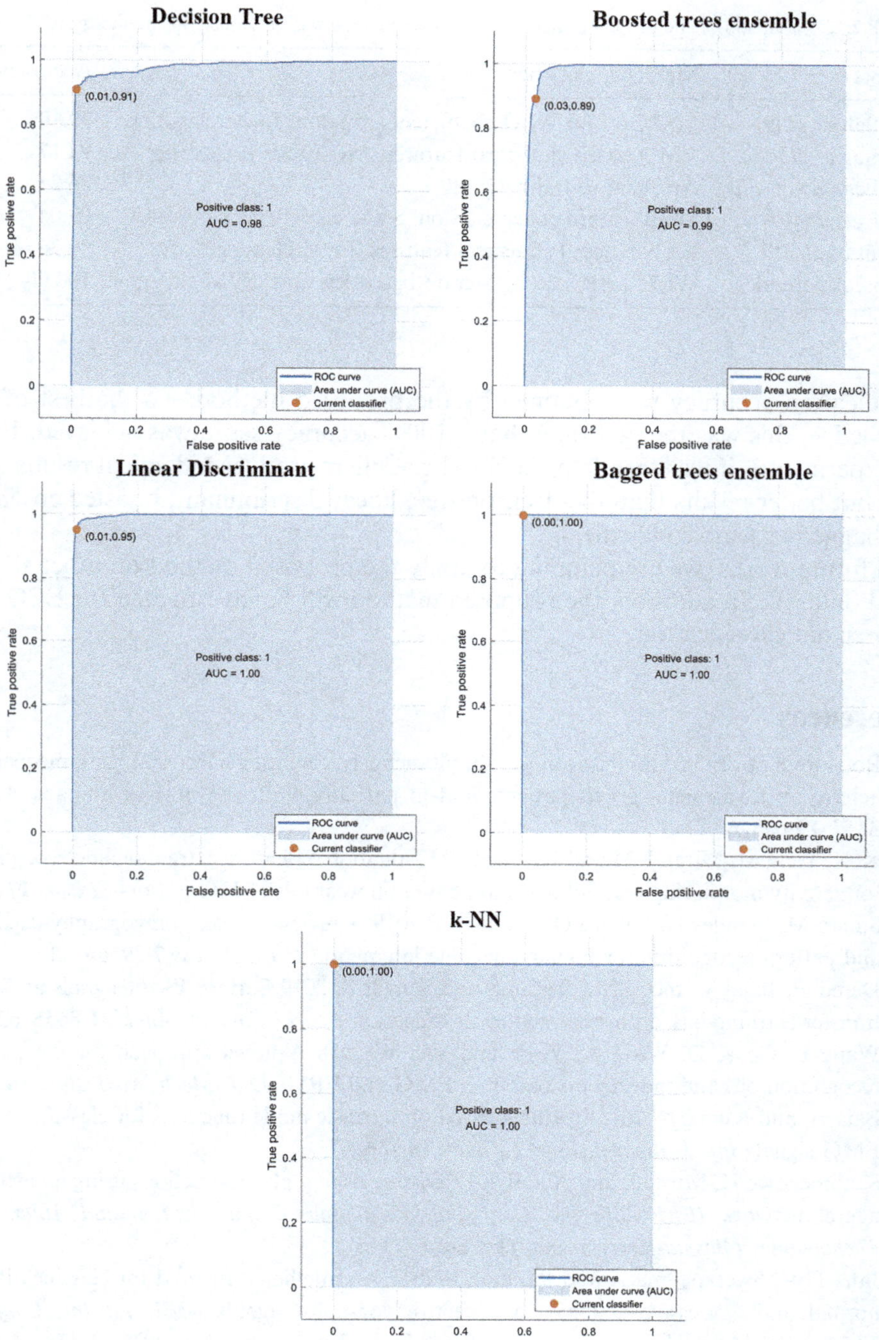

Figure 8.9. The ROC curves and AUC values of electrode 6 for all classifiers.

Table 8.2. Comparison of the proposed method to the results reported previously in the literature.

Authors	Methods, classifier	Accuracy (%)
Abdullah *et al* [9]	MSPCA and WPD coefficients, random forest classifier	99.10
Sukumar *et al* [62]	VMD based statistical features, MC-LSSVM classifier	98.17
Sahinbegovic *et al* [7]	Artificial neural network	86.25
Demir *et al* [4]	Deep feature concatenation, SVM classifier	99.04
Sravani *et al* [19]	FAWT based statistical features, ELM classifier	99.36
Proposed method	WPT, LBP, and histogram based features, SVM classifier	100.00

classification accuracy was obtained by the proposed method. To the best of our knowledge, this was the first time that a 100% accuracy score was achieved. From the experiments, it was seen that the SVM classifiers produced identical results to *k*-NN, but better results than the decision tree, linear discriminant, boosted ensemble and bagged ensemble classifiers.

In future works, we are planning to apply the proposed method on other sEMG based datasets. In addition, the proposed method will be investigated for EEG and ECG signal classification.

References

[1] Botelho S Y 1955 Comparison of simultaneously recorded electrical and mechanical activity in myasthenia gravis patients and in partially curarized normal humans *Am. J. Med.* **19** 693–6

[2] Xi X, Tang M, Miran S M and Luo Z 2017 Evaluation of feature extraction and recognition for activity monitoring and fall detection based on wearable sEMG sensors *Sensors* **17** 1–20

[3] Simão M, Mendes N, Gibaru O and Neto P 2019 A review on electromyography decoding and pattern recognition for human–machine interaction *IEEE Access* **7** 39564–82

[4] Demir F, Bajaj V, Ince M C, Taran S and Şengür A 2019 Surface EMG signals and deep transfer learning-based physical action classification *Neural Comput. Appl.* **31** 8455–62

[5] Wang L, Ge K D, Wu J Y, Ye Y and Wei W 2018 A novel approach for the pattern recognition of hand movements based on EMG and VPMCD *J. Mech. Med. Biol.* **18** 1–15

[6] Bajaj V and Kumar A 2015 Features based on intrinsic mode functions for classification of EMG signals *Int. J. Biomed. Eng. Technol.* **18** 156–67

[7] Sahinbegovic H, Music L and Alic B 2017 Distinguishing physical actions using an artificial neural network *IEEE 26th Int. Conf. on Information, Communication and Automation Technologies (Bosnia-Herzegovina, Oct. 2017)* 1–5

[8] Jana G C, Swetapadma A and Pattnaik P 2017 An intelligent method for classification of normal and aggressive actions from electromyography signals *IEEE 1st Int. Conf. on Electronics, Materials Engineering and Nano-Technology (India, Apr. 2017)* 1–5

[9] Abdullah A A, Subasi A and Qaisar S M 2017 Surface EMG signal classification by using WPD and ensemble tree classifiers *Proc. of the Int. Conf. on Medical and Biological Engineering (Bosnia-Herzegovina, Mar. 2017)* A Badnjevic (Singapore: Springer) pp 475–81

[10] Bai D, Chen S and Yang J 2019 Upper arm motion high-density sEMG recognition optimization based on spatial and time–frequency domain features *J. Healthc. Eng.* **2019** 1–16

[11] Tuncer T, Dogan S and Subasi A 2020 Surface EMG signal classification using ternary pattern and discrete wavelet transform based feature extraction for hand movement recognition *Biomed. Signal Process. Control* **58** 1–12

[12] Kumari P, Narayan Y and Mathew L 2016 Binary movement classification of sEMG signal using linear SVM and wavelet packet transform *IEEE 1st Int. Conf. on Power Electronics, Intelligent Control and Energy Systems (India, July 2016)* pp 1–4

[13] Amma C, Krings T, Böer J and Schultz T 2015 Advancing muscle–computer interfaces with high-density electromyography *33rd Annual ACM Conf. on Human Factors in Computing Systems (Korea, Apr. 2015)* 929–38

[14] Wei W, Wong Y, Du Y, Hu Y, Kankanhalli M and Geng W 2019 A multi-stream convolutional neural network for sEMG-based gesture recognition in muscle–computer interface *Pattern Recognit. Lett.* **119** 131–8

[15] Chaya N A, Bhavana B R, Anoogna S B and Hiranmai M 2019 Real-time replication of arm movements using surface EMG signals *Procedia Comput. Sci.* **154** 186–93

[16] Turlapaty A C and Gokaraju B 2019 Feature analysis for classification of physical actions using surface EMG data *IEEE Sens. J.* **19** 12196–204

[17] Ouyang G, Zhu X, Ju Z and Liu H 2013 Dynamical characteristics of surface EMG signals of hand grasps via recurrence plot *IEEE J. Biomed. Health Inform.* **18** 257–65

[18] Sapsanis C, Georgoulas G, Tzes A and Lymberopoulos D 2013 Improving EMG based classification of basic hand movements using EMD *35th Annual Int. Conf. of the IEEE Engineering in Medicine and Biology Society (Japan, July 2013)* pp 5754–7

[19] Sravani C, Bajaj V, Taran S and Sengur A 2020 Flexible analytic wavelet transform based features for physical action identification using sEMG signals *IRBM* **41** 18–22

[20] Qi J, Jiang G, Li G, Sun Y and Tao B 2019 Intelligent human–computer interaction based on surface EMG gesture recognition *IEEE Access* **7** 61378–87

[21] Malat S G 1989 A theory for multiresolution signal decomposition *IEEE Trans. Pattern Anal. Mach. Intell.* **11** 674–93

[22] Mohamed I M, Moawad I M D, Salah D and Fatma H E 2008 Signal denoising by wavelet packet transform on FPGA technology *Ubiquit. Comput. Commun. J.* **3** 54–8

[23] Khushaba R N, Kodagoda S, Lal S and Dissanayake G 2010 Driver drowsiness classification using fuzzy wavelet-packet-based feature-extraction algorithm *IEEE Trans. Biomed. Eng.* **58** 121–31

[24] Pancholi S and Joshi A M 2019 Improved classification scheme using fused wavelet packet transform based features for intelligent myoelectric prostheses *IEEE Trans. Ind. Electron.* **14** 1–9

[25] Ojala T, Pietikäinen M and Mäenpää T 2000 Gray scale and rotation invariant texture classification with local binary patterns *European Conference on Computer Vision (Ireland, July 2000)* (Lecture Notes on Computer Science) D Vernon (Berlin: Springer) pp 404–20

[26] Huang D, Shan C, Ardabilian M, Wang Y and Chen L 2011 Local binary patterns and its application to facial image analysis: a survey *IEEE Trans. Syst. Man Cybern.* C **41** 765–81

[27] Chatlani N and Soraghan J J 2010 Local binary patterns for 1-D signal processing *IEEE 18th European Signal Processing Conf. (Denmark, Aug. 2010)* pp 95–9

[28] Güner A, Alçin Ö F and Şengür A 2019 Automatic digital modulation classification using extreme learning machine with local binary pattern histogram features *Measurement* **145** 214–25
[29] Widodo A and Yang B S 2007 Support vector machine in machine condition monitoring and fault diagnosis *Mech. Syst. Signal Process.* **21** 2560–74
[30] Qi Z, Tian Y and Shi Y 2013 Robust twin support vector machine for pattern classification *Pattern Recognit.* **46** 305–16
[31] Adaminejad H, Shayegani I, Ohammadi M and Farjah E 2013 An algorithm for power quality events core vector machine-based classification *Modares J. Electr. Eng.* **12** 50–9
[32] Acı M, Avcı M and Acı Ç 2017 Reducing simulation duration of carbon nanotube using support vector regression method *J. Faculty Eng. Archit. Gazi Univ.* **32** 901–7
[33] Cheng L and Bao W 2014 Remote sensing image classification based on optimized support vector machine *Telkomnika Indonesian J. Electr. Eng.* **12** 1037–45
[34] Shen L, Chen H, Yu Z, Kang W, Zheng B, Li H, Yang B and Liu D 2016 Evolving support vector machines using fruit fly optimization for medical data classification *Knowl.-Based Syst.* **96** 61–75
[35] Cıbuk M, Budak U, Guo Y, Ince M C and Sengur A 2019 Efficient deep features selections and classification for flower species recognition *Measurement* **137** 7–13
[36] Toğaçar M, Ergen B and Cömert Z 2020 Detection of lung cancer on chest CT images using minimum redundancy maximum relevance feature selection method with convolutional neural networks *Biocybern. Biomed. Eng.* **40** 23–39
[37] Liu Q, Xu X, Tao Y and Wang X 2016 An improved decision tree method base on RELIEFF for medical diagnosis *IEEE 6th Int. Conf. on Digital Home (China, Dec. 2016)* pp 133–8
[38] Lavanya K, Bajaj S, Tank P and Jain S 2017 Handwritten digit recognition using Hoeffding tree, decision tree and random forests—a comparative approach *IEEE Int. Conf. on Computational Intelligence in Data Science (India, June 2017)* pp 1–6
[39] Loh W Y 2011 Classification and regression trees *Wiley Interdiscip. Rev.: Data Min. Knowl. Discov.* **1** 14–23
[40] Quinlan J R 1987 Simplifying decision trees *Int. J. Man–Mach. Stud.* **27** 221–34
[41] Quinlan J R 1986 Induction of decision trees *Mach. Learn.* **1** 81–106
[42] Quinlan J R 1993 *C4.5: Programming for Machine Learning* 1st edn ed P Langley (San Francisco, CA: Morgan Kaufmann) pp 1–56
[43] Brodley C E and Utgoff P E 1995 Multivariate decision trees *Mach. Learn.* **19** 45–77
[44] Kavzoğlu T and Çölkesen İ 2010 Classification of satellite ımages using decision trees: Kocaeli case *Electr. J. Map Technol.* **2** 36–45
[45] Rokach L 2010 Ensemble-based classifiers *Artif. Intell. Rev.* **33** 1–39
[46] Polikar R 2006 Ensemble based systems in decision making *IEEE Circuits Syst. Mag.* **6** 21–45
[47] Dietterich T G 2000 An experimental comparison of three methods for constructing ensembles of decision trees: bagging, boosting, and randomization *Mach. Learn.* **40** 139–57
[48] Breiman L 1996 Bagging predictors *Mach. Learn.* **24** 123–40
[49] Zhou Z H 2009 Ensemble learning *Encyclopedia of Biometrics* vol 1 (Berlin: Springer) 270–3
[50] Altman N S 1992 An introduction to kernel and nearest-neighbor nonparametric regression *Am. Stat.* **46** 175–85

[51] Hoyos-Barcelo C, Monge-Alvarez J, Shakir M Z, Alcaraz-Calero J M and Casaseca-de-La-Higuera P 2017 Efficient *k*-NN implementation for real-time detection of cough events in smartphones *IEEE J. Biomed. Health Inform.* **22** 1662–71

[52] Zhao X, Zhang J and Qin X 2017 *k*NN-DP: handling data skewness in *k*NN joins using map reduce *IEEE Trans. Parallel Distrib. Syst.* **29** 600–13

[53] Jiao L, Geng X and Pan Q 2019 BP*k*NN: *k*-nearest neighbor classifier with pairwise distance metrics and belief function theory *IEEE Access* **7** 48935–47

[54] Zhang S, Li X, Zong M, Zhu X and Wang R 2017 Efficient *k*NN classification with different numbers of nearest neighbors *IEEE Trans. Neural Netw. Learn. Syst.* **29** 1774–85

[55] Mohammed M A, Ghani M K A, Hamed R I, Mostafa S A, Ibrahim D A, Jameel H K and Alallah A H 2017 Solving vehicle routing problem by using improved *k*-nearest neighbor algorithm for best solution *J. Comput. Sci.* **21** 232–40

[56] Liew S H, Choo Y H, Low Y F and Yusoh Z I M 2017 EEG-based biometric authentication modelling using incremental fuzzy-rough nearest neighbour technique *IET Biometrics* **7** 145–52

[57] Aslan M, Akbulut Y, Şengür A and İnce M C 2017 Skeleton based efficient fall detection *J. Faculty Eng. Archit. Gazi Univ.* **32** 1025–34

[58] Agarwal D and Chen B C 2010 fLDA: matrix factorization through latent Dirichlet allocation *Third ACM Int. Conf. on Web Search and Data Mining (USA, Feb. 2010)* pp 91–100

[59] Moghaddam B, Weiss Y and Avidan S 2007 Spectral method for sparse linear discriminant analysis *US Patent Application* 11/440,825

[60] Bekios-Calfa J, Buenaposada J M and Baumela L 2010 Revisiting linear discriminant techniques in gender recognition *IEEE Trans. Pattern Anal. Mach. Intell.* **33** 858–64

[61] Dheeru D and Taniskidou K E 2017 *UCI Machine Learning Repository* https://archive.ics.uci.edu/ml/index.php

[62] Sukumar S, Marsadek M, Ramasamy A and Mokhlis H 2018 Grey Wolf optimizer based battery energy storage system sizing for economic operation of microgrid *IEEE Int. Conf. on Environment and Electrical Engineering and IEEE Industrial and Commercial Power Systems Europe (Italy, June 2018)* 1–5

[63] Şengür D and Turhan M 2018 Prediction of the action identification levels of teachers based on organizational commitment and job satisfaction by using *k*-nearest neighbors method *Fırat Univ. Turkish J. Sci. Technol.* **13** 61–8

[64] Şengür D and Tekin A 2013 Prediction of student's grade point average by using the data mining methods *Int. J. Inform. Technol.* **6** 7–16

IOP Publishing

Modelling and Analysis of Active Biopotential Signals in Healthcare, Volume 1

Varun Bajaj and G R Sinha

Chapter 9

Empirical wavelet transform based classification of surface electromyogram signals for hand movements

Anurag Nishad[1], Abhay Upadhyay[2]

[1]*Department of Electrical and Electronics Engineering, BITS Pilani, KK birla Goa Campus, 403726, India*

[2]*Department of Electronics and Communication Engineering, Institute of Engineering and Technology Bundelkhand, Jhansi 284003, India*

A hand amputee patient will require a prosthetic arm to perform basic hand movements. A prosthetic arm which operates according to input electromyogram (EMG) signals recorded from the surface of the forearm is comfortable to use. However, the prosthetic arm must perform the correct hand movement task when it receives the input surface EMG (sEMG) signals. Therefore, the controller in the prosthetic arm must be able to classify different hand movements when the input sEMG signals are received. In this chapter, we have proposed an algorithm for the classification of six basic hand movements. Our methodology is tested on sEMG signals recorded from five people. For each hand movement, the sEMG signals are recorded from two channels. These sEMG signals are decomposed using the empirical wavelet transform (EWT). In the next step, the cross information potential (CIP) features are computed from the EWT coefficients of the decomposed sEMG signals. Then, these features are ranked and fed into the classifier for classification. For feature ranking, three methods, namely the local learning-based clustering feature selection (LLCFS), multi-cluster feature selection (MCFS) and RELIEFF methods, are tested in the proposed model. Similarly, in the classification stage, the naïve Bayes (NB), *k*-nearest neighbors (*k*-NN), bootstrap aggregating (BA) and random forest (RF) classifiers are tested. Through simulation, it is observed that RELIEFF for feature ranking and RF for classification are the most effective in the proposed model. This combination in the proposed model achieves an average accuracy (ACC) of 87.44%. The result suggests that our proposed method should be tested on a larger dataset for further validation.

9.1 Introduction

There are six primary motions of the human hand which a hand amputee can perform using a specially designed prosthetic hand (PH) [1]. When designing a PH, the objective is to maintain the precise functioning of the PH based on the input physiological signals. In the literature, the physiological signals suggested for the input of the PH are electromyogram (EMG) signals or electroencephalogram (EEG) signals. EMG and EEG signals represent the electrical activity of the muscles and the brain, respectively [2, 3]. The signal can be input to the PH through multiple channels or electrodes. Between EMG and EEG, PHs controlled by EMG signals are preferred. This is because EMG signals can be recorded from the forearm by the electrodes present inside the PH. Such designed PHs are comfortable in use [4]. In the case of a PH controlled by EEG signals, the recording is done from the scalp, which is not comfortable to use [4].

The EMG signals are usually recorded from the surface of the forearm since the characteristics in the EMG signal differ for different muscle movements [1, 3]. The recorded signals from the surface of the forearm are called surface EMG (sEMG) signals. The reason for choosing the forearm to record sEMG signals is that the muscles are in the rest position while performing different hand movements [5–7].

The objective here is to design a PH that can classify different hand movements accurately based on the acquired sEMG signals [8]. However, the input sEMG signals may be acquired from many electrodes [9]. Having many electrodes in a PH should be avoided as it would be costly and uncomfortable in use [1]. To design an accurate PH, research on biosensors, pattern recognition methods and advanced biosignal processing is available in the literature [10, 11].

In the literature, methods for the classification of different numbers of hand movements have been proposed. In [12], eight hand movements were classified using the wavelet thresholding technique. The wavelet threshold method was applied to determine the sEMG signals. The wavelet parameters were first determined using only three hand movements. The bi-orthogonal wavelet, six decomposition levels and SURE thresholding were selected as the proper parameters. Then, using these parameters, the proposed model was tested to classify eight hand movements [12]. A method for a four-class classification problem was proposed in [13]. A neuro-fuzzy interference system was proposed to classify four hand grasp movements, namely free, ball, can and dumbbell [13]. The authors in [13] first examined a recurrence plot based nonlinear estimate to compute the dynamic properties of the sEMG signals which are usually hidden. Then through experimentation, it was found that the application of recurrence quantification analysis (RQA) discriminates the sEMG signals effectively. Also, it is shown in [13] that the classification result improved when both the RQA and linear measures were applied together in the proposed model. In [14] models based on a fuzzy Gaussian mixture were presented to classify ten hand movements. These models were developed using Willison amplitude and determinism [14]. During experimentation, the authors in [14] applied three sensors together in order to determine the sEMG signals, the angle trajectory of the fingers and the contact forces by the hand. Then the correlation of information from different sensors was examined and it was shown that there was an important relation between the trajectories of the finger and the signals of the muscle. Similarly, the relationship between the contact forces and muscle signals was significant as well [14].

Similarly, in [1, 3] the sEMG signals acquired from two electrodes were decomposed into intrinsic mode functions (IMFs) through empirical mode decomposition (EMD). The authors in [1, 3] addressed the six hand movements classification problem. In [1] features such as the Willison amplitude, skewness, kurtosis, integrated EMG, variance, zero-crossing, waveform length and slope sign change were used. Then, a simple linear classifier was employed to classify the sEMG signals corresponding to six hand movements. In [3], features such as the median, standard deviation (SD) and kurtosis of the instantaneous frequencies of the IMFs, in addition to the features used in [1], were applied for hand movement classification. Then, after computing the features, principal component analysis (PCA) for a reduction in dimensionality and the RELIEFF method for feature ranking were tested [3]. For classification, a simple classifier was applied [3].

Similarly, in [15–21] the six-class classification problem is addressed. In [15], the classifications of six classes, namely power, pinch, palm-up, precision, oblique and hook, were studied. The features proposed were time/frequency based [15]. The dimensionality of these features was reduced through PCA. In the last stage, the radial basis function (RBF) kernel support vector machine (SVM) was applied to classify the six classes [15]. In [16], the classification problem of six hand movements was based on a singular value decomposition (SVD) applied on the sEMG signals and then PCA. The features were extracted from the sub-frame of the sEMG data [16]. The singular values computed from the sub-frames and the variance and mean (MN) of the top five principal eigenvectors were used as features [16]. In the classification stage, the authors applied the hierarchical *k*-nearest neighbors (*k*-NN) classifier [16]. The hand movements belonging to six classes were first classified into two classes [16]. The two classes were defined based on the engagement of the motor unit and neural firing [16]. After classifying the hand movements into two classes, the hand movements in each class were further classified into three classes [16].

In [17] the energy histogram correlation obtained from each data electrode was applied as a feature to classify six hand movements. For the classification stage, a classifier based on a cascaded structure was used [17]. SVD was also applied in [18] to extract the features for a six-class classification problem, after which many classifiers were tested, namely the *k*-NN, naïve Bayes (NB), decision tree, sequential minimal optimization (SMO) and radial basis function classifiers. SMO produced the best result in this work. Recently, in [19] six hand movements were classified by acquiring sEMG signals from three electrodes and decomposing them through variational mode decomposition (VMD). Here, the composite permutation entropy index (CPEI) features were computed and after ranking, three classifiers, namely NB, *k*-NN and bootstrap aggregating (BA), were tested. In [20], six hand movements were classified by applying the discrete wavelet transform (DWT) and wavelet neural network (WNN) methods on sEMG signals recorded from three sensors.

Another work where the wavelet was applied is [21]. In this work, a tunable-Q wavelet transform (TQWT) based filter bank was used to decompose the cross-covariance of the sEMG signals. The TQWT based filter bank consisted of many constant bandwidth sub-bands obtained from the different values of the quality (Q) factor [21]. The decomposed sub-band signals from the TQWT based filter bank had a constant bandwidth but different oscillatory natures [21]. From the decomposed sub-band signals, Kraskov entropies were computed and ranked. In the last stage [21], the k-NN classifier was used to classify six basic hand movements.

More advance classification problems are addressed in [22–24]. In [22], the classification of hand, wrist and the finger movements is presented. The authors applied windowing techniques and multi-layer perceptron (MLP) to classify 52 classes. A similar problem was tackled in [23], where the variance, waveform length, MN absolute value, cepstral coefficients, histogram of sEMG, marginal discrete wavelet transform and short-time Fourier transform were applied as the features in the proposed method. In the classification stage, the SVM produced the most consistent result. In [24], the classification problem of 52 classes was again addressed. Time-domain features, namely zero crossings, slope sign change, mean absolute value and waveform length, were used. Then, in the classification stage, linear discriminant analysis (LDA) and an artificial neural network (ANN) were tested. Through simulation, it was shown that the performance of the ANN classifier was better than that of LDA.

In this chapter, we address the six-class classification problem. Here, the classification of six basic hand movements from sEMG signals being presented. The sEMG signals used in this method are acquired from two electrodes. Therefore, the proposed method is suitable for developing low cost and comfortable PHs. In the proposed method, the sEMG signals acquired from both electrodes are decomposed separately by empirical wavelet transform (EWT). Through decomposition, we obtain the EWT coefficients for both sEMG signals. Then these EWT coefficients are used to compute cross information potential (CIP) features. After this, these computed features are ranked and then classified. For feature ranking, three methods are tested. Similarly, in the classification stage, four classifiers are tested. Through simulation, the best performance of the proposed model is obtained when RELIEFF method is used for ranking and random forest (RF) classifier is applied for classification. The efficacy of the proposed model is tested in terms of average accuracy (ACC) since the proposed model is tested on data from five people. The ACC is defined as the number of correctly identified instances out of the total number of instances.

The rest of this chapter is organized as follows. The dataset used in this work is explained in section 9.2. Then an overview of EWT is presented in section 9.3 and the proposed model is described in section 9.4. The simulation result and discussion are presented in sections 9.5 and 9.6, respectively. Finally, this chapter is concluded in section 9.7.

9.2 Dataset

In this work, we have used a basic hand movement dataset [1]. The dataset is available online at https://archive.ics.uci.edu/ml/datasets/sEMG+for+Basic+Hand

+movements#. The dataset is a collection of sEMG signals obtained from five people (P1, P2, P3, P4 and P5). Two of them are male and the rest female, and their age at the time of recording was between 20 to 22 years. In this work, P1, P2, and P3 represents the first female, second female, and third female, respectively, whose data in the form of sEMG signals are provided in the dataset. Similarly, P4 and P5 represent the first male and second male, respectively, whose data are given in the dataset.

The sEMG signals were recorded at a 500 Hz sampling rate and with the help of two electrodes, namely electrode 1 and electrode 2, which were placed on the forearm of the subject. Hence for each recording, a pair of sEMG signals is available.

Each person performed six hand movement tasks during the recording of the sEMG signals. The performed hand movement tasks are cylindrical (CY), hook (HO), lateral (LA), palmar (PA), spherical (SP) and tip (TI). The pictorial representation of these hand movements is shown in figure 9.1. Also, example sEMG signals corresponding to each hand movement are shown in figure 9.2. While performing these hand movement tasks, objects of different sizes and weights were held. Further detailed information on this dataset is available in [1].

During the recording of the sEMG signals, each person performed each hand movement task 30 times. Therefore, each person performed 180 hand movement tasks in total. Each task was performed for 6 s. Hence the recorded signal consists of 3000 samples. The amount of force applied and the speed to perform a task depend on the individual.

9.3 Overview of empirical wavelet transform

The signal decomposition techniques proposed in the literature are either rigid or adaptive. The rigid methods, such as wavelets and Fourier transforms, apply basis functions which are designed independent of the nature of the signal. On the other hand, the basic functions of adaptive methods are developed according to the nature of the signal. The EWT falls into the category of adaptive decomposition techniques

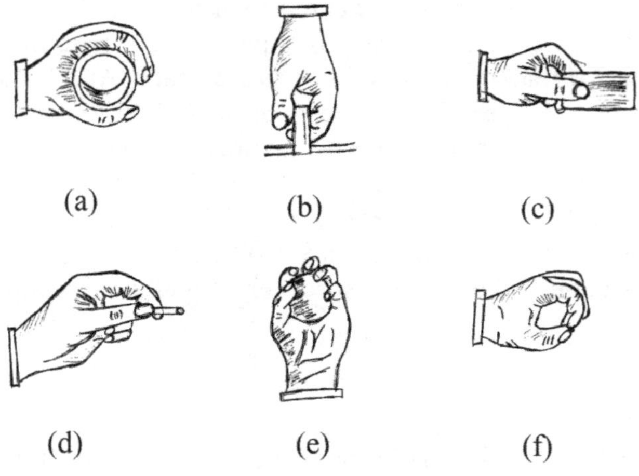

(a) (b) (c) (d) (e) (f)

Figure 9.1. The basic hand movements classified in this work: (a) CY, (b) HO, (c) LA, (d) PA, (e) SP and (f) TI.

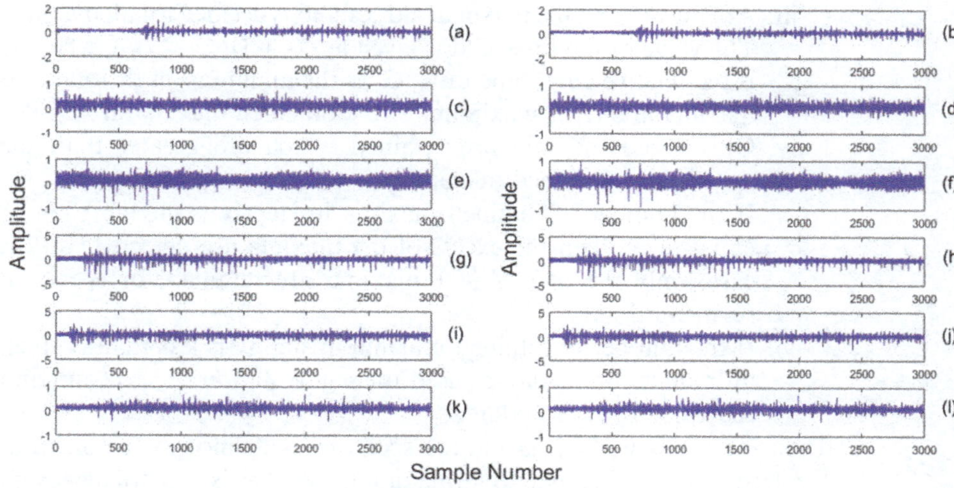

Figure 9.2. The sEMG signal corresponding to hand movement (a) CY recorded from electrode 1, (b) CY recorded from electrode 2, (c) HO recorded from electrode 1, (d) HO recorded from electrode 2, (e) LA recorded from electrode 1, (f) LA recorded from electrode 2, (g) PA recorded from electrode 1, (h) PA recorded from electrode 2, (i) SP recorded from electrode 1, (j) SP recorded from electrode 2, (k) TI recorded from electrode 1 and (l) TI recorded from electrode 2.

for non-stationary signals. It was introduced in [25]. The EWT method generates adaptive wavelets in order to extract various modes of the signal. The extracted modes of the signals are compact in the Fourier spectrum and hence these modes can also be termed as amplitude modulated (AM)–frequency modulated (FM) components.

The steps involved in EWT based decomposition of the signal are as follows [25]:

1. The EWT method begins with the computation of the Fourier transform of an input signal in order to obtain the spectrum of the signal in the range from 0 to π.
2. The next step corresponds to the segmentation of the obtained Fourier spectrum. In this step, each segment of the spectrum will produce an AM–FM component which is centered over a particular frequency. In order to achieve this goal, the following algorithm is used.
    a. First, it is assumed that the spectrum will be divided into N number of segments [25]. Then the $N + 1$ boundary is required for segmentation. Since 0 and π are also boundaries present at the extremes of the spectrum, the algorithm will have to compute $N - 1$ boundaries.
    b. After assuming N segments, there is an algorithm in the EWT to find the actual number of segments. This algorithm works as follows:
        i. The algorithm first detects the G number of significant maxima in the spectrum of the signal.
        ii. Then these G maxima (G_1, G_2, … G_G) are arranged in decreasing order such that $G_1 \geqslant G_2 \geqslant \ldots \geqslant G_G$ and are then normalized between 0 and 1.

iii. After arranging the maxima, a threshold is decided for all maxima. The value of the threshold is given as $G_G + \Omega(G_1 - G_G)$, where Ω is the relative amplitude ratio, G_G is the minimum of the obtained maxima and G_1 is maximum of the obtained maxima in step i.
iv. Only those maxima are retained which are greater than the threshold mentioned in step iii.
v. If the number of retained maxima in step iv is greater than or equal to N, then the excess of the maxima are detected. In this case, only the first $N - 1$ maxima are retained for spectrum segmentation.
vi. If the number of retained maxima in step iv is less than N, then all retained maxima are used for segmentation of spectrum and the value of N is updated.

c. Then the nth boundary, ω_n, in the spectrum is defined as the mean of the frequencies of consecutive maxima. Here, ω is the frequency in radians per second.

3. For segments in the spectrum, the empirical wavelets-based set of band-pass filters is defined. To implement filters, the scaling function ($\zeta(\omega)$) and wavelet functions ($\xi(\omega)$) are used. The concept of Littlewood–Paley and Meyer's wavelets is applied to obtain filters based on wavelets [25, 26]. Mathematically, $\zeta_n(\omega)$ and $\xi_n(\omega)$ for the nth boundary are expressed as [25]:

$$\zeta_n(\omega) = \begin{cases} 1, & \text{if } |\omega| \leqslant (1-\alpha)\omega_n \\ \cos\left(\frac{\pi}{2}\mu\left(\frac{1}{2\alpha\omega_n}(|\omega| - (1-\alpha)\omega_n)\right)\right), & \text{if } (1-\alpha)\omega_n \leqslant |\omega| \leqslant (1+\alpha)\omega_n \\ 0, & \text{otherwise} \end{cases} \tag{9.1}$$

$$\xi_n(\omega) = \begin{cases} 1, & \text{if } (1+\alpha)\omega_n \leqslant |\omega| \leqslant (1-\alpha)\omega_{n+1} \\ \cos\left(\frac{\pi}{2}\mu\left(\frac{1}{2\alpha\omega_{n+1}}(|\omega| - (1-\alpha)\omega_{n+1})\right)\right), & \text{if } (1-\alpha)\omega_{n+1} \leqslant |\omega| \leqslant (1+\alpha)\omega_{n+1} \\ \sin\left(\frac{\pi}{2}\mu\left(\frac{1}{2\alpha\omega_n}(|\omega| - (1-\alpha)\omega_n)\right)\right), & \text{if } (1-\alpha)\omega_n \leqslant |\omega| \leqslant (1+\alpha)\omega_n \\ 0, & \text{otherwise} \end{cases} \tag{9.2}$$

Here, the arbitrary function $\mu(x)$ is expressed as [25]

$$\mu(x) = \begin{cases} 0, & \text{if } x \leqslant 0 \\ \text{and } \mu(x) + \mu(1-x) = 1 & \\ 1, & \text{if } x \geqslant 1 \end{cases}. \tag{9.3}$$

The parameter α in equations (9.1) and (9.2) ensures that $\zeta(\omega)$ and $\xi(\omega)$ provide a tight frame within $L^2(R)$. The condition for the tight frame is expressed as [25]:

$$\alpha < \min_n\left(\frac{\omega_{n+1} - \omega_n}{\omega_{n+1} + \omega_n}\right). \tag{9.4}$$

4. After obtaining $\xi(\omega)$ and $\zeta(\omega)$ from equations (9.1) and (9.2), the detail coefficients ($\rho_\xi(n, t)$) and approximate coefficients ($\rho_\zeta(0, t)$) of wavelets are determined. $\rho_\xi(n, t)$ and $\rho_\zeta(0, t)$ are determined from the inner product of signal $z(t)$ with $\xi(\omega)$ and $\zeta(\omega)$, respectively. The mathematical expressions to compute coefficients for signal $z(t)$ are as follows [25]:

$$\rho_{z,\xi}(n, t) = \int z(\tau)\overline{\xi_n(\tau - t)} \tag{9.5}$$

$$\rho_{z,\zeta}(0, t) = \int z(\tau)\overline{\zeta_1(\tau - t)}. \tag{9.6}$$

The EWT has been used in the literature for the analysis of non-stationary signals. In [27, 28] the EWT is applied for the detection of seizures from EEG signals. Also, the EWT has been applied for removing ocular artifacts from EEG signals. The EWT has also been applied for decomposing non-stationary signals in order to obtain the time–frequency representation of the signal. In [29] a new time–frequency plane based on the EWT and Fourier–Bessel series expansion (FBSE) representation is proposed. Similarly, in [30] the EWT is used to reduce the cross-terms in the Wigner–Ville distribution (WVD) of a non-stationary signal.

9.4 The proposed method

The method proposed for the classification of basic hand movements consists of four stages. These are signal decomposition, feature extraction, feature ranking, and classification. These four stages are shown in the form of a block diagram in figure 9.3.

Each stage of the block diagram is explained in the following sub-sections.

9.4.1 EWT based decomposition

For each hand movement task, the dataset provides sEMG signals from two electrodes, namely electrode 1 and electrode 2. Hence the data provided by the dataset are bivariate in nature. As shown in figure 9.3, the first stage of the proposed model is EWT decomposition. The sEMG signals from the two electrodes are decomposed separately by the EWT. For decomposition, the parameters of the EWT are set as described in section 9.5. The spectrum of the sEMG signal is segmented into several bands through the EWT decomposition. The EWT

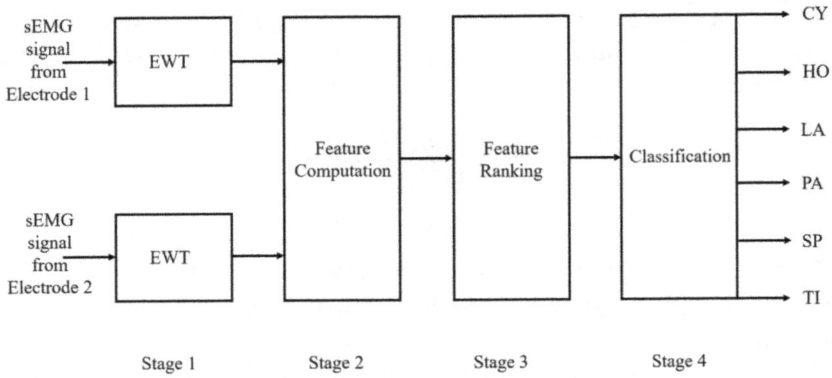

Figure 9.3. Block diagram of proposed model for classification of basic hand movements.

coefficients obtained from equations (9.5) and (9.6), from each segmented band, are then fed into the feature computation stage.

9.4.2 Feature computation

In the proposed model, the features are computed through the cross information potential (CIP). The EWT coefficients obtained from the EWT decomposition are used for the computation of the CIP. The term information potential (IP) in CIP computes Renyi's quadratic entropy. Mathematically, it can be expressed as [31]:

$$\mathrm{IP}(E_i) = \frac{1}{C^2}\sum\nolimits_{ix=1}^{C}\sum\nolimits_{iy=1}^{C} F(E_{ix} - E_{iy}). \tag{9.7}$$

Equation (9.7) shows the expression to compute IP from EWT coefficients (E_i) of the ith band of the segmented spectrum of a signal. There are a total of C coefficients and F is the kernel function whose argument is the difference between E_{ix} and E_{iy}. E_{ix} and E_{iy} are the xth and yth EWT coefficients of E_i, respectively. The CIP computes the closeness between probability distributions [32]. Hence, the CIP computes the similarity between two sets of EWT coefficients E_i and E_j. Here, E_i and E_j are the EWT coefficients obtained from the ith and jth segmented band, respectively. It is mathematically expressed as [31]:

$$\mathrm{CIP}(E_i,\ E_j) = \frac{1}{C^2}\sum\nolimits_{ix=1}^{C}\sum\nolimits_{iy=1}^{C} F(E_{ix} - E_{jy}), \tag{9.8}$$

where E_{ix} and E_{iy} are the xth and yth EWT coefficients of E_i and E_j, respectively. In the proposed method, the CIP is computed from the EWT coefficients of the ith band of the segmented spectrum of the sEMG signal obtained from electrode 1 and the EWT coefficients of the ith band of the segmented spectrum of the sEMG signal obtained from electrode 2. Therefore in equation (9.8) it may seem that we could replace $i = j$, since we are considering only the ith band of both spectrums and therefore equation (9.8) would be converted into equation (9.7). This is an incorrect conclusion. Hence, we must replace E_i and E_j by E_{i1} and E_{i2}, respectively, in equation (9.8). Here, E_{i1} and E_{i2} represents the EWT coefficients obtained from the ith band of the segmented spectrum of the sEMG signal obtained from electrode 1 and electrode 2, respectively.

The CIP has been used for feature computation in the literature. In [33], the CIP was used as a feature to classify seizure, seizure-free, and normal EEG signals. The CIP was applied to the sub-band signals obtained after the decomposition of EEG signals by a TQWT based filter bank. Similarly, in [34], the CIP was applied on the coefficients of a flexible analytic wavelet transform (FAWT) to compute features to classify coronary artery disease (CAD) electrocardiogram (ECG) signals from normal ECG signals. In [35], the CIP feature, along with other features, was used to classify normal and amyotrophic lateral sclerosis (ALS) EMG signals.

9.4.3 Feature ranking

Among the extracted features, some have more significance than others. Therefore, the features extracted in the previous stage need to be ranked according to their significance level. By choosing a few significant features, good classification ACC can be achieved. In this work, we have tested the following feature ranking methods:

a. *Local learning-based clustering feature selection (LLCFS).* The LLCFS method, as proposed in [36], first assigns a weight to each feature. Then the features with assigned weights are added in local learning-based clustering (LLC) in order to find the importance of every feature in the clustering. This process is repeated in the form of several iterations. Therefore, the assigned weights are also computed in each iteration. In the end, the weights corresponding to insignificant features will approach zero [36].
b. *Multi-cluster feature selection (MCFS).* This method was proposed in [37] and computes an MCFS score for each feature for ranking. The features in this method are selected in such a way that the multi-cluster pattern of data does not change. The method then determines the solution of an optimization problem containing a sparse eigen and L1 regulation least squares problem in order to compute the MCFS score.
c. *The RELIEFF method.* The RELIEFF method was proposed in [38, 39]. In this method, if a feature is capable of distinguishing close instances, then the feature is considered as significant. The method first selects some random instances, and then it searches k-NNs of the same class and k-NNs from different classes [38, 39]. After this, a quality parameter is computed from k-NNs of similar and different classes and selected instances. The value of the quality parameter determines the significance of the class. In this chapter, k_r represents the variable k in the k-NN used in RELIEFF.

These feature ranking methods were used in [21] for the classification of hand movements. In [21], these ranking methods were applied to the computed Kraskov entropies, which were used as features. In addition, the RELIEFF method was also applied in [33] on CIP features for the classification of epileptic EEG signals.

9.4.4 Classification

Classification is the last stage of the proposed method to classify six different hand movements. In this work, we have tested the following classifiers:

a. *Bootstrap aggregating (BA)*. This classifier was proposed in [40] and is also referred to as bagging. In this method, a training set is formed whose size is the same as that of the original instances. That is, if there are T original instances, then the size of the training set would also be T. However, there may be some instances in the training set which are present multiple times, and there may be instances that are not present in the training set. After generating a training set, a weak classifier is developed by the learning system which performs the classification. This entire process is repeated several times and then a strong classifier is generated from the combination of weak classifiers.
b. *k-nearest neighbors (k-NN)*. The nearest neighbors algorithm is the basis for this classifier [41]. Its performance depends on two factors: k neighbors and the distance. The classification of the test data depends on their *k-NN* which act as a training set. The k neighbors are determined from the given distance computation method. This classifier is computationally fast.
c. *Naïve Bayes (NB)*. In [42] this classifier was proposed, which applies Bayes' theorem for classification. This classifier makes use of probabilistic knowledge for classification. The algorithm in the classifier considers that no hidden attribute exists which can change the process of prediction [42].
d. *Random forest (RF)*. The performance of the RF classifier as proposed in [43] depends on several classification trees. These trees are produced through a random tree algorithm [44]. Each generated tree performs the classification task. Then the overall classification result depends on the classification performance of each tree. For classification, a random vector is allotted to each tree. The allotted vectors produce different trees that have similar distributions and are independent of each other. Let say that r_i is the vector allotted to the ith vector. Then a margin function (M) for the training set, selected randomly from a distribution of random vectors X, Y, is created in order to determine the class of the instance. The expression to compute M is [43]

$$M(X,\ Y) = AV(I[TC(x,\ r_i) = Y]) - \max\{AV(I[TC(x,\ r_i) = j])\}, \tag{9.9}$$

where x is the training data, $TC(x,\ r_i)$ is a tree classifier, AV is the average, $I[.]$ is the indicator function and $j \neq Y$ in [43]. If the value of M is greater, then it indicates better classification ACC.

These above-mentioned classifiers have been applied for classifying physiological signals in the literature. The k-NN classifier was used in [21] to classify basic hand movements. Also, k-NN was applied in [45, 46] to classify focal and non-focal EEG signals. Similarly, the RF classifier was used in [33] to classify epileptic EEG signals. In [47], the BA and RF classifiers were tested in a model which classifies apneic and non-apneic ECG segments. Also, in [48] the BA, k-NN, NB, and RF classifiers and some other classifiers were used to screen for sleep apnea. In [49], different sleep stages were classified where NB, k-NN, RF classifiers along with some other classifiers are tested.

9.5 Simulation results

In this work, the proposed model is tested separately on the data of P1, P2, P3, P4 and P5. The sEMG signal is first decomposed by the EWT. The sEMG signals are available from electrode 1 and electrode 2. Hence, the sEMG signals from both electrodes are decomposed separately by the EWT, as shown in figure 9.3. The decomposition is performed in MATLAB using the toolbox of the EWT [25]. To perform the decomposition, the sEMG signal along with some parameters of the EWT are required as input to the MATLAB function of the EWT. These parameters are global trend remover, regularization, detection, maximum number of bands, equidistant partitioning, detection on log spectrum and sampling frequency [50].

The global trend remover parameter determines the pattern in the spectrum globally and then removes it [50]. By default, this parameter is set to 'none' which deactivates this parameter. The regularization parameter can be applied when the signal is contaminated by noise [50]. This parameter is also disabled by setting the value of this parameter equal to 'none'. The detection parameter selects the algorithm for detecting boundaries, modes, maxima and/or minima in the spectrum [50]. In this work, this parameter is set to 'scalespace'. This 'scalespace' setting determines the significant modes in the spectrum under scalespace representation [50].

The next parameter is N as defined in section 9.3. This parameter is a positive integer which indicates the maximum number of bands into which the spectrum of the signal should be decomposed. In this work, this parameter is set to fifty. The equidistant partitioning parameter is set to zero [50]. This parameter is required for equidistant partitioning of the spectrum if the number of maxima detected by the EWT algorithm is less than the maximum predicted bands. The next parameter is detection on the log spectrum. This parameter decides whether the boundaries of the spectrum will be detected in a log scale or not. It is set to zero [50] in this work. The last parameter is sampling frequency, which is set to 500 Hz, as mentioned for the database. The values of the EWT parameters are kept the same for decomposing the data from all five subjects.

After decomposing the sEMG signals by setting the EWT parameters, the next stage in the proposed work is to extract the features. The EWT based decomposition may result in different numbers of segmented bands in the spectrum for different signals. This is because, in the EWT decomposition stage, we have set the maximum number of bands parameter as equal to 50. However, the actual number of bands into which signal is decomposed may differ from the value set for the maximum number of bands. As mentioned in section 9.3, the boundaries of the bands in the EWT are formed based on the number of maxima detected in the spectrum. This is because the EWT aims towards extracting modes centered around a particular frequency. The number of maxima detected by the EWT may differ from the value of the maximum number of bands [25]. Due to this reason, the different signals may decompose into different numbers of bands since their spectrum may have a different number of maxima and characteristics.

Hence, the spectrum of sEMG signals obtained from electrode 1 and electrode 2 of the same pair may be decomposed into different bands. Similarly, if we examine all the sEMG signals from all the classes in the dataset, then their numbers of segmented bands obtained after decomposition may vary. Let us say B_i is the number of bands into which the spectrum of the ith sEMG signal is decomposed. Also, let B be the minimum of all obtained B_i. Then for all sEMG signals, only the first B bands from their spectrum are used for feature computation. If we do not apply this rule, then the number of features corresponding to a class may be more than the number of features corresponding to another class. In such a case, it would be difficult for the classifier to classify different classes. When the data of P1, P2, P3, P4, and P5 are considered in the proposed model, then the obtained values of B are 19, 19, 15, 17 and 17, respectively.

To compute the features, the CIP is used, as mention in section 9.4.2. The CIP is implemented using the ITL toolbox (http://www.sohanseth.com/Home/codes). The EWT coefficients of the jth band from the spectrum of the sEMG signal of electrode 1 and the EWT coefficients of the jth band from the spectrum of the sEMG signal of electrode 2 are used to compute the CIP feature. Here, the $1 \leqslant j \leqslant B$ and sEMG signals from both electrodes belong to the same pair. Therefore, B number of features are available and the feature ranking method will rank these B features.

In the feature ranking stage of the proposed model different ranking methods are tested, as mentioned in section 9.4.3. These ranking methods are implemented through a feature selection library [51]. These ranked features are given as input to the classification stage, which is the last stage of the proposed method. In the classification stage different classifiers are tested, as mentioned in section 9.4.4. These classifiers are available from the Waikato environment for knowledge analysis (WEKA) [52] software. For classification, the best rank feature is first fed in the classifier and then the obtained ACC is recorded. Then, the two best ranked features are fed into the classifier and the ACC is recorded and so on. In this way, the number of ranked features (NRF) is different in each iteration. In the last iteration, all B ranked features are used in the classification. During classification, 'leave one out' cross-validation is used [53]. In this type of cross-validation, if there are D data points, then $D - 1$ data points will train the classifier and the remaining one data point will be the test point for the classifier. Then in the next iteration another data point will be the test point for the classifier and the remaining $D - 1$ data points will be the training set. In this way, the classifier will perform classification tasks for D iterations and no data point will be repeated as a test point more than once.

Tables 9.1–9.5 show the obtained ACC in percentage from the proposed model for different subjects. In these tables, we present the results of ACC obtained for each NRF. Among them, we select the maximum ACC and consider it as the ACC achieved by a particular classifier. Also, these tables show the ACC (%) when different combinations of feature ranking methods and classifiers are used in the proposed model. Since we have tested three feature ranking methods and four classifiers, a total of 12 combinations or cases are possible.

Table 9.1. ACC (%) obtained through the proposed method for P1 data.

	BA			*k*-NN			NB			RF		
NRF	LLCFS	MCFS	RELIEFF	LLCFS	MCFS	RELIEFF	LLCFS	MCFS	RELIEFF	LLCFS	MCFS	RELIEFF
1	50.55	48.33	45	46.67	39.44	40.56	48.88	50	43.33	47.22	45.55	41.66
2	50	48.33	52.77	45.56	46.67	46.67	52.22	53.88	54.44	48.88	48.88	52.77
3	55.55	58.33	54.44	53.89	51.67	52.22	59.44	66.66	59.44	53.33	62.77	56.66
4	57.22	61.66	53.88	54.44	52.22	57.22	61.11	70.55	67.77	53.33	65	56.11
5	52.22	63.33	52.77	54.44	54.44	58.89	60	71.66	70	58.33	65.55	60.55
6	57.77	68.88	60	58.89	54.44	60.00	62.22	73.88	70.55	58.33	73.33	64.44
7	57.22	**72.77**	58.33	57.22	54.44	58.33	66.66	75.55	70	60	72.77	65.55
8	57.77	67.22	58.88	56.67	52.78	58.89	67.77	75	68.88	59.44	76.66	63.88
9	58.33	70.55	59.44	61.11	52.22	61.11	68.33	77.22	72.77	61.11	73.33	67.77
10	56.66	70.55	65.55	59.44	58.33	**63.89**	70	78.33	70.55	58.88	76.11	68.33
11	62.22	71.11	67.22	63.89	61.11	63.89	73.88	79.44	71.66	65.55	78.33	70.55
12	62.22	69.44	66.66	64.44	58.89	62.78	74.44	77.77	72.77	64.44	76.66	70.55
13	61.66	68.88	64.44	65.56	58.89	63.33	73.33	78.33	68.88	66.66	78.88	72.22
14	65	69.44	64.44	64.44	60.56	62.78	72.22	80	70	66.66	77.22	73.88
15	61.66	70.55	65.55	**67.22**	61.67	63.33	71.11	79.44	72.77	63.33	**79.44**	68.33
16	62.22	68.33	63.33	65.56	61.67	62.22	72.22	80.55	72.22	66.11	78.33	68.33
17	62.77	68.88	63.88	66.67	61.67	62.22	70.55	**81.11**	70.55	64.44	75.55	72.77
18	63.33	70	62.77	62.22	**63.33**	62.22	71.11	80.55	71.11	73.88	76.66	65.55
19	**68.88**	69.44	**70**	62.22	62.22	62.22	**79.44**	79.44	**79.44**	**80.55**	78.33	**80**

Table 9.2. ACC (%) obtained through the proposed method for P2 data.

	BA			*k*-NN			NB			RF		
NRF	LLCFS	MCFS	RELIEFF	LLCFS	MCFS	RELIEFF	LLCFS	MCFS	RELIEFF	LLCFS	MCFS	RELIEFF
1	65.55	33.88	50.55	48.33	26.67	42.22	56.66	38.33	55	51.11	33.33	40.55
2	65.55	62.22	56.11	59.44	49.44	59.44	63.33	64.44	63.88	57.77	58.33	58.88
3	68.33	72.22	63.88	65.56	65.00	62.78	66.11	75.55	68.88	66.11	75	63.88
4	70.55	75	62.22	67.78	70.56	67.22	75	85	68.88	70	77.77	66.11
5	68.33	75	74.44	72.22	72.22	75.00	77.77	81.11	80.55	69.44	76.11	75
6	70.55	73.33	73.88	71.67	70.00	75.00	79.44	80.55	81.66	71.66	76.11	77.77
7	65.55	**77.77**	77.22	71.11	80.00	77.78	77.22	**86.11**	82.22	72.22	82.77	80.55
8	70.55	76.11	72.77	74.44	81.11	81.67	77.77	83.88	84.44	75	82.77	80
9	66.66	75	72.77	73.89	80.00	82.78	78.88	82.77	84.44	72.22	83.88	77.77
10	72.77	77.22	77.77	72.78	81.11	**83.89**	81.11	81.66	84.44	76.66	83.33	80.55
11	69.44	76.66	76.11	73.89	81.11	81.11	79.44	83.88	85	77.22	82.22	81.66
12	70	76.11	71.66	76.67	82.78	82.78	80.55	85.55	84.44	79.44	81.66	82.22
13	73.33	76.66	72.22	77.78	81.67	79.44	81.11	86.11	85	81.66	**87.22**	82.22
14	74.44	76.66	74.44	78.89	82.78	76.11	81.11	86.11	85	82.22	84.44	82.77
15	75.55	75	77.77	78.33	83.33	77.78	84.44	85.55	83.88	82.22	85	**87.22**
16	75.55	76.11	79.44	79.44	**83.89**	81.67	85	85.55	**86.11**	**86.66**	84.44	83.88
17	**78.33**	76.11	**82.22**	80.00	81.67	82.22	**85.55**	83.88	85.55	85.55	85.55	84.44
18	78.33	77.22	82.22	**81.11**	81.67	81.67	85	84.44	85.55	85.55	85	85
19	78.33	77.22	79.44	81.11	81.11	81.11	85.55	85.55	85.55	84.44	85.55	85

Table 9.3. ACC (%) obtained through the proposed method for P3 data.

	BA			k-NN			NB			RF		
NRF	LLCFS	MCFS	RELIEFF	LLCFS	MCFS	RELIEFF	LLCFS	MCFS	RELIEFF	LLCFS	MCFS	RELIEFF
1	57.22	38.33	54.44	43.33	29.44	46.11	50	38.33	53.33	45	37.77	43.33
2	67.22	72.22	65.55	58.33	62.22	55.00	64.44	67.22	66.66	63.33	62.77	61.66
3	72.77	78.88	72.77	72.78	63.89	61.11	74.44	70	71.11	75.55	76.66	70
4	81.11	77.77	66.11	77.22	65.56	67.78	78.33	70	75.55	86.66	75.55	70.55
5	77.77	79.44	75.55	77.78	65.56	74.44	77.77	78.88	80.55	84.44	80	81.66
6	77.22	**81.66**	75.55	76.11	72.22	76.67	81.11	81.66	82.77	82.22	79.44	85
7	78.33	81.66	75	79.44	72.78	76.11	83.88	85	80.55	84.44	82.77	81.11
8	77.77	81.11	75.55	78.33	76.67	77.22	82.22	86.11	78.88	84.44	81.66	82.22
9	76.11	79.44	79.44	79.44	76.11	78.89	85.55	87.22	78.88	84.44	84.44	86.66
10	75.55	79.44	78.88	**80.56**	73.33	77.78	82.22	86.11	77.77	83.88	83.88	86.66
11	77.77	77.77	80.55	78.33	79.44	78.33	83.88	87.22	87.77	82.22	85.55	**87.77**
12	80.55	80.55	80.55	79.44	79.44	81.11	81.11	88.33	86.66	83.33	**86.11**	86.66
13	78.33	80	81.11	78.89	79.44	**82.78**	82.22	88.88	**91.66**	85	84.44	87.77
14	78.88	80.55	**82.22**	79.44	**80.56**	81.11	82.22	89.44	89.44	83.88	84.44	86.66
15	**84.44**	81.66	81.66	79.44	79.44	79.44	**90**	**90**	90	**88.33**	85.55	87.22

Table 9.4. ACC (%) obtained through the proposed method for P4 data.

	BA			*k*-NN			NB			RF		
NRF	LLCFS	MCFS	RELIEFF	LLCFS	MCFS	RELIEFF	LLCFS	MCFS	RELIEFF	LLCFS	MCFS	RELIEFF
1	45.55	38.88	38.88	43.33	36.11	36.11	38.33	20.55	20.55	42.77	43.88	43.88
2	66.11	50	51.66	63.89	50.56	42.78	52.22	47.77	44.44	66.66	44.44	47.77
3	72.77	52.77	58.33	76.11	55.56	50.56	65.55	47.77	53.33	80.55	56.66	55.55
4	**75.55**	71.11	62.22	77.22	66.11	58.33	70.55	61.66	65.55	78.88	74.44	68.88
5	73.33	75	63.33	78.33	67.22	62.22	74.44	63.33	64.44	77.77	76.11	68.33
6	72.22	75	67.77	79.44	68.89	63.89	76.66	65	72.77	81.11	78.33	72.77
7	72.22	77.77	68.33	78.33	70.56	72.22	75.55	66.11	75	79.44	78.33	77.22
8	73.33	76.11	75	81.11	71.67	81.11	73.33	67.77	80.55	78.88	78.88	81.11
9	73.88	**79.44**	**76.11**	81.11	79.44	81.67	74.44	76.66	78.33	82.22	81.11	83.33
10	72.22	78.88	75.55	**82.22**	78.89	83.33	73.88	73.88	79.44	**83.88**	80	**84.44**
11	73.88	76.66	75.55	80.56	79.44	**83.89**	75	75.55	80	82.22	80.55	82.77
12	72.22	76.11	73.88	80.56	80.56	83.33	75.55	76.66	80	81.11	79.44	82.77
13	71.11	75.55	73.33	80.00	78.89	81.67	76.11	78.33	78.33	81.11	82.22	82.77
14	71.66	72.77	70	79.44	80.56	82.22	76.66	**80**	77.77	81.11	82.22	83.33
15	71.11	73.33	70	79.44	**81.67**	82.78	77.77	79.44	**82.22**	82.22	**83.33**	82.22
16	69.44	72.22	72.22	78.89	78.89	81.67	77.77	78.33	79.44	82.22	82.22	82.22
17	69.44	72.77	71.66	78.89	78.89	78.89	**79.44**	79.44	79.44	82.22	81.66	82.22

Table 9.5. ACC (%) obtained through the proposed method for P5 data.

	BA			*k*-NN			NB			RF		
NRF	LLCFS	MCFS	RELIEFF	LLCFS	MCFS	RELIEFF	LLCFS	MCFS	RELIEFF	LLCFS	MCFS	RELIEFF
1	37.77	45	43.88	30.56	39.44	37.22	37.77	41.11	36.11	31.11	38.88	36.11
2	60	71.66	57.77	56.11	51.67	55.56	57.22	67.22	53.88	57.77	68.33	58.33
3	59.44	82.77	65	65.00	58.33	59.44	62.22	77.22	54.44	61.66	82.77	61.11
4	65.55	83.88	69.44	73.89	67.22	67.22	63.33	81.66	60	72.22	87.77	67.22
5	66.11	**87.77**	68.88	75.00	71.67	66.67	63.33	84.44	65	70	90	66.66
6	62.22	85.55	68.88	75.56	75.56	65.00	62.77	88.33	65.55	67.22	90.55	72.22
7	65	85	70	77.78	73.89	68.89	70.55	87.77	65.55	73.88	90.55	76.11
8	76.66	84.44	71.66	75.56	77.22	72.78	78.88	87.77	71.11	80	89.44	77.77
9	76.66	87.22	73.88	80.00	76.67	73.33	83.33	87.77	72.22	81.11	92.22	77.77
10	76.66	87.22	89.44	83.89	81.11	73.33	86.11	88.33	85	80	**93.33**	**93.33**
11	77.22	87.22	**90**	84.44	80.56	73.33	86.66	87.77	85	83.88	91.66	92.77
12	75.55	86.11	89.44	**87.78**	81.11	81.11	88.33	87.22	88.33	83.88	91.66	92.22
13	76.66	83.33	89.44	87.22	77.78	**87.22**	88.33	87.22	90.55	82.77	90.55	91.66
14	74.44	83.33	90	86.67	77.22	86.67	87.22	86.11	**91.11**	83.33	91.11	91.66
15	75.55	83.33	88.33	86.67	76.67	85.56	86.11	86.66	88.88	82.77	89.44	91.11
16	76.11	83.33	89.44	86.67	81.11	86.11	86.11	87.77	89.44	82.22	91.11	91.11
17	**87.77**	87.22	89.44	86.67	**86.67**	86.67	**90.55**	**90.55**	90.55	**91.11**	91.66	92.77

In tables 9.1–9.5, the maximum ACC achieved with the minimum NRF in each case is highlighted in bold. From these tables, it can be observed that the maximum ACC achieved by the BA and k-NN classifiers is usually lower than the maximum ACC obtained by the NB and RF classifiers. Also, in some cases, the maximum ACC achieved by the NB classifier is greater than that of the RF classifier and vice versa. For example, in tables 9.1 and 9.3, the maximum ACCs achieved by NB are 81.11% and 91.66%, respectively, which are greater than the maximum ACCs achieved by the RF classifier. In contrast, in tables 9.2, 9.4 and 9.5, the maximum ACC obtained by the RF is greater than that of the NB classifier. However, the maximum ACCs obtained by both classifiers are generally close to each other. A similar conclusion can be made for feature ranking methods as well. We can observe from tables 9.1–9.5 that there is not much difference in the maximum obtained ACC by a particular classifier when any of the three ranking methods is used.

Therefore, in order to choose the best classifier between NB and RF, we compute the average of the maximum ACC obtained by them when the RELIEFF method is used for feature ranking in the proposed model. The reason for choosing RELIEFF over MCFS is that the parameter k_r in RELIEFF can be changed. Due to changes in k_r, the ranking of features may change, which can further improve the maximum ACC obtained by the classifier. In tables 9.1–9.5, the value of k_r is set to one initially to evaluate the performance of different classifiers in the proposed model.

In this condition, the average ACCs obtained by the NB classifier and RF classifier are 86.1% and 86.55%, respectively. Hence, we consider RELIEFF for feature ranking and RF for classification in the proposed model. Then the value of k_r is changed from one to five to evaluate the performance of the RF classifier. The ACC obtained by the RF classifier for different values of k_r for the data of each subject is shown in tables 9.6–9.10.

The maximum ACC corresponding to the minimum NRF is highlighted in bold in table 9.6–9.10. This information is summarized in table 9.11, where the corresponding maximum ACC, minimum NRF and k_r are listed for the data of P1, P2, P3, P4 and P5.

For the given value of NRF and *kr* in table 9.11, the variation of the probability (p)-values of ranked features with respect to NRF is shown in table 9.12 for the data of P1, P2, P3, P4 and P5. The p-values are computed through the Kruskal–Wallis test [54].

In table 9.12, the ranked features are represented by *Rj* where *j* is the rank of the feature and NA indicates not applicable.

Also, the statistical analysis of these ranked features for the P1, P2, P3, P4 and P5 data are shown in figures 9.4–9.8, respectively.

The confusion matrices for the conditions mentioned in table 9.11 are shown in tables 9.13–9.17.

To evaluate the efficacy of the proposed model, the average of ACCs listed in table 9.11 is computed. For the proposed model, the average ACC is found to be 87.44%. The obtained ACC is greater than the ACC obtained in [16]. The authors in [16] reported an average ACC of 86.71% using the same dataset.

Table 9.6. ACC (%) at different values of k_r for P1 data.

NRF	$k_r = 1$	$k_r = 2$	$k_r = 3$	$k_r = 4$	$k_r = 5$
1	41.66	41.66	47.77	47.77	47.77
2	52.77	53.33	53.88	53.88	53.88
3	56.66	56.11	55.55	59.44	59.44
4	56.11	60	64.44	58.333	58.33
5	60.55	63.88	69.44	60	70
6	64.44	64.44	70.55	58.88	75
7	65.55	66.11	71.66	62.77	77.22
8	63.88	70	72.22	61.66	73.33
9	67.77	68.33	71.11	63.33	76.11
10	68.33	69.44	73.33	67.22	75.55
11	70.55	70.55	70.55	65.55	80
12	70.55	73.33	73.88	70.55	77.77
13	72.22	70.55	73.33	72.22	82.77
14	73.88	70.55	71.11	80.55	78.88
15	68.33	68.88	71.11	82.22	80
16	68.33	81.66	70.55	81.66	82.77
17	72.77	81.11	73.33	82.22	80
18	65.55	**83.33**	81.11	78.33	81.66
19	80	78.88	80	76.66	80

Table 9.7. ACC (%) at different values of k_r for P2 data.

NRF	$k_r = 1$	$k_r = 2$	$k_r = 3$	$k_r = 4$	$k_r = 5$
1	40.55	42.22	33.33	33.33	57.77
2	58.88	58.33	60.55	67.22	65
3	63.88	66.11	73.33	73.33	70
4	66.11	72.77	76.11	75	70
5	75	72.77	77.77	77.22	73.33
6	77.77	77.77	78.33	79.44	76.66
7	80.55	77.77	80	81.66	78.88
8	80	80	78.33	79.44	84.44
9	77.77	78.88	83.33	80.55	84.44
10	80.55	80	85.55	85.55	85.55
11	81.66	82.77	85.55	85	86.66
12	82.22	86.11	86.11	85	85.55
13	82.22	85	86.11	**87.22**	87.22
14	82.77	83.33	85.55	86.66	83.33
15	87.22	85.55	86.11	85.55	86.11
16	83.88	83.88	85	84.44	83.88
17	84.44	85	85.55	85.55	85
18	85	85	86.11	85	85.55
19	85	86.11	87.22	85	86.11

Table 9.8. ACC (%) at different values of k_r for P3 data.

NRF	$k_r = 1$	$k_r = 2$	$k_r = 3$	$k_r = 4$	$k_r = 5$
1	43.33	48.33	53.88	53.88	42.77
2	61.66	75	63.33	72.77	65.55
3	70	77.77	70.55	77.22	71.66
4	70.55	75	73.33	75.55	75.55
5	81.66	81.11	75.55	82.77	79.44
6	85	82.22	81.11	85.55	83.88
7	81.11	87.22	85.55	87.22	86.66
8	82.22	87.77	86.66	87.77	87.77
9	86.66	86.11	85.55	88.33	87.22
10	86.66	86.11	87.77	87.22	87.22
11	87.77	86.11	88.33	86.11	86.66
12	86.66	**88.88**	86.66	87.77	86.66
13	87.77	87.77	84.44	86.66	85.55
14	86.66	87.22	85	86.66	85
15	87.22	86.11	86.66	86.66	85

Table 9.9. ACC (%) at different values of k_r for P4 data.

NRF	$k_r = 1$	$k_r = 2$	$k_r = 3$	$k_r = 4$	$k_r = 5$
1	43.88	43.88	43.88	43.88	43.88
2	47.77	47.77	51.66	51.66	51.66
3	55.55	55.55	61.11	61.11	61.11
4	68.88	62.77	63.88	63.88	74.44
5	68.33	71.11	66.66	68.33	72.77
6	72.77	73.88	73.88	76.11	76.66
7	77.22	72.22	70.55	78.88	76.11
8	81.11	76.11	73.33	75	80.55
9	83.33	75	77.77	76.66	76.11
10	**84.44**	77.22	75.55	78.33	78.88
11	82.77	78.88	76.11	76.66	76.66
12	82.77	78.33	79.44	82.22	81.11
13	82.77	77.77	80	81.11	80.55
14	83.33	76.66	80	81.66	81.66
15	82.22	81.66	81.66	82.22	81.66
16	82.22	83.33	82.22	81.66	82.77
17	82.22	82.22	83.33	81.66	83.33

Table 9.10. ACC (%) at different values of k_r for P5 data.

NRF	$k_r = 1$	$k_r = 2$	$k_r = 3$	$k_r = 4$	$k_r = 5$
1	36.11	45.55	45.55	46.11	46.11
2	58.33	70.55	70.55	73.33	73.33
3	61.11	76.66	82.77	81.11	81.11
4	67.22	83.88	84.44	86.11	84.44
5	66.66	85	86.11	86.66	83.88
6	72.22	85	86.66	90.55	85
7	76.11	87.22	87.22	90.55	91.66
8	77.77	88.88	87.22	91.11	90
9	77.77	89.44	87.22	91.11	88.88
10	**93.33**	92.22	90.55	90.55	90.55
11	92.77	91.11	92.22	91.66	92.22
12	92.22	91.66	91.66	92.22	92.22
13	91.66	91.11	92.77	90.55	92.22
14	91.66	91.66	90.55	90	91.66
15	91.11	91.11	92.22	92.22	91.11
16	91.11	91.66	92.77	90.55	91.11
17	92.77	91.11	91.11	90.55	91.66

Table 9.11. Obtained average ACC (%).

Person	Maximum ACC	Minimum NRF	Minimum k_r
P1	83.33	18	2
P2	87.22	13	4
P3	88.88	12	2
P4	84.44	10	1
P5	93.33	10	1

9.6 Discussion

The work proposed in this chapter shows the algorithm for an automated system developed for hand movement detection by a PH. The developed system includes advance signal processing and machine learning techniques to classify six basic hand movements. The EWT method first decomposes the sEMG signals from electrode 1 and electrode 2, separately. Then, the CIP features are computed. A CIP feature is computed from the EWT coefficients of the jth band of the sEMG signals from electrode 1 and electrode 2. Here, the value of j is greater than one and less than B. The spectrum of every sEMG signal will have at least B number of bands obtain from the EWT decomposition. After obtaining B number of features, these features are ranked and then fed into the classifier. Three ranking methods, namely LLCFS,

Table 9.12. p-values of ranked features for all data.

Ranked feature	p-value for P1 data	p-value for P2 data	p-value for P3 data	p-value for P4 data	p-value for P5 data
R1	1.10×10^{-28}	4.32×10^{-21}	7.42×10^{-32}	3.89×10^{-21}	1.71×10^{-23}
R2	1.88×10^{-30}	4.94×10^{-32}	5.24×10^{-31}	6.50×10^{-26}	1.09×10^{-24}
R3	5.58×10^{-29}	3.87×10^{-31}	1.10×10^{-29}	1.55×10^{-25}	4.90×10^{-26}
R4	7.61×10^{-30}	1.00×10^{-30}	4.44×10^{-32}	1.79×10^{-30}	4.23×10^{-26}
R5	6.03×10^{-30}	1.44×10^{-32}	6.85×10^{-31}	1.77×10^{-25}	2.47×10^{-22}
R6	1.06×10^{-29}	2.32×10^{-30}	8.94×10^{-33}	3.46×10^{-28}	8.78×10^{-24}
R7	1.39×10^{-31}	1.08×10^{-31}	3.68×10^{-16}	1.22×10^{-27}	1.51×10^{-27}
R8	6.81×10^{-30}	2.93×10^{-30}	1.35×10^{-31}	3.51×10^{-28}	2.59×10^{-22}
R9	1.81×10^{-29}	1.06×10^{-30}	9.10×10^{-23}	1.91×10^{-23}	1.93×10^{-22}
R10	4.75×10^{-30}	2.29×10^{-31}	1.82×10^{-32}	4.96×10^{-29}	1.75×10^{-24}
R11	6.02×10^{-10}	1.04×10^{-29}	1.49×10^{-31}	NA	NA
R12	8.39×10^{-30}	1.45×10^{-30}	5.27×10^{-32}	NA	NA
R13	7.37×10^{-30}	1.04×10^{-30}	NA	NA	NA
R14	6.24×10^{-30}	NA	NA	NA	NA
R15	1.46×10^{-29}	NA	NA	NA	NA
R16	1.37×10^{-25}	NA	NA	NA	NA
R17	4.63×10^{-30}	NA	NA	NA	NA
R18	3.17×10^{-30}	NA	NA	NA	NA

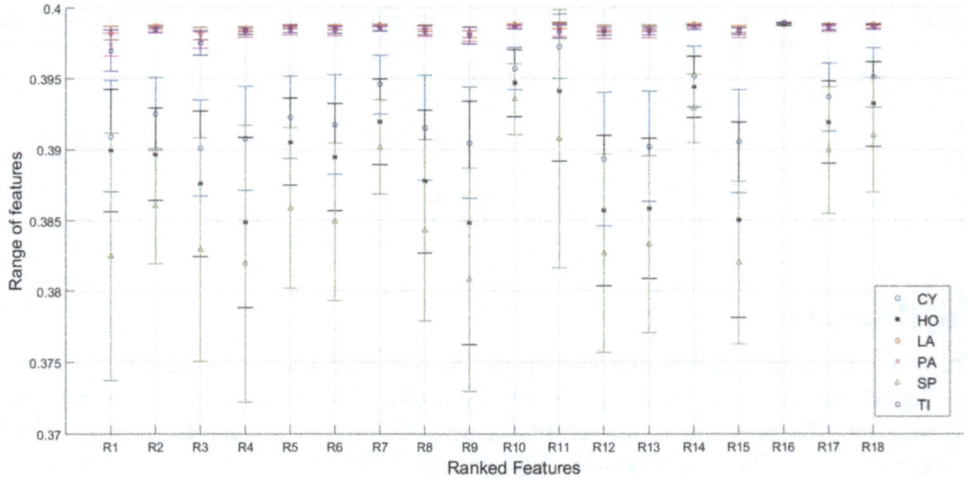

Figure 9.4. Statistical analysis (MN and SD) of ranked features for P1 data.

MCFS, and RELIEFF are tested for ranking, and the NB, k-NN, BA, and RF classifiers are tested in the classification stage. The proposed model is tested on the data of five people separately. Then, the obtained ACC for each person's data is averaged to obtain an average ACC of 87.44%.

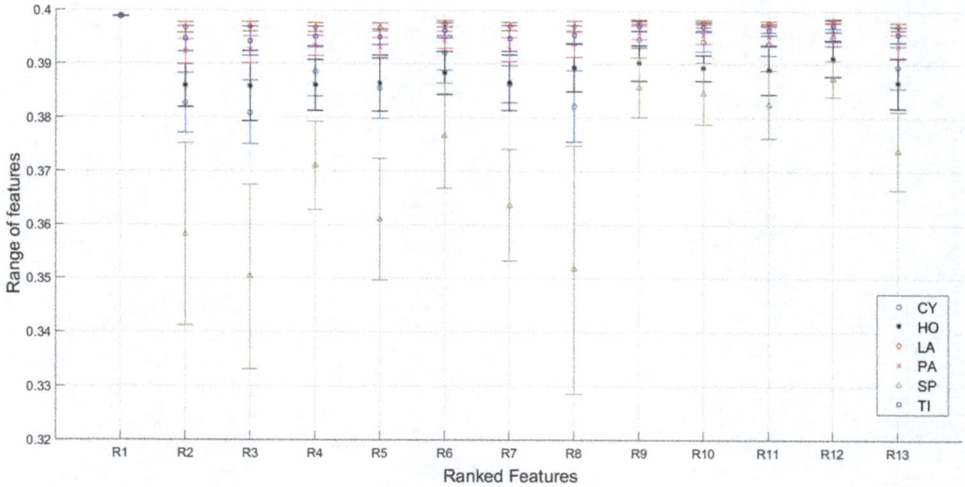

Figure 9.5. Statistical analysis (MN and SD) of ranked features for P2 data.

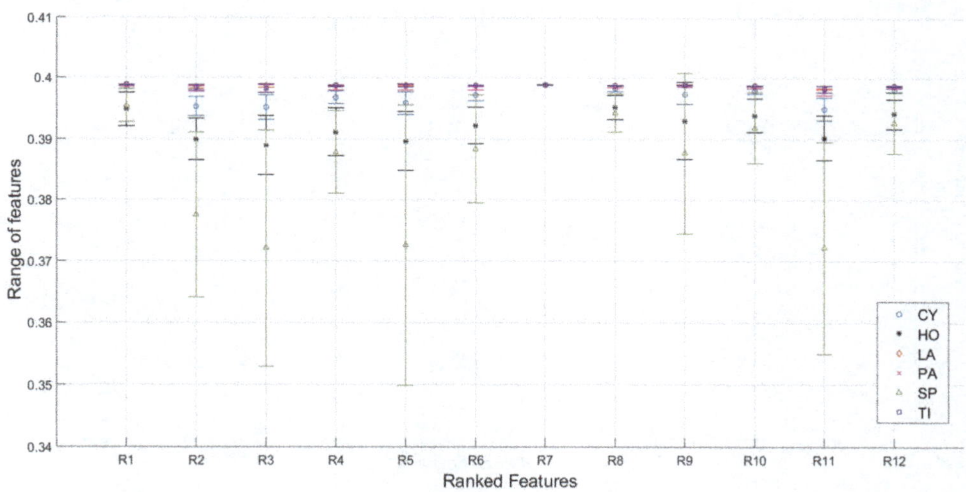

Figure 9.6. Statistical analysis (MN and SD) of ranked features for P3 data.

From tables 9.1–9.5, we can observe that there is no single model that will provide the best result for all the subject's data. However, we can say that the ACCs obtained by the NB and RF classifiers are usually greater than that of the BA and *k*-NN classifier. In contrast, in the feature ranking stage, one method may be better than the other two methods in some cases, but not in other cases. Still, we can observe from tables 9.1–9.5 that there is not much difference in the obtained maximum ACC by a classifier when the feature ranking method is changed in the proposed model. Therefore, RELIEFF is chosen for feature ranking since we have the option to vary k_r, which can affect the obtain ACC. Similarly, in the

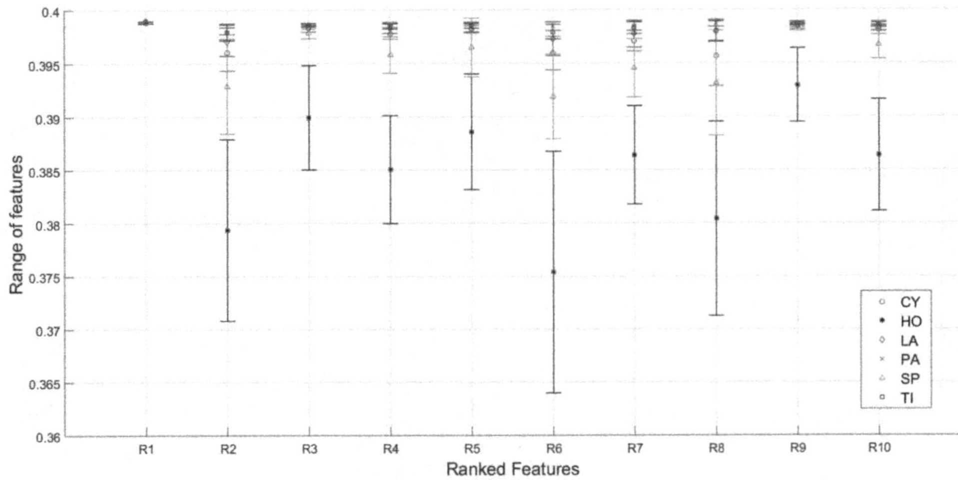

Figure 9.7. Statistical analysis (MN and SD) of ranked features for P4 data.

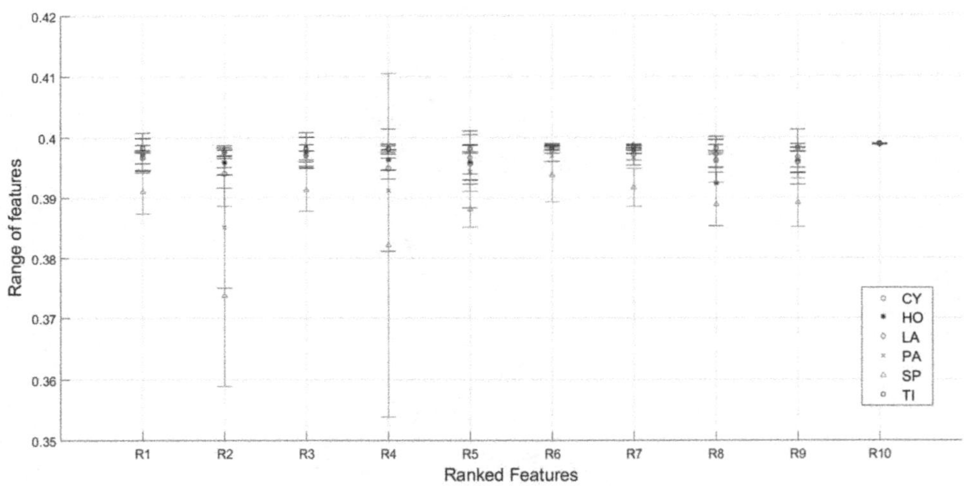

Figure 9.8. Statistical analysis (MN and SD) of ranked features for P5 data.

Table 9.13. Obtained confusion matrix using P1 data.

		Predicted					
		CY	HO	LA	PA	SP	TI
Actual	**CY**	29	0	0	0	1	0
	HO	1	26	0	0	3	0
	LA	0	0	27	2	0	1
	PA	0	0	2	24	0	4
	SP	2	3	0	0	25	0
	TI	0	0	1	10	0	19

Table 9.14. Obtained confusion matrix using P2 data.

		Predicted					
		CY	HO	LA	PA	SP	TI
Actual	**CY**	27	2	0	1	0	0
	HO	1	27	0	1	1	0
	LA	0	0	26	0	0	4
	PA	0	0	0	25	0	5
	SP	0	1	0	0	29	0
	TI	0	0	3	4	0	23

Table 9.15. Obtained confusion matrix using P3 data.

		Predicted					
		CY	HO	LA	PA	SP	TI
Actual	**CY**	27	1	0	2	0	0
	HO	2	26	0	0	2	0
	LA	0	0	26	3	0	1
	PA	1	0	2	27	0	0
	SP	1	2	0	0	27	0
	TI	0	0	3	0	0	27

Table 9.16. Obtained confusion matrix using P4 data.

		Predicted					
		CY	HO	LA	PA	SP	TI
Actual	**CY**	28	0	0	0	2	0
	HO	0	30	0	0	0	0
	LA	2	0	21	2	1	4
	PA	0	0	4	22	0	4
	SP	1	0	0	0	29	0
	TI	0	0	4	4	0	22

classification stage, RF is chosen since the average ACC obtain by RF is greater than that of the NB classifier.

The optimum values of the NRF and k_r for which maximum ACC is obtained for each person's data by the proposed method are shown in table 9.11. In this table, the NRF indicates the number of ranked features used to produce a classification ACC by the classifier. Table 9.12 shows the p-values of these ranked features. It can be

Table 9.17. Obtained confusion matrix using P5 data.

		Predicted					
		CY	HO	LA	PA	SP	TI
Actual	**CY**	29	0	0	1	0	0
	HO	0	25	3	2	0	0
	LA	0	1	29	0	0	0
	PA	2	1	0	27	0	0
	SP	0	0	0	0	30	0
	TI	1	1	0	0	0	28

observed that all the *p*-values are less than 0.001 for each person's data. This indicates that the ranked features are statistically significant.

The statistical analysis of these ranked features can be seen in figures 9.4–9.8. In these figures, there are different colored bars which correspond to different classes or hand movements. The length of the bar indicates twice the SD and the mid-point of the bar indicates the MN. We can also observe from these figures that the range of ranked features is small. Therefore, the MN of some ranked features of some classes may be very close. In figures 9.4–9.6, and figure 9.8, we can observe that the SD of the ranked features corresponding to the SP class is much greater than that of the other classes of ranked features. In contrast, in figure 9.7 the SD of the HO class ranked feature is greater than that of the other classes of ranked features. In figure 9.4–9.8, the MN of the SP class ranked features is usually distinguishable. Also, in some cases, the MNs of the CY and HO class ranked features are also dissimilar to those of the other classes of ranked features. These conclusions cannot be made for the ranked features belonging to the LA, PA and TI classes.

Tables 9.13–9.17 show the confusion matrices obtained using the proposed method for P1, P2, P3, P4, and P5 data when the parameters mentioned in table 9.11 are chosen in the proposed model. The confusion matrix shows how many actual samples of a particular class are predicted correctly and incorrectly. In the case of an incorrect prediction, we also have the information of which sample of a class is wrongly predicted in other classes. For example, in table 9.16, all 30 samples of the HO class are correctly predicted into the HO class. This can be related to figure 9.7, where the MN of the HO class ranked features is very far from the MN of the other class ranked features. A similar conclusion can be made for the SP class from table 9.17 and figure 9.8. Other than these cases, all 30 samples of a class are never predicted correctly. In some cases, the prediction is poor. For example, in table 9.13, only 19 samples of TI are predicted correctly. Ten of the TI samples are predicted as PA and one is predicted as LA. As seen in figures 9.4–9.8, the MNs of the ranked features of LA, PA, and TI are close to each other, therefore the incorrect predictions of the samples of the LA, PA, and TI classes lie within these classes.

From table 9.11, we can observe that the value of NRF never goes above 19 for any subject's data. Therefore, the proposed model provides decent results in terms of average ACC, with a smaller number of features. If the efficacy of the proposed method is compared to [16], the average ACC obtained by the proposed method is greater. The method proposed in [16] achieves an average ACC equal to 86.71%. In [16] the data are framed and then SVD and PCA are applied.

9.7 Conclusion and future scope

In this chapter, a methodology for basic hand movement classification is proposed. The proposed model classifies six basic hand movements from input sEMG signals. The sEMG signals are first decomposed through EWT and then the obtained EWT coefficients are used to compute the CIP features. Through simulations, it is observed that the RELIEFF method for feature ranking and the RF classifier for classification are the most suitable in the proposed model. Due to RELIEFF, the obtained ACC depends on the k_r value as well. The proposed model is tested on the data of five people and the average ACC obtained is 87.44%. Through simulations, it is shown that not many NRFs are required to obtain a decent ACC. The maximum NRFs required is 19 for the P1 and P2 data. Also, the p-values of the ranked features are computed, which are below 0.001. This indicates the statistical significance of the obtained ranked features.

The input to the proposed model is bivariate sEMG signals which are obtained from two electrodes only. Therefore, the proposed model is suitable for designing a low cost prosthetic arm which can perform hand movement tasks with fewer electrodes.

The dataset on which the proposed model is tested consists of 30 samples from each of six classes. In the future, the proposed model can be tested on a larger dataset which includes more data from more individuals. Also, some more features can be added in the proposed model to improve its performance in terms of average ACC. In addition to this, deep learning techniques can be employed to classify different hand movements, since deep learning techniques automatically extract good features from input signals. Hence the deep learning technique can improve the obtained average ACC.

References

[1] Sapsanis C, Georgoulas G, Tzes A and Lymberopoulos D 2013 Improving EMG based classification of basic hand movements using EMD *Proc. IEEE 35th Annual Int. Conf. of the Engineering in Medicine and Biology Society (Osaka, Japan)* pp 5754–7

[2] Caton R 1875 The electric currents of brain *J. Nerv. Ment. Dis.* **2** 610

[3] Sapsanis C, Georgoulas G and Tzes A 2013 EMG based classification of basic hand movements based on time-frequency features *Proc. IEEE 21st Mediterranean Conf. on Control and Automation (Chania, Greece)* pp 716–22

[4] Kiguchi K and Hayashi Y 2012 A study of EMG and EEG during perception-assist with an upper-limb power-assist robot *Proc. IEEE Int. Conf. on Robotics and Automation (Saint Paul, MN, USA)* pp 2711–6

[5] Kurita Y, Tada M, Matsumoto Y and Ogasawara T 2002 Simultaneous measurement of the grip/load force and the finger EMG: effects of the grasping condition *Proc. 11th IEEE Int. Workshop on Robot and Human Interactive Communication (Berlin, Germany)* pp 217–22

[6] Finneran A and O'Sullivan L 2013 Effects of grip type and wrist posture on forearm EMG activity, endurance time and movement accuracy *Int. J. Ind. Ergon.* **43** 91–9

[7] Liu J and Zhou P 2013 A novel myoelectric pattern recognition strategy for hand function restoration after incomplete cervical spinal cord injury *IEEE Trans. Neural Syst. Rehabil. Eng.* **21** 96–103

[8] Andrianesis K and Tzes A 2008 Design of an anthropomorphic prosthetic hand driven by shape memory alloy actuators *Proc. 2nd IEEE RAS & EMBS Int. Conf. on Biomedical Robotics and Biomechatronics (Scottsdale, AZ, USA)* pp 517–22

[9] Oskoei M A and Hu H 2007 Myoelectric control systems—a survey *Biomed. Signal Process. Control* **2** 275–94

[10] Carroll D and Alwarappan S 2012 Recent advances in biosensors and biosensing protocols *J. Biosens. Bioelectron.* **3** e112

[11] Kaniusas E 2012 Fundamentals of biosignals *Biomedical Signals and Sensors I* (Berlin: Springer) pp 1–26

[12] Khezri M and Jahed M 2008 Surface electromyogram signal estimation based on wavelet thresholding technique *Proc. IEEE 30th Annual Int. Conf. of the IEEE Engineering in Medicine and Biology Society (Vancouver, Canada)* pp 4752–5

[13] Ouyang G, Zhu X, Ju Z and Liu H 2014 Dynamical characteristics of surface EMG signals of hand grasps via recurrence plot *IEEE J. Biomed. Health Inform.* **18** 257–65

[14] Ju Z and Liu H 2014 Human hand motion analysis with multisensory information *IEEE/ASME Trans. Mechatron.* **19** 456–66

[15] Kakoty N M and Hazarika S M 2011 Recognition of grasp types through principal components of DWT based EMG features *Proc. IEEE Int. Conf. on Rehabilitation Robotics (Zurich, Switzerland)* pp 1–6

[16] Iqbal O, Fattah S A and Zahin S 2017 Hand movement recognition based on singular value decomposition of surface EMG signal *Proc. IEEE Region 10 Humanitarian Technology Conf. (Dhaka, Bangladesh)* pp 837–42

[17] Akben S B 2017 Low-cost and easy-to-use grasp classification, using a simple 2-channel surface electromyography (sEMG) *Biomed. Res.* **28** 577–82

[18] Ruangpaisarn Y and Jaiyen S 2015 sEMG signal classification using SMO algorithm and singular value decomposition *Proc. 7th IEEE Int. Conf. on Information Technology and Electrical Engineering (Chiang Mai, Thailand)* pp 46–50

[19] Xiao F, Yang D, Lv Z, Guo X, Liu Z and Wang Y 2020 Classification of hand movements using variational mode decomposition and composite permutation entropy index with surface electromyogram signals *Future. Gener. Comput. Syst.* **110** 102–36

[20] Duan F, Dai L, Chang W, Chen Z, Zhu C and Li W 2016 sEMG-based identification of hand motion commands using wavelet neural network combined with discrete wavelet transform *IEEE Trans. Ind. Electron.* **63** 1923–34

[21] Nishad A, Upadhyay A, Pachori R B and Acharya U R 2019 Automated classification of hand movements using tunable-Q wavelet transform based filter-bank with surface electromyogram signals *Future Gener. Comput. Syst.* **93** 96–110

[22] Nazemi A and Maleki A 2014 Artificial neural network classifier in comparison with LDA and LS-SVM classifiers to recognize 52 hand postures and movements *Proc. 4th IEEE Int. Conf. on Computer and Knowledge Engineering (Mashhad, Iran)* pp 18–22

[23] Kuzborskij I, Gijsberts A and Caputo B 2012 On the challenge of classifying 52 hand movements from surface electromyography *Proc. IEEE Annual Int. Conf. of the Engineering in Medicine and Biology Society (San Diego, CA, USA)* pp 4931–7

[24] Saeed B, Gilani S O, ur Rehman Z, Jamil M, Waris A and Khan M N 2019 Comparative analysis of classifiers for EMG signals *Proc. IEEE Canadian Conf. of Electrical and Computer Engineering (Edmonton, Canada)* pp 1–5

[25] Gilles J 2013 Empirical wavelet transform *IEEE Trans. Signal Process* **61** 3999–4010

[26] Daubechies I 1992 *Ten Lectures on Wavelets* (Philadelphia, PA: SIAM)

[27] Bhattacharyya A and Pachori R B 2017 A multivariate approach for patient-specific EEG seizure detection using empirical wavelet transform *IEEE Trans. Biomed. Eng.* **64** 2003–15

[28] Bhattacharyya A, Gupta V and Pachori R B 2017 Automated identification of epileptic seizure EEG signals using empirical wavelet transform based Hilbert marginal spectrum *Proc. 22nd IEEE Int. Conf. on Digital Signal Processing (London, UK)* pp 1–5

[29] Bhattacharyya A, Singh L and Pachori R B 2018 Fourier–Bessel series expansion based empirical wavelet transform for analysis of non-stationary signals *Digital. Signal. Process.* **78** 185–96

[30] Sharma R R, Kalyani A and Pachori R B 2020 An empirical wavelet transform-based approach for cross-terms-free Wigner–Ville distribution *Signal Image Video Process.* **14** 249–56

[31] Xu D and Erdogmuns D 2010 Renyi's entropy, divergence and their nonparametric estimators *Information Theoretic Learning* (Berlin: Springer) pp 47–102

[32] Xu J W, Paiva A R C, Park I and Principe J C 2008 A reproducing kernel Hilbert space framework for information-theoretic learning *IEEE Trans. Signal Process.* **56** 5891–902

[33] Nishad A and Pachori R B 2020 Classification of epileptic electroencephalogram signals using tunable-Q wavelet transform based filter-bank *J. Ambient Intell. Human. Comput.* 1–15

[34] Kumar M, Pachori R B and Acharya U R 2017 Characterization of coronary artery disease using flexible analytic wavelet transform applied on ECG signals *Biomed. Signal Process. Control* **31** 301–8

[35] Singh R and Pachori R B 2020 Iterative filtering-based automated method for detection of normal and ALS EMG signals *Recent Trends in Image and Signal Processing in Computer Vision* (Berlin: Springer) pp 33–53

[36] Zeng H and Cheung Y 2011 Feature selection and kernel learning for local learning-based clustering *IEEE Trans. Pattern Anal. Mach. Intell.* **33** 1532–47

[37] Cai D, Zhang C and He X 2010 Unsupervised feature selection for multi-cluster data *Proc. 16th ACM SIGKDD Int. Conf. on Knowledge Discovery and Data Mining* pp 333–42

[38] Kononenko I, Simec E and Robnik-Sikonja M 1997 Overcoming the myopia of inductive learning algorithms with RELIEFF *Appl. Intell.* **7** 39–55

[39] Robnik-Sikonja M and Kononenko I 2003 Theoretical and empirical analysis of ReliefF and RReliefF *Mach. Learn.* **53** 23–69

[40] Breiman L 1996 Bagging predictors *Mach. Learn.* **24** 123–40

[41] Aha D W, Kibler D and Albert M K 1991 Instance-based learning algorithms *Mach. Learn.* **6** 37–66

[42] John G H and Langley P 1995 Estimating continuous distributions in Bayesian classifiers *Proc. Eleventh Conf. on Uncertainty in Artificial Intelligence* pp 338–45

[43] Breiman L 2001 Random forests *Mach. Learn.* **45** 5–32

[44] Fraiwan L, Lweesy K, Khasawneh N, Wenz H and Dickhaus H 2012 Automated sleep stage identification system based on time frequency analysis of a single EEG channel and random forest classifier *Comput. Methods Programs Biomed.* **108** 10–9

[45] Gupta V, Priya T, Yadav A K, Pachori R B and Acharya U R 2017 Automated detection of focal EEG signals using features extracted from flexible analytic wavelet transform *Pattern Recognit. Lett.* **94** 180–8

[46] Das A B and Bhuiyan M I H 2016 Discrimination and classification of focal and nonfocal EEG signals using entropy-based features in the EMD-DWT domain *Biomed. Signal Process. Control* **29** 11–21

[47] Nishad A, Pachori R B and Acharya U R 2018 Application of TQWT based filter-bank for sleep apnea screening using ECG signals *J. Ambient Intell. Human. Comput.* 1–12

[48] Hassan A R 2015 A comparative study of various classifiers for automated sleep apnea screening based on single-lead electrocardiogram *Proc. IEEE Int. Conf. on Electrical Electronic Engineering (Rajshahi, Bangladesh)* pp 45–8

[49] Sharma R, Pachori R B and Upadhyay A 2017 Automatic sleep stages classification based on iterative filtering of electroencephalogram signals *Neural Comput. Appl.* **28** 2959–78

[50] Gilles J 2013 *Empirical Wavelet Transform Toolbox Version 4.0* pp 1–22

[51] Roffo G, Melzi S, Castellani U and Vinciarelli A 2017 Infinite latent feature selection: a probabilistic latent graph-based ranking approach *Proc. IEEE Int. Conf. on Computer Vis. (Venice, Italy)* pp 1398–406

[52] Hall M, Frank E, Holmes G, Pfahringer B, Reutemann P and Witten I H 2009 The WEKA data mining software: an update *SIGKDD Explor.* **11** 10–8

[53] Kohavi R 1995 A study of cross-validation and bootstrap for accuracy estimation and model selection *Proc. 14th Int. Joint Conf. on Artificial Intelligence* pp 1137–43

[54] McKight P E and Najab J 2010 Kruskal–Wallis test *Corsini Encyclopedia of Psychology* (Hoboken, NJ: Wiley)

Chapter 10

Analysis of the muscular activity pattern of recurring physical action

Ajay Somkuwar[1], Vandana Somkuwar[2]

[1]*Maulana Azad National Institute of Technology, Bhopal (MP) India*

[2]*National Institute of Technical Teachers Training and Research, Bhopal (MP) India*

The pedalling action of riding a bicycle is a repeated task that is carried out for a long time. Thus a few muscles are subjected to excessive repetitive use. This creates many short- and long-term health problems for bicycle operators due to the accumulation of local muscle fatigue which leads to many short- and long-term ailments. These ailments necessitate information about the efforts human beings have to apply in different tasks and work situations. Hence a reliable method for measuring muscular activity and a mode of relating these measures to fatigue is required. Therefore, the purpose of this study is to present an integral investigation of the biomechanics of the human–bicycle system, keeping the focus on the practical role of the significant leg muscles. Surface electromyograms (sEMGs) are collected from eight muscles of the leg. The adverse effects of power-line, crosstalk and low level myoelectric signal reception from other sources, which obscure the description of the muscle's activation timing and the comparative intensity of its activity, are separated via an independent component analysis technique. The integrated myoelectric signal (IEMG) is calculated for muscular activity analysis and a Kalman filter is used to predict the pattern of joint forces exerted at the ankle, knee and hip. Based on the biomechanical model of the leg as a five-bar linkage model, the kinematics of leg and joint moments at the ankle, knee and hip joint are calculated and correlate well with the predicted normalized joint moments.

10.1 Introduction

Performance optimization using muscular activity in bicycling is very well understood via a simplified physiological model. The earliest work on this subject was by

doi:10.1088/978-0-7503-3279-8ch10

Houtz and Fischer [7] who used an ergometer to associate the electromyograms (EMGs) of 14 muscles and the simultaneous motion of the joints during bicycle riding. Faria and Cavanagh [13] discussed the functions of the muscles and joints involved in pedalling. Despite their presentation of the muscular activity regions for the duration of pedalling activity, their experimental procedure and methods were not presented in detail. The effect of seat position and alignment on the activity of the associated muscles was predicted by Despires [5], but the human–machine linkage was not mentioned. In an experiment carried out by Ericson *et al* [14], the muscular activity of eleven muscles of the leg was recorded and quantified using myoelectrical signals. They set up load movements acting around the ankle, knee and hip joint axes that were lower than the loads normally exerted during walking activity. Using an empirical model, Jorge and Hull [8] analysed the muscular activity of various pedalling conditions and concluded that the level of muscular activity is significantly affected by the pedalling conditions. They showed that using a biomechanical model of the lower extremities, the joint moments developed during pedalling could be computed easily based on some cost function of the force exerted. Also, using some variables incorporated in the model of the cost function, an average power 200 W was obtained in their experimental work [15]. Yoshihuku and Herzog optimized the design parameters which maximized the output power from the muscles of the human lower limb while pedalling. However, they assumed that the muscles exerted force according to the Hill equation. The average power and its dependence on the design parameters were computed [16]. The present work has resorted to an approach that is partly model and partly experimental. On the basis of the latest ideas in physiology, models are proposed for better articulation of individual problems and, wherever possible, the results are compared to the experimental evidence in the existing literature. Because of the complexities involved, the muscular activity can be determined easily only when the physiological model is expressed in basic terms. Because of the involvement of complex muscular activity, it has been decided to explore the possibility of using an experimental model to simplify the computation of muscular activity while pedalling. The results of the model have to be verified with the help of suitable experimental as well as analytical methods.

10.2 Analytical expressions of joint moments

The aim of this section is to describe the analytical procedure in order to explore the effects of the mechanical aspects of the pedalling task on the kinematic patterns described by the hip, knee and ankle joints. Considering that the moments developed at the joint are a result of the effort exerted by the associated muscles, the joint moments can serve in the examination of optimal physiological cost as well as muscular effort. In a repetitive task, it is necessary make certain assumptions to simplify the complex system in order to achieve an analytical solution. In general, it is always best to begin with a simple model which represents the human leg and crank system. If the task is to analyse the force exerted at a particular joint, the best approach is to set the model while neglecting all other parts of the system. Hence, inspired by the work of Jorge and Hull [8], the leg–crank system is modelled as a four

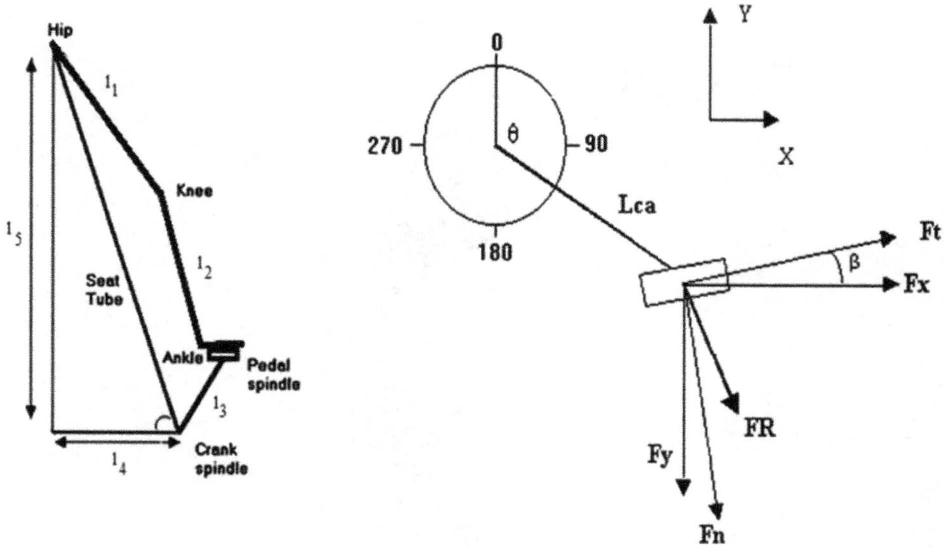

Figure 10.1. The four bar linkage model and the forces acting on the pedal and crank arm.

bar linkage model that is fixed at both the hip joint and the crank axis, as shown in figure 10.1. The links are the thigh (T), shank (S), crank (C) and the saddle tube, which provides a stationary link. The crank axis, pedal spindle, hip joint and knee joint serve as the pivot points. The foot is assumed as a rigid body and acts coplanar to the pedal surface, hence pedal spindle rotation is precisely equal to the movements of the ankle joint [11]. We assume that the bones are a rigid body and rely on the method of lumping muscles into functional groups to create a deterministic system for biomechanical analysis. To restrain the model, one geometric constraint and five kinematic inputs are required to specify the linkage configuration. Regarding the geometric constraints, the knee joint is not allowed to extend beyond the straight position. The relative pedal angle is approximated as a sinusoidal pattern. The normal and tangential forces are measured using a dynamometer. The cadence is set as constant (40 rpm) by following a metronome. The biomechanical analysis needs to specify the values of the anthropometric parameters of the subjects considered for analysis. The operator's body segments, mass and length are calculated according to Dempster's formula to obtain the model parameters [4].

As illustrated in figure 10.1, the lower limb and saddle tube may be modelled as closed loop four bar linkages with the tube serving as a fixed link. The angle of the moving links, thigh, shank and the crank arm form a closed model, hence according to the loop equation the following condition must be fulfilled:

$$l_1 e^{i\varepsilon 1} + l_2 e^{i\varepsilon 2} + l_3 e^{i\varepsilon 3} + l_4 e^{i\varepsilon 4} + l_5 e^{\pi/2} = 0. \tag{10.1}$$

In order to specify a complete linkage configuration, the measured crank angle and relative pedal angle (θ_2) are required to validate the following constraint equation:

$$\varepsilon_4 = \varepsilon_3 + \frac{\pi}{2} - \theta_2. \tag{10.2}$$

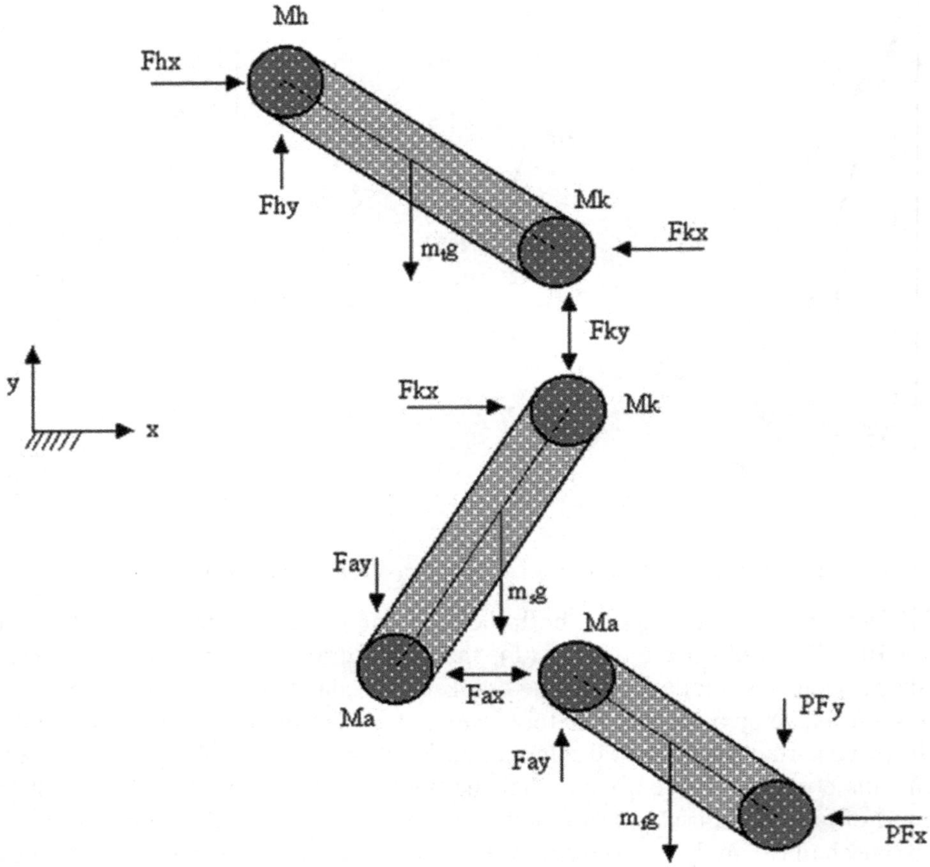

Figure 10.2. The forces acting on the segments of the lower limb.

The sinusoidal pattern of the foot angle correlates with the data very well, with the frequency equal to the pedalling rate. The first and second derivatives of the obtained data represent the angular velocity and acceleration, respectively. Angular acceleration is defined as positive for counter-clockwise joint moments, positive if it acts clockwise on the body segments and negative if it acts counter-clockwise on the segment of the distal joint, as shown in figure 10.2.

The joint moment's equations are

$$M_{\mathrm{H}} = -I_{\mathrm{TCG}}\,\ddot{\varepsilon}_1 + M_{\mathrm{K}} + (F_{\mathrm{HY}}\sin - F_{\mathrm{HX}}\cos)d_1 + (F_{\mathrm{KY}}\sin - F_{\mathrm{KY}}\cos)(l_1 - d_1) \tag{10.3}$$

$$M_{\mathrm{K}} = -I_{\mathrm{SEG}}\,\ddot{\varepsilon}_2 + M_{\mathrm{A}} + (F_{\mathrm{AY}}\sin - F_{\mathrm{AX}}\cos)(l_2 - d_2) + (K_{\mathrm{KY}}\sin - F_{\mathrm{KX}}\cos)\cdot d_2 \tag{10.4}$$

$$M_{\mathrm{A}} = I_{\mathrm{FCG}}\,\ddot{\varepsilon}_3 + (F_{\mathrm{AY}}\sin - F_{\mathrm{AX}}\cos)\,l_3, \tag{10.5}$$

where

$$\left.\begin{aligned} F_X &= F_t\cos\theta_2 - F_n\sin\theta_2 \\ F_Y &= F_t\sin\theta_2 + F_n\cos\theta_2 \end{aligned}\right\} \tag{10.6}$$

$$\left.\begin{aligned} F_{AX} &= m_f a_{fx} + Fx \\ F_{AY} &= m_f(a_{fy} + g) + Fy \end{aligned}\right\} \tag{10.7}$$

$$\left.\begin{aligned} F_{KX} &= m_s a_{sx} + F_{AX} \\ F_{KY} &= m_s(a_{sy} + g) + F_{AY} \end{aligned}\right\} \tag{10.8}$$

$$\left.\begin{aligned} F_{HX} &= m_t a_{tx} + F_{KX}, \\ F_{HX} &= m_t a_{tx} + F_{KX}, \end{aligned}\right\} \tag{10.9}$$

Table 10.1 lists the symbols used in the above equations. The two performance variables (seat height and seat tube angle) are obtained for optimal results.

10.2.1 Data collection and joint moment analysis

The average absolute joint moment was calculated for constant average pedal forces. The pedal force was recorded from all ten subjects (table 10.2) for ten consecutive crank cycles and the most appropriate data were utilized for the

Table 10.1. Symbols used in the biomechanical analysis.

Sr No.	Symbol used	Description
1	l_1, l_2, l_3	Lengths of the thigh, shank and crank arm
2	l_4, l_5	Horizontal and vertical seat distances
3	d_1, d_2	Distances of the thigh and shank centres of gravity from the proximal joint
4	F_t, F_n	Tangential and normal pedal forces
5	M_H, M_K, M_A	Moments at the hip, knee and ankle joints
6	I_{TCG}, I_{SCG}, I_{FCG}	Moments of inertia of the thigh, shank and foot about the centre of gravity
7	F_{HX}, F_{HY}, F_{KX}, F_{KY}, F_{AX}, F_{AY}	Reaction forces at the hip, knee and ankle in the *x*- and *y*-directions
8	θ_1, θ_2	Measured crank and relative pedal angles
9	ε_1, $\dot{\varepsilon}_1$, $\ddot{\varepsilon}_1$ ε_2, $\dot{\varepsilon}_2$, $\ddot{\varepsilon}_2$ ε_3, $\dot{\varepsilon}_3$, $\ddot{\varepsilon}_3$	Angles, velocities and accelerations of the thigh, shank, pedal spindle and crank, respectively
10	ω_1, ω_2	Angular velocities of the crank and rear cog
11	m_f, m_s, m_t	Masses of the foot, shank and thigh
12	a_t, a_s, a_f	Absolute accelerations of the thigh, shank and foot CG

Table 10.2. Anthropometric data of the subjects.

Sr No.	Name	Age (years)	Height (trochanter leg length) (m)	Weight (kg)	Thigh data Length (m)	Thigh data Mass (kg)	Thigh data Moments of inertia (kg m^2)	Shank data Length (m)	Shank data Mass (kg)	Shank data Moments of inertia (kg m^2)	Foot data Length (m)	Foot data Mass (kg)	Foot data Moments of inertia (kg m^2)
1	MG	35	1.57 (0.72)	43	0.3459	4.30	53.7×10^{-3}	0.3820	1.9995	26.6×10^{-3}	0.1790	0.6200	4.48×10^{-3}
2	NR	50	1.62 (0.74)	48	0.3569	4.80	64.2×10^{-3}	0.3942	2.2320	31.6×10^{-3}	0.1847	0.6921	5.33×10^{-3}
3	PL	29	1.61 (0.74)	48	0.3547	4.80	63.1×10^{-3}	0.3918	2.2320	31.1×10^{-3}	0.1835	0.6921	5.27×10^{-3}
4	MK	32	1.66 (0.76)	51	0.3657	5.10	71.7×10^{-3}	0.4039	2.3715	35.1×10^{-3}	0.1892	0.7353	5.96×10^{-3}
5	SS	28	1.68 (0.77)	52	0.3701	5.20	74.7×10^{-3}	0.4088	2.4180	36.9×10^{-3}	0.1915	0.7498	6.24×10^{-3}
6	MY	26	1.64 (0.75)	53	0.3613	5.30	72.9×10^{-3}	0.3991	2.4645	35.7×10^{-3}	0.1870	0.7642	6.08×10^{-3}
7	RK	24	1.65 (0.759)	54	0.3635	5.40	75.1×10^{-3}	0.4015	2.5110	37.0×10^{-3}	0.1881	0.7786	6.24×10^{-3}
8	DN	37	1.63 (0.74)	57	0.3591	5.70	77.1×10^{-3}	0.3966	2.6505	37.9×10^{-3}	0.1858	0.8219	6.48×10^{-3}
9	MP	24	1.72 (0.79)	58	0.3789	5.80	87.4×10^{-3}	0.4185	2.6970	43.2×10^{-3}	0.1961	0.8363	7.20×10^{-3}
10	SK	20	1.68 (0.77)	58	0.3701	5.80	83.3×10^{-3}	0.4088	2.6970	41.1×10^{-3}	0.1915	0.8363	6.90×10^{-3}
Average		30.5 ± 8.1 SD	1.646 ± 4.6 SD	52.2 ± 0.46 SD	0.363± 0.009 SD	5.22 ± 0.464 SD	$72. \times 10^{-3}$ ± 9.3 SD	0.400 ± 0.01 SD	2.43 ± 0.21 SD	35.6×10^{-3} ± 4.6 SD	0.188 ± 0.005 SD	0.755 ± 0.069 SD	6.02×10^{-3} ± 0.767 SD

calculation of the joint moments. A one-way analysis of variance (ANOVA) was executed for the repeated evaluation, using the value obtained for the pedal forces (both tangential and normal) at each value of the biomechanical variable considered. The *F*-ratio obtained was significant at the 0.05 level. Similarly, the *F*-ratio for seat angle variation was also significant.

Understanding the biomechanical aspects of pedalling is important for improving performance as well as avoiding overuse injuries in the leg joints acquired due to pedalling activity. The information required to understand the pedalling process includes the load on the pedal and the kinematics of the leg segments. The average absolute joint moment is calculated for constant average pedal forces. At steady cadence equal to 40 rpm, the pedal force is recorded for ten cycles of the crank assembly using a pedal dynamometer, as shown in figure 10.3. The data are stored using BNC2080 (National Instruments, Austin, TX) in a TEAC-R60 data recorder (four channels). The average vertical and horizontal force is computed for one rotation of the crank. These are utilized to compute the average joint moment at the hip, knee and ankle joints, given in figure 10.4. A one-way analysis of variance (ANOVA) is carried out for the repeated calculation. The value obtained for the pedal forces (both tangential and normal) at each value of the biomechanical variable are considered for the average force calculation.

The *F*-ratio obtained is significant at the 0.05 level. Similarly, the *F*-ratio for seat angle variation is also significant. Hence relying on the above findings, with a constant pedalling rate (= 40 rpm) and crank arm length (= 170 mm), the average forces may be computed using the ten subjects given in table 10.2. All the processing is performed on a personal computer, using the optimization tool box for the Matlab™ software. Univariate optimizations, using the Davidon–Fletcher–Powell method, are performed for each of the variables. The value of the seat tube angle is found to be 76°, while 97% of trochanter leg length is found for minimum joint excursion. The obtained kinematic pattern of the leg moments indicates that the

Figure 10.3. The pedal dynamometer.

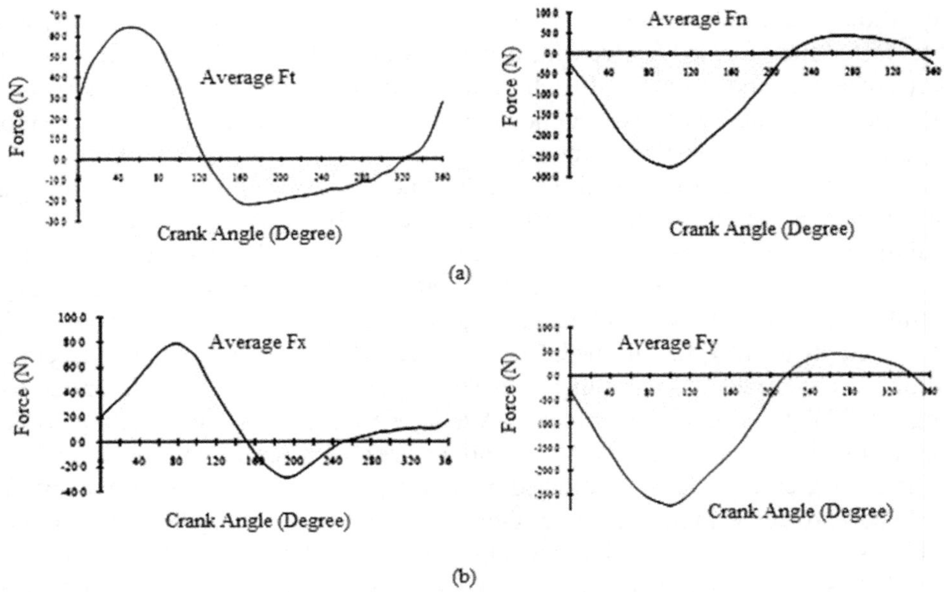

Figure 10.4. Average tangential, normal, horizontal and vertical forces over one crank revolution.

joint moments at the ankle, knee and hip do vary with the biomechanical variables considered, and these results are presented in table 10.3.

10.3 Myoelectric signals during recursive work

The attractive feature of electromyography signals is that they represent the state of contraction of an individual muscle and therefore are the best indicators of the force or energy generating inputs in biomechanical systems. There are number of tasks and work situations or conditions in which human effort needs to be evaluated on the basis of muscular activity. Hence a reliable technique which relates muscular fatigue to workload is required, and EMG based analysis is crucial to meet this requirement [9]. The changes in activity or changes in the interaction between muscles or muscle groups can be estimated easily via recorded EMG, either at rest or during movement of the human body. Work using machines has profoundly reduced the physical load on the human operator, but at the same time it has lead to an increase in the duration and number of situations where stresses are concentrated on a single part or set of muscles of the human body. Such a repetitive nature of tasks increases muscles tension. This may induce muscle degradation or lower human performance or, even worse, result in severe injuries to muscles. Sometimes this results in the reduction of human capacity as well. Currently there are three main applications of EMG signals which include:

- Determining muscle activation timings, i.e. the beginning and end of muscle activity.
- Estimating the force exerted by muscles or groups of muscles.

Table 10.3. Joint moments calculated by biomechanical analysis and normalized joint moments estimated using EMG activity over one crank cycle.

Joint moments calculated by pedal dynamometer and biomechanical analysis	Normalized joint moments estimated by EMG analysis
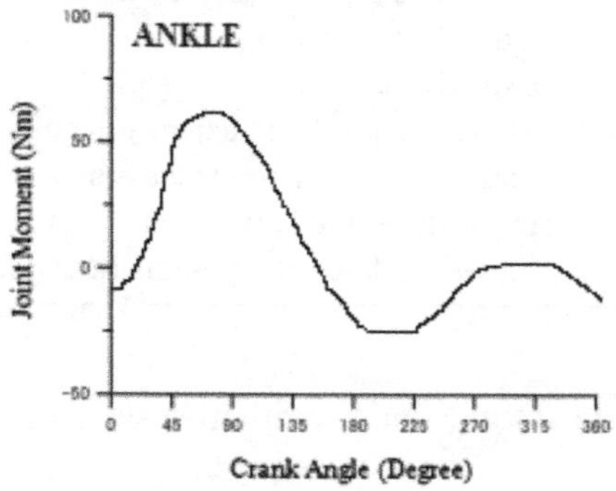	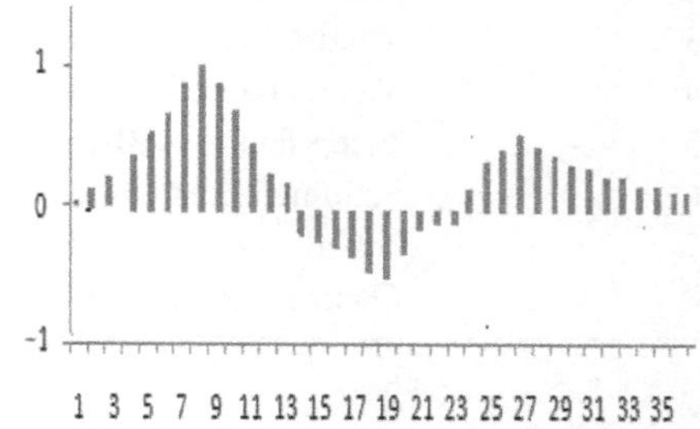
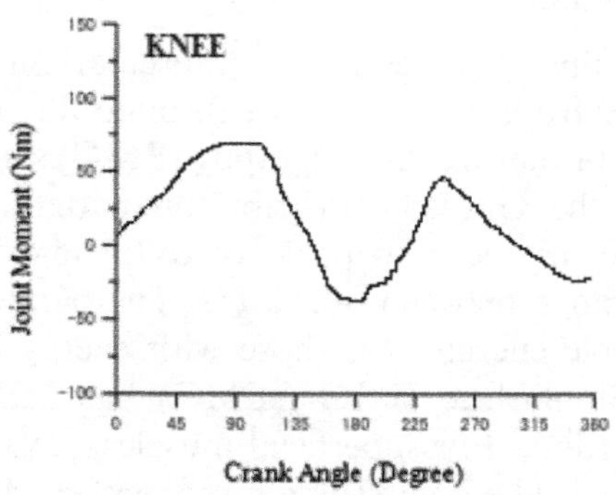	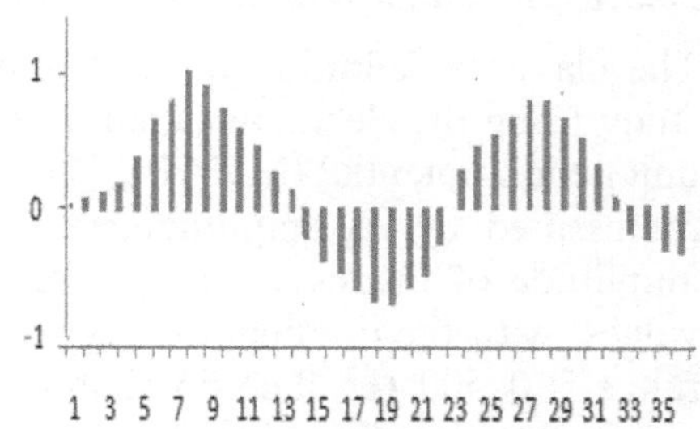
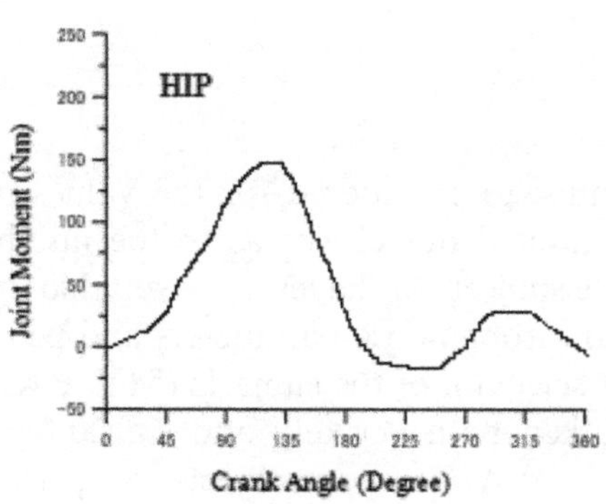	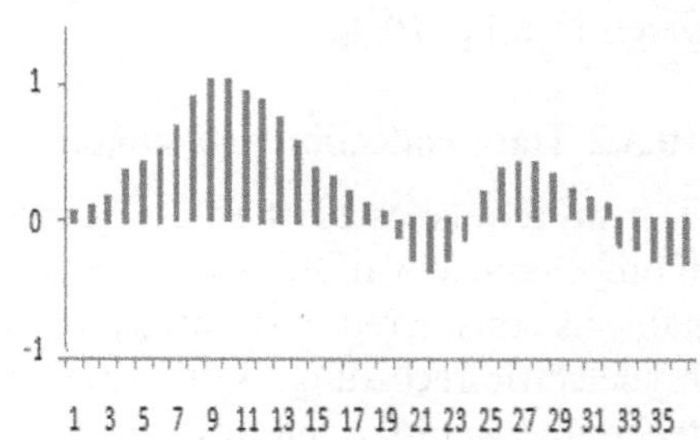

Table 10.4. The major muscles of the lower limb selected for this study and their functions.

Sr No.	Muscle name	Function
1	Gluteus maximus (GM)	Extends the hip
2	Rectus femoris (RF)	Flexes the hip and extends the knee
3	Vastus medialis (VM)	Extends the knee
4	Vastus lateralis (VL)	Extends the knee
5	Bicep femoris (BF)	Flexes the knee and extends the hip
6	Semimembranous (S)	Flexes the knee and extends the hip
7	Tibialis anterior (TA)	Dorsi-flexes the foot
8	Gastrocnemius (G)	Flexes the knee and plantar flexes the foot

- Obtaining the fatigue rate or muscle fatigue index with the support of its frequency spectrum.

10.3.1 The major muscles of the lower extremity

The classical definition of electromyography describes it as a muscular function study using the electrical signal originating from the muscles or through the motor unit action potential (MUAP). The nature of the electromyography (EMG) signal is represented by its amplitude, similar to the Gaussian distribution pattern. The amplitude of the signal ranges from 0–10 mV peak to peak, or 0–1.5 mV RMS values, with the dominant energy signal range between 0–150 Hz. The bandwidth range is 0–500 Hz. The dominant or usable energies are those with energy levels above the electrical signal noise level [3]. Surface EMG (sEMG) has achieved widespread use because of its ease of recording. For superficial muscles, it yields a more representative and reliable EMG signal. The EMG signals were collected from eight major muscles from the right leg of the subjects, the functions of which are given in table 10.4.

10.3.2 Data collection and subjects

The prerequisites of the data, time and muscles considered for the values of the biomechanical variables of concern have limited our choice as to the number of subjects considered in the analysis. Only five subjects of the ten were suitable for this myoelectric recording. Similarities in the anthropometric parameters and pedalling technique were the major determinants for selection of the subjects (MY, RK, DN, MP and SK, table 10.2). Consent was taken both verbally and signed by each person. A one-way analysis of variance (ANOVA) between subjects was performed for each muscle dataset. The *F*-ratio obtained (for the within subjects effect) was significant at the 0.05 level.

10.3.3 The myoelectrical signal, electrodes and recording

Noise can be defined as unwanted electrical signals which act as a disturbance in the desired EMG signals. The two main issues which affect the fidelity of the signal during detection and recording of the EMG signal are the signal to noise ratio and distortion. The signal to noise ratio is the ratio of the energy of the EMG signal to the energy present in the noise signal. The distortion of the signal means the change in the relative contribution of the frequency component in the EMG signals. Any change in the frequency component of the EMG signal should be avoided. The new processing technique suggested by Crowninshield and Brand [2] emphasizes the importance of the configuration of electrodes in extracting quantitative information from the EMG signal. The value of the bandwidth and the amplitude of the EMG signal are affected by surface distance. A small change in the distance shifts the bandwidth to higher frequencies and lowers the amplitude of the signal. Thus to avoid these changes the distance should be fixed so that quantitative comparisons between recordings can be made within and among muscles, and among individuals. To detect representative electrical activity of a muscle during a contraction, a inter-detection surface with a 1.0 cm configuration is sufficient. By limiting the distance to 1.0 cm, the signals from small muscles can also be detected easily.

By setting a larger size for the detection surfaces, the amplitude of the signal will be greater and can be detected easily, but the electrical noise generated in the signal will be low at the skin–detection surface interface. To detect the amplitude and noise of the EMG signal the electrode can be placed between single motor points, between two motor points and the tendon insertion, and along the longitudinal midline of the muscles. The length of the muscle fibres should be aligned parallel to the longitudinal axis of the electrode. To fix the common reference to the differential input of the preamplifier in the electrode, a reference electrode is necessary and it should be placed as far away as possible and on electrically neutral tissue. Power-line interference noise may be reduced by judicious placement of the ground electrode.

10.3.4 Crosstalk and muscle movement artefacts

Crosstalk is one of the serious confounding factors of dynamic EMG signals when the time and intensity of the activation of muscles is to be considered. Crosstalk is detected by selectively stimulating a specific muscle and observing the signal generated on neighbouring muscles. A signal separation technique, FAST-ICA, provided by Hyvärinen, is a technique which identifies and removes the crosstalk signals in clinical measurements. During dynamic contraction the muscles moves and this cause a change in the geometry of the electrode–muscle system. Due to this sliding of the innervation zone near or below a bipolar electrode, the dataset shows marked reduction of signal amplitude. This reduction of signal amplitude might be interpreted as a reduction of muscle activation. To avoid this, the EMG electrodes were placed over the distal half of the muscle belly so that the contact surfaces were aligned longitudinally to the muscle fibres. The steps involved in conducting clinical measurement are as follows:

a. Prepare the electrode sites by cleaning the skin with isopropyl alcohol and shave the hair, if necessary, to ensure good contact.
b. Attach silver/silver chloride electrodes using an adhesive pad and electrolytic gel (centre-to-centre electrode spacing = 30 mm, diameter = 8 mm).

c. Position the common reference electrode over the distal end of the muscles concerned.
d. Keep the amplifier gain constant at 10 000 with an input impedance > $25M^*$ at DC and > $15M^*$ at 100 Hz, with an EMG amplifier (common mode rejection ratio = 85 dB at 50 Hz, bandwidth = 20–500 Hz).
e. Repeat the measurement protocol for each subject in each test session.
f. Maintain a 40 rpm cadence by following a metronome.
g. Achieve a steady cadence, collect 15 s (i.e. ten cycles of the crank assembly at 40 rpm) of EMG signal and store it using a BNC connector board in a data recorder (four channels).
h. Identify the crank top dead centre. The circular disc attached to the crank assembly and crank arm length will produce spikes in the EMG signals from which the crank top dead centre can be identified (figure 10.1).
i. Process the data obtained using the software further in the computer.
j. Sample the final obtained data with a sampling rate for each channel of 1000 Hz and stored in ASCII formats.
k. Use the Fast-ICA technique to reduce the crosstalk and power-line frequency disturbance [1].

10.3.5 The measured electromyogram

The normalized EMG for eight important muscles over one crank cycle is given in figure 10.5 for the optimum value of the angle of seat tube and seat height acquired in section 10.2.1. The muscle activity pattern shows that there is little overlap in the region of greatest activity of the quadriceps and hamstring muscles, and the simultaneous activities of these muscles crossing opposite sides of a joint are known to be incompressible. Overlap in the muscular activity regions of potentially synergistic muscle pairs (the gluteus maximus, quadriceps and gastrocnemius) has been observed, due to the activity of the different muscles being diverse during different stages of the crank cycle. In an analysis of the EMG signal and its correlation with pedal force data, three different stages of the crank rotation cycle can be identified. The first stage (TDC to 100°) includes the greater activity of the tibialis anterior, quadriceps and gluteus maximus. The peaks of the vastus lateralis and vastus medialis occur in this region. The second stage is the region from 100° to –260° degrees of the crank cycle, where the active hamstring and gastrocnemius muscles are largely considered. However, the gluteus maximus co-acts with the hamstring muscles for a short time (after 100°) but the peak activity levels are not overlapping. The muscles active in the third stage (from 260° to the TDC) are the tibialis anterior and quadriceps. The peak activity of the gluteus maximus is very clear during the power stroke. However, it surprisingly co-contracts with both the hamstring and quadriceps groups. The activity of the gastrocnemius starts slightly after cessation of the tibialis anterior. It can be concluded that these two muscles are activated on opposite sides of the pedal cycle. The onset of the vastus lateralis and vastus medialis is close to 260° and terminates at about 100°–110°. Although these muscles form part of the same muscle group as the rectus femoris, it is interesting to note that they are active prior to the rectus femoris.

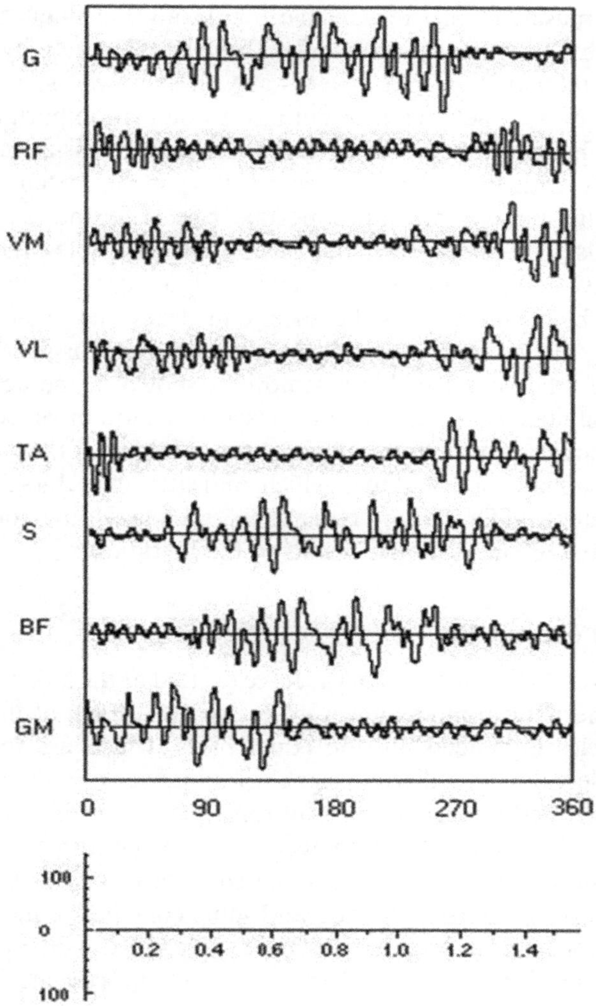

Figure 10.5. The normalized EMG for the eight muscles of the lower limb over one crank cycle.

10.3.6 Muscle activity pattern

Examining the activation pattern of each muscle, it was found that the regions for each subject are not significantly different (figure 10.5). During pedalling, the vastus lateralis, vastus medialis, gastrocnemius and biceps femoris were activated synergistically during the power phase of the pedalling action, which points towards these muscles pushing the crank downward. The vastus lateralis extends the knee in support of the RF and the vastus medialis. In the same activity the gastrocnemius muscle plantar flexes the ankle joint to transfer this exerted force to the pedal. When the knee is bent, the semimembranosus supports the tibia inward for turning the foot. The activity of the quadriceps is to extend the leg leading from the thigh, while the movements of the pelvis and trunk upon the femur are maintained by the rectus femoris. The same muscle is also responsible for flexing the thigh on the pelvis. However, if the thigh is fixed the action is to flex the pelvis. The activity of the vastus medialis is to draw the patella media inwards as well as upwards. The role of the bicep femoris is not only to flex the knee but also to play an important role in the extension of the hip, hence astonishingly both the vastus medialis and vastus lateralis co-act synergistically with the bicep femoris resulting in hip extension and knee flexion. The hamstring muscles play an important role in flexing the leg upon the thigh. If the knee is semi-flexed then the bicep femoris rotates the leg slightly outward. The activity of the semimembranous is such as to rotate the leg inwards and is dissimilar to bicep activity. As the knee joint flexes due to shortness of the hamstring muscles, the complete flexion of the hip remains unchanged. The responsibility of the dual action of bicep femoris is shown in the activity pattern when the leg is straightened, similar to the pushdown motion of pedalling. The tibialis anterior is activated through the opposite range to that of the gastrocnemius, during the recovery phase of pedalling, which indicates that dorsiflexion occurs during pedalling near the TDC. Many researchers, such as Houtz and Fischer [7], and Despires [5], have reported similar results in the literature.

10.4 Joint force estimation

The normalized EMG is used to estimate the force exerted at the ankle, knee and hip joint using the biomechanical model given in section 10.1. Each of the normalized EMGs of the muscles is divided into 36 equal parts, which are integrated and averaged for a single value as a discrete signal of number, $N = 36$. These discrete integrated EMGs are used to decode the joint force pattern at the ankle, knee and hip joint, using the Kalman estimation technique. When some conditions are satisfied, the Kalman filter minimizes the estimated error covariance. The Kalman filter model depends on a recursive method and some consecutive measured inputs taken separately to calculate an accurate observed value [6]. The filter endeavours to discover an estimation of the state x_k, where the discrete process attains time steps equal to k:

$$X_{k+1} = A_k x_k + w_k, \tag{10.10}$$

where matrix A_k is a matrix which is linearly related to the next state and the present state, and $w_k \sim N\,(0,\, w_k)$ represents the incorporated noise. The equation from the measurement (in our case discrete IEMG, $n = 36$) is defined as

$$z_k = H_k x_k + q_k, \tag{10.11}$$

where H_k stands for the relation between the measurement and the process state. $q_k \sim N\,(0,\, q_k)$ is the noise value during measurement. The Kalman filter is applied to the recorded EMG z_k, for estimating the joint forces (Fx, Fy) of the hip, knee and ankle joints. The size of A_k is $N \times N$, and the size of H_k is $N \times M$, where N is the number of muscles exerting force on a particular joint (table 10.5), and M represents the joint forces (horizontal and vertical forces for the ankle, knee and hip joints).

In our analysis we assume that H_k, A_k, q_k and w_k do not vary with time while computing their value during the training phase:

$$AX_2X_2^{\mathrm{T}}(X_1X_1^T)^{-1} \tag{10.12}$$

$$H = ZX^T(XX^T)^{-1}, \tag{10.13}$$

where X, X_1, X_2 and Z are defined as

$$X = \begin{bmatrix} X_{11} & .. & X_{1M} \\ .. & .. & .. \\ X_{21} & .. & X_{2M} \end{bmatrix}$$

$$X_1 = X = \begin{bmatrix} X_{11} & .. & X_{1M-1} \\ .. & .. & .. \\ X_{21} & .. & X_{2M-1} \end{bmatrix}$$

$$X = \begin{bmatrix} X_{12} & .. & X_{1M} \\ .. & .. & .. \\ X_{22} & .. & X_{2M} \end{bmatrix}$$

$$Z = \begin{bmatrix} Z_{11} & .. & Z_{1M} \\ .. & .. & .. \\ Z_{N1} & .. & Z_{2NM} \end{bmatrix},$$

where N is the number of muscle involved.

After obtaining the values of A and H, we work out the values for W and Q with the help of the expressions [6]

$$W = (X_2 - AX_1)(X_2 - AX_1)^T/(M - 1) \tag{10.14}$$

and

$$Q = (Z - HX)(Z - HX)^T/M. \tag{10.15}$$

The joint forces Fx and Fy at the ankle, knee and hip joints are estimated using the measured (discrete signal $n = 36$) EMG signal z_k. For every step k, the *a priori* state $\hat{X}_k^-$ is estimated using the error covariance matrix P_k^- and the previous state:

$$\hat{X}_k^- = \mathrm{A}\hat{X}_{k^{k-1}} \tag{10.16}$$

$$P_k^- = \mathrm{A}P_{k-1}A^T + W, \tag{10.17}$$

where in the first iteration $\hat{x}_k^-$ and P_k^- are initialized by one. We then obtain the Kalman gain (K) and use this new value with the estimation $\hat{X}_k^-$ and measured electromyogram, z_k, to update the estimate and the error covariance matrix:

$$K_k = P_k^- H^T(HP_k^- H^T + Q)^{-1} \tag{10.18}$$

$$\hat{X}_k = \hat{X}_k^- + K_k\left(z_k - H\hat{X}_k^-\right), \tag{10.19}$$

$$P_k = (I - K_k H)P_k^-. \tag{10.20}$$

Table 10.5. Normalized estimated joint force.

Body segment	Muscle involved	Activity pattern	Normalized estimated force using Kalman filter
Foot	**TA**		$F_{AX} = m_f\, a_{fx} + Fx$ 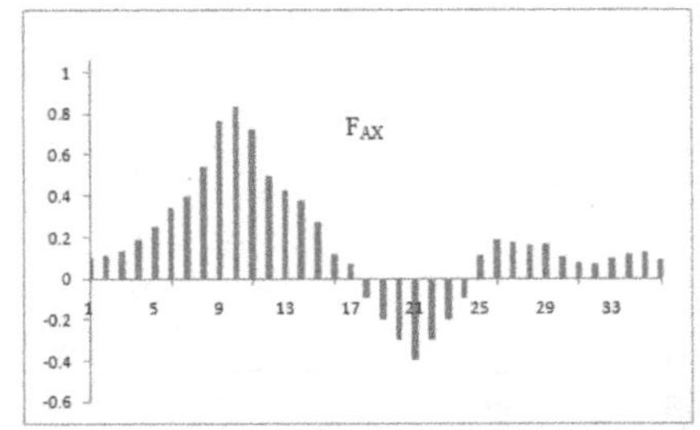
	G	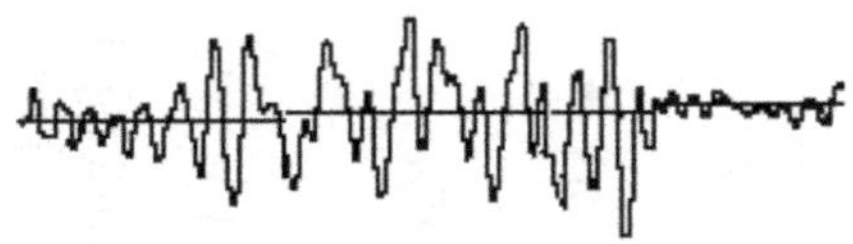	$F_{AY} = m_f\,(a_{fy} + g) + Fy$ 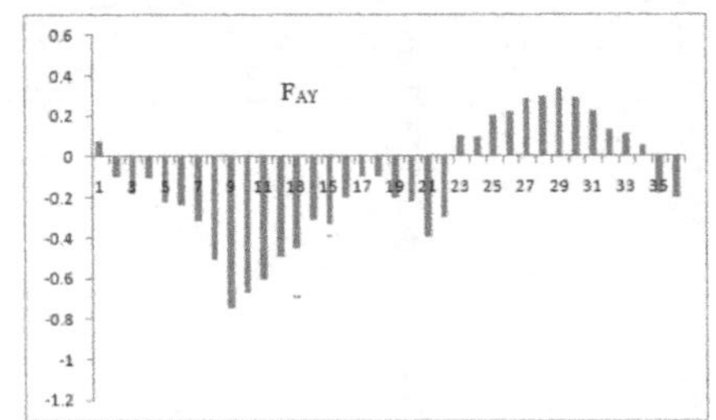

Knee	**RF**	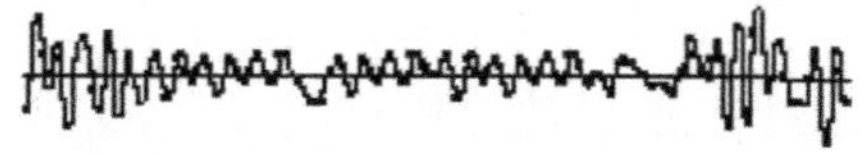	$F_{KX} = m_s\, a_{sx} + F_{AX}$ 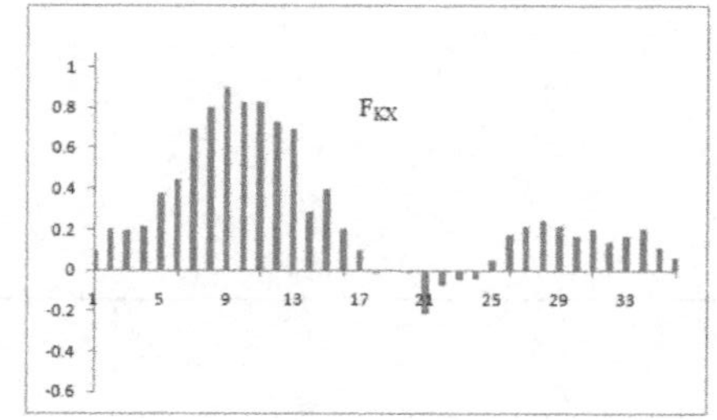
	VM	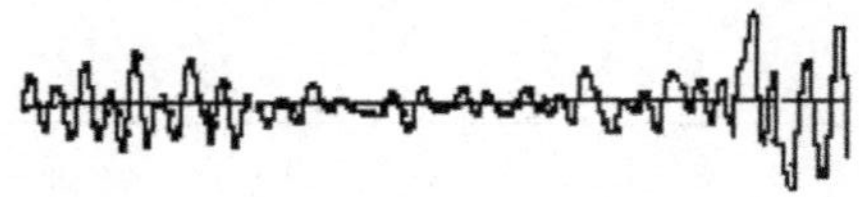	
	VL	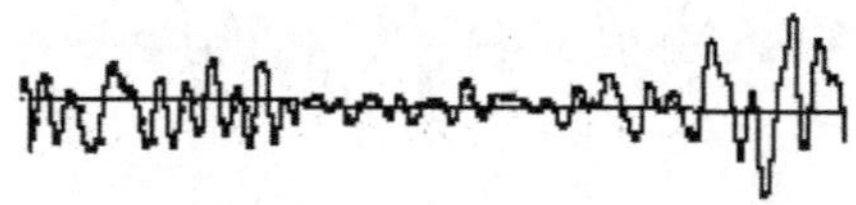	$F_{KY} = m_s\,(a_{sy} + g) + F_{AY}$ 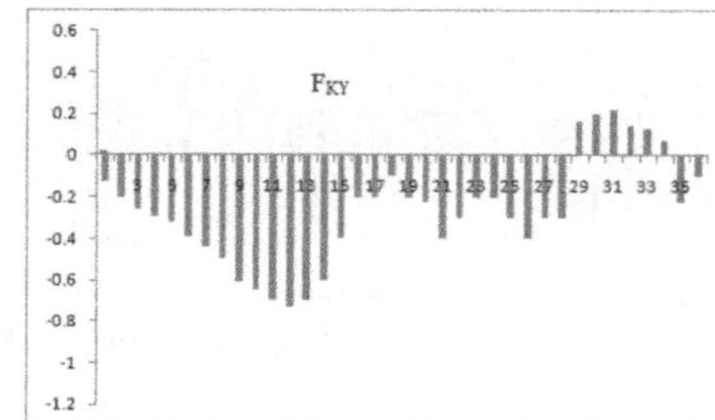
	BF	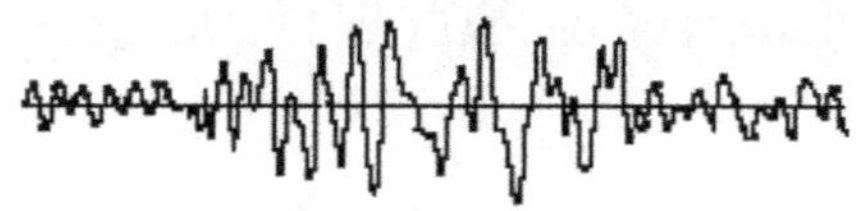	
	S	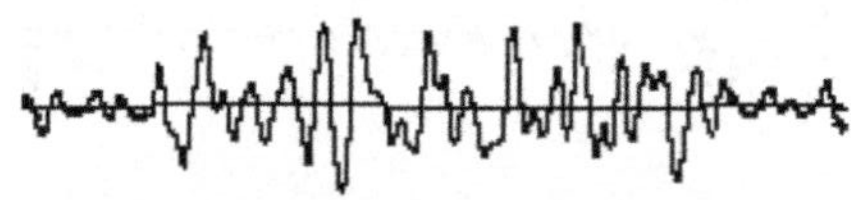	
	G	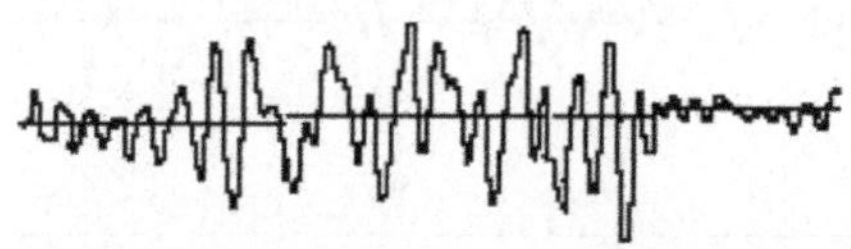	

(*Continued*)

Table 10.5. *(Continued)*

Body segment	Muscle involved	Activity pattern	Normalized estimated force using Kalman filter
Hip	**GM**	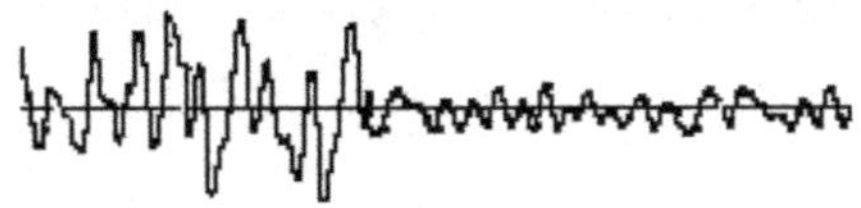	$F_{HX} = m_t\, a_{tx} + F_{KX}$ 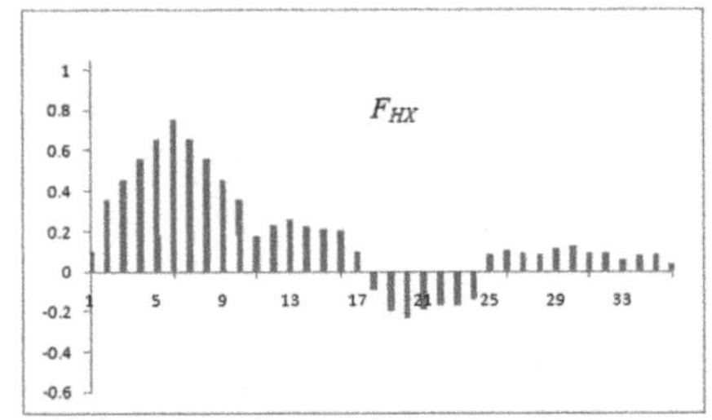
	RF	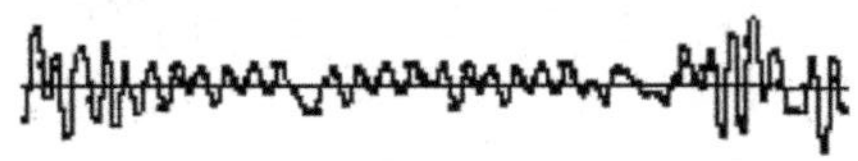	
	BF	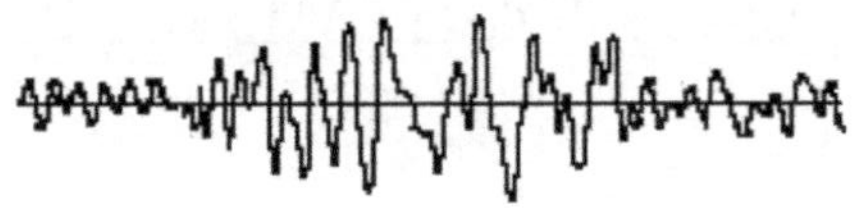	$F_{HY} = m_t\,(a_{ty} + g) + F_{KY}$ 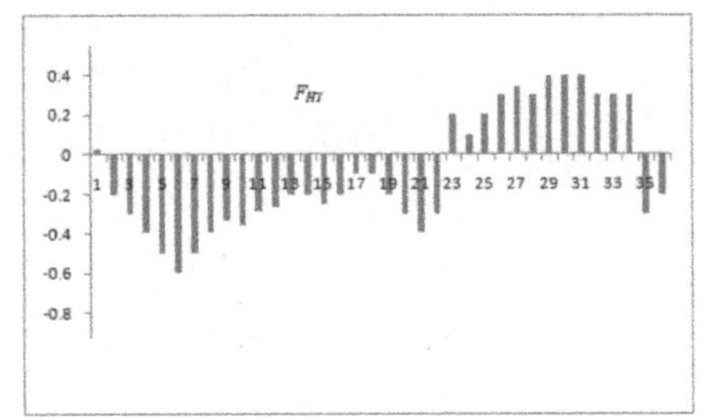
	S	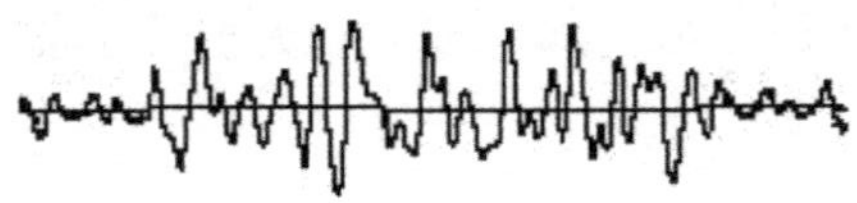	

The above algorithm replicates lone iteration only, where joint force is interpreted for one crank cycle (i.e. 1.5 s, table 10.5).

Table 10.5 illustrates the predicted normalized force using the Kalman filter and the major muscles accountable for the force generated [12]. The horizontal and vertical force pattern in the foot–pedal segment is developed by the tibialis anterior and gastrocnemius muscle activity. The last column indicates the Kalman estimation of the force using associated muscles. Since the computation of force is intricate and using normalized muscular activity we can predict only the force pattern applied for complete crank cycle. Similarly to the knee, the hip joint force (normalized) is also presented using a normalized activity pattern of the contributing muscles.

10.4.1 Joint moment pattern and performance measurement

The normalized joint force obtained using the Kalman filter is used to compute the joint moment pattern. The actual force measured by the pedal dynamometer is utilized to calculate the joint moments. The normalized estimated force for all three joints is utilized to obtain the value of the moments at the hip, knee and ankle joints. Table 10.3 clearly indicates that the patterns obtained by estimated values are similar in pattern to the actual joint moments calculated by biomechanical analysis. The actual joint moments calculated and the joint moments calculated by the predicted joint forces were normalized using the Pearson formula [17]:

$$X_{\text{Normalized}} = \frac{X - X_{\min}}{X_{\max} - X_{\min}}. \tag{10.21}$$

The performance of the proposed approached is analysed using a correlation coefficient (CC) matrix [17]. This is an indicator of the similarity between the expected joint moments given by the EMG analysis and given by the mathematical model. The values of both moments were taken at an increment of 10° of crank rotation,

$$r = \frac{n\sum(xy) - \sum(x)\sum(y)}{\sqrt{[n\sum x^2 - (\sum x)^2]}\sqrt{[n\sum y^2 - (\sum y)^2]}}, \tag{10.22}$$

where n represent the samples, which are 36 in our case. Figure 10.6 shows the correlation between the normalized joint moments calculated using biomechanical analysis and EMG estimation

The computationally complex biomechanical analysis of the lower limb demonstrates the often counterintuitive role of joint moments in recurring activity. Each joint moment can produce linear and angular acceleration of all body segments due to the dynamic coupling inherent in the interconnected musculoskeletal system. Table 10.3 shows the joint moments calculated by the pedal dynamometer and the biomechanical analysis and moments estimated by EMG analysis of the corresponding muscle activity only. The second column presents the normalized predicted moments over one crank cycle using Kalman estimation.

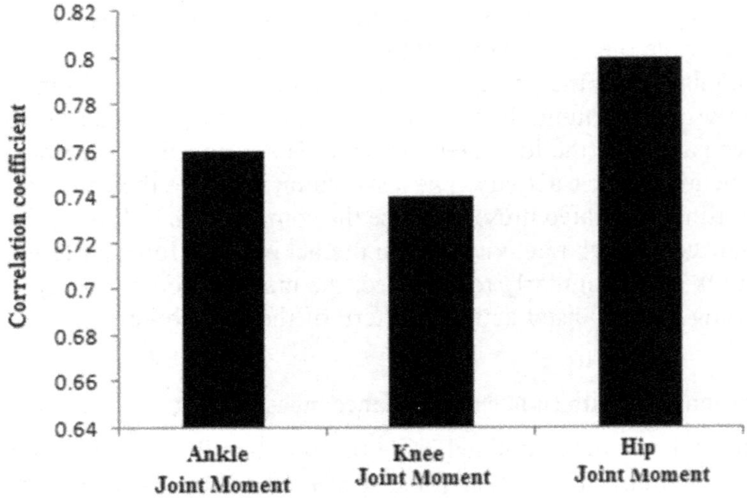

Figure 10.6. Correlation between the joint moments using biomechanical analysis and predicted by EMG analysis.

10.5 Conclusions

The estimation of muscle forces and activity patterns is crucial for defining many physical ailments, if the analysis correlates with joint activity and its load pattern to avoid excessive load on a particular joint. In our study an optimum clarification of the inside allocation problem of muscle load is given which encourages the prediction of synergistic muscle action with biomechanically reasonable individual muscle forces. These methods utilize a limitation on muscle potency which is a set segment of small muscle strengths that contribute at a specific joint moment. Simulated muscle activity often fails to represent experimentally recorded muscle activity. In contrast the given method is shown to correlate well with recorded EMG activity. Bicycle operators carry out the repetitive action of pedalling for long periods, thus a few muscles are subjected to excessive repetitive use. This creates many short- and long-term health problems. The adverse effects of power-line, crosstalk and low level myoelectric signal reception from other sources, which obscure the description of a muscle's activation timing and the comparative intensity of its activity, is emphasized. Raw signals have been separated from unwanted components via an independent component analysis (ICA) technique. This led to predicting the joint forces in the vertical and horizontal directions. Based on the biomechanical model of the leg as a five-bar linkage model, the kinematics of the leg and joint moments at the ankle, knee and hip joint were calculated. A saddle angle equal to 76° and a seat height equal to 97% of trochanter leg length are found to be optimal for the activity of the muscles considered. Correlation coefficients equal to 0.80, 0.73 and 0.76 are obtained between the measured and estimated joint forces at the hip, knee and ankle joint, respectively. Thus the activity pattern of muscles correlates well with the force applied. The activity regions of the major muscles are illustrated, which also play an important role in reducing physical ailments.

References

[1] Hyvärinen A 1999 Survey on independent component analysis *Neural Comput. Surv.* **2** 94–128

[2] Crowninshield R D and Brand R A 1981 A physiologically based criterion of muscle force prediction in locomotion *J. Biomech.* **14** 793–801

[3] DeLuca C J 1984 Myo-electrical manifestations of localized muscular fatigue in humans *CRC Crit. Rev. Biomed. Eng.* **11** 251–79

[4] Dempster W T 1956 The anthropometry of body action *Ann. NY Acad. Sci.* **63** 559–85

[5] Desiprés M 1974 An electromyographic study of competitive road cycling conditions simulated on a treadmill *Biomechanics IV* (Baltimore, MD: University Park Press) pp 349–55

[6] Bishop G and Welch G 2001 An introduction to the Kalman filters *Proc. of SIGGRAPH, Course* **8** 41–53

[7] Houtz S J and Fischer F J 1959 An analysis of muscle action and joint excursion on a stationary bicycle *J. Bone Joint Surg.* **41A** 123–31

[8] Hull M L and Jorge M 1985 A method for biomechanical analysis of bicycle pedaling *J. Biomech.* **18** 307–16.

[9] Kautz S A, Feltner M E, Coyle E F and Baylor A M 1991 The pedaling technique of elite endurance cyclists changes with increasing workload at constant cadence *Int. J. Sports Biomech.* **7** 29–53

[10] Mukaka M 2012 A guide to appropriate use of correlation coefficient in medical research *Malawi Med. J.* **24** 69–71

[11] Merletti R and Torino P D 1999 Standards for reporting EMG data *J. Electromyogr. Kinesiol.* **9** 3–4

[12] Wu W, Black M J, Gao Y, Bienenstock E, Serruya M and Donoghue J P 2002 Inferring hand motion from multi-cell recordings in motor cortex using a Kalman filter *SAB'02-workshop on Motor Control in Humans and Robots: On the Interplay of Real Brains and Artificial Devices* pp 66–73

[13] Faria I E and Cavanagh P R 1978 *The Physiology and Biomechanics of Cycling* (New York: Wiley) p 175

[14] Ericson M O, Bratt A, Nisell R, Nemeth G and Ekholm J 1986 Load moments about the hip and knee joints during ergometer cycling *Scand. J. Rehabil. Med.* **18** 165–72

[15] Gonzales H and Hull M L 1989 Multivariable optimization of cycling biomechanics *J. Biomech.* **22** 1151–61

[16] Yoshihuku Y and Herzog W 1990 Optimal design parameters of the bicycle-rider system for maximal muscle power output *J. Biomech.* **23** 1069–990.

[17] Pearson K 1895 Notes on regression and inheritance in the case of two parents *Proc. R. Soc. Lond.* **58** 240–2

IOP Publishing

Modelling and Analysis of Active Biopotential Signals in Healthcare, Volume 1

Varun Bajaj and G R Sinha

Chapter 11

Cloud-based cardiac health monitoring using event-driven ECG processing and ensemble classification techniques

Saeed M Qaisar, Abdulhamit Subasi

Effat University, College of Engineering, Jeddah, 21478, Saudi Arabia

Electrocardiogram (ECG) signals have been widely utilized to detect heart arrhythmias. Cloud-based health monitoring is beneficial in tracking, avoiding and treating these abnormalities. In this chapter, a method for the monitoring of cardiovascular health is presented. It processes and classifies the electrocardiogram (ECG) signals to realize an automated cloud-based diagnosis of chronic heart disease. To achieve real-time compression and efficient signal processing and transmission, event-driven ECG signal acquisition is utilized. The experimental results reveal that a substantial gain in compression and usage of bandwidth is achieved by the designed solution, in contrast to the classical counterpart, due to the event-driven nature. It results in a significant reduction of the computational complexities of the post-denoising, feature extraction and classification. The denoised signal features are extracted by employing the autoregressive (AR) Burg method. Ensemble classifiers are then used to classify the different types of ECG signals. The precision of the system is studied regarding the performance of the classification.

11.1 Introduction

Recent developments in the smart device industry have gradually increased the deployment of biomedical implants to provide users with a virtual healthcare framework. The role of wearable devices in monitoring biomedical signals is to regulate the smart sensor data. Over the last decade, clinically related wearable devices have included instruments such as stethoscopes, blood pressure monitors,

doi:10.1088/978-0-7503-3279-8ch11

Holter ECG readers and portable EEG systems. While this trend still persists today, wearable design firms have also developed revolutionary tools and techniques to capture and analyse body signals. Thus biomedical signals obtained from wearable sensors enable patients with chronic diseases to be remotely monitored continuously and in real time. In addition, several of these biomedical signals have clinical significance in monitoring the stage of the condition or disease [1].

Advanced and sophisticated implants are also utilized for processing biomedical signals. This is particularly popular in the case of contemporary hospitals and homes for the elderly. Such implants augment the precision and accelerate the procedure of diagnosis and decision making. New biomedical implants offer the future potential of using wireless sensing methods for remote cardiovascular system healthcare monitoring. This is particularly beneficial for the elderly and seriously ill cardiac patients [2]. The effective realization and embedding of wireless sensing technology has attracted various fields, for example biomedical implant based fabric design and computer based health diagnosis [3].

Decreased fertility and increased lifespan have led to a high elderly population ratio in developed countries. This has subsequently increased medical needs. Mobile health monitoring, diagnosis and treatment offer the possibility of making medical care cheaper and thus easily affordable. It will be possible to obtain quality medical treatment without the need to go to hospitals or healthcare centers. There is also a significant population who cannot take appropriate care of their health because of transportation issues and limitations of mobility. There might be a lack of transportation and/or healthcare infrastructure. This is particularly valid for developing countries with limited health budgets and is also true for rural and remote communities. Their level of healthcare can be improved tremendously by connecting them with medical specialists virtually. This can be realized in this modern technological era by smartly combining information and wireless communication technologies. Such a development will be highly beneficial for rural and remote communities in general and for the developing world in particular. In this context, promising studies are ongoing to develop a cloud-based healthcare infrastructure, that is capable of gathering and managing the data of potential patients [4].

Recently, there have been many advances in surgical devices and clinical technology to meet the needs of existing diagnosis and treatment in the healthcare field. These modern healthcare tools and devices provide precise and accelerated health analysis. Additionally, thanks to wireless sensing technology and contemporary networking strategies, it is becoming technically convenient to realize the real-time continuous observation of elderly people and patients with critical health conditions.

The continuous and live surveillance of patients and elderly people, particularly those with chronic illnesses, is essential [4]. Cloud-based mobile healthcare monitoring devices play a key role in tracking and analysing the patient condition. Real-time patient monitoring systems gather people's biomedical signals on an ongoing basis to examine the risks associated with vital health conditions in real time [5]. Thanks to the integration of cloud servers and sensing devices, sensor-based mobile patient monitoring has become increasingly common in personalized healthcare frameworks in recent years [6].

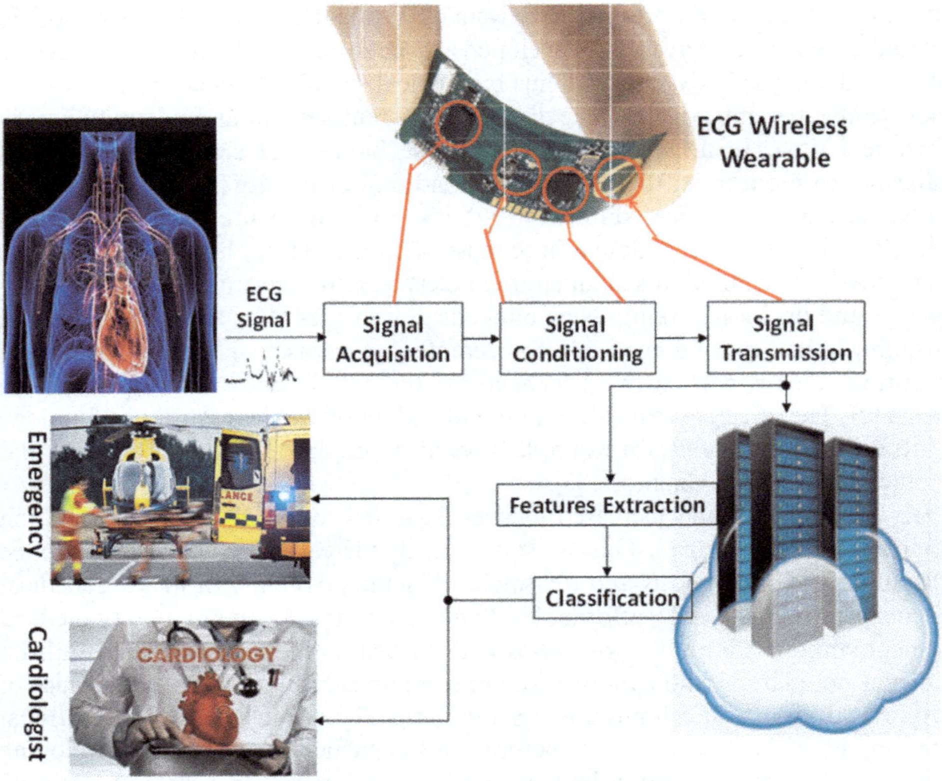

Figure 11.1. A framework for cloud-based cardiovascular disease monitoring.

Figure 11.1 presents a cloud-based patient tracking framework for medical diagnosis and care. In this framework, mobile devices constantly gather biomedical signals from different intelligent wireless wearables, conditioning the signals and transmitting them to a cloud-based application. The outcomes are transmitted to the relevant health centers. Notifications are sent in the case of emergency to alert the emergency services, cardiologists and the relatives of the patient [6].

In order to analyse biomedical signals for signal recognition, numerous signal processing and machine learning algorithms can be employed. Wang *et al* [7] utilized a field programmable gate array (FPGA) based tactic for cloud-based ECG signal analysis. In [8] an ECG monitoring device is utilized to conduct machine intelligence-based ECG signal analysis and send the findings to the cloud. Cloud-based methods for the detection of coronary heart failure have become common to track patients with heart disease [8].

Remote healthcare management services based on cloud servers are supported by smart wireless wearables, personal digital assistants and cloud computing [7, 8]. They involve the use of services offered by cellular network providers, the Global Positioning System (GPS) and general packet radio services (GPRSs). Sensors can acquire biomedical signals on the basis of these capabilities but cannot store, interpret or transmit these signals without using a mobile network such as 5G [9].

Thanks to the remarkable advances in intelligent biomedical wireless implants, remote and mobile processing modules, and wireless communication networks, the latest developments of cloud-based mobile patient tracking services have become possible. There is a large number of accurate, wireless and smart sensors for collecting different health-related biomedical data. These sensors can easily be incorporated into a smartphone or smartwatch. These smart devices can contribute crucially and significantly to the realization of contemporary and efficient cloud-based mobile healthcare solutions [9]. An effective wireless implant records the vital health condition signals. These recordings are made directly from the relevant patient. Then, the implant preconditions, treats and transmits the conditioned vital health information to the smartphone. Then the smartphone sends the collected data to the cloud via a wireless cellular network to verify the patient's status continuously. In the case of an emergency, the cloud service application sends the details to the emergency services, clinician and the patient's family [10].

This chapter introduces an original event-based architecture for smart acquisition and processing of ECG signals via front-end wireless wearables connected to a cloud-based application in an arrhythmia detection system. The goal is to increase device performance and the wearable power efficiency by achieving real-time compression, computational effectiveness and reduction of transmission activity, while ensuring adequate classification accuracy. The method is promising and leads to an effective ECG wearable implementation along with precise decision support for arrhythmia identification [11–13].

11.2 Background and literature review

Heart disease is one of the main causes of death worldwide [14]. Ventricular arrhythmias cause irregular heartbeats and are the cause of nearly 80% of sudden heart failures. The electrocardiogram (ECG) is a complicated signal and provides essential information about cardiovascular system functionality [15]. Interference and physiological artifacts alter the ECG signal by decreasing the effectiveness of post-processing, such as feature extraction and classification. To overcome these deficiencies, numerous signal processing techniques have been proposed [15]. Several techniques have been implemented for the preprocessing of ECG signals, such as quadratic filtering [16], Kalman filtering [17] and principal component analysis [18]. The ECG signals are processed and analysed using effective methods of signal processing to accurately classify cardiac diseases [15, 19]. The short time Fourier transform (STFT), discrete wavelet transformation (DWT), wavelet packet decomposition (WPD) and autoregressive (AR) Burg method are the ECG feature extraction tools commonly used [15, 20]. Features, obtained with the extraction process, are used for automatic diagnosis of cardiac arrhythmia. Numerous data mining techniques are utilized for the classification of ECG signals [20]. Some key ECG classification approaches are the neural networks [20], support vector machines (SVMs) [15], exhaustive k-means clustering [21], k-nearest neighbors (k-NN) [20] and fuzzy logic [15].

The successful treatment of heart failure can be achieved with a prompt diagnosis of arrhythmia disorders. Therefore, patients with chronic heart disorders need constant monitoring. A wearable ECG system is one of the best instruments for the continuous monitoring of heart disorders [22]. Smart wearables acquire the ECG signal and forward their findings to a testing center [22]. In addition, the embedded application can be used in the case of emergency to call the relevant emergency center or clinician with specifics of the patient's GPS tracking location [22].

Precise and accurate notification to healthcare experts of adverse conditions is important for emergency management. In this sense, a cloud-based mobile healthcare system could broaden and improve healthcare outreach, decision making, chronic illness treatment and the tackling of emergencies. Such a strategy could allow healthcare professionals to make timely decisions and provide the elderly with timely treatment and care.

The smart ECG wearable architecture remains challenging due to the strict size, weight and power consumption constraints. In these circumstances, self-powered smart sensors are usually preferred, as they allow a continuous acquisition of ECG signals without causing the patient too many restrictions. Important power savings in a wireless ECG system can only be accomplished by reducing the use of wireless transceivers, i.e. the transmission of processed data is favored over raw data [22]. To this end, several studies have been performed on the compression of the ECG signal [23–25], non-uniform resampling [26] and event-driven data acquisition [12, 22].

11.3 ECG in healthcare

An electrocardiogram (ECG) is recorded non-invasively from the body surface of the patient. It presents the activities of the cardiovascular system of the patient in question in the form of an electrical signal. It is recorded with the help of appropriate electrodes which are placed in particular positions on the patient's body. A particular ECG recording zone is the patient's chest area [13]. Einthoven devised a variety of approaches and ideas in this regard. He invented certain useful concepts for the conditioning, processing and automatic categorization of different types of heart pulses. He also investigated the impact of electrode placement, at various body locations, on the performance of an automatic heartbeat identifier. During this process, various locations on the arms, legs and chest were considered. He also presented a theoretical model of the measurement and identification mechanism based on a single time-varying dipole [27]. Additionally, a differential lead based ECG signal recording was also outlined in [27]. In this case, the measurement was carried out between two predefined positions on the patient's body surface.

The ECG signal is the most important element in clinical practice. Through the introduction of computers, integrated circuits and mobile networks, the technological advances in electrocardiography have accelerated greatly. In addition, ECG enhancement turns out to be feasible in the presence of wireless connectivity and cloud computing. A variety of advanced materials have been investigated for ECG sensors. There are currently numerous nano-devices available which can acquire, analyse and convey ECG signals to the cloud. These are tiny, wireless and very lightweight devices and allow the effective monitoring of the cardiovascular functionality of the people concerned. Because of their tiny dimensions and light weight, the patient does not feel too much inconvenience while using these implants. Most of these implants allow non-invasive measurement of the ECG. Thanks to low-power architectures and electronics technology, the autonomous working life of these devices is extended. They allow the uninterrupted and live monitoring of a patient's cardiovascular system functionality for a much longer period compared to the traditional Holter recorders. These implants are not limited only to the monitoring of the cardiovascular system, but are also employed for other potential biomedical applications, such as the monitoring of patients with epilepsy, Alzheimer's, diabetes, high blood pressure and temperature. Additionally, they can also provide the geographical locality of the person concerned. This information is particularly useful for emergency cases [9].

A known clinical standard is to utilize the vital biomedical signs of the patient concerned for monitoring and diagnosis of health disorders. These signals are recorded using the smart nano-implants. The acquisition process introduces noise and artifacts into the signal. This can affect and corrupt the information which is contained by the signal and can result in an erroneous diagnosis. Therefore, it is important to prepare and precondition these signals before passing them to the analysing physicians. This task is performed by trained and qualified medical technicians. They denoise and condition the raw signal in order to prepare it for the medical consultants. This is realized using noise and artifact removal methods such as filtering and principle component analysis.

Clinical biomedical indications are still considered a common norm for the treatment of illnesses and diseases. Artifact-free and noise-free signals are prepared by qualified technicians before they are sent to the consulting physician for the diagnosis procedure. It is important to use wearable biomedical implants for the uninterrupted live surveillance, monitoring, analysis and diagnosis of cardiovascular system functionality. Multi-channel ECG signals are acquired via smart biomedical implants to automatically detect cardiac functionality disorders. An uninterrupted and long-duration ECG analysis is necessary for precise diagnosis and health condition identification of the cardiovascular system of the patient concerned. Overall it is a very complex and sophisticated mechanism and involves a variety of multidisciplinary advanced techniques and technologies. The system's findings should be effective and precise, otherwise harm to the patient could result and incorrect judgments could be made during diagnosis. Through using cloud infrastructure and software, this can be avoided through creating a tracking system between patients and physicians. In addition, some of the currently available wearable devices that acquire biomedical signals from patients are rarely used by physicians. This is due to the quality of the acquired data and its difficulty of interpretation. Different academic organizations and research and development teams have recognized these shortfalls and have begun designing applications for analysing critical signals. Some popular wearables already commercialized for analysing ECG signals are Muse [28], Epoc++ [29] and General Electrics's Holter

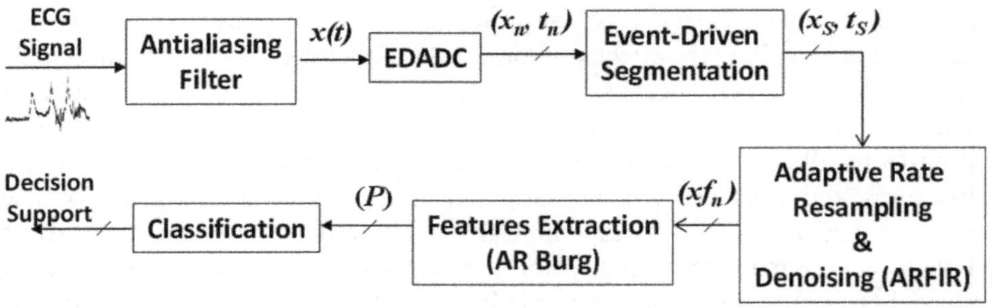

Figure 11.2. A block diagram of the proposed system.

ECG [30]. There are several exciting studies in the literature of smart health surveillance systems that have led to the development of many platforms, approaches and frameworks. Smart monitors and mobile applications provide solutions for delivering healthcare services to remote patients in digital health monitoring systems.

Salvador *et al* [31] developed a system for the provision of portable acquisition tools and cellular phones to cardiac patients that enable data transmission. However, the system only took into account patients in a stable state and deliberately omitted emergency conditions. Herscovici *et al* [32] investigated existing technologies for the construction of smart healthcare networks that identify emergencies in ambulances and rural health centers for mobile electrocardiography, and are completely combined with mobile telechography. Shih *et al* [33] established an integrated smart ECG system for the detection and monitoring of older patients using mobile devices. Ren *et al* [34] designed a smart healthcare network for patient tracking, with emphasis on security and sustainable wireless communication links. Xia *et al* [35] designed a cloud-based healthcare framework to track and evaluate ECG quality assessment, ECG enhancement and ECG parameter extraction in real time.

11.4 The proposed approach

The classical ECG diagnostic systems are static in nature, leading to device design constraints for the worst possible case [36, 37]. They capture the signal at a fixed frequency and then analyse it using fixed parameter based post modules. Thus they store and process a large amount of surplus samples. This increases the overall cost of computation, the volume of data transmission and the power consumption [37, 38]. To alleviate these shortcomings, this chapter proposes an event-driven acquisition, processing and classification system for ECG (see figure 11.2).

11.4.1 Dataset

The functionality of the proposed framework is verified using the MIT-BIH Arrhythmia database [39]. A collection of 24 h outpatient ECG recordings is utilized in this analysis. The dataset contains arrhythmias which are clinically significant. The signal is band-limited to 60 Hz and is captured at a sampling frequency of 360 Hz with standard 11 bit resolution analog-to-digital conversion (ADC). Every ECG recording considered is divided into short time length windows. The windows lengths are selected as equal to 0.9 s. In order to study the performance of the proposed method in terms of identifying various categories of heart pulses, ECG recordings of five different categories are employed. The categories considered are right bundle branch block (RBBB), normal (N), premature ventricular contraction (PVC), left bundle branch block (LBBB) and atrial premature contraction (APC). A total of 1500 ECG instances are considered with 300 instances considered for each of the five classes. This equal representation of each class avoids any possible bias during the classification process.

11.4.2 The event-driven acquisition

In the proposed system, an analog bandpass filter is used to filter the incoming analog ECG signal $y(t)$. The band-limited signal, $x(t)$, is conveyed to an event-driven analog-to-digital converter (EDADC). The traditional ECG based cardiovascular system disorder identification and diagnosis systems are founded on the basis of the Nyquist theory. They utilize traditional ADCs and record and process the ECG signals at a fixed rate. Therefore, the choice of processing rate is made for the worst possible case. As a result, these solutions are not efficient, which is particularly true in the case of time-varying intermittent signals such as ECG. These disadvantages can be mitigated by embedding signal-piloted EDADCs in the system [37, 40]. The functional mechanism of EDADCs is completely different to traditional ADCs. EDADCs do not follow the Nyquist sampling criterion and the acquisition of a new sample is not triggered by a fixed rate clock. Instead, they are based on the mechanism of level crossing sampling (LCS) and respect the Bernstein sampling theorem [37]. The sampling acquisition is triggered by the phenomenon of threshold crossing. A new sample is acquired once the band-limited analog signal $x(t)$ traverses any of the preset threshold levels. As a result, the incoming analog signal adapts the system's acquisition activity and sampling rate.

The signal discretized with LCS is repartitioned in a non-uniform time frame. The instants of sampling are defined by

$$t_n = t_{n-1} + \mathrm{d}t_n, \tag{11.1}$$

where t_n is the present sampling instant, t_{n-1} is the prior instant, and $\mathrm{d}t_n$ is the time difference between the present and the prior sampling instants [41]. For a selected EDADC amplitude dynamics, ΔV, and resolution, M, its quantum, q, can be calculated as $q = \frac{\Delta V}{2^{M-1}}$ [37, 39, 42]. The EDADC focuses on the relevant ECG information and avoids the unwanted information. They obtain a smaller number of samples relative to the classical equivalents. This greatly decreases the operations of the post-processing and transmission modules and achieves power-efficient implementation.

11.4.3 The event-driven segmentation

The EDADC output is segmented by the activity selection algorithm (ASA) [41]. It utilizes the non-uniformity of the sampling method to select only the important sections of the signal. The traditional windowing functions [40, 41] do not offer suitable features for the ASA. Only the necessary information of the signal is selected. Furthermore, the form and length of the window function are modified according to the local characteristics of the windowed signal [37, 41]. In addition, an efficient reduction of the phenomenon of leakage is also achieved [41].

11.4.4 The adaptive rate resampling and denoising

For a given resolution M, the EDADC sampling rate changes with the temporal variations of $x(t)$. The highest sampling frequency of EDADC [37] is determined by

$$Fs_{\max} = 2 \cdot f_{\max} \cdot (2^M - 1) \cdot \frac{A_{\text{in}}}{\Delta V}, \tag{11.2}$$

where $f_{\max}$ is the $x(t)$ highest frequency component, A_{in} is the input signal amplitude and $Fs_{\max}$ is the maximum sampling frequency of the EDADC. Let W^i represent the ith chosen segment which is produced by the ASA. If Fs^i is the frequency of sampling for W^i, then mathematically it can be computed by the relation $Fs^i = N^i/L^i$. L^i is measured in seconds and it is the time length of W^i. The count of samples which exist in W^i is denoted by N^i. There are plenty of mature and sophisticated signal treatment tools which are based on the traditional sampling mechanism [40, 41]. Therefore, to benefit from such methods and to integrate them in the suggested solution, each W^i is resampled uniformly. The resampling frequency is tuned by following the locally extracted parameters [41]. It resamples and processes the selected signal sections at the same or lower rates compared to to the conventional equivalents. Consequently, the proposed solution's computational advantage is greatly improved relative to conventional counterparts.

Let Frs^i represents the resampling rate for the selected segment W^i; then Frs^i is chosen specifically for each W^i. A reference rate of sampling is selected in the system and is denoted as F_r. This Frs^i selection mechanism is a function of F_r and Fs^i. Then, W^i is resampled at a rate of Frs^i and this results in a new count of samples, namely Nr^i. The online resampling is obtained using interpolation. The interpolation method updates the resampled signal characteristics with respect to the original signal [43]. For the devised method, the interpolation error depends on ΔV, M and the interpolation technique used [43].

Qaisar *et al* [43] showed that for the case of the EDADC, the maximum resampling error is bounded by q. However, it could be equal to ΔV in the traditional counterparts. This demonstrates the appropriateness of using quick and low-complexity interpolation approaches, rather than the conventional equivalents, in our proposed method. In addition, this study focuses on increasing the computational performance of wearable ECG devices and cloud-based computing. Consequently, the simplified linear interpolation (SLI) approach is considered for online resampling of the segmented signal parts. In SLI the value of xr_n is set equal to the mean of its prior and next non-uniform samples and the highest error per interpolation is limited by $\frac{q}{2}$ [43].

It is necessary to condition the biomedical signals to reduce the impact of artifacts and noise. The conditioning process is frequently achieved by utilizing digital filters [44]. Traditional digital filtering is static in nature. The sampling rate is kept fixed and the digitized signal is conditioned using a fixed order filter. Therefore, the parameters, such as sampling rate and the order of filters, are selected for the worst possible case [45]. Such a choice could result in a remarkable but unnecessary increase in the processing load and power consumption of the conditioning stage. This performance limitation can be improved by using multirate processing concepts [46]. In this context, novel techniques of digital filtering are proposed which can adjust their rate of conditioning and the order of filter according to the incoming signal changes [47].

Keeping in mind the targeted application, an appropriate bank of reference digital filters is realized offline. The implementation is achieved for a set of reference sampling frequencies, denoted as *Fref*. Different elements of the set *Fref* are computed by utilizing a fixed step of frequency, denoted as Δ. This leads towards a uniformly spaced placement of various elements of the set *Fref*. The first element in the set *Fref* is $Fs_{\min} \geqslant 2F_{C\max}$ and it defines the lower limit on the elements of set *Fref*. It assures an effective digital filtering process. The last element in the set *Fref* is F_r and it defines the upper bound on the elements of the set *Fref*.

For each selected segment one appropriate predesigned digital filter is selected during the real-time online signal conditioning. This selection is a function of the values of various elements of the set *Fref* and the real-time calculated value of Fs^i. Depending on the signal variations, the value of Fs^i can be specified and therefore an appropriate mechanism of the reference filter choice is adopted to obtain an effective and efficient denoising. For the case when $Fs^i \geqslant F_r$, the offline designed reference filter built for F_r is used for W^i. However, if $Fs^i < F_r$, then the reference filter whose corresponding value of $Fref_c$, which is neighboring the Fs^i is selected for W^i. Here, the index '*c*' represents the real-time selected element of the set *Fref* for online conditioning of W^i.

This adaptation of Frs^i causes W^i to be resampled closer to the Nyquist rates or at sub-Nyquist rates [47]. Hence, during data resampling and denoising procedures, it prevents unwanted interpolations and filtering operations. As a result, the proposed solution increases the computational and power efficiencies.

11.4.5 Extraction of features

The interesting features of the denoised signal are extracted using the autoregressive (AR) Burg method [15]. In this case, xr_n is modeled as the output of an all-pole causal filter. The method for an *O*th order model can be presented as

$$z_n = \sum_{j=1}^{O} a_j xr_{n-j} + b_n, \tag{11.3}$$

where xr_{ni} is the input, a_j are the model's weighing coefficients and b_n is the variance of the white noise σ^2. An Oth order model could estimate the power spectral density (PSD) using the following equation, where F_s is the sampling frequency:

$$P_{\text{AR Burg}}(f) = \frac{\sigma^2}{\left| 1 + \sum_{j=1}^{O} a_j \cdot e^{\frac{-i2\pi fj}{F_s}} \right|^2}. \tag{11.4}$$

11.4.6 Machine learning methods

The technological advancements in computing and electronics have evolved the utilization of machine learning approaches in a variety of applications in our modern era. The same trend is being followed in the domain of biomedical signal processing. A variety of mature machine learning algorithms have been introduced for the classification and diagnosis of biomedical signals. Certain practical examples are the identification of myocardial ischemia, epilepsy, heart attacks and cardiac hypertrophy. Such an approach is particularly beneficial in the case of emergencies. It can accelerate the process of analysis and decision making and can lead towards timely and effective treatment of the patient. Accurate and timely care measures can be taken through effective use of sophisticated machine learning approaches. Currently, various treatment approaches are being considered by medical practitioners. Certain common examples are electrocardiography, electroencephalography, magnetic resonance imaging and tomographic scanners [48].

Real-time and uninterrupted biomedical signal analysis is very important for tracking the health conditions of patients with critical conditions. The overall process consists of different stages of signal acquisition, conditioning, analysis and classification. It involves a smart combination of a variety of sophisticated techniques and technologies. An effective diagnosis can be realized by extracting pertinent features from the acquired and conditioned biomedical signals. This makes it convenient to identify the divergence of the intended signal from its standard parameters. The monitoring process should be robust and accurate in identifying the type of abnormality occurring. In addition, the doctor should be well trained in interpreting the detected abnormality as an accurate health disorder. Manual analysis of such data needs a lot of training and experience. This is very difficult, in particular in the case of big and multidimensional health data. Therefore, computer based machine learning methods can contribute a great deal in this respect [48].

Machine learning techniques may automate the process of biomedical signal analysis and classification between normal and pathological patterns by producing decisions to classify these patterns. A crucial feature of clinical surveillance has been the automated identification and recognition of physiological signals using various signal processing techniques [48]. Because biomedical signals provide such a large amount of data, the main problem for classification is how to classify the biomedical signal records. Next, the relevant features must be derived from the acquired biomedical signals, then the size of this feature set must be decreased and in next step the decreased size feature set is employed in the final classification stage. An algorithm that implements classification is called a classifier. A classifier knows how to use training sets to categorize the class of a function variable. The algorithms used for predicting categorical labels in the classification process are *the k*-nearest neighbors (*k*-NN), artificial neural network (ANNs), support vector machine (SVM), classification and regression tree (CART), random tree, REP tree, random forest and LAD tree [49] classifiers.

The *artificial neural network* (ANN) is a set of linked input and output units. Each link can have a specific weight. The learning is accomplished by adjusting these weights. It allows proper classification of test data. The ANN is inherently parallel in nature. Therefore, parallelization approaches can be used to speed up the classification [50]. The *support vector machine* (SVM) performs a nonlinear projection to maximize the dimension of the training data. It enables the most effective classifier to be found and can accurately differentiate the test data for each class. *k-nearest neighbors* (k-NN) learns and describes the test sample by contrasting it to the testing samples. *k* training instances are used to classify the considered instance. The decision is based on distance calculation. The *classification and regression tree* (CART) recognizes statistical reference variables, regression, but does not calculate rule sets. It constructs binary trees using the function and threshold that generate the maximum intelligence at each node. The *REP tree* develops a tree of prediction to manage the missing labels by splitting the given dataset into sections. The test set is defined by using models with a similarity exclusion effect applied to the nodes [50]. The *random tree* is based on the principle of an ensemble learning approach which generates distinct learners. A random set of data is generated by employing the bagging technique. It is later employed to raise the decision tree. Its classification principle is to pass the input variables to a collection of predictors. The classification vote of each predictor is employed to make the final decision. The label of the class is decided on the basis of the majority of votes. The *LAD tree* seeks to characterize data by splitting it with a linear plane, which implies that the data are generated independently from two different normal distributions with equivalent co-variance models for each group. To have different means, it finds the best position to place a line between the two distributions. It allows minimizing the approximation error. The *random forest* is based on a combination of predictors for a tree. The tree is based on values of a random vector which is sampled independently. Nonetheless, for all trees in the forest a common distribution is practiced. The classification error is a function of the accuracy and the connotation between different trees [50].

An ensemble of robust classifiers is used in the ensemble learning approaches. Such an approach is beneficial in terms of classification precision and can outperform single classifier based methods. It is a frequently used technique in a variety of machine learning domains. Recent technological advancements have expanded the use of ensemble classification approaches in biomedical applications. Ensemble classifications centering on the completely complex-valued networks of neural relaxation have been extended to the classification of images. They apply the same classification problem to many classifiers. Classifier tests with different accuracy ratings are paired with various methods such as average and voting in this form of study. Therefore, better predictive results can be obtained compared to a single classifier. In this chapter each considered single classifier, such as k-NN, ANN, SVM, etc, is also used with the bagging, AdaBoost and random subspace models in order to attain a higher arrhythmia classification accuracy [51].

In the case of *bagging*, different identically sized training datasets are chosen at random from the broader problem domain. Then using a different machine learning method, a decision tree can be created for each dataset. These trees may be expected to be identical in substance and to enforce the same classification for each new test case. This hypothesis, however, is usually incorrect, particularly if the training datasets are very small, as induction of the decision tree is not a stable process. Small changes in the training data therefore simply generate a different attribute that is selected at a specific node, with significant consequences for the subtree structure under that node. It shows that there are certain instances where multiple decision trees produce specific forecasts and others do not. Bagging attempts to balance the uncertainty of learning methods by simulating the procedure described before using a given training set, instead of obtaining independent domain datasets [52]. *AdaBoost* can be utilized for some bagging-like learning. To simplify the method, the learning algorithm is presumed to be able to modify weighted instances that are positive numbers. The presence of instance weight affects the way an error is measured for a classifier. Typically, in this case it is determined by dividing the weights accumulated by the cumulative weights of all instances into the misclassified instances. The weighting of instances allows the learning algorithm to concentrate on a given set of higher weighted instances. Such higher weighted instances are more critical than others as there is a greater inducement to identify them correctly [52]. *Random subspace* is an ensemble learning approach that aims to reduce the correlation among estimators in an ensemble by making them conditioned on random samples of features rather than the whole set of features. It is comparable to bagging except that the attributes for each learner are collected randomly, with substitution. This informally allows individual learners not to over-focus on features that tend to be incredibly predictive in the training set, but fail to be as reliable for points beyond that set. Random subspace is thus an appealing alternative for problems where the number of features is much greater than the number of training points.

11.5 The performance evaluation measures

The following evaluation measures are used.

11.5.1 Compression ratio

One criterion is to measure the system output in terms of the gain in compression. Classically, $x(t)$ is captured at a fixed rate. The total sample count for a given time period L_T is obviously calculated as $N = F_{\text{ref}} \times L_T$. The discrepancies of the incoming signal tune the sampling rate of the EDADC [37]. When the signal is smoother the sampling rate remains low and when the signal varies rapidly the sampling rate increases accordingly. This is the reason behind the change in the collected number of samples for each selected segment, even if the temporal length of each segment is kept equal to L_T. The sampling density of the EDADC also depends on the resolution of the EDADC alongside the characteristics of the incoming signal [37]. Let us suppose that the designed approach collects N_{ED} samples for the time length of L_T. Then, it is straightforward to calculate the compression ratio, R_{COMP}. It can be computed by calculating the ratio between N and N_{ED}. This process can be mathematically described as

$$R_{\mathrm{COMP}} = \frac{N}{N_{\mathrm{ED}}}. \tag{11.5}$$

11.5.2 Computational complexity

The computational complexity of the front-end processing module is studied in depth up to the denoising stage. The complexity of the feature extraction and classification modules is evaluated at an intuitive level by taking into account the compression ratio of the proposed solution. The adaptive rate FIR (ARFIR) method is utilized for denoising the resampled signal [41].

The numerical complexity of a classical K order FIR filter is well known, and the entire computational complexity C_{FIR} can be measured using

$$C_{\mathrm{FIR}} = \underbrace{K \cdot N}_{\text{Additions}} + \underbrace{K \cdot N}_{\text{Multiplications}}. \tag{11.6}$$

The online adaptation requires additional operations in the case of an invented solution. First, a reference filter hc_k is selected for W^i. This real-time filter selection is performed using the successive approximation algorithm. It performs $\log_2(Q)$ comparisons for the worst case. Here, Q is the length of set $Fref$, in other words it is the count of elements in this set. It is also necessary to resample the selected segments at the real-time computed resampling rate Frs^i. The live resampling is achieved by using the SLI. For SLI, the count of operations for each W^i is Nr^i additions and Nr^i binary weighted divisions. The binary weighted divisions are realized by employing the one-bit right shift mechanism. The complexity of such an implementation is insignificant in contrast to the addition and multiplication operations [53]. With this background, the computational cost of SLI is approximated as Nr^i additions for W^i. After uniform resampling, the W^i is conditioned by utilizing a K^ith order digital filter. The count of operations during the conditioning process is $K^i{\cdot}Nr^i$ multiplications and K^iNr^i additions for W^i. The complexity of a comparison operation is assumed to be equal to that of an addition operation. This assumption allows us to evaluate the processing cost of the devised ARFIR with respect to that of the classical one. In this context the computational complexity of the ARFIR can be represented as

$$C_{\text{ARFIR}} = \underbrace{K^i \cdot Nr^i + Nr^i + \log_2(Q)}_{\text{Additions}} + \underbrace{K^i \cdot Nr^i}_{\text{Multiplications}}. \tag{11.7}$$

11.5.3 Classification accuracy

The pertinent parameters of the denoised signal are extracted using the AR Burg method. Several robust classification algorithms use these derived parameters to identify the ECG categories. The effectiveness of the entire chain is evaluated in terms of the classification performance. A limited collection of data causes difficulties in estimating classification results and a cross-validation scheme is often used in this situation [50]. Therefore, this analysis utilizes the ten-fold cross-validation. For each classifier, the accuracy, *F*-measure, ROC field and Kappa values are evaluated using the same test set.

Accuracy is defined as the ratio of properly categorized instances against all instances and is given as

$$\text{Accuracy} = \frac{\text{TP} + \text{TN}}{\text{TP} + \text{FP} + \text{FN} + \text{TN}}, \tag{11.8}$$

where TP is true positives, TN is true negatives, FP is false positives and FN is false negatives.

The *F*-measure, defined as the harmonic mean of the accuracy and recall indicators, can be calculated as

$$F\text{-measure} = \frac{2 \times \text{TP}}{2 \times \text{TP} + \text{FP} + \text{FN}}. \tag{11.9}$$

Receiver operating characteristic (ROC) analysis is one of the helpful tools that allows the evaluation of classifiers at multiple operating points, the collection of operating points and the correlation of operating points. The ROC curve displays the full range of different operating points in a single graph, with the corresponding varying rates of the true positives and false positives.

The Kappa statistic is a metric which takes into account the predicted figure by deducing it from the successes of the predictor. For a good predictor it expresses the outcome as a proportion of the sum. Kappa statistics are used to calculate the agreement between expected and actual classifications of a dataset, and to correct the agreement that occurs by chance.

11.6 Experimental results and discussion

11.6.1 Experimental results

The efficiency of the proposed solution is studied for the collection, processing and classification of the MIT-BIH arrhythmia dataset signals. In the classical case, each considered record is divided into fixed length windows of 0.9 s length. The signal is recorded at a fixed sampling rate of 360 Hz. Therefore, it results in 330 samples for each window. In the case of the suggested approach, the ECG signal is acquired with an EDADC of $M = 5$. Since the ECG signals are band-limited by $f_{\max} = 60$ Hz, the maximum EDADC sampling frequency is $Fs_{\max} = 3.72$ kHz (see equation (11.2)).

The output of EDADC is irregularly spaced in time. Therefore, it cannot be segmented by utilizing the standard windowing operations. In this context, the ASA is utilized for segmenting the EDADC output. The reference window length L_{ref} is chosen as equal to 1 s [41]. L_{ref} is tuned for each selected segment by the ASA. It mainly allows focusing on and treating the active signal portion while neglecting the unwanted and redundant baseline. This activity selection process augments the processing gain and the power effectiveness of the suggested method. By extracting the parameters of each selected segment the ASA permits adjusting the sampling frequency in a real-time manner and the denoising stage filter order as a function of the incoming signal disparities. This allows the process to concentrate only on the interesting sections of the signal and achieves a reduced number of samples for each selected window, in contrast to the traditional method. However, in the case of typical equivalents, despite incoming signal variations, the window length and the count of samples per window remain identical, 0.9 s and 330 samples, respectively. Consequently, this induces an increase in the processing and power efficiencies of the suggested approach.

For all 300 instances of each class the average compression ratios are determined using equation (11.8). These are 2.76, 2.25, 2.59, 2.61 and 2.69, respectively, for classes N, RBBB, LBBB, APC and PVC. The average compression gain obtained overall for all five classes is 2.6-fold.

The ASA's output is resampled uniformly using the SLI, and is conditioned using the ARFIR. A bank of bandpass filters is built offline. The cut-off frequencies of each filter are $[F_{C\min} = 0.5; F_{C\max} = 30]$ Hz. It enables the emphasis to be on the ECG band of interest while attenuating the undesirable parts. It results in an increased precision of the extraction and classification modules for the post applications. The filter bank is implemented for a range of sample frequencies, $Fref$, between 76.25 Hz $> 2F_{C\max}$ to $F_r = 320$ Hz. In this case $\Delta = 32.5$ Hz is chosen. It realizes a bank of $Q = 8$ bandpass FIR filters. A description of the bank of filters is shown in table 11.1.

Depending on the incoming signal variations and the employed EDADC resolution, the real-time calculated resampling frequency, Frs^i, could be specific for each W^i and it is aligned in a real-time manner with the chosen reference filter's sampling frequency, $Fref_c$. The conditioning stage improves the acquired signal's signal to noise ratio (SNR) and allows us to achieve an improved classification performance [12, 13]. The employed adaptive rate denoising tactic obtains the

Table 11.1. Summary of the reference filter bank parameters.

hc_k	$h1_k$	$h2_k$	$h3_k$	$h4_k$	$h5_k$	$h6_k$	$h7_k$	$h8_k$
$\boldsymbol{Fref_c}$ **(Hz)**	132.5	165	197.5	230	262.5	295	327.5	360
$\boldsymbol{K_c}$	43	53	64	75	85	96	107	117

conditioning of the signal with a reduced number of operations in contrast to the traditional counterparts [41]. The average reductions, compared to the classical method, in the number of operations of the devised *ARFIR* are calculated for all 300 instances of each class considered. In terms of additions these are, respectively, 5.95, 4.62, 5.47, 5.64 and 5.93 for classes N, RBBB, LBBB, APC and PVC. In terms of multiplications these are, respectively, 6.08, 4.71, 5.59, 5.77 and 6.06 for classes N, RBBB, LBBB, APC and PVC. The average gain for all five classes is also computed. The gain in the additions and multiplications are 5.52-fold and 5.64-fold, respectively.

The AR Burg method is employed to extract the discriminative and informative features of the conditioned signal. These attributes are then used for classification. For the classical case, summaries of the classification results obtained using the different classifiers are described in tables 11.2 and 11.3, and the classification results obtained by applying the proposed event-driven acquisition and processing-based approach are presented in tables 11.4 and 11.5.

Table 11.2 shows the results for the five-class ECG dataset in terms of the classification precision and the *F*-measure values for the classical case. Among the single classifiers, the highest classification accuracy of 95.7% is achieved by the random forest. A minor increase in performance is achieved by the ensemble of random forest with AdaBoost, with an accuracy of 95.8%. Among the single classifiers, the highest *F*-measure value of 0.957 is also achieved by the random forest. Among the ensemble classifiers, the random forest with AdaBoost secures the highest *F*-measure value of 0.958. It is apparent that the accuracies of the weak classifiers are improved by the ensemble classifier models.

Table 11.3 shows the results for the five-class ECG dataset in terms of the ROC area and the kappa statistics for the classical sampling case. Among the single classifiers, the highest ROC area value of 0.996 is achieved by the random forest. A

Table 11.2. Summary of the classification accuracies and *F*-measures for the classical case.

	Accuracy				*F*-measure			
	Single	Bagging	AdaBoost	Random subspace	Single	Bagging	AdaBoost	Random subspace
SVM	0.881	0.876	0.891	0.847	0.882	0.877	0.893	0.848
***k*-NN**	0.9	0.893	0.9	0.907	0.899	0.892	0.899	0.907
ANN	0.868	0.891	0.868	0.883	0.866	0.89	0.866	0.882
Random forest	0.957	0.957	0.958	0.951	0.957	0.957	0.958	0.951
CART	0.901	0.929	0.948	0.935	0.901	0.929	0.948	0.935
REP tree	0.91	0.929	0.949	0.933	0.91	0.929	0.949	0.934
Random tree	0.891	0.942	0.897	0.952	0.891	0.942	0.897	0.952
LAD tree	0.887	0.917	0.935	0.914	0.887	0.917	0.935	0.915

Table 11.3. Summary of ROC areas and kappa statistics for the classical case.

	ROC area				Kappa			
	Single	Bagging	AdaBoost	Random subspace	Single	Bagging	AdaBoost	Random subspace
SVM	0.952	0.966	0.98	0.959	0.8508	0.845	0.8642	0.8083
***k*-NN**	0.975	0.978	0.946	0.989	0.875	0.8667	0.875	0.8842
ANN	0.97	0.983	0.966	0.982	0.835	0.8642	0.835	0.8533
Random forest	0.996	0.997	0.996	0.996	0.9458	0.9458	0.9475	0.9383
CART	0.964	0.993	0.995	0.993	0.8767	0.9117	0.935	0.9183
REP tree	0.965	0.992	0.994	0.993	0.8875	0.9117	0.9367	0.9167
Random tree	0.932	0.992	0.935	0.992	0.8642	0.9275	0.8708	0.94
LAD tree	0.978	0.988	0.992	0.99	0.8583	0.8958	0.9192	0.8925

Table 11.4. The summary of classification accuracies and *F*-measures for the proposed approach.

	Accuracy				*F*-measure			
	Single	Bagging	AdaBoost	Random subspace	Single	Bagging	AdaBoost	Random subspace
SVM	0.937	0.929	0.93	0.927	0.938	0.93	0.93	0.927
***k*-NN**	0.931	0.92	0.928	0.93	0.931	0.92	0.928	0.93
ANN	0.93	0.941	0.93	0.945	0.93	0.941	0.93	0.945
Random forest	0.943	0.943	0.943	0.947	0.943	0.943	0.943	0.947
CART	0.894	0.914	0.93	0.927	0.894	0.914	0.93	0.927
REP tree	0.873	0.909	0.925	0.914	0.873	0.909	0.925	0.914
Random tree	0.868	0.931	0.88	0.935	0.868	0.931	0.88	0.935
LAD tree	0.79	0.851	0.901	0.855	0.793	0.852	0.901	0.856

similar performance is achieved by the ensemble of random forest with Adaboost and random forest with random subspace. However, a minor performance increase is obtained by the random forest with bagging, achieving an ROC area value of 0.997. Among the single classifiers, the highest kappa value of 0.946 is also achieved by the random forest. Among the ensemble classifiers, the random forest with bagging secures a similar performance. A minor increase is obtained by the random forest with AdaBoost by obtaining a kappa value of 0.948. Moreover, it is clearly seen that the accuracies of the weak classifiers are improved by the ensemble classifier models.

Table 11.5. The summary of ROC areas and kappa statistics for the proposed approach.

	ROC area				Kappa			
	Single	Bagging	AdaBoost	Random subspace	Single	Bagging	AdaBoost	Random subspace
SVM	0.981	0.988	0.988	0.987	0.9217	0.9117	0.9125	0.9083
***k*-NN**	0.985	0.989	0.971	0.992	0.9142	0.9	0.91	0.9125
ANN	0.989	0.995	0.987	0.995	0.9125	0.9258	0.9125	0.9317
Random forest	0.996	0.996	0.996	0.997	0.9283	0.9283	0.9292	0.9333
CART	0.96	0.991	0.991	0.994	0.8675	0.8925	0.9125	0.9092
REP tree	0.962	0.991	0.992	0.992	0.8408	0.8858	0.9058	0.8925
Random tree	0.918	0.992	0.925	0.994	0.835	0.9133	0.85	0.9183
LAD tree	0.942	0.977	0.99	0.978	0.7375	0.8142	0.8767	0.8192

The above results (tables 11.2 and 11.3) show that for the classical case the chain of classical ADC, windowing, FIR filter and AR Burg achieves the best classification performance with the random forest. A minor increase in the performance is achieved by using a smart ensemble of random forest with AdaBoost. In addition, the accuracies of the weak classifiers are improved by the ensemble classifier models.

Table 11.4 shows results for the five-class ECG dataset in terms of the classification precision and the *F*-measure values for the devised chain. Among the single classifiers, the highest classification accuracy of 94.3% is achieved by the random forest. A similar performance is achieved by the ensembles of random forest with bagging and random forest with AdaBoost. A minor increase is obtained by the ensemble of ANN with random subspace by achieving a precision of 94.5%. Overall, the highest classification accuracy of 94.7% is achieved by the ensemble of random forest with random subspace. Among the single classifiers, the highest *F*-measure value of 0.943 is achieved by the random forest. A similar performance is achieved by the ensembles of random forest with bagging and random forest with AdaBoost. A minor increase is obtained by the ensemble of ANN with random subspace by achieving an *F*-measure of 0.945. Overall, the highest *F*-measure value of 0.947 is secured by the ensemble of random forest with random subspace. In this case, the accuracies of the weak classifiers are improved by the ensemble classifier models as well.

Table 11.5 shows the results for the five-class ECG dataset in terms of the ROC area and the kappa statistics for the devised chain. Among the single classifiers, the highest ROC area value of 0.996 is achieved by the random forest. A similar performance is achieved by the ensembles of random forest with bagging and random forest with AdaBoost. A minor increase is achieved by the ensemble of random forest with random subspace, achieving an ROC area value of 0.997. Among the single classifiers, the highest Kappa value of 0.928 is achieved by the random forest. A similar performance is achieved by the ensemble of random forest with bagging. A minor increase is obtained by the ensemble of random forest with AdaBoost, achieving a kappa value of 0.929. A further increase is achieved by the ensemble of ANN with random subspace, yielding a kappa value of 0.932. Overall, the highest kappa value of 0.933 is achieved by the ensemble of random forest with random subspace. One more point that should be mentioned is again that the accuracies of the weak classifiers are improved by the ensemble classifier models.

The above results (tables 11.4 and 11.5) show that, for the suggested event-driven chain of the EDADC, ASA, ARFIR and AR Burg method, random forest achieves the best classification performance. A minor increase in the performance is achieved using a smart ensemble of random forest with random subspace.

11.6.2 Discussion

Recent advancements in wireless sensor techniques and technologies enhance the development and deployment of smart implants in general and the wireless and self-powered biomedical implants in particular. A variety of biomedical implants have been introduced in the literature. The remote monitoring and diagnosis of patients is a beneficial concept and is particularly helpful for the remote and rural communities of the developing word. Inspired by its promising benefits this domain of research has attracted many researchers and industries for different potential domains, such as computation, information science and electronics. A recent trend is the integration of cloud-based applications in modern mobile healthcare solutions. By considering the developments and results in the cloud-based mobile healthcare sector, it can be imagined how much this approach will expand and be deployed further in the near future. It will revolutionize the sensor, computer and networking fields of research and development.

Cloud-based biomedical signal processing for the automatic diagnosis of health condition monitoring is a novel trend. It is evident that it is going to play an important role in future mobile healthcare solutions. One key application is to automatically detect cardiovascular system disorders to avoid any possible heart failures. One of the aims of this chapter is to reflect many of these developments and also to recognize other essential research issues in the mobile healthcare framework. It is based on smart wearables that are linked to cloud-based applications.

The benefits of the proposed method are evident from the results. It employs an intelligent combination of the event-driven acquisition and the adaptive rate processing mechanisms. It is capable of adjusting L^i, Frs^i and N^i as a function of the incoming signal variations. This ensures the dynamic parameter adaptation of the suggested solution in accordance with the incoming signal disparities. It is shown how the adaptation of L^i and N^i automates the real-time tuning of Frs^i. This real-time adjustment of Frs^i enhances the processing effectiveness and the power efficiency of the designed approach in contrast to the traditional equivalent. It is obtained by omitting needless operations while performing the real-time and online interpolation of the active signal portions. It also avoids the collection and conditioning of irrelevant information and leads towards an effective denoising mechanism. Additionally, the real-time adoption of filter orders also adds to the effectiveness of the suggested conditioning mechanism [11].

One crucial function is the consistent and effective mobile supervision of chronically ill patients. In this chapter we developed an event-driven and adaptive rate cloud-based remote health monitoring system. It utilizes ECG signals for arrhythmia categorization. The primary objective of this chapter is to illustrate how wireless biomedical instruments can be used to anticipate and detect chronic disorders in real time, using an automated, smart and scalable cloud-based mobile healthcare monitoring framework. The suggested approach was checked as being practical, accurate and successful in the measurement and evaluation of cardiac health monitoring. The advances in mobile device and sensor technologies include innovative ways to improve the quality of care for patients with chronic disorders. The proposed approach is new and has the ability to improve healthcare technology and medical systems for the coming age.

11.7 Conclusion

In this chapter we developed an event-driven and adaptive rate cloud-based remote cardiac health monitoring system. The proposed approach is new, and has the ability to improve the healthcare technology and medical systems in future.

The results show that, as a whole, the average reductions in the count of operations in terms of additions and multiplications, respectively, for the adaptive rate filtering in contrast to the classical equivalent are 5.52-fold and 5.64-fold. In addition, the proposed approach also achieves a notable reduction in the acquired count of samples, for the studied case it is 2.6-fold. This ensures an important enhancement in the effectiveness of the post-processing stages, such as conditioning, transmission and classification. Moreover, a more than 50% reduction in bandwidth use is also ensured by these findings. On the side of cloud-based processing, the transformation of a 2.6-fold reduced count of samples is realized with reduced order Auto Regressive Burg models in contrast to the fixed rate classical equivalents. It ensures a more than 50% decrease in the real-time count of operations of the feature extraction process in contrast to the traditional methods. A similar amount of gain in terms of processing effectiveness is also expected for the classification stage. Furthermore, a simpler 5 bit resolution EDADC is employed in the proposed approach for the recording of ECG signals. In contrast a relatively complex 11 bit resolution ADC is utilized for the ECG signals recorded in the case of traditional equivalents. This ensures a simpler circuit level realization of the proposed method in contrast to the conventional equivalents.

The highest accuracy obtained by the proposed method is 1.1% lower than the classical counterpart's highest classification accuracy. This shows that the suggested solution achieves comparable classification accuracy compared with the traditional counterpart while visibly out-performing it in terms of processing, transmission and hardware simplicity.

System output is a function of the utilized resampling, denoising, feature extraction and classification approaches. The study of device effectiveness in terms of processing cost and precision while utilizing alternative methods of feature mining, such as the round cosine transform and the discrete wavelet transform, is a future work. Certain other research opportunities are to explore the miniaturization, optimization and integrated development of the proposed solution.

Acknowledgments

This work is funded by Effat University under the grant number UC#7/28Feb 2018/ 10.2–44g.

References

[1] Arunkumar K and Bhaskar M 2019 Heart rate estimation from photoplethysmography signal for wearable health monitoring devices *Biomed. Signal Process. Control* **50** 1–9

[2] Baig M M, Afifi S, GholamHosseini H and Mirza F 2019 A systematic review of wearable sensors and IoT-based monitoring applications for older adults—a focus on ageing population and independent living *J. Med. Syst.* **43** 233

[3] Atalay O, Atalay A, Gafford J and Walsh C 2018 A highly sensitive capacitive-based soft pressure sensor based on a conductive fabric and a microporous dielectric layer *Adv. Mater. Technol.* **3** 1700237

[4] Majumder S, Chen L, Marinov O, Chen C H, Mondal T and Deen M J 2018 Noncontact wearable wireless ECG systems for long-term monitoring *IEEE Rev. Biomed. Eng.* **11** 306–21

[5] Venkatesan C, Karthigaikumar P and Satheeskumaran S 2018 Mobile cloud computing for ECG telemonitoring and real-time coronary heart disease risk detection *Biomed. Signal Process. Control* **44** 138–45

[6] Catarinucciet L *et al* 2015 An IoT-aware architecture for smart healthcare systems *IEEE Internet Things J.* **2** 515–26

[7] Wang X, Zhu Y, Ha Y, Qiu M and Huang T 2016 An FPGA-based cloud system for massive ECG data analysis *IEEE Trans. Circuits Syst. II Express Briefs* **64** 309–13

[8] Wang X, Gui Q, Liu B, Jin Z and Chen Y 2013 Enabling smart personalized healthcare: a hybrid mobile-cloud approach for ECG telemonitoring *IEEE J. Biomed. Health Inform.* **18** 739–45

[9] Guzik P and Malik M 2016 ECG by mobile technologies *J. Electrocardiol.* **49** 894–901

[10] Serhani M A, Menshawy M E and Benharref A 2016 SME2EM: smart mobile end-to-end monitoring architecture for life-long diseases *Comput. Biol. Med.* **68** 137–54

[11] Qaisar S M 2019 Efficient mobile systems based on adaptive rate signal processing *Comput. Electr. Eng.* **79** 106462

[12] Qaisar S M and Subasi A 2018 An adaptive rate ECG acquisition and analysis for efficient diagnosis of the cardiovascular diseases *Presented at the 2018 IEEE 3rd Int. Conf. on Signal and Image Processing (ICSIP)* pp 177–81

[13] Subasi A, Bandic L and Qaisar S M 2020 Cloud-based health monitoring framework using smart sensors and smartphone *Innovation in Health Informatics* (Amsterdam: Elsevier) pp 217–43

[14] Mozaffarianet D *et al* 2016 Executive summary: heart disease and stroke statistics—2016 update: a report from the American Heart Association *Circulation* **133** 447–54

[15] Alickovic E and Subasi A 2015 Effect of multiscale PCA de-noising in ECG beat classification for diagnosis of cardiovascular diseases *Circuits Syst. Signal Process.* **34** 513–33

[16] Phukpattaranont P 2015 QRS detection algorithm based on the quadratic filter *Expert Syst. Appl.* **42** 4867–77
[17] Hesar H D and Mohebbi M 2016 ECG denoising using marginalized particle extended Kalman filter with an automatic particle weighting strategy *IEEE J. Biomed. Health Inform.* **21** 635–44
[18] Rodríguez R, Mexicano A, Bila J, Cervantes S and Ponce R 2015 Feature extraction of electrocardiogram signals by applying adaptive threshold and principal component analysis *J. Appl. Res. Technol.* **13** 261–9
[19] Linh T H 2018 A solution for improvement of ECG arrhythmia recognition using respiration information *Vietnam J. Sci. Technol.* **56** 335
[20] da S. Luz E J, Schwartz W R, Cámara-Chávez G and Menotti D 2016 ECG-based heartbeat classification for arrhythmia detection: a survey *Comput. Methods Programs Biomed.* **127** 144–64
[21] Li P *et al* 2016 High-performance personalized heartbeat classification model for long-term ECG signal *IEEE Trans. Biomed. Eng.* **64** 78–86
[22] Zhang X and Lian Y 2014 A 300-mV 220-nW event-driven ADC with real-time QRS detection for wearable ECG sensors *IEEE Trans. Biomed. Circuits Syst.* **8** 834–43
[23] Rezaii T Y, Beheshti S, Shamsi M and Eftekharifar S 2018 ECG signal compression and denoising via optimum sparsity order selection in compressed sensing framework *Biomed. Signal Process. Control* **41** 161–71
[24] Shaw L, Rahman D and Routray A 2018 Highly efficient compression algorithms for multichannel EEG *IEEE Trans. Neural Syst. Rehabil. Eng.* **26** 957–68
[25] Chen S, Hua W, Li Z, Li J and Gao X 2017 Heartbeat classification using projected and dynamic features of ECG signal *Biomed. Signal Process. Control* **31** 165–73
[26] Niederhauser T, Haeberlin A, Jesacher B, Fischer A and Tanner H 2017 Model-based delineation of non-uniformly sampled ECG signals *Presented at the 2017 Computing in Cardiology (CinC)* pp 1–4
[27] Einthoven W 1903 The string galvanometer and the human electrocardiogram *Presented at the KNAW Proc.* vol 6 pp 107–15
[28] MUSE Meditation Made Easy Muse: the Brain Sensing Headband http://choosemuse.com/ (accessed 19 Oct. 2018)
[29] Emotiv https://emotiv.com/ (accessed 19 Oct. 2018)
[30] GE Healthcare https://gehealthcare.co.uk/ (accessed 19 Oct. 2018)
[31] Salvador C H *et al* 2005 Airmed-cardio: a GSM and Internet services-based system for out-of-hospital follow-up of cardiac patients *IEEE Trans. Inf. Technol. Biomed.* **9** 73–85
[32] Herscoviciet N *et al* 2007 m-health e-emergency systems: current status and future directions [Wireless corner] *IEEE Antennas Propag. Mag.* **49** 216–31
[33] Shih D H, Chiang H S, Lin B and Lin S B 2010 An embedded mobile ECG reasoning system for elderly patients *IEEE Trans. Inf. Technol. Biomed.* **14** 854–65
[34] Ren Y, Werner R, Pazzi N and Boukerche A 2010 Monitoring patients via a secure and mobile healthcare system *IEEE Wirel. Commun.* **17** 59–65
[35] Xia H, Asif I and Zhao X 2013 Cloud-ECG for real time ECG monitoring and analysis *Comput. Methods Programs Biomed.* **110** 253–9
[36] Qaisar S M *et al* 2017 Time-domain characterization of a wireless ECG system event driven A/D converter *Presented at the 2017 IEEE Int. Instrumentation and Measurement Technology Conf. (I2MTC)* pp 1–6

[37] Qaisar S M, Yahiaoui R and Gharbi T 2013 An efficient signal acquisition with an adaptive rate A/D conversion *Presented at the 2013 IEEE Int. Conf. on Circuits and Systems (ICCAS)* pp 124–9
[38] Jin S W, Li J J, Li Z N and Wang A X 2017 A hysteresis comparator for level-crossing ADC *Presented at the 2017 29th Chinese Control And Decision Conf. (CCDC)* pp 7753–7
[39] Moody G B and Mark R G 2001 The impact of the MIT-BIH arrhythmia database *IEEE Eng. Med. Biol. Mag.* **20** 45–50
[40] Welch T B, Wright C H and Morrow M G 2016 *Real-time Digital Signal Processing from MATLAB to C with the TMS320C6x DSPs* (Boca Raton, FL: CRC Press)
[41] Qaisar S M, Fesquet L and Renaudin M 2014 Adaptive rate filtering a computationally efficient signal processing approach *Signal Process.* **94** 620–30
[42] Allier E, Sicard G, Fesquet L and Renaudin M 2003 A new class of asynchronous A/D converters based on time quantization *Presented at the Ninth Int. Symp. on Asynchronous Circuits and Systems, 2003* pp 196–205
[43] Qaisar S M, Akbar M, Beyrouthy T, Al-Habib W and Asmatulah M 2016 An error measurement for resampled level crossing signal *Presented at the 2016 Int. Conf. on Event-based Control, Communication, and Signal Processing (EBCCSP)* pp 1–4
[44] Kohler B U, Hennig C and Orglmeister R 2002 The principles of software QRS detection *IEEE Eng. Med. Biol. Mag.* **21** 42–57
[45] Tan L and Jiang J 2018 *Digital Signal Processing: Fundamentals and Applications* (New York: Academic)
[46] Noh D and Katsianos T 2018 Multi-rate system for audio processing *US Patent Application No.* 10/008,217
[47] Qaisar S M, Fesquet L and Renaudin M 2006 Spectral analysis of a signal driven sampling scheme *Presented at the 2006 14th European Signal Processing Conf.* pp 1–5
[48] Begg R, Lai D T and Palaniswami M 2007 *Computational Intelligence in Biomedical Engineering* (Boca Raton, FL: CRC Press)
[49] Siuly S, Li Y and Zhang Y 2016 *EEG Signal Analysis and Classification* (Berlin: Springer)
[50] Subasi A 2019 *Practical Guide for Biomedical Signals Analysis Using Machine Learning Techniques: A MATLAB Based Approach* (New York: Academic)
[51] Saraswathi D and Srinivasan E 2014 An ensemble approach to diagnose breast cancer using fully complex-valued relaxation neural network classifier *Int. J. Biomed. Eng. Technol.* **15** 243–60
[52] Hall M, Witten I and Frank E 2011 *Data Mining: Practical Machine Learning Tools and Techniques* (Burlington, MA: Kaufmann)
[53] Cavanagh J 2017 *Computer Arithmetic and Verilog HDL Fundamentals* (Boca Raton, FL: CRC Press)

IOP Publishing

Modelling and Analysis of Active Biopotential Signals in Healthcare, Volume 1

Varun Bajaj and G R Sinha

Chapter 12

Electrocardiogram beat classification using deep convolutional neural network techniques

Zafer Cömert[1], Yaman Akbulut[2], Muhammed H Akpinar[3], Ömer F Alçin[4], Ümit Budak[5], Muzaffer Aslan[6], Abdulkadir Şengür[3]

[1]*Samsun University, Engineering Faculty, Software Engineering Department, Samsun, Turkey*
[2]*Firat University, Informatics Department, Elazig, Turkey*
[3]*Firat University, Technology Faculty, Electrical and Electronics Engineering Department, Elazig, Turkey*
[4]*Malatya Turgut Ozal University, Faculty of Engineering and Natural Sciences, Department of Electrical Engineering, Malatya, Turkey*
[5]*Bitlis Eren University, Engineering Faculty, Electrical and Electronics Engineering Department, Bitlis, Turkey*
[6]*Bingol University, Engineering Faculty, Electrical-Electronics Engineering Department, Bingol, Turkey*

The electrocardiogram (ECG) is a useful method which enables the monitoring of various cardiac conditions, such as arrhythmia and heart rate variability (HRV). ECG beats help to determine various heart failures such as cardiac disease and ventricular tachyarrhythmia. In the literature, it can be seen that various advanced signal processing and machine learning techniques and deep learning algorithms have been employed for ECG beat categorization. These methods were generally based on either the time domain or frequency domain. Time–frequency (T–F) based techniques have also been proposed for ECG beat classification. In this chapter, a different model is proposed for the ECG beat classification task. In the proposed approach, the ECG beats are initially represented by images. Instead of using a time–frequency approach for converting the ECG beats to ECG images, we opt to use the ECG beats directly to construct the ECG images. In other words, the ECG beat values are directly saved as ECG images. Three deep convolutional neural network (CNN) approaches are considered in ECG beat classification. These approaches ensure end-to-end learning schema, fine-tuning of pre-trained CNN models, extraction of deep features and their classification using a traditional classifier, such as the support vector machine (SVM) or deep machine learning approaches. The well-known

MIT-BIH arrhythmia database is considered in the evaluation of the proposed deep learning approaches. The database is separated into two sets, the training and test dataset in proportions of 75% and 25%, respectively. The experimental results are evaluated using the classification accuracy score. The results show that the proposed methods have potential for use in ECG beat classification.

12.1 Introduction

Electrocardiography (ECG) is utilized as a standard means for monitoring and diagnosing cardiovascular diseases and it has an important role in routine clinical practice. Using this non-invasive tool, the electrical cardiac activities covering the depolarization and repolarization patterns of the heart can be observed on an ECG trace [1]. An ECG trace reflects the time-varying signal changes that depend on the ionic current flow that causes the cardiac fibers to contract and subsequently relax. The changes in potential between the electrodes placed on the surface of the skin can be observed via ECG [2]. A single and routine period of an ECG presents the sequential atrial and ventricular depolarization as well as repolarization. These continuous cycles consist of significant peaks that reflect the morphological structures underlying specific peaks, denoted by P, Q, R, S and T [3]. This means that a heartbeat creates a specific pattern on the ECG trace. This specific pattern occurs when the signal moves away from the baseline heart rate and then returns to the baseline heart rate level. A single sinus rhythm of the ECG traditionally includes the above-mentioned peaks, as shown in figure 12.1.

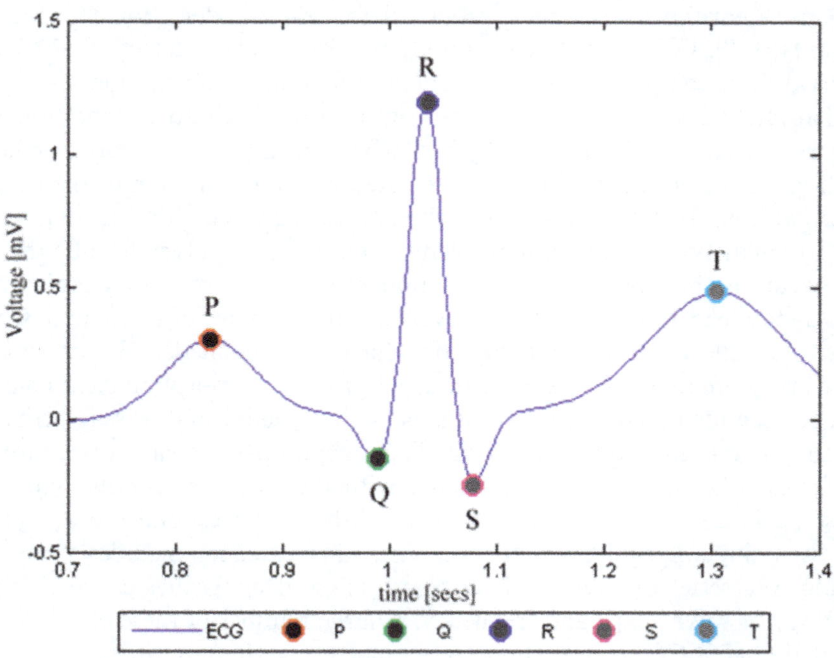

Figure 12.1. The morphological structure of one PQRST-complex of a single ECG cycle.

The depolarization of the atria, which causes depolarization waves to spread from the sinoatrial (SA) node to the atria, increases the voltage measured from the heart surface, causing it to diverge from the baseline heart rate, and this wave is referred to as a P-wave [4]. The time between the events of atrial depolarization and ventricular depolarization is the PQ-interval. The QRS-complex is one of the most vital patterns in computerized heart rate analysis and it is frequently used for basic heart rate analysis. The QRS-complex produced the largest amplitude in an ECG cycle [5]. The ventricles become depolarized before their contractions and this depolarization forms the QRS-complex. Atrial repolarization appears just before the QRS-complex. However, the following wave pattern, which is referred to as the QRS-complex, has a wider amplitude. Therefore, atrial repolarization cannot be observed on an ECG trace [6]. The time between the beginning of the ventricular depolarization and end of the ventricular repolarization overlaps the QT-interval on the ECG trace. It has been shown that the QT-interval increases due to the linear increase of the RR-interval in experimental studies. Also, a prolonged QT-interval has been related to late ventricular repolarization and this cardiac ailment may lead to sudden death [7]. The time interval between the S-wave and T-wave is referred to as the ST-interval. High or depressive patterns far from the baseline are consistently considered as symptoms that indicate heart diseases. The relief time elapsed until the next ECG cycle, in other words, the ventricular repolarization, is shown as a T-wave on the ECG signal [8]. ECG signals can have undesired artefacts for different reasons, such as patient or electrode movements, displacements of the electrodes and power-line noise [9]. To achieve useful clinical information, noisy ECG signals must be converted into reliable signals using preprocessing procedures [10, 11]. To this aim, many algorithms have been proposed for R-peak detection [12–14], QRS detection [15–18], QT-interval detection [3, 19], RR-interval detection [20–23] and ST-interval detection [8, 24, 25] to provide useful clinical information.

Cardiac arrhythmia is a medical term and corresponds to a condition which reflects irregularities in heart rate rhythm, where it may be quite slow or fast [1]. Disruptions of the electrical impulses associated with coordination of the heartbeat cause arrhythmia. In other words, arrhythmia is defined as discomfort that includes rhythm irregularity or conduction problems in electrical impulses through the heart. It can occur in the upper or lower chamber of the heart. Arrhythmias of the ventricles may lead to life-threatening events. Various types of arrhythmias, which demonstrate quite different characteristics, have been defined [26]. Each of these types of arrhythmias is associated with a pattern whose morphological characteristics have been identified. Therefore, it is possible to define and classify arrhythmia types using a computer-aided approach. The arrhythmias are categorized into two basic classes. The irregularities associated with the electrical conductivity of the heart is the first class. This category of arrhythmia corresponds to arrhythmias caused by a single irregular vessel shot. The other category includes arrhythmias consisting of a series of irregular heartbeats. These irregularities cause changes in morphology or wave shape and frequency. A large number of these changes can be observed though ECG examination. A normal heart rhythm where each specific peak can be observed properly is defined as the normal sinus rhythm (NSR). During

NSR, the triggering pulses from the sinoatria spread in a coordinated manner across the four chambers of the heart. In addition to the NSR, there are also different common arrhythmia types. The source of the NSR beats is the SA node. The regular heart rate can vary due to autonomous inputs occurring in the SA node. Premature beats are a widely observed cardiac disease and are harmless. Patients with this type of arrhythmia feel a beating or missing heartbeat in the chest. Premature beats also include premature atrial contractions (PACs) and premature-ventricular contractions (PVCs) [27]. Supraventricular arrhythmias occur in the atria, with fast heart rates, and some of them are atrial fibrillation (AF), atrial flutter, paroxysmal supraventricular tachycardia (PSVT) and Wolff–Parkinson–White syndrome [28]. Ventricular arrhythmias (VAs) include ventricular flutter, ventricular tachycardia (VT) and ventricular fibrillation (VF). These types of arrhythmias are life-threatening and therefore require quick intervention [29]. Sinus node dysfunction is related to problems associated with the SA node. A slow heart rhythm is observed in this arrhythmia and patients use a pacemaker to tackle this type of arrhythmia [30]. Heart block problems may appear in the AV node or HIS Purkinje system because of incomplete electrical activity [31]. Slow and irregular heartbeats are then observed. To tackle this condition, using a pacemaker is recommended. Arrhythmia diagnosis based on ECG beats has great importance for cardiologists [32]. A broad range of cardiac conditions covering arrhythmia and HRV can be observed on ECG traces [33]. However, even for specialists in the field, it is very difficult to examine, define and categorize a large number of ECG beats. In addition, the possibility of missing vital information is high due to the large scale of the data. As a practical solution, computer-aided systems have been adopted and have become quite important in daily clinical practice [1]. In this view, automatic ECG beat classification has been considered as an important issue in addressing cardiac ailments.

In computer-aided approaches, the two basic procedures of feature extraction and classification are common. The performance of such systems is sensitive and depends on specifically designed features as well as the fine-tuning of the classifiers. Time-domain features have been used for ECG analysis and have been adopted as a useful approach. In addition, the frequency domain holds great clues regarding HRV. Furthermore, the time–frequency domain have been used to describe large-scale ECG data [34]. The morphological and higher-order statistic [35] diagnostic indices have been measured using advanced digital signal processing techniques in the feature extraction stage, such as principal component analysis (PCA) [3, 36, 37], wavelet transforms (WT) [38, 39], Hilbert transforms [40], cross-correlation approaches [41] and the Kalman filter [42]. Support vector machines (SVMs) [43] have ensured high general performance. The extreme learning machine (ELM) has contributed to shortening the training time [44]. Similarly, the *k*-nearest neighbors (*k*-NN) method [38], linear discriminant analysis (LDA) [45], artificial neural network (ANNs) [46] and probabilistic neural network (PNN) [47] have also been used for classification purposes. Optimization techniques such as particle swarm optimization (PSO) [48], artificial bee colony [49] and genetic algorithms [50] have been embedded into the proposed models for automatic ECG beat classification tasks [33]. The permutation entropy (PE) and the conditional entropy of ordinal patterns (CEOP) have been applied to ECG signals to realize a binary classification [51]. Deterministic learning schema have been applied to the ECG beat classification task. To this aim, the proposed model exploited the dynamics structure of the ECG beats as a unique feature. In this approach, the classification was carried out without using any feature extraction procedure for the test beats. Also, the model presented good generalization capacity [33]. The enormous differences between individual patients were addressed in another study focused on ECG beat classification. An optimization mechanism was proposed to enhance the achievements of the model. Recurrent neural networks (RNNs) have been employed to learn features automatically and promising results have been reported [52]. To prevent sudden cardiac death, a computer-aided diagnosis model based on spectral entropy and a deep convolutional neural network (DCNN) was introduced. The time–frequency transform and entropy measurement were dealt with in this study. Furthermore, two-directional two-dimensional PCA was utilized to reduce the number of features used. The study points out that the spectral entropy can provide good separation between different ECG beats [34]. A novel model was built on the SVM with rejection schema. Traditional procedures, such as preprocessing, feature extraction and classification, applied in machine learning approaches were followed in the proposed model. As a result, a cost-sensitive classifier was provided [53]. Vector quantization (VQ) has been used in ECG beat classification, since this method increases the classification accuracy while reducing the dimension of the feature set. To this aim, a novel dictionary learning algorithm was presented. This study resulted in an accelerated algorithm and a model with increased accuracy [54]. The ECG beat classification task has also been handled as a binary classification task between normal and abnormal. A method relying on the radial basis function (RBF) system for canceling ectopic beats was proposed. The model yielded good results in terms of the low distortion of the signal recordings [55]. A neuro-fuzzy network model fed with autoregressive model coefficients, higher-order cumulants and the WT has been used for ECG beat classification. The proposed model provided a significant performance enhancement in ECG beat classification [56]. In addition to the normal beat type, four different beat types were addressed in a novel model exploring RNNs trained with the Levenberg–Marquart algorithm. The combined eigenvector method and RNN was reported as a useful tool for ECG beat classification [57]. A combination of PSO and RBF neural networks was used to separate six different ECG beat types. The morphological features were taken into account in the model. As a result, an efficient diagnostic model was provided [58]. Six discriminative ECG beat classification problems were addressed using a PNN and WT. The model was fed with only eleven features and an effective and efficient model was achieved [59]. A novel model, called a modified mixture of experts (MME) network, was used in ECG beat classification with diverse features and a supervised learning approach with the expectation–maximization (EM) algorithm was used in the model. The model used handcrafted features extracted using wavelet coefficients and Lyapunov exponents. The proposed MME approach was found to be a useful tool in categorizing ECG signals for early heart abnormality detection [60]. An automated medical diagnostic tool based on a cross-correlation and least square support vector machine (LS-SVM) was developed for ECG beat classification which covers normal beats, PVC beats and other beats. The performance of the proposed model in this three-class classification scheme was reported to have 95.51% accuracy [61]. In another study, ECG beat classification was performed using a novel independent components (ICs) arrangement strategy and independent component analysis (ICA). The PNN and SVM classifiers were employed to evaluate the proposed method. Eight different ECG beat types were considered in the experiment and a 98.7% classification accuracy was achieved [62].

The ECG beat classification task has mainly been addressed in two different ways: the global-classifier and patient-adapting approaches. In the first approach, the experience of specialists is not considered and the classification is performed automatically [52]. The desired performance has not been achieved in this approach since a large-scale inter-individual variability has been observed in ECG waveforms. The second approach focuses on the ECG beat classification task by considering the specific characteristics of each patient's ECG recording to enhance the accuracy of automated systems [51]. However, this enhancement has remained limited since there are different temporal and physical conditions for patients as well as variations in the morphological characteristics of ECG waveforms. ECG signals generally display their characteristic features periodically. Sometimes differences arise in periodic ECG beats and sometimes in the morphological structure of the ECG. These enormous time-varying patterns are marked as temporal or dynamic [63]. For this reason, the recognition of this temporal pattern is very difficult. Also, the presence of notable similarities between different temporal ECG beats is a challenge that remains to be solved [64].

As mentioned above, many different combinations of advanced signal processing and machine learning techniques have already been proposed for the ECG beat classification task [65–67]. Recently, a novel approach called the deep convolution neural network (CNN), a sub-branch of machine learning, has become attractive due to its generalizability on large-scale real-world problems [68]. The great advantage of CNN models is the end-to-end learning scheme [69]. In this way, the tedious and complex feature extraction step is eliminated and the features representing the characteristic and discriminative features of the input data can be obtained using activation maps [70]. Although CNN models require large-scale data and high calculation costs for the training, they have been adopted as an efficient method due to the highly generalizable performance provided.

In the proposed approach, pre-trained CNN models are used and the fine-tuning procedure is performed to transfer domain knowledge. Then, the deep features extracted from the activation maps in the deep CNN architecture are input to one of the traditional machine learning algorithms called SVMs. In the training of the CNN models, raw ECG beat images are considered and five ECG beat types, namely normal (N), left-bundle-branch-block (LBBB), right-bundle-branch-block (RBBB), premature-ventricular-contraction (PVC) and paced beats (PBs) are taken into account in the experiments [71]. As a result, a novel ECG beat classification diagnostic tool with high sensitivity is produced via the proposed model.

The rest of this chapter is organized as follows: the dataset description and methods are provided in section 12.2; the experimental work and results are reported in section 12.3; a comparison is presented in section 12.4; and, lastly, concluding remarks are presented in section 12.5.

12.2 Material and methods

12.2.1 The MIT-BIH database

The MIT-BIH ECG arrhythmia database was the first publicly available database on arrhythmia and was collected by Beth Israel Hospital (BIH) between 1975 and 1979 and has been distributed by MIT since 1980 [72]. The MIT-BIH database contains 48 half-hour ECG records obtained from 47 patients, including 25 men aged 32–89 years and 22 women aged 23–89 years. The database consists of two parts. There are 23 patient records in the first part and 25 patients in the second part. The database contains 116 137 QRS complexes. The ECG signals were sampled at 360 samples per second with an 11 bit resolution over a 10 mV range and were band-pass filtered at 0.1–100 Hz. Each record in the database has been annotated independently by two or more cardiologists.

12.2.2 Producing ECG beat images

In this study, as one-dimensional (1D) ECG record signals are converted into two-dimensional (2D) images, we need to crop each ECG beat region and create ECG beat images as a separate input into the machine learning approaches. To do this, we fix the ECG beat image size in a rectangular shape. The first step is to obtain the signal values of the ECG beats by reading the record files. The format of the record files is comma-separated values (CSVs). Each record file has nearly 650 000 records for the 30 min duration and 360 records per second. To label the ECG beat type, we need to read annotate files that have the 'txt' file format. The second step is to determine what the ECG beat image size should be. While producing the ECG beat image, we need a reference point in the signal. The peak value of each signal was selected as a reference point and the width of the ECG beat image is determined by the sum of two parts of the signals. The first part is the points which are before the peak value and the second part is the points which are after the peak value. While producing the ECG beat image, 90 signal values are used for the part before the peak value and 150 signal values for the part after the peak value to define the signal as lossless.

The flowchart for producing the ECG beat images is shown in figure 12.2. The preprocessing block of the flowchart in figure 12.2 has parameters such as sig1, wden, sym8, peak values, beat types and size of the images. sig1 is the raw signal of a patient record, wden is a function that denoises the raw signal using WT, sym8 is the name of the wavelet function, lev is the level of wavelet function and its value is six in this process, peak value and beat type are obtained through annotate files. The peak value is the highest point in each beat and the beat type is the label.

There are 23 symbols including beat and non-beat types in the MIT-BIH ECG database. The symbols of the beat types and non-beat types are !,], |, [, ~, +, A, a, E, e, F, f, J, j, L, . or N, / or p, Q, q, R, S, V and x. The beat types used in this study are L, N, p, R and V. They are called the left-bundle-branch-block (LBBB) beats, normal (N) beats, paced beats (PBs), right-bundle-branch-blocks (RBBB) beats and

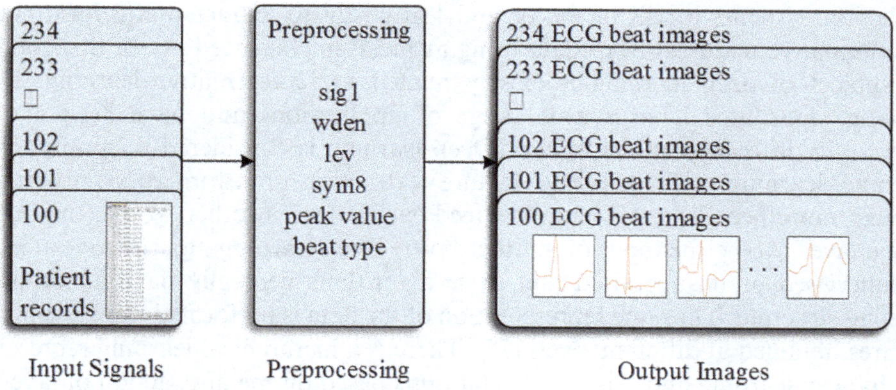

Figure 12.2. The flowchart for producing ECG beat images from MIT-BIH ECG database patient records.

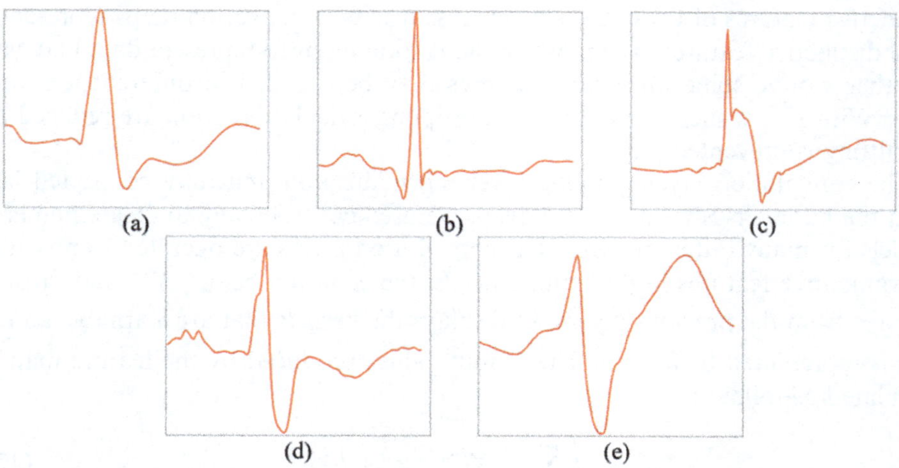

Figure 12.3. Produced ECG beat images after preprocessing: (a) LBBB beat images, (b) N beat images, (c) PB images, (d) RBBB beat images and (e) PVC beat images.

premature-ventricular-contraction (PVC) beats. These five beat types are selected due to the huge number of their occurrence. Sample beat type images that are produced using the preprocessing block are shown in figure 12.3.

In total, 8001 LBBB (symbol L) beat samples, 8067 N (symbol N) beat samples, 7018 PB (symbol P) samples, 7247 RBBB (symbol R) beat samples and 7125 PVC (symbol V) beat samples were used in the experiments to evaluate the proposed method.

12.2.3 Convolutional neural networks (CNNs)

The deep learning approach consists of multiple processing layers combined with multiple abstraction structures to learn the representations of the data. Various preprocessing procedures, dimension reduction and feature selection are performed to reveal these features [73]. To reduce the cost at this stage, it is necessary to avoid the dependence on features. In this way, when designing classifiers and other prediction systems, it will be easier and less costly to extract useful information (discriminative features) from data using artificial intelligence [74]. In this context, the subject of deep learning is closely related to representative learning. Deep learning algorithms have a wide range of applications and have been applied successfully in many different fields. Deep learning is considered a sub-branch of machine learning and performs feature extraction transformation procedures through non-linear layers. The layers feed each other, in other words, the output of one layer can be the input of another layer. Deep learning algorithms can either be supervised or unsupervised and these algorithms generally have a non-linear learning structure. Thus, the representation of the data is performed using distinctive features obtained at different levels [75]. There is a hierarchical learning representation in deep learning and this representation is based on the abstraction of layers at different levels. The data applied as an input in deep learning are passed through a series of processes and learning is realized. Convolution is performed to reveal the distinctive features of the data. This process results in activation maps representing local distinctive features. The power of activation maps to represent data may vary. In other words, some distinctive features may be retained in different activation maps.With layers such as pooling and dropout, calculation costs are reduced and overfitting is prevented [76].

The convolution layer, pooling layer, normalization and fully connected layer form the basic CNN model [77]. Layers are used consecutively to create numerous models for many different tasks. The convolution layers are operated to reveal the discriminative features in the input data. In the equation below, X_i^{l-1} indicates the features from the previous layers, k_{ij}^l displays the kernels that are learnable and the bias is represented by b_j^l [78]. The output values produced by the feature map are calculated as follows:

$$X_j^l = f\left(\sum\nolimits_{i \in M_j} X_i^{l-1} * k_{ij}^l + b_j^l\right), \tag{12.1}$$

where the input map selection represented by M_j and $f(.)$ indicates an activation function. With the pooling layer, the calculation cost is reduced and different approaches, such as average pooling or maximum pooling, can be used for this purpose. In CNN models, another advantage of pooling layers is to prevent overfitting. This process is also supported by layers such as dropout layers. In these layers, the downsampling is performed as defined in

$$\mathrm{X}_j^l = \mathrm{down}\left(X_j^{l-1}\right). \tag{12.2}$$

Here in the related procedure is symbolized by the function 'down'. It can be stated that the number of local distinctive features is reduced [79]. The activation maps processed in previous layers are combined in fully connected (FC) layers. The deep features obtained as a result of all the operations applied in the deep architecture are combined in the FC layers [80]. Optimization algorithms play an important role in the training of deep networks, and their use is inevitable for proper learning to be achieved. Equation (12.3) below represents an optimization algorithm. For optimization purposes, stochastic gradient descent with momentum (SGDM) and adaptive moment estimation (ADAM) are preferred frequently due to their performance on real-world problems [81]:

$$V_t = \beta V_{t-1} + a\nabla_w L(W, X, y), \tag{12.3}$$

where W denotes the weights, and L and α show the loss function and learning rate, respectively. During the training of the CNN, the new weights are computed as follows:

$$W = W - aV_t. \tag{12.4}$$

The ADAM optimization algorithm performs weights within each iteration, taking into account the average of the second momentum. It also uses the average first moment in the RMSProp method [82].

12.2.4 Deep transfer learning (DTL)

Transfer learning is a learning approach that examines the use of artificial learning systems with different or similar solutions to problems. The transfer learning approach is based on the learning model of human nature. In other words, human beings find solutions to a different event that they encounter by knowing or not knowing their past experiences. Deep transfer learning is an effective approach to solve image classification problems, in particular when there is a finite number of training images. Achieving network weights that represent learning in the deep learning model requires considerable time and computational costs. In the cases where the data are limited, updating the previously obtained weights is adopted as an efficient solution instead of performing network training from scratch, because the computational node numbers mentioned here are at the level of millions. As part of this solution, a modification is made, in particular in the last layers of the network [83].

As mentioned above, the transfer learning approach allows a pre-trained network to be used for a new task. For this purpose, it is based on updating the weights of the network, which is trained on the general problem, providing the connections within the network [84]. Except for the last few layers of the network, all are defined as a new network. Later, the connections and weights are updated considering the number of classes in the studied problem. Training of the network is performed to enable the network to reveal specific information in the new area. Fine-tuning is required to enable deep network learning. In this context, an initial learning coefficient is determined first. The choice of this parameter is crucial to ensure appropriate and valid learning. The initial learning coefficient should be neither too large nor too small [85]. Another important parameter of the transfer learning approach is the mini-batch size. The number of samples that is used to train the network in each epoch is determined by this parameter. Similarly, the maximum number of epochs is one of the important parameters to be specified for transfer learning. Optimization algorithms are used to provide better generalization performance and adapt deep networks to a new area through transfer learning. These algorithms can use various learning schemes. These schemes allow the learning coefficient to be updated as the epoch progresses. Therefore, the initial learning coefficient can be updated during transfer learning and thus more efficient learning can be realized [86].

12.2.5 Support vector machines

The SVM is the most popular traditional supervised machine learning approach and is used by researchers all over the world [87]. Due to its easy implementation and robustness, it is possible to come across an SVM in almost all machine learning-based classification studies. The basic principle of the SVM is to linearly separate two classes from each other by drawing a margin (a line, a plane or a hyperplane) between them. The goal of the SVM method is to maximize the margin between two classes [88]. When the datasets become non-linear, researchers solved this issue by adding some new kernel structures to the SVM to overcome these non-linear datasets. Some of these SVM kernels are linear, quadratic, cubic, fine Gaussian, medium Gaussian and coarse Gaussian.

12.2.5.1 The linear SVM kernel

$$K(x, y) = x^T y + c, \tag{12.5}$$

where K denotes the kernel, x is the input vector (the feature vector of one class), y is the input vector (the feature vector of another class), c is a constant term and T denotes the transpose. The terms defined in equation (12.5), i.e. K, x, y, c, T, are also the same for equations (12.6)–(12.9).

12.2.5.2 The quadratic SVM kernel

$$K(x, y) = (\alpha x^T y + c)^2, \tag{12.6}$$

where α denotes the slope term. The quadratic SVM kernel is a type of polynomial SVM that has a polynomial degree of 2.

12.2.5.3 The cubic SVM kernel

$$K(x, y) = (\alpha x^T y + c)^3, \tag{12.7}$$

cubic SVM is a type of polynomial SVM that has a polynomial degree of 3.

12.2.5.4 The fine Gaussian SVM kernel

$$K(x, y) = \exp\left(-\frac{\|x - y\|^2}{2\sigma^2}\right) \tag{12.8}$$

$$K(x, y) = \exp(-\gamma\|x - y\|^2), \tag{12.9}$$

where γ depends on σ which is the Gaussian term. The fine Gaussian SVM kernel scale set is the result of the square root of P divided by 4, where P is the number of predictors. Equation (12.9) includes all types of Gaussian SVM kernels, such as fine, medium and coarse.

12.2.5.5 The medium Gaussian SVM kernel

The medium Gaussian SVM kernel scale set is the square root of P, where P is the number of predictors.

12.2.5.6 The coarse Gaussian SVM kernel

The coarse Gaussian SVM kernel scale set is the result of the square root of P multiplied by 4, where P is the number of predictors.

12.2.6 Performance metrics

The most basic tool used to assess the performance of a model is the confusion matrix, also called the error matrix. This matrix includes the true-positive, false-positive, false-negative and true-negative parameters. Using these indices, various performance metrics such as accuracy can be calculated. In this study, only the accuracy metric is taken into account and this reflects the system's ability to accurately estimate the ECG beats. It can be calculated as follows:

$$\text{Accuracy} = \frac{\text{NoTCB}}{\text{NoACB}} \tag{12.10}$$

$$\text{NoTCB} = \text{NoACB} - \text{NoFCB}, \tag{12.11}$$

where NoTCB is the number of true classified beats, NoACB is the number of all classified beats and NoFCB is the number of false classified beats. All the numbers of classified beats belong to only one class.

12.3 Experimental work and results

A workstation equipped with the NVIDIA Quadro M4000 GPU was used in the experiments. The MATLAB software on the MIT-BIH arrhythmia database was considered in all coding. Five ECG beat types namely normal (N) beats, left-bundle-branch-block (LBBB) beats, right-bundle-branch-block (RBBB) beats, premature-ventricular-contraction (PVC) beats and paced beats (PBs) were used in the classification stage of the work. The MIT-BIH arrhythmia database contains 40 ECG records and all records were used in the experiments. In the experiments, 8067 N beat samples, 8001 LBBB beat samples, 7247 RBBB beat samples, 7125 PVC beat samples and 7018 PB samples were used.

The ECG signals were converted into ECG images by taking into account the annotations prepared by the experts. The signal-to-image conversion was handled using a code that was written by the chapter authors. First, the resolutions of the constructed images were adjusted to 224 × 224 px and then were converted to gray-scale images. Five well-known pre-trained deep CNNs, called VGG16, VGG19,

ResNet18, ResNet50 and ResNet101, were considered in the experiments. In addition, five SVM methods, namely the linear SVM, quadratic SVM, cubic SVM, medium Gaussian SVM and coarse Gaussian SVM were used in the classification. In the linear SVM, the linear kernel function was used and the epsilon value of the SVM method was set to 0.01. In the quadratic SVM, the quadratic function was used as the kernel function and the epsilon value was adjusted to 0.2. The cubic kernel function was used in the cubic SVM and the epsilon value was assigned as 0.02. For both the medium and coarse Gaussian SVMs, the Gaussian kernel function was used and the epsilon parameters were adjusted to 0.1 and 0.2. The performance of the proposed method was evaluated by using the accuracy score. A ten-fold cross-validation procedure was also performed in the experiment to yield objective results. At the end of the ten-fold cross-validation procedure, the mean accuracy scores were calculated. Table 12.1 shows the obtained accuracy scores. The rows of table 12.1 show the SVM classifier types, while the columns show the pre-trained CNN models. The last row and column show the average accuracy scores.

From table 12.1, it is seen that only the VGG16 and VGG19 models produced average accuracy scores above 99.0%. VGG16 achieved 99.32% and VGG19 produced a 99.27% average accuracy score. In addition, the ResNet18 model obtained a 98.98% average accuracy score, and the ResNet50 and ResNet101 models both produced 98.65% average accuracy scores. The highest accuracy score of 99.6% was also produced by both the VGG16 and VGG19 models. When the obtained results are evaluated from the viewpoint of the classifiers, it is seen that the best mean accuracy score of 99.42% was yielded by the cubic SVM classifier. The second best mean accuracy score of 99.40% was yielded by the quadratic SVM, and the fine and medium Gaussian SVMs produced 99.18% and 99.16% average accuracy scores, respectively. Linear SVM produced a 99.00% average accuracy score and, finally, the worst average accuracy score of 97.68% was produced by the coarse Gaussian SVM classifier.

Figure 12.4 shows the confusion matrix for the ResNet50 features and the linear SVM classifier. The labels 1, 2, 3, 4 and 5 show N, LBBB, RBBB, PVC and PB,

Table 12.1. The results for the pre-trained deep CNN models and SVM classifiers in ECG beat classification.

	Accuracy (%)					
Method	ResNet18	ResNet50	ResNet101	VGG16	VGG19	Average
Linear SVM	99.5	98.4	98.5	99.3	99.3	99.00
Quadratic SVM	99.3	99.2	99.3	99.6	99.6	99.40
Cubic SVM	99.4	99.2	99.3	99.6	99.6	99.42
Fine Gaussian SVM	99.2	99.0	99.0	99.4	99.3	99.18
Medium Gaussian SVM	99.2	98.9	99.0	99.4	99.3	99.16
Coarse Gaussian SVM	97.3	97.2	96.8	98.6	98.5	97.68
Average	98.98	98.65	98.65	99.32	99.27	

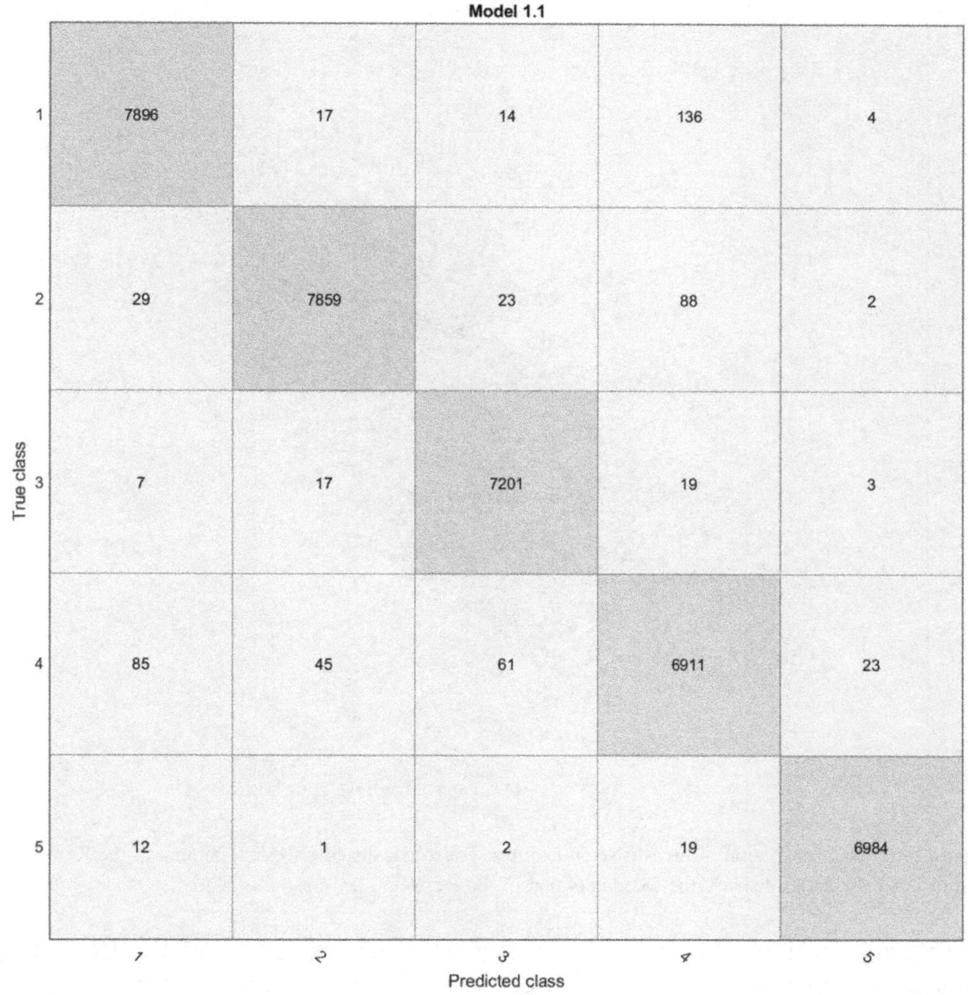

Figure 12.4. The confusion matrix obtained for ResNet50 and the linear SVM classifier.

respectively. From figure 12.4, it is observed that 171 N beat samples were classified incorrectly. The correct classification rate of the N beat samples was 97.88%. 142 LBBB beat samples were misclassified and the accuracy of the LBBB beats was 98.23%.

The obtained correct classification rate for RBBB was 99.37% and only 46 misclassified RBBB samples can be seen in figure 12.4. The highest misclassification was obtained for PVC beats. In total, 214 PVC beat samples were classified incorrectly and the accuracy of the PVC beat was 97.00%. Finally, the highest accuracy score of 99.52% was obtained for the PB class. Only 34 samples were misclassified. Figure 12.5 shows the distribution of ResNet50 features. The *x*-axis shows the first column of the ResNet50 features and the *y*-axis shows the second column of the ResNet50 features.

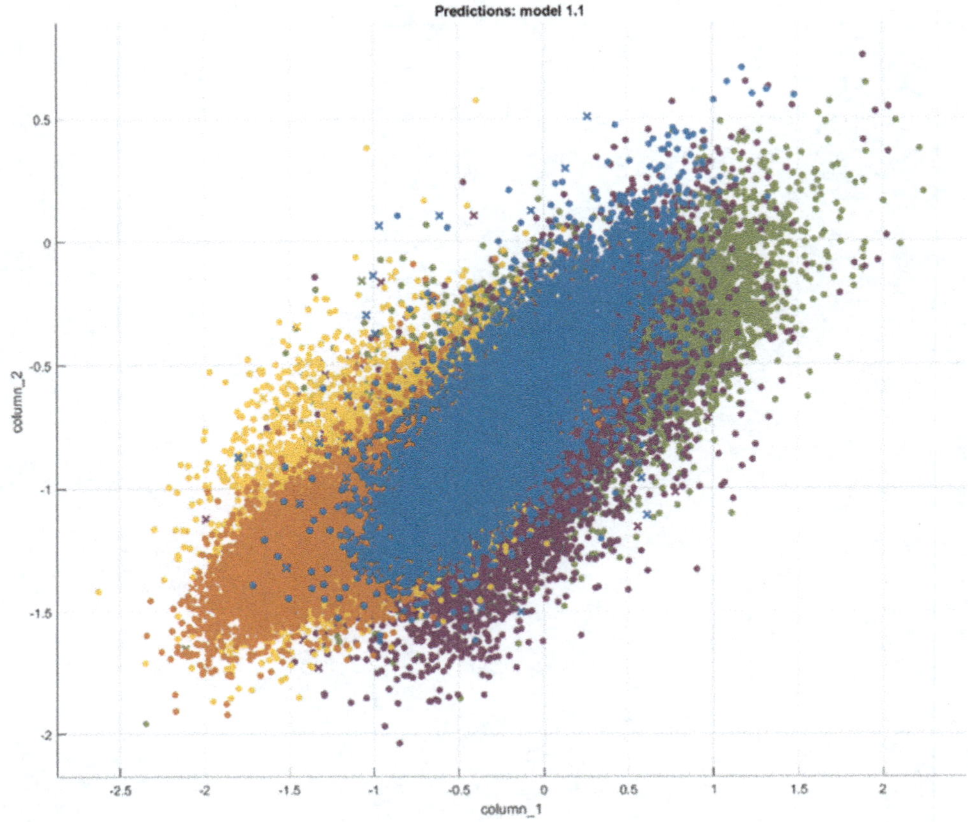

Figure 12.5. The distribution of the ResNet50 features. The x-axis displays the first column of the ResNet50 features and the y-axis displays the second column of the ResNet50 features.

Figure 12.6 illustrates the training period of the fine-tuning of the ResNet50 model. The top panel of figure 12.6 displays the training and testing accuracies (blue for training and black for testing) and the bottom panel of figure 12.6 shows the loss values for both the training and testing samples (orange for training and black for testing). Table 12.2 provides the results of the fine-tuning of the pre-trained deep CNNs on ECG beat categorization.

As seen in table 12.2, all the fine-tuned deep CNN models achieved classification accuracy scores above 99.00%. The highest accuracy score of 99.66% was yielded by the VGG19 model and a 99.49% accuracy score, which was the second best result, was obtained by the ResNet101 model. Accuracy scores of 99.39%, 99.36% and 99.04% were obtained by the ResNet18, ResNet50 and VGG16 models, respectively.

In the final experiments, a novel CNN model was constructed and trained in an end-to-end fashion by the chapter authors and used in ECG beat classification. The developed CNN model is depicted in figure 12.7 and, as can be seen, the developed CNN model was composed of 15 layers. The network started with the input layer

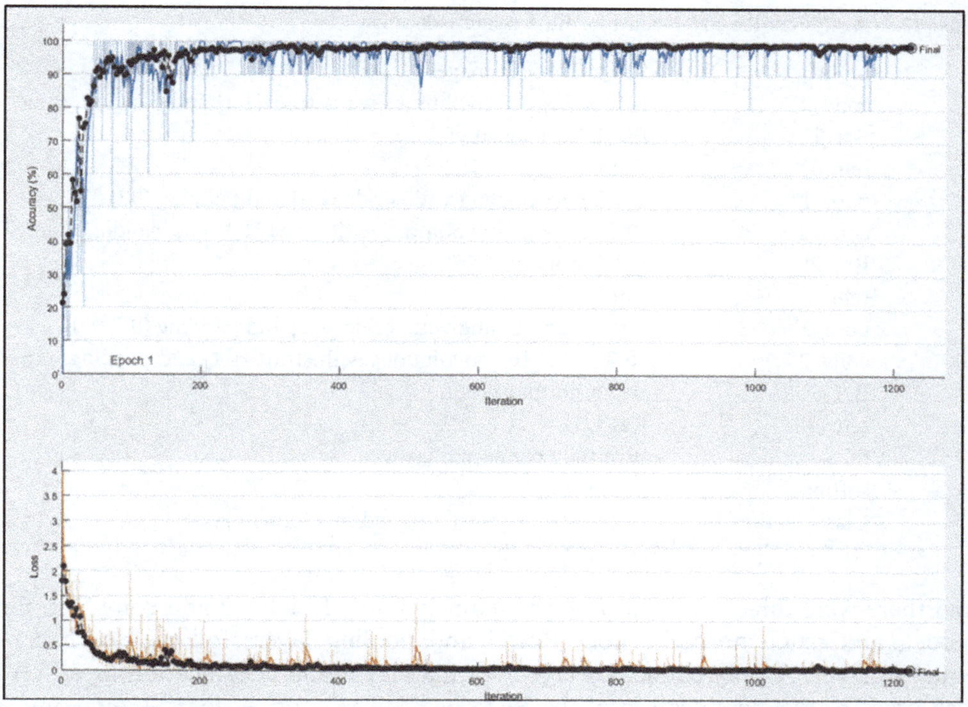

Figure 12.6. The training of the developed CNN model for ECG beat classification.

Table 12.2. The results of fine-tuning pre-trained deep CNN models on ECG beat classification.

Fine-tuning	Accuracy (%)
VGG16	99.04
ResNet18	99.39
ResNet50	99.36
ResNet101	99.49
VGG19	99.66

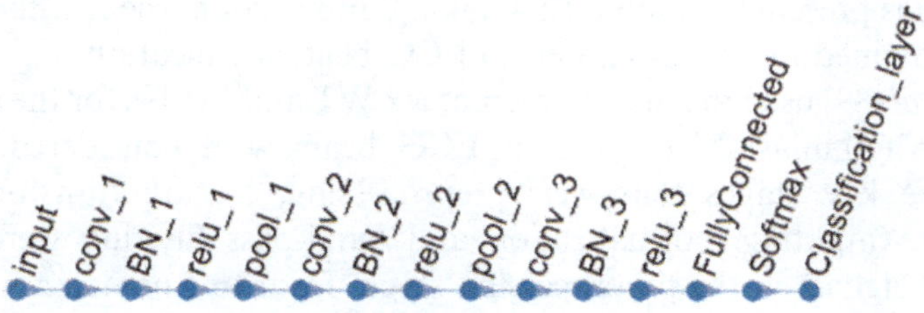

Figure 12.7. The developed CNN model.

Table 12.3. Analysis of the CNN model for ECG beat classification.

1	'input'	224 × 224 × 3 images
2	'conv_1'	64 3 × 3 × 3 convolutions with stride [1 1] and padding 'same'
3	'BN_1'	Batch normalization
4	'relu_1'	ReLU
5	'pool_1'	2 × 2 max pooling with stride [2 2] and padding [0 0 0 0]
6	'conv_2'	32 3 × 3 × 8 convolutions with stride [1 1] and padding 'same'
7	'BN_2'	Batch normalization
8	'relu_2'	ReLU
9	'pool_2'	2 × 2 max pooling with stride [2 2] and padding [0 0 0 0]
10	'conv_3'	16 3 × 3 × 16 convolutions with stride [1 1] and padding 'same'
11	'BN_3'	Batch normalization
12	'relu_3'	ReLU
13	'fc'	Five fully connected layers
14	'softmax'	Softmax
15	'Classification'	Crossentropyex with '0' and nine other classes

and there were three convolution layers namely conv_1, conv_2 and conv_3 in the model, and batch normalization, ReLU and pooling layers followed each convolution layer. A fully connected layer, softmax layer and classification layer were also used for the classification of the ECG beats. The conv_1, conv_2 and conv_3 layers contained 64, 32 and 18 filters of size 3 × 3. The max operator was used in the pooling layers.

The layers of the proposed architecture and the parameters and values used with the layers are given in table 12.3. The SGDM was chosen as the optimization algorithm. The initial learning coefficient was set to 0.01 and 20 was used for the batch size value. The network trained more than 1200 iterations.

The end-to-end learning scheme of the proposed CNN model is given in figure 12.8. The top panel of figure 12.8 shows the training and testing accuracies (blue for training and black for testing) and the bottom panel of figure 12.8 demonstrates the loss values for both the training and test samples (orange for training and black for testing). The obtained accuracy score was 98.68% and the training procedure was completed in 876 iterations.

12.4 Discussion

A comparison is presented in table 12.4, taking into account the methods, databases and results obtained in studies related to ECG beat classification.

Thomas *et al* [89] used the dual-tree-complex WT and MLNs for the classification of cardiac arrhythmias. Five types of ECG beats were considered. The model followed three key stages that wer: preprocessing, feature transformation and classification. Amplitude normalization and band-pass filtering were applied to the raw ECG signals in the preprocessing stage. Dual-tree complex wavelet-based features and four other extracted features were used in the feature set. Finally, the MLNs were employed to determine the five different ECG beat types. The model

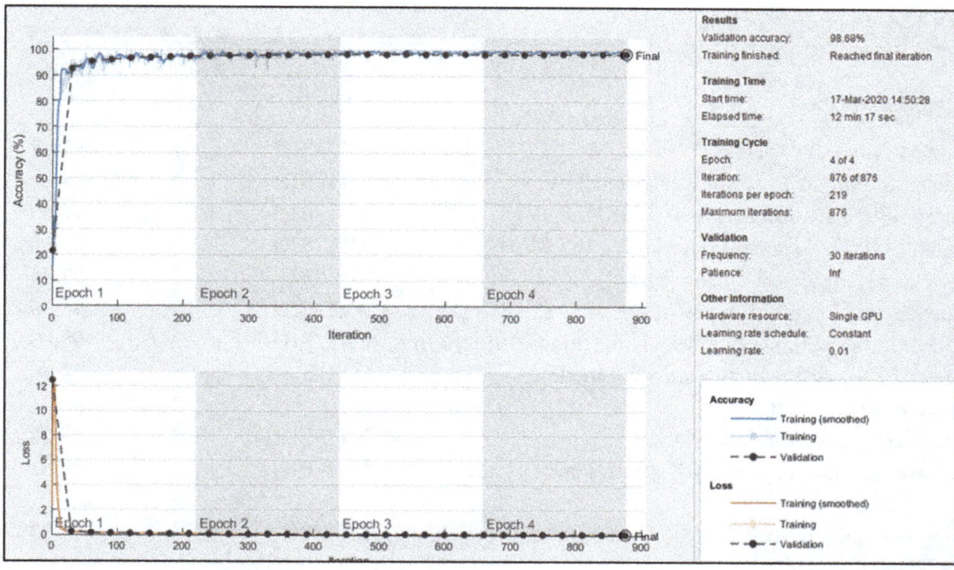

Figure 12.8. The training of the developed CNN model for ECG beat classification.

achieved 97.68% classification accuracy. Kaur *et al* [90] used wavelet coefficients and feature reduction methods for efficient cardiac dysfunction classification. The WT was computed with db4. All WT coefficients at levels 4, 5, 6 and 7 were taken into account and a total of 43 coefficients constituted the feature vector. Then, factor analysis was utilized to reduce the size of the feature set without rotation and with orthogonal rotation. In this way, the number of features was decreased and 23 features were obtained. The model achieved 99.06% accuracy. Rai *et al* [91] used Daubechies WT and RBFNN for five types of ECG beat classification. A preprocessing step covering noise elimination, baseline wander removal, R-peak detection and QRS-complex extraction was applied to the raw ECG signals. Each ECG sample was represented by 21 points, which were input to the model. In this study, the performances of the BPNN, MLP and SVM classifiers were also examined and the model yielded a 99.60% classification score. Khalaf *et al* [92] used spectral correlation and the SVM classifier for cardiac arrhythmia classification. For this purpose, the spectral correlation was utilized. The proposed system used Otsu's thresholding, and statistic features which included PCA and Fisher scores. The SVM classifier was used to determine the ECG beat types and the model produced a 98.60% classification accuracy. Dong *et al* [33] introduced a new approach based on the deterministic learning algorithm. This system ensured different features and the model was performed on the MIT-BIH database. The main advantage of the system was that no feature extraction procedure was required for the test beats. The model yielded a 97.78% classification score. Osowski and Ling [35] carried out ECG beat classification based on higher-order statistical features and a fuzzy neural network. The model produced an accuracy of 98.00% using a multi-layer perceptron and Fourier analysis. Zidelmal *et al* [53] used an SVM with

Table 12.4. Comparison of the related studies.

Reference, method	Database	# of beat types	Accuracy (%)
Thomas *et al* [89], DTCWT	MIT-BIH	5	97.68
Kaur *et al* [90], WT, factor analysis, LDA	MIT-BIH	5	99.06
Rai *et al* [91], Daubechies WT, RBFNN	MIT-BIH	5	99.60
Khalaf *et al* [92], spectral correlation and SVM	MIT-BIH	5	98.60
Dong *et al* [33], deterministic learning, RBFNN	MIT-BIH	5	97.78
Osowski and Ling [35], fuzzy hybrid neural network	MIT-BIH	7	98.00
Zidelmal *et al* [53], support vector machine with rejection	MIT-BIH	2	98.80
Liu *et al* [54], vector quantization, k-medoids	QTDB	8	94.50
Mateo *et al* [55], RBFNN	A	2	98.71
Engin [56], neuro-fuzzy network	MIT-BIH	4	93.50
Asgharzadeh-Bonab *et al* [34], spectral entropy, deep CNN technique	MIT-BIH	5	99.33
Wang *et al* [52], morphological and temporal vector, global RNN	MIT-BIH INCARTDB SVDB	4	99.80
Li *et al* [93], deep residual CNN technique	MIT-BIH	5	99.06
Acharya *et al* [94], CNN technique	B	4	94.90
Yao *et al* [95], CNN technique	C	9	81.20
The proposed model, deep features and support vector machines	**MIT-BIH**	**5**	**99.66**

A: PhysioNet database (MIT-BIH Atrial Fibrillation Database, Long-Term AF Database, MIT-BIH Arrhythmia Database, AF Termination Challenge Database, MIT-BIH Normal Sinus Rhythm Database, St Petersburg Institute of Cardiological Technics 12-lead Arrhythmia Database, Normal Sinus Rhythm RR Interval Database).

INCARTDB: St Petersburg Institute of Cardiological Technics 12-lead Arrhythmia Database.

SVDB: The MIT-BIH Supraventricular Arrhythmia Database.

B: MIT-BIH Atrial Fibrillation Database (AFDB), MIT-BIH Arrhythmia Database (MITDB), Creighton University Ventricular Tachyarrhythmia Database (CUDB).

C: China Physiological Signal Challenge.

rejection. The frequency information, RR intervals, QRS morphology and AC power of QRS detail coefficients were also considered in the analysis. The beats were collected into two classes and the model achieved 98.80% classification. Liu *et al* [54] used the vector quantization method for classifying eight different ECG beat types. To de-noise the raw ECG signal, a median filter was used in the preprocessing step. Wave boundary detection was also performed. The dictionary learning algorithm was employed to describe the signals. As a result, the model achieved 94.50% classification success using VQ k-medoids. Mateo *et al* [55] introduced a system using RBFNN to cancel out ectopics.

As seen in table 12.4 and mentioned in the studies [35, 53–56, 89–92], ECG beat classification has been realized using advanced signal processing techniques that include WT, high-ordered statistical features, PCA, ICA, VQ, factor analysis, PSO and specific morphological feature extraction processes. These approaches mainly followed three key stages: preprocessing, feature transformation (feature extraction and feature selection) and finally classification. The quality of both the preprocessing and feature extraction steps is critical to ensuring a consistent and robust diagnostic tool, since the performance of the next steps of the model depends on these inital steps. In other words, the selected features have a great impact on the performance of the classifiers.

Asgharzadeh-Bonab *et al* [34] proposed a model based on the spectral entropy and deep CNN. The noise reduction and QRS segmentation were applied as the preprocessing procedures in the proposed model. Then the model followed time–frequency analysis, feature reduction, and classification steps. The images obtained using the two-directional two-dimensional principal component analysis were applied as the input to deep CNN. The model ensured an average accuracy of 98.33%. Wang *et al* [52] proposed a model fed with morphological and temporal vectors and uses a global recurrent neural network. The model applied a single RNN for automatic feature learning and classification. An end-to-end learning schema was ensured in this study. As a result, a diagnostic tool with high accuracy was ensured. Li *et al* [93] offered a deep residual CNN model from 2 lead ECG. The efficient features were learned directly ECG samples. The 1D convolutional layer was used in the architecture. As a result, a successful model was ensured. Acharya *et al* [94] introduced two architectures of CNN performed on the three different databases. They did not consider the QRS-complex in the experiment. Also, feature extraction and feature selection were not involved. The model achieved an accuracy of 94.90%. Yao *et al* [95] suggested an attention-based time-incremental CNN model. The model performed on a database consisting of a normal and eight different beat types and ensured 81.20% classification accuracy.

Considering the related studies [34, 52, 93–95], it is seen that in addition to the models following traditional preprocessing, feature transformation and classification processes, deep CNN models have also been used to classify different ECG beat types, thanks to their efficient end-to-end learning schemes. In this context, it has been observed that both deep CNN models, LSTM-based models, ResNet models and RNN models have been used efficiently. Deep CNN networks push the raw data, which is received as inputs, forward to many non-linear layers throughout the network. In this process, the deep CNN models perform the learning and classification processes directly on the input data using functions such as convolution, pooling and dropout. Thanks to this end-to-end learning scheme, the labor-intensive extraction process can be eliminated from the diagnostic models. Another great feature of the deep CNN models is that they offer a hierarchical learning scheme. Activation maps obtained from different layers can reveal many distinctive features of the input data. Sharing weights in the network also provides stronger generalization performance. In the proposed method, the performance of the pretrained CNN models was first examined to classify the ECG beats. Later, instead of extracting features manually, the deep features obtained from the activation maps were inputted into the SVM classifier and quite efficient results were obtained. In addition to all their advantages, deep CNN models need a lot of data for proper training. As this experimental study focused on ECG beats, sufficient data were available. However, for many clinical applications, accessing such a large-scale dataset may not be possible or may take a lot of time. In addition, sufficient hardware resources with high capacity must be available for the training of deep CNN models. Moreover, compared to traditional machine learning methods, deep CNN models take a longer time to train.

The vast majority of studies were carried out on the MIT-BIH database. In addition, different databases are also available through PhysioNet for ECG beat classification. In the studies, it can be observed that five classes are generally used for ECG beat classification problems, and binary classification has also been realized by separating the beats as normal and abnormal. Moreover, some researchers evaluated the problem as having four, seven or eight classes. These differences also show that it is not possible to make a one-to-one comparison. In related studies, it is seen that the classification performance is generally over 95%. It is also worth noting that there is a significant increase in classification success with the RNN and CNN based methods that use end-to-end learning schemes. The classification success has been over 99% using these models. Asgharzadeh-Bonab *et al* [34] achieved a 99.33% classification score while Wang *et al* [52] reported an accuracy of 99.80%. Similarly, for the proposed model we achieved 99.66% classification. As a result, from table 12.4, it is observed that the proposed method outperformed most other state-of-the-art methods.

12.5 Conclusion

In this chapter, three deep CNN approaches are considered in ECG beat classification. These approaches achieve an end-to-end learning scheme of a new CNN model, fine-tuning of pre-trained CNN models, and extraction of deep features and their classification using a traditional or deep machine learning approach such as SVM. The obtained results present the following conclusions:

1. Saving ECG beats as images and applying the CNN approach to them for classification purposes produced quite efficient results. All deep CNN based approaches produced reasonable accuracy scores.
2. Fine-tuning of a pre-defined model and deep feature extraction produced better results than end-to-end training of the new CNN model. This might have happened because the pre-trained models were trained on 25 million images.
3. The cubic kernel function produced better accuracy scores than the other kernel functions.

In future works, dense CNN models will be developed and trained on ECG beat classification. In addition, time–frequency images are considered for comparison with the presented results.

References

[1] Sahoo S, Dash M, Behera S and Sabut S 2020 Machine learning approach to detect cardiac arrhythmias in ECG signals: a survey *IRBM* in press

[2] McSharry P E, Clifford G D, Tarassenko L and Smith L A 2003 A dynamical model for generating synthetic electrocardiogram signals *IEEE Trans. Biomed. Eng.* **50** 289–94
[3] Cömert Z, Kocamaz A F and Kocamaz A F 2015 Determination of QT interval on synthetic electrocardiogram *23rd Signal Processing and Communications Applications Conf. (SIU)* pp 1–4
[4] Diker A, Cömert Z, Avci E and Velappan S 2018 Intelligent system based on genetic algorithm and support vector machine for detection of myocardial infarction from ECG signals *2018 26th Signal Processing and Communications Applications Conf. (SIU)* pp 1–4
[5] Thomas I C, Cork D P, Bhatia A K, Nayak H M, Burke M C and Moss J D 2015 QRS complex fragmentation and survival following left ventricular assist device implantation *Int. J. Cardiol. Heart Vasc.* **7** 165–9
[6] Chiladakis J, Kalogeropoulos A, Zagkli F and Chouchoulis K 2015 Effect of heart rate on the intrinsic and the ventricular-paced QRS duration *J. Electrocardiol.* **48** 689–95
[7] Ünal S *et al* 2018 Tp–e interval and Tp–e/QT ratio in patients with human immunodeficiency virus *J. Infect. Public Health* **11** 35–8
[8] Correa R, Arini P D, Correa L S, Valentinuzzi M and Laciar E 2014 Novel technique for ST-T interval characterization in patients with acute myocardial ischemia *Comput. Biol. Med.* **50** 49–55
[9] Diker A, Cömert Z and Avcı E 2017 A diagnostic model for identification of myocardial infarction from electrocardiography signals *Bitlis Eren Univ. J. Sci. Technol.* **7** 132–9
[10] Guyot P, Djermoune E-H and Bastogne T 2019 Assessment of non-negative matrix factorization for the preprocessing of long-term ECG *J. Pharmacol. Toxicol. Methods* **99** 106595
[11] Jain S, Bajaj V and Kumar A 2018 Riemann Liouvelle fractional integral based empirical mode decomposition for ECG denoising *IEEE J. Biomed. Health Inform.* **22** 1133–9
[12] Gu X, Hu J, Zhang L, Ding J and Yan F 2020 An improved method with high anti-interference ability for R peak detection in wearable devices *IRBM* **41** 172–83
[13] Gupta V, Mittal M and Mittal V 2019 R-peak detection using chaos analysis in standard and real time ECG databases *IRBM* **40** 341–54
[14] Zalabarria U, Irigoyen E, Martinez R and Lowe A 2020 Online robust R-peaks detection in noisy electrocardiograms using a novel iterative smart processing algorithm *Appl. Math. Comput.* **369** 124839
[15] Bajaj A and Kumar S 2019 QRS complex detection using fractional Stockwell transform and fractional Stockwell Shannon energy *Biomed. Signal Process. Control* **54** 101628
[16] Lee J S, Lee S J, Choi M, Seo M and Kim S W 2019 QRS detection method based on fully convolutional networks for capacitive electrocardiogram *Expert Syst. Appl.* **134** 66–78
[17] Malik J, Soliman E Z and Wu H-T 2020 An adaptive QRS detection algorithm for ultra-long-term ECG recordings *J. Electrocardiol.* **60** 165–71
[18] Jain S, Ahirwal M K, Kumar A, Bajaj V and Singh G K 2017 QRS detection using adaptive filters: a comparative study *ISA Trans.* **66** 362–75
[19] Christov I *et al* 2006 Dataset of manually measured QT intervals in the electrocardiogram *Biomed. Eng. Online* **5** 31
[20] Wang L, Lin Y and Wang J 2019 A RR interval based automated apnea detection approach using residual network *Comput. Methods Programs Biomed.* **176** 93–104
[21] Pachori R B, Avinash P, Shashank K, Sharma R and Acharya U R 2015 Application of empirical mode decomposition for analysis of normal and diabetic RR-interval signals *Expert Syst. Appl.* **42** 4567–81

[22] Thuraisingham R A 2006 Preprocessing RR interval time series for heart rate variability analysis and estimates of standard deviation of RR intervals *Comput. Methods Programs Biomed.* **83** 78–82
[23] Hsu C F, Lin P-Y, Chao H-H, Hsu L and Chi S 2019 Average entropy: measurement of disorder for cardiac RR interval signals *Phys. A Stat. Mech. Appl.* **529** 121533
[24] Schneider M A *et al* 2017 Reducing door-in door-out intervals in helicopter ST-segment elevation myocardial infarction interhospital transfers *Air Med. J.* **36** 244–7
[25] Bouaguel I, Foudad H and Trichine A 2018 Corrected QT interval prolongation: a new predictor of cardiovascular risk inpatients with non-ST-elevation acute coronary syndrome Algerian cohort *Arch. Cardiovasc. Dis. Suppl.* **10** 49
[26] Lanatá A, Valenza G, Mancuso C and Scilingo E P 2011 Robust multiple cardiac arrhythmia detection through bispectrum analysis *Expert Syst. Appl.* **38** 6798–804
[27] Kondo H, Shinohara T and Takahashi N 2015 A case of short-coupled premature ventricular beat-induced ventricular fibrillation with early repolarization in the inferolateral leads *J. Arrhythmia* **31** 60–3
[28] Shah J *et al* 2020 Prevalence of cardiovascular risk factors and financial burden in younger adults hospitalized with atrial fibrillation *Heart Lung* **49** 393–7
[29] Hulleman M *et al* 2020 Ventricular fibrillation waveform characteristics in out-of-hospital cardiac arrest and cardiovascular medication use *Resuscitation* **151** 173–80
[30] Sangaiah A K, Arumugam M and Bian G-B 2020 *Artif. Intell. Med.* **103** 101788
[31] Wainwright B *et al* 2020 Autoimmune-mediated congenital heart block *Best Pract. Res. Clin. Obstet. Gynaecol.* **64** 41–51
[32] Chen Q *et al* 2019 The value of baseline and arrhythmic ECG in the interpretation of arrhythmic mechanisms *Card. Electrophysiol. Clin.* **11** 219–38
[33] Dong X, Wang C and Si W 2017 ECG beat classification via deterministic learning *Neurocomputing* **240** 1–12
[34] Asgharzadeh-Bonab A, Amirani M C and Mehri A 2020 Spectral entropy and deep convolutional neural network for ECG beat classification *Biocybern. Biomed. Eng.* **40** 691–700
[35] Osowski S and Linh T H 2001 ECG beat recognition using fuzzy hybrid neural network *IEEE Trans. Biomed. Eng.* **48** 1265–71
[36] Castells F, Laguna P, Sörnmo L, Bollmann A and Roig J M 2007 Principal component analysis in ECG signal processing *EURASIP J. Adv. Signal Process.* **2007** 74580
[37] Monasterio V, Laguna P and MartÍnez J P 2009 Multilead analysis of T-wave alternans in the ECG using principal component analysis *IEEE Trans. Biomed. Eng.* **56** 1880–90
[38] Banerjee S and May M M 2013 ECG beat classification based on discrete wavelet transformation and nearest neighbour classifier *J. Med. Eng. Technol.* **37** 264–72
[39] Daamouche A, Hamami L, Alajlan N and Melgani F 2012 A wavelet optimization approach for ECG signal classification *Biomed. Signal Process. Control* **7** 342–9
[40] Yan J and Lu L 2014 Improved Hilbert–Huang transform based weak signal detection methodology and its application on incipient fault diagnosis and ECG signal analysis *Signal Process.* **98** 74–87
[41] Dutta S, Chatterjee A and Munshi S 2011 Identification of ECG beats from cross-spectrum information aided learning vector quantization *Measurement* **44** 2020–7
[42] Akhbari M, Ghahjaverestan N M, Shamsollahi M B and Jutten C 2018 ECG fiducial point extraction using switching Kalman filter *Comput. Methods Programs Biomed.* **157** 129–36

[43] Bhagyalakshmi V, Pujeri R V and Devanagavi G D 2018 GB-SVNN: genetic BAT assisted support vector neural network for arrhythmia classification using ECG signals *J. King Saud Univ.—Comput. Inf. Sci.* in press

[44] Zhou S and Tan B 2020 Electrocardiogram soft computing using hybrid deep learning CNN-ELM *Appl. Soft Comput.* **86** 105778

[45] Yeh Y-C, Wang W-J and Chiou C W 2009 Cardiac arrhythmia diagnosis method using linear discriminant analysis on ECG signals *Measurement* **42** 778–89

[46] Joo S, Choi K-J and Huh S-J 2012 Prediction of spontaneous ventricular tachyarrhythmia by an artificial neural network using parameters gleaned from short-term heart rate variability *Expert Syst. Appl.* **39** 3862–6

[47] Haseena H H, Joseph P K and Mathew A T 2011 Classification of arrhythmia using hybrid networks *J. Med. Syst.* **35** 1617–30

[48] Zeng N, Wang Z, Li Y, Du M and Liu X 2012 A hybrid EKF and switching PSO algorithm for joint state and parameter estimation of lateral flow immunoassay models *IEEE/ACM Trans. Comput. Biol. Bioinform.* **9** 321–9

[49] Jain S, Bajaj V and Kumar A 2016 Efficient algorithm for classification of electrocardiogram beats based on artificial bee colony-based least-squares support vector machines classifier *Electron. Lett.* **52** 1198–200

[50] Diker A, Avci D, Avci E and Gedikpinar M 2018 A new technique for ECG signal classification genetic algorithm wavelet kernel extreme learning machine *Optik* **180** 46–55

[51] à Mougoufan J B B, Fouda J S A E, Tchuente M and Koepf W 2020 Adaptive ECG beat classification by ordinal pattern based entropies *Commun. Nonlinear Sci. Numer. Simul.* **84** 105156

[52] Wang G *et al* 2019 A global and updatable ECG beat classification system based on recurrent neural networks and active learning *Inf. Sci.* **501** 523–42

[53] Zidelmal Z, Amirou A, Ould-Abdeslam D and Merckle J 2013 ECG beat classification using a cost sensitive classifier *Comput. Methods Programs Biomed.* **111** 570–7

[54] Liu T, Si Y, Wen D, Zang M and Lang L 2016 Dictionary learning for VQ feature extraction in ECG beats classification *Expert Syst. Appl.* **53** 129–37

[55] Mateo J, Torres A M, Aparicio A and Santos J L 2016 An efficient method for ECG beat classification and correction of ectopic beats *Comput. Electr. Eng.* **53** 219–29

[56] Engin M 2004 ECG beat classification using neuro-fuzzy network *Pattern Recognit. Lett.* **25** 1715–22

[57] Übeyli E D 2009 Combining recurrent neural networks with eigenvector methods for classification of ECG beats *Digit. Signal Process.* **19** 320–9

[58] Korürek M and Doğan B 2010 ECG beat classification using particle swarm optimization and radial basis function neural network *Expert Syst. Appl.* **37** 7563–9

[59] Yu S-N and Chen Y-H 2007 Electrocardiogram beat classification based on wavelet transformation and probabilistic neural network *Pattern Recognit. Lett.* **28** 1142–50

[60] Güler İ and Übeyli E D 2005 A modified mixture of experts network structure for ECG beats classification with diverse features *Eng. Appl. Artif. Intell.* **18** 845–56

[61] Dutta S, Chatterjee A and Munshi S 2010 Correlation technique and least square support vector machine combine for frequency domain based ECG beat classification *Med. Eng. Phys.* **32** 1161–9

[62] Yu S-N and Chou K-T 2009 Selection of significant independent components for ECG beat classification *Expert Syst. Appl.* **36** 2088–96

[63] Wang K, Yang G, Huang Y and Yin Y 2020 Multi-scale differential feature for ECG biometrics with collective matrix factorization *Pattern Recognit.* **102** 107211

[64] Mondéjar-Guerra V, Novo J, Rouco J, Penedo M G and Ortega M 2019 Heartbeat classification fusing temporal and morphological information of ECGs via ensemble of classifiers *Biomed. Signal Process. Control* **47** 41–8

[65] Librenza-Garcia D *et al* 2017 The impact of machine learning techniques in the study of bipolar disorder: a systematic review *Neurosci. Biobehav. Rev.* **80** 538–54

[66] Toğaçar M, Cömert Z and Ergen B 2020 Classification of brain MRI using hyper column technique with convolutional neural network and feature selection method *Expert Syst. Appl.* **149** 113274

[67] Toğaçar M, Ergen B and Cömert Z 2020 Application of breast cancer diagnosis based on a combination of convolutional neural networks, ridge regression and linear discriminant analysis using invasive breast cancer images processed with autoencoders *Med. Hypotheses* **135** 109503

[68] Butun E, Yildirim O, Talo M, Tan R-S and Acharya U R 2020 1D-CADCapsNet: one dimensional deep capsule networks for coronary artery disease detection using ECG signals *Phys. Med.* **70** 39–48

[69] Daldal N, Sengur A, Polat K and Cömert Z 2020 A novel demodulation system for base band digital modulation signals based on the deep long short-term memory model *Appl. Acoust.* **166** 107346

[70] Cömert Z 2020 Fusing fine-tuned deep features for recognizing different tympanic membranes *Biocybern. Biomed. Eng.* **40** 40–51

[71] Goldberger A L *et al* 2000 PhysioBank, PhysioToolkit, and PhysioNet components of a new research resource for complex physiologic signals *Circulation* **101** e215–20

[72] Moody G B and May M R G 2001 The impact of the MIT-BIH arrhythmia database *IEEE Eng. Med. Biol. Mag.* **20** 45–50

[73] Toğaçar M, Ergen B and Cömert Z 2019 Waste classification using AutoEncoder network with integrated feature selection method in convolutional neural network models *Measurement* **2019** 107459

[74] Yu Y, Lin H, Meng J, Wei X, Guo H and Zhao Z 2017 Deep transfer learning for modality classification of medical images *Information* **8** 91

[75] Traore B B, Kamsu-Foguem B and Tangara F 2018 Deep convolution neural network for image recognition *Ecol. Inform.* **48** 257–68

[76] Bakator M and Radosav D 2018 Deep learning and medical diagnosis: a review of literature *Multimodal Technol. Interact.* **2** 47

[77] Başaran E, Cömert Z and Çelik Y 2020 Convolutional neural network approach for automatic tympanic membrane detection and classification *Biomed. Signal Process. Control* **56** 101734

[78] Toğaçar M, Ergen B and Cömert Z 2020 Detection of lung cancer on chest CT images using minimum redundancy maximum relevance feature selection method with convolutional neural networks *Biocybern. Biomed. Eng.* **40** 23–39

[79] Budak Ü, Cömert Z, Rashid Z N, Şengür A and Çıbuk M 2019 Computer-aided diagnosis system combining FCN and Bi-LSTM model for efficient breast cancer detection from histopathological images *Appl. Soft Comput.* **85** 105765

[80] Liu H, Li B, Lv X and Huang Y 2017 Image retrieval using fused deep convolutional features *Procedia Comput. Sci.* **107** 749–54

[81] Zeiler M D 2012 *ADADELTA: An Adaptive Learning Rate Method* http://doi.acm.org.ezproxy.lib.ucf.edu/10.1145/1830483.1830503

[82] Shindjalova R, Prodanova K and Svechtarov V 2014 Modeling data for tilted implants in grafted with bio-oss maxillary sinuses using logistic regression *AIP Conf. Proc.* **1631** 58–62

[83] Deniz E, Sengür A, Kadiroglu Z, Guo Y, Bajaj V and Budak Ü 2018 Transfer learning based histopathologic image classification for breast cancer detection *Health Inf. Sci. Syst.* **6** 18

[84] Altuntaş Y, Cömert Z and Kocamaz A F 2019 Identification of haploid and diploid maize seeds using convolutional neural networks and a transfer learning approach *Comput. Electron. Agric.* **163** 104874

[85] LeCun Y, Bengio Y and Hinton G 2015 Deep learning *Nature* **521** 436–44

[86] Talo M 2019 Automated classification of histopathology images using transfer learning *Artif. Intell. Med.* **2019** 101743

[87] Cortes C and Vapnik V 1995 Support-vector networks *Mach. Learn.* **20** 273–97

[88] Burges C J C 1998 A tutorial on support vector machines for pattern recognition *Data Min. Knowl. Discov.* **2** 121–67

[89] Thomas M, Das M K and Ari S 2015 Automatic ECG arrhythmia classification using dual tree complex wavelet based features *AEU—Int. J. Electron. Commun.* **69** 715–21

[90] Kaur M and Arora A S 2012 Classification of ECG signals using LDA with factor analysis method as feature reduction technique *J. Med. Eng. Technol.* **36** 411–20

[91] Rai H M, Trivedi A, Chatterjee K and Shukla S 2014 R-peak detection using Daubechies wavelet and ECG signal classification using radial basis function neural network *J. Inst. Eng. Ser.* B **95** 63–71

[92] Khalaf A F, Owis M I and Yassine I A 2015 A novel technique for cardiac arrhythmia classification using spectral correlation and support vector machines *Expert Syst. Appl.* **42** 8361–8

[93] Li Z, Zhou D, Wan L, Li J and Mou W 2020 Heartbeat classification using deep residual convolutional neural network from 2-lead electrocardiogram *J. Electrocardiol.* **58** 105–12

[94] Acharya U R, Fujita H, Lih O S, Hagiwara Y, Tan J H and Adam M 2017 Automated detection of arrhythmias using different intervals of tachycardia ECG segments with convolutional neural network *Inf. Sci.* **405** 81–90

[95] Yao Q, Wang R, Fan X, Liu J and Li Y 2020 Multi-class arrhythmia detection from 12-lead varied-length ECG using attention-based time-incremental convolutional neural network *Inf. Fusion* **53** 174–82

IOP Publishing

Modelling and Analysis of Active Biopotential Signals in Healthcare, Volume 1

Varun Bajaj and G R Sinha

Chapter 13

ECG signal watermarking to enhance the security of telecardiology

Siddharth Bhalerao, Irshad A Ansari, Anil Kumar

PDPM-Indian Institute of Information Technology Design and Manufacturing, Jabalpur - 482005 MP, India

The telemonitoring of physiological signals is a recent and major development in the medical domain. Telemedicine requires the transmission and storage of biological signals and medical images over a network. Electrocardiogram (ECG) signals are involved in telecardiology and are used to identify various cardiovascular ailments. These signals are sensitive, therefore before using these signals for diagnosis their integrity must be ensured. Watermarking is emerging as a tool to safeguard ECG signals. Here, watermarking serves two purposes: it provides ownership information for the signal and it helps in detecting signal tampering. In addition, the reversibility of the watermarking algorithm is also of major concern in the case of medical signals. Therefore, in this chapter, a comprehensive survey of ECG watermarking algorithms is presented and a detailed description of a deep neural network (DNN) based reversible ECG data hiding method is provided. The proposed DNN based reversible ECG data hiding uses prediction error expansion (PEE) and histogram shifting (HS) to achieve reversibility. This chapter analyses different embedding schemes and also examines the impact of signal abnormalities on the watermarking scheme. It provides a detailed description of different aspects such as data preparation, training and testing of the model.

13.1 Introduction

Telemedicine [1] has become a reality due to advancements in information and communication technology. It is intended to change the healthcare system in a positive way, as it has the capability of reducing the pressures on existing medical services. With telemedicine, it is possible to provide services to remote locations, where it can achieve wonders with a limited number of resources. Telemedicine is

doi:10.1088/978-0-7503-3279-8ch13

helpful in every stage of an ailment, that is, pre-hospitalization, in-hospital and post-hospital care. Given the large proportion of heart-related ailments in medical fatalities, cardiology is becoming one of the successful applications of telemedicine [2] by transmitting a patient's medical data and electrocardiogram (ECG), thus telecardiology facilitates the real-time evaluation of a patient's health. The patient can obtain a consultation with a doctor without any need for either to travel. Telecardiology is emerging as an area of significant research. Telemedicine and telecardiology consist of multiple components, which include medical devices, electronic patient records (EPRs) and communications set-up. With technological advancements wearable devices are being developed [3] and such devices generate large amounts of data. Artificial intelligence (AI) and machine learning (ML) are also major players in the medical domain. Similarly, deep learning (DL) is emerging as a groundbreaking technology in healthcare and medicine, and these approaches are achieving positive improvements in telemedicine and telecardiology. It is a well known fact that technologies such as DL and ML are data-hungry methods and require a large amount of data to work. A large amount of data is generated in telecardiology applications.

Telemedicine requires the transmission of a person's medical data over the Internet. This medical data may comprise images and biomedical signals [4]. The presence of medical data being transmitted over the Internet exposes them to a variety of security threats. Techniques such as data encryption and watermarking are used to establish data integrity. Using cryptography, the form of the data is changed; an attacker may see the communication but cannot understand it. In contrast, in watermarking a signal called the watermark is embedded in another signal called the host signal [5]. The main aim of watermarking is to detect ownership and authenticate the data. Watermarking is extensively used for securing medical signals and images [6]. Watermarking can be done on any signal having different attributes. Based on its perceptivity, watermarking can be classified as visible or invisible watermarking. In invisible watermarking, the presence of a watermark is indiscernible in the watermarked signal. Due to this, invisible watermarking is preferred for medical signals and images.

Robustness is another characteristic of watermarking. In the case of robust watermarking the watermark is supposed to be unaffected even in the presence of attacks [7, 8]. Robust watermarking is used for copyright protection and ownership detection. In contrast to robust watermarking there is fragile watermarking, where the watermark is destroyed when the signal is attacked. Fragile watermarking is used for tamper detection or authentication of the host signal. Based on recovery, watermarking can be classified as blind and non-blind. Blind watermarking only requires watermarked data during extraction, whereas non-blind watermarking requires additional information (original host/watermark/both) during extraction [5]. When it is possible to separate the host signal from the watermarked signal without loss, then such a method is called reversible watermarking [9, 10]. Reversible watermarking is preferable for medical signals, as non-reversible methods may deform the signals, making them unsuitable for diagnosis.

Extensive research is ongoing in the field of physiological signal watermarking and some notable works are discussed here. One of the earliest works on biological signal watermarking was reported by Kong and Feng [6] for EEG signals. In this study, three different procedures were discussed for EEG watermarking and the performance of the algorithm was tested for a wide variety of communication channel errors. Embedding was performed using the patchwork method, least significant bit (LSB) watermarking and a quantization based method, and the results were compared thoroughly. All the tests were performed on 22 different EEG signals with multiple medical conditions. In [11], an initial ECG watermarking method using the discrete wavelet transform (DWT) is reported. In this method, additive white Gaussian noise (AWGN) generated using a known seed was added to the wavelet decomposition of the ECG signal. By reversing the embedding, watermarked AWGN could be extracted, subject to the presence of the original signal. This method, therefore, was a non-blind method. They used signals from the MIT-BIH arrhythmia database for performance evaluation of their method. In [12], a 15 bit unique identification code was embedded in the ECG recording of a patient using a chirp signal. The use of the chirp signal spreads the patient identification code all over the ECG. The signal guarantees blind recovery for high signal to noise ratio (SNR) values. Although their method can identify tampering with the signal, it fails at low SNR values.

Similar to [11], another wavelet based method was proposed by Ibaida and Khalil [13]—an ECG steganography technique for point-of-care systems. The patient's secret data were encrypted using XOR-based encryption and then embedded in the wavelet transform of the ECG signal. The embedding process was secured using scrambling. Their method was fast and introduced little change in the original signal. The algorithm was tested for a wide variety of signals. An approach similar to the above system is mentioned in [14]. Another transform domain method is proposed in [15] by Chen *et al.* In their work, the embedding is performed using quantization and the performances of three different transforms are compared. Signals from the MIT-BIH arrhythmia database were used for the experiments. They obtained good performance but only ten signals from that database were used for performance computation.

In [16–19], an innovative idea was proposed where the ECG signal was reshaped to form a 2D array before watermarking. All of these methods are transform domain techniques. By reshaping the ECG signal, different image transforms could be applied to ECG signals. In [19, 20] a reshaped ECG was decomposed using DWT and patient data in the form of a quick response (QR) code was embedded in it. Transform domain based ECG steganography techniques were proposed in [21] and [22]. In [21] curvelet transform based steganography is implemented and a transform was applied to the 2D converted ECG signal. In both works, signals from the MIT-BIH arrhythmia database were used.

One of the important characteristics of medical signal watermarking is reversibility. The reversible watermarking methods can be classified as compression-based, histogram shifting (HS) [23] and difference expansion (DE) methods. The difference expansion method was developed by Tian [24], and was originally proposed for reversible data hiding in images. Thodi and Rodríguez [25] improved DE and developed prediction error expansion (PEE). PEE is the heart and soul of many successful watermarking algorithms. It can embed large amounts of data in the host signal. In [26–28], PEE and HS were used for image watermarking. An improvement for PEE was proposed in [29], where the correlation between prediction errors was exploited to obtain better performance. Similarly, in [30–32] an adaptive PEE was proposed for reversible image watermarking. In [33–35] PEE and HS were used for watermarking ECG signals.

In [36] a reversible data hiding method for ECG recordings was proposed by Zheng and Qian. It was a transform based technique and ensured reversibility using a lifting based Haar wavelet transform. The watermark data were scrambled through an Arnold transform before embedding. To improve the quality of the watermarked signal, no embedding was performed in the QRS segment of the ECG signal. They used signals from the MIT-BIT arrhythmia database, and results were provided for only five signals chosen from the database. Due to no embedding in the QRS segment, the capacity of this scheme was low. A local prediction-based reversible image watermarking algorithm was proposed in [37] which used PEE and HS for data embedding in digital images. Their technique was extended to ECG signals by Wang *et al* in [34], which resulted in a high capacity reversible ECG data hiding method. This method was a progressive embedding technique and three different predictors were used for ECG sample prediction. Based on the position of the ECG sample, an appropriate predictor was chosen. Due to the progressive technique, extraction needed to be done in reverse order compared to the embedding.

In [38] Shiu *et al* discussed a reversible watermarking method for biomedical signals based on the Hamming code. The algorithm was tested for ECG and electromyogram (EMG) signals. They used ECG signals from the MIT-BIH arrhythmia database and EMG signals from PhysioNet. Their method was a robust and blind watermarking technique based on a shared key. Knowledge of the key was required during embedding and extraction. In addition to the key, a parity-check matrix and decision and syndrome codes were also required during embedding and extraction. In [39–41] transform based reversible ECG data hiding techniques were discussed.

In [35] a DNN based predictor was proposed for reversible ECG data hiding. It was a high capacity technique and signals from the MIT-BIH arrhythmia database were used to evaluate the performance. The embedding was performed using PEE and the patient's confidential data were used as the watermark. The SHA-1 hashing algorithm was used for tampering detection and localization. In this method, the ECG signal was fragmented and SHA-1 hash was calculated for each fragment and the hash was embedded in the fragment itself. As the calculated hash of a fragment was embedded in the fragment itself, this method failed to identify the tampering when any fragment was completely replaced by a watermarked fragment from a different signal or the same signal.

The organization of the rest of the chapter is as follows. In section 13.2 the reversible data hiding method preliminary, i.e. PEE, is elaborated. Section 13.3 details the prediction scheme and the ECG database used for reversible data hiding. In section 13.4 the training procedure, different embedding procedures and their performance are discussed in detail. In section 13.5 an improved embedding scheme is proposed as a solution to remove the weakness of the method proposed in [35]. This work also analyses the performance of the proposed technique on different types of ECG signals. Finally, the chapter is completed by the conclusion and bibliography.

13.2 Preliminaries

In the case of watermarking or data hiding in medical signals, reversibility is a desirable feature. For achieving reversibility, PEE is one of the most successful methods. Therefore, in the proposed method watermarking is performed using PEE and histogram shifting. PEE was proposed by Thodi and Rodríguez [25] as an improvement over the DE method. In PEE, the host signal values are predicted using an appropriate predictor. In the case that the host signal is an image, its pixels are predicted based on the values of neighboring pixels. In the proposed method the host is an ECG signal. The values are predicted using a DNN based prediction mechanism. The prediction model is trained based on neighboring samples of the sample to be predicted. The error is computed between the predicted value and the actual value, and the watermark bit is embedded in the difference between them. When the error is smaller, the similarity between the actual signal and the watermarked signal will be greater. In this section PEE is discussed in detail, the prediction scheme and embedding procedure are explained in the following sections.

13.3 Prediction error expansion

The idea behind PEE and watermark embedding is explained through equations (13.1)–(13.3). Initially, the sample value is predicted using an appropriate mechanism, and the difference (e) is calculated between the actual (y) and predicted ($\hat{y}$) value (13.1). This error is modified by doubling the error and adding a watermark bit to it, and this step is equivalent to left-shifting the error and inserting a watermark bit into its LSB (13.2). The updated error (ee) is then added to the predicted value and this gives the modified sample value (yy) (13.3). When all the samples are modified the new signal generated is the watermarked signal:

$$e = y - \hat{y} \tag{13.1}$$

$$ee = 2^{*}e + b \tag{13.2}$$

$$yy = \hat{y} + ee \tag{13.3}$$

$$yy = y + e + T, \qquad e > 0 \tag{13.4}$$

$$yy = y + e - T + 1, \quad e < 0. \tag{13.5}$$

When the error between the predicted and actual values is greater, the quality of the resultant signal deteriorates. The deterioration of the output signal can be kept under check by introducing a threshold (T) on the error value. When the error is less than the threshold, the watermark is embedded according to (13.1)–(13.3). In the case when the error exceeds the threshold, in place of embedding the watermark in the sample the error and the threshold are added and the resultant is added to the sample; this process is equivalent to HS as discussed in [23]. The modification of the sample in the case of thresholding is performed using either (13.4) or (13.5). The presence of thresholding helps in controlling the amount of information embedded in the signal and its quality.

During extraction, the watermark bit and original sample value are recovered. Recovery is possible only when the predicted value for any particular sample during extraction is the same as the predicted value obtained during embedding. To understand extraction, assume that the sample value is (yy) which was updated during the embedding process using equation (13.3). The prediction for this sample is the same as obtained during embedding, i.e. ($\hat{y}$). The same process is followed and error is calculated as shown in equation (13.6), where er is the error during extraction. By substituting the value of (yy) from equation (13.3), the error at extraction becomes equal to the modified error, as calculated in equation (13.2). The LSB of the modified error has watermark information. The original sample can be obtained as shown in equation (13.8):

$$er = yy - \hat{y} \tag{13.6}$$

$$er = \hat{y} + ee - \hat{y} = 2{*}e + b \tag{13.7}$$

$$y = yy - e - b. \tag{13.8}$$

The samples where the error at extraction comes out to be greater than twice the threshold are identified as HS samples, with no watermark embedded in them. The extraction process does not recover watermarks from HS samples, just the original sample is restored. The threshold-based watermarking process requires knowledge of the threshold at both ends.

13.4 Prediction scheme and ECG database

In this section, the DNN based prediction scheme employed for reversible ECG data hiding is discussed. The imperceptibility of the PEE algorithm depends on better prediction. In this chapter, the details of the database used are also provided.

13.4.1 Deep neural network

A deep neural network (DNN) is essentially an artificial neural network (ANN) with multiple hidden layers. Each hidden layer consists of numerous independent units called neurons. All the neurons are similar and perform the same task. The output of the neurons is the weighted sum of all the inputs to it, passed through an activation function. During the training of a neural network, the training data are processed by the network and in doing so it learns the relation between the data and weights are updated. The weights are updated based on the error between the actual output and the target output. The algorithm for updating weights is called back-propagation.

The error calculated at the output layer is used to update the weights and biases of the network. An optimizer is used to update the weights, which results in minimizing the error. Stochastic gradient descent (SGD) and 'ADAM' are a few examples of optimizers used in ANN. An example of a layered DNN is present in figure 13.1. DNNs are extensively used for image processing applications and also find applications in biomedical signal processing [42–44].

The DNN architecture used for the proposed work is presented in figure 13.1. It has three hidden layers and a 40:60:40 neuron configuration. The weights and biases are learned using a stochastic gradient descent algorithm. The error function used is the mean square error (MSE).

13.4.2 ECG database

In the proposed work, a prediction error expansion based on a reversible data hiding technique for the ECG signal is implemented. The reversible data hiding method will ensure the integrity of ECG signals in telecardiology applications. ECG signals from the MIT-BIH arrhythmia database [45, 46] are used for performance evaluation. The database contains 48 ECG signals recorded at Beth Israel Hospital Arrhythmia Laboratory from 1975 to 1979. The recordings comprise two ECG signals each obtained from 47 subjects, and all recordings are around 30 min in duration. The subjects were 25 men and 22 women, aged between 23 to 89 years. In 46 recordings,

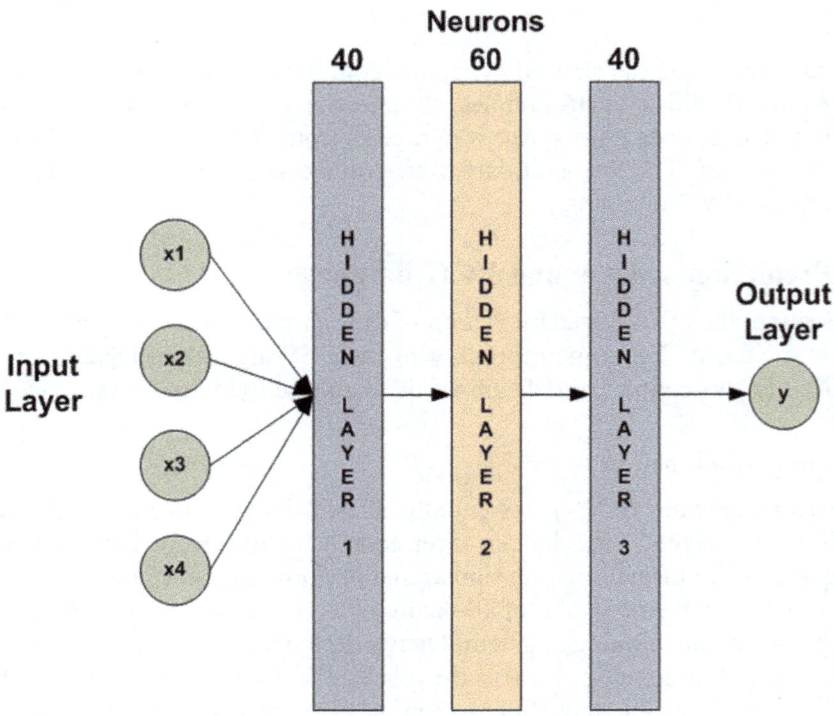

Figure 13.1. Deep neural network architecture.

the upper signal is a modified lead limb II (ML II) signal; therefore these 46 signals were used in this study. For all purposes, i.e. training, testing and performance evaluation, 10 s signals were extracted from the original signals. The signals used during testing are different to the signals used for training as they are extracted from different parts of the original signal.

To evaluate the performance of the prediction model, three different quantities are measured, which are percentage residual difference (PRD), SNR and normalized cross-correlation (NCC). The performance of the algorithm depends on sample prediction by the prediction model during PEE; accurate prediction will lead to improvement of the values of the metrics. The metrics are expressed in equations (13.9)–(13.11). Here, x represents the original signal and y is the watermarked signal:

$$\mathrm{PRD}(X,\ Y) = \sqrt{\frac{\sum_{i=1}^{N}(x_i - y_i)^2}{\sum_{i=1}^{N}{x_i}^2}} \times 100\% \tag{13.9}$$

$$\mathrm{SNR}(X,\ Y) = 10\log_{10}\sqrt{\frac{\sum_{i=1}^{N}{x_i}^2}{\sum_{i=1}^{N}(x_i - y_i)^2}} \tag{13.10}$$

$$\mathrm{NCC}(X,\ Y) = \frac{\sum \overline{x_i\ xor\ y_i}}{N*N}. \tag{13.11}$$

13.5 Training and embedding

To embed the patient's data in the ECG signal, different training and embedding schemes are tested, and the scheme providing the highest bits per sample (BPS) and consuming the least resources is chosen. The data hiding method is based on PEE and during watermarking data are fragmented in the same manner as used during the training. The embedding scheme is expected to be fast during embedding as well as extraction. In addition to being fast, the scheme should provide large embedding strength and should introduce minimal changes in the signal. Therefore, of all the schemes tested, the one which had all the qualities discussed here is chosen. The training and embedding process is explained through figure 13.2, in which the

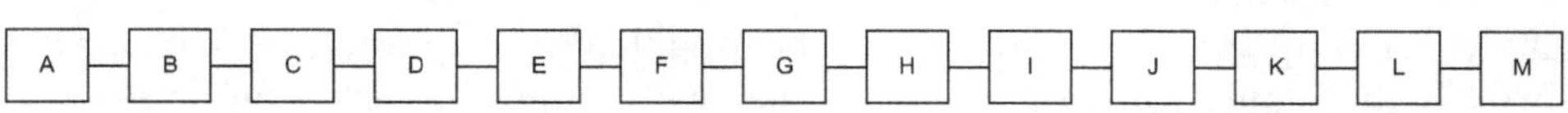

Figure 13.2. ECG signal snippet.

squares represent the ECG samples. The training procedure and schemes investigated for data embedding are described next.

13.5.1 Training

The ECG signal is required to be fragmented before training and embedding. The fragmentation and division of the ECG signal are similar for both the training and embedding. Initially, for training the model the ECG signal is first fragmented into non-overlapping groups of three consecutive samples. The center sample is the one that will be predicted by the prediction scheme based on its neighboring samples. Considering figure 13.2, A–B–C and D–E–F will be the fragmented groups, here 'B' and 'E' act as the target samples and A–C act as the training samples for sample B and D–F for sample E, and similarly for other groups. Thus, a 10 s ECG signal with 360 Hz sampling frequency will have 3600 samples in total and it will be fragmented into 1200 non-overlapping groups. The models can be trained on such groups. The fragmentation is further changed by adding more samples to it. This greatly increases the prediction accuracy. The scheme is explained next.

To increase the accuracy of the prediction, two immediate left and right neighbors of the center sample are provided to the model during training and embedding. The training is performed in the same manner as mentioned in [35]. Although the DNN architecture used in the present work is different from [35], there is no significant improvement in the performance. The training vector consists of four samples as input and the center sample as the target sample. Therefore, the training vector for the target sample C will be A–B–D–E. The three samples (B, C, D) in the center are unique for each group, whereas every group shares the extreme samples with neighboring groups. The 10 s snippets of all 46 ECG signals are divided to form such groups and the fragmented data are provided to the DNN for training. To effectively embed the data in ECG signals, different embedding strategies are tested and the rest of this section discusses various embedding strategies. In the next section a modified embedding strategy and its performance are discussed. The results of all the schemes are also provided after discussion of the schemes.

13.5.2 Embedding scheme 1

For embedding the watermark bit in the first scheme, only one sample (the central one) is chosen from the group of unique non-overlapping ECG samples to embed the watermark bit in it. The embedding scheme is explained in detail using figure 13.2, the blocks in the figure representing the ECG samples. In the figure, samples, B–C–D, E–F–G and so on, form non-overlapping groups of three samples. Here, C and F are the samples which will carry the watermark bit. The input to the DNN predictor will be A–B–D–E for the prediction of C, similarly, for sample F the input to the predictor will be D–E–G–H. In this manner, fragmentations are performed and the central sample is chosen to embed the watermark bit, thus after completion samples C, F, I and L would be modified to carry watermark bits. Thus, it is clear that going forward in this manner will obtain the maximum embedding strength of about 33% or 0.33 bits per sample. The predicted value for any sample must be equal during embedding and extraction, therefore unique and non-overlapping groups become a necessity. Having such groups ensures that the predicted samples have the same neighbors during embedding and extraction.

The output of embedding scheme 1 is shown in figure 13.3. In figure 13.3(a), the host ECG signal is shown. The maximum embedding strength obtained from this scheme is 0.33 BPS—this is the case when embedding is done in every suitable pixel. By selecting an appropriate threshold the embedding strength could be varied. Figures 13.3(b) and (c) show the watermarked signal for different strengths of embedding. In figure 13.3(b) the embedding strength is 0.21 BPS and the PRD of the watermarked signal is 0.43. Similarly, figure 13.3(c) shows the watermarked signal with 0.33 BPS. The PRDs of the watermarked signals are written on the images. It can be examined from the figures that the deformation introduced by data embedding is almost negligible, and the watermarked signal resembles the host signal. In figure 13.4(a) the performance of scheme 1 can be examined through the plot of PRD and NCC for different BPS. Similarly, in figure 13.4(b) the SNR of the watermarked signal is plotted for different BPS.

This scheme displays a fast run time but has lower embedding strength. Given its lower embedding strength, less distortion is introduced in the resultant ECG signal. To increase the embedding strength different schemes are explored, as described in the following sections.

13.5.3 Embedding scheme 2

To increase the embedding strength from 33%, more samples must carry watermark bits in them. This can be achieved using scheme 2, which is a progressive embedding scheme. The training procedure is the same as that discussed in the previous scheme, but the embedding is performed differently in scheme 2. In this scheme, the watermark is first embedded in sample C using its four neighbors. Then the modified neighbor (which is carrying a watermark) and other side neighbors are used to predict the value for the new sample. Therefore, after embedding the watermark bit in C, the modified value of C along with other neighbors will be used to embed the watermark bit in D. In this way, almost the complete signal is modified and an embedding strength of about 99% is obtained.

The output of the embedding scheme 2 is shown in figure 13.5. In figure 13.5(a) the host ECG signal is illustrated. The maximum embedding strength that can be achieved using scheme 2 is 0.99 BPS. Similarly to scheme 1, by selecting an appropriate threshold the embedding strength can be varied. Figures 13.5(b) and (c) shows the watermarked signal for different strengths of embedding. In table 13.1 the performance of scheme 1 can be examined.

This method provides good embedding strength but the output signal quality is degraded, as is evident from table 13.1. The main disadvantage of this method is its large time consumption. As it requires that the previous samples should be watermarked before updating the current sample, a delay is introduced. Thus, to embed the watermark in the nth sample the process requires that the watermark should be present in $(n-1)$ preceding samples. Therefore, due to the unavailability of the

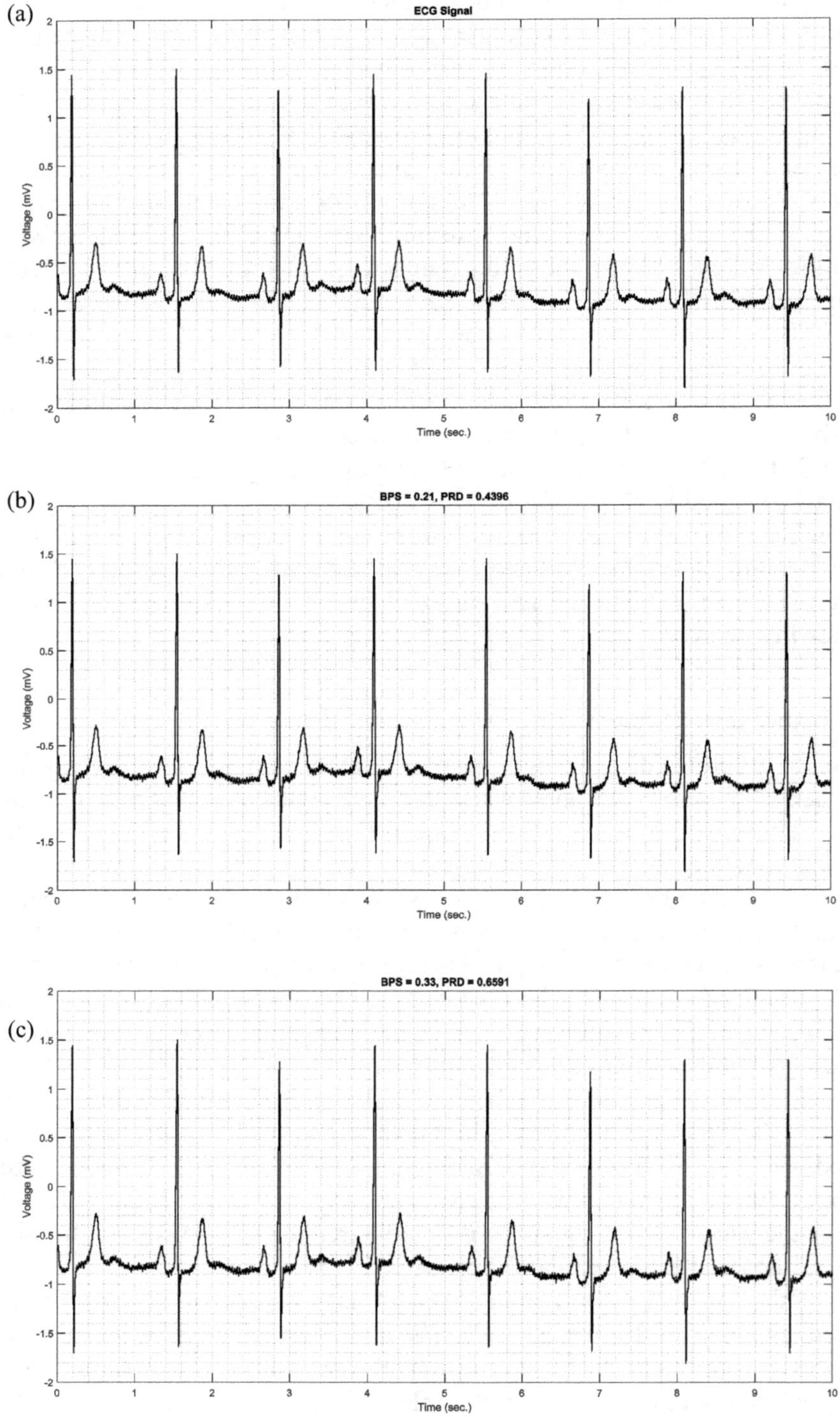

Figure 13.3. Embedding scheme 1: (a) host ECG signal, (b) watermarked ECG signal BPS = 0.21 and (c) watermarked ECG signal BPS = 0.33.

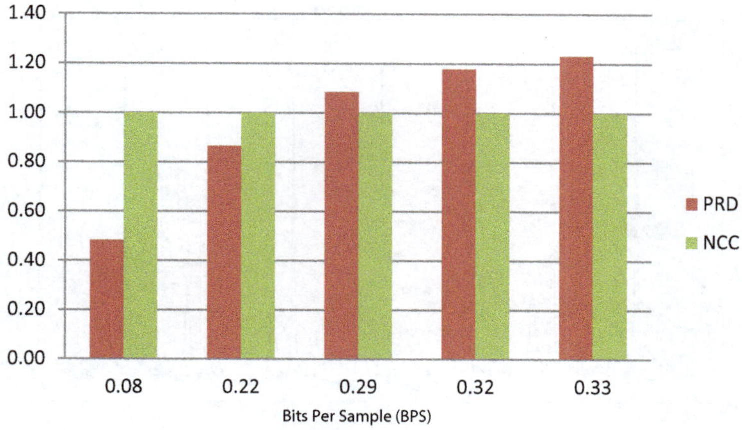

(a)

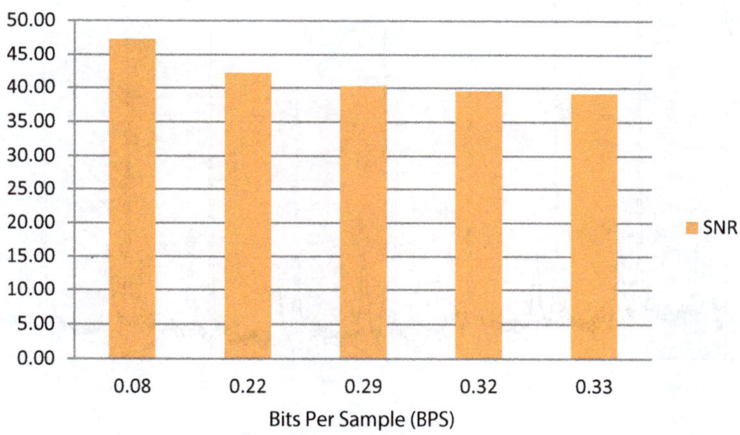

(b)

Figure 13.4. Embedding scheme 1: (a) PRD and NCC comparison for different BPS and (b) SNR comparison for different BPS.

previous neighbor's modified values at the same instant, this process cannot run in parallel, resulting in its large time consumption. In this progressive scheme, extraction is performed in reverse order. To produce a faster scheme with large embedding strength, a modified version of scheme 1 was proposed. It works in three phases and provides a maximum embedding strength of 99% or 0.99 BPS.

13.5.4 Embedding scheme 3

Embedding scheme 3 is a high capacity reversible data hiding method and 0.99 BPS embedding is obtained. As explained in the above section, the data are fragmented in

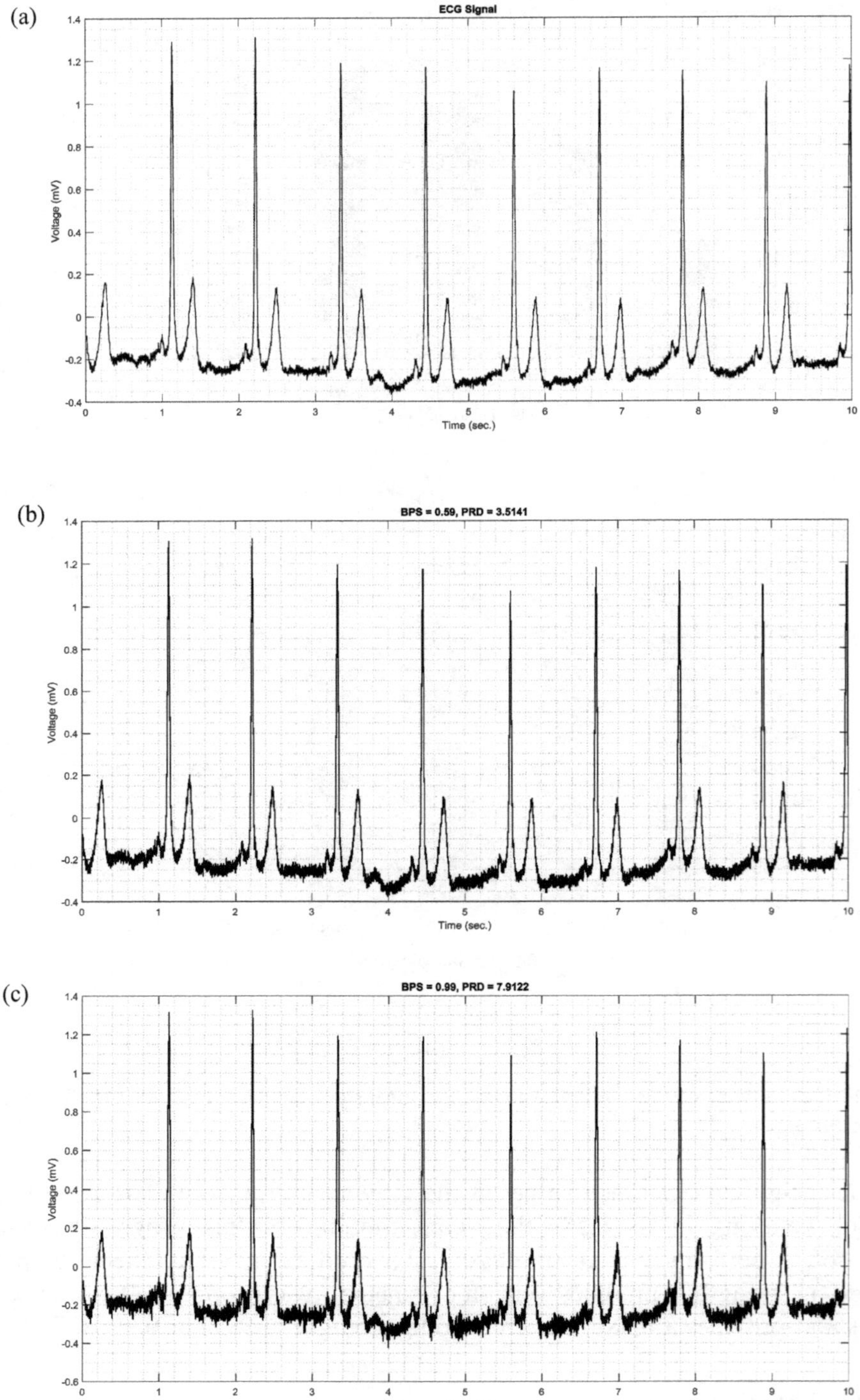

Figure 13.5. Embedding scheme 2: (a) host ECG signal, (b) watermarked ECG signal BPS = 0.59 and (c) watermarked ECG signal BPS = 0.99.

Table 13.1. Performance of scheme 2 for different BPS.

BPS	PRD	SNR	NCC	Time (s)
0.21	0.88	41.14	1.00	37.83
0.47	1.73	35.30	1.00	37.49
0.63	2.53	31.98	1.00	39.91
0.89	4.16	27.66	1.00	37.38
0.98	5.07	25.94	1.00	37.46
1.00	5.70	24.98	1.00	37.44

groups of three non-overlapping samples. In the first phase, the central sample value is predicted by providing its two immediate left and right neighbors to the prediction model. After prediction, the watermark is embedded using PEE. At the end of the first phase, watermark embedding is completed on a single sample from each group, achieving 0.33 BPS embedding. In the second and third phases, the watermark is embedded in the remaining samples of the non-overlapping group. Thus, at the end of the second phase watermarking of 0.66 BPS is obtained and similarly, 0.99 BPS watermarking is observed after the third phase.

In the case of the second and third phases, the samples modified during the previous phase take part in prediction. Therefore, during extraction, it is necessary to extract the watermark bits in reverse order. Thus, the third phase watermark is extracted first, followed by the second phase watermark and the first phase watermark is extracted last. This requirement is necessary as PEE needs the same predicted value during embedding and extraction, as explained in section 13.2. This scheme was used for ECG data hiding in [35] and is explained in detail there. However, the tampering location algorithm of [35] suffered when a specific attack was carried out, thus in the next section an improved embedding scheme is discussed.

The output of embedding scheme 3 is presented in figure 13.6. In figure 13.6(a) the host ECG signal is present and in figures 13.6(b) and (c) watermarked signals are present. The embedding strengths of the watermarked signals in images are 0.66 and 0.99 BPS.

13.6 Improved embedding scheme

This section focuses on the improvement of the tampering localization ability of the scheme proposed in [35]. The effect of doubling the embedding strength (BPS) of the above discussed scheme is also examined here. The highest embedding strength that the above discussed scheme offers is 0.99 BPS. In the improved scheme, the embedding strength is increased beyond 0.99 BPS and this is achieved by increasing the embedding phases from the three-phase embedding mentioned in [35]. Thus, when the number of embedding phases is increased to four, the embedding strength of 1.33 BPS is obtained. Similarly, using a fifth and sixth embedding phase, the embedding strength can be increased to 1.66 and 1.99 BPS, respectively. Increasing

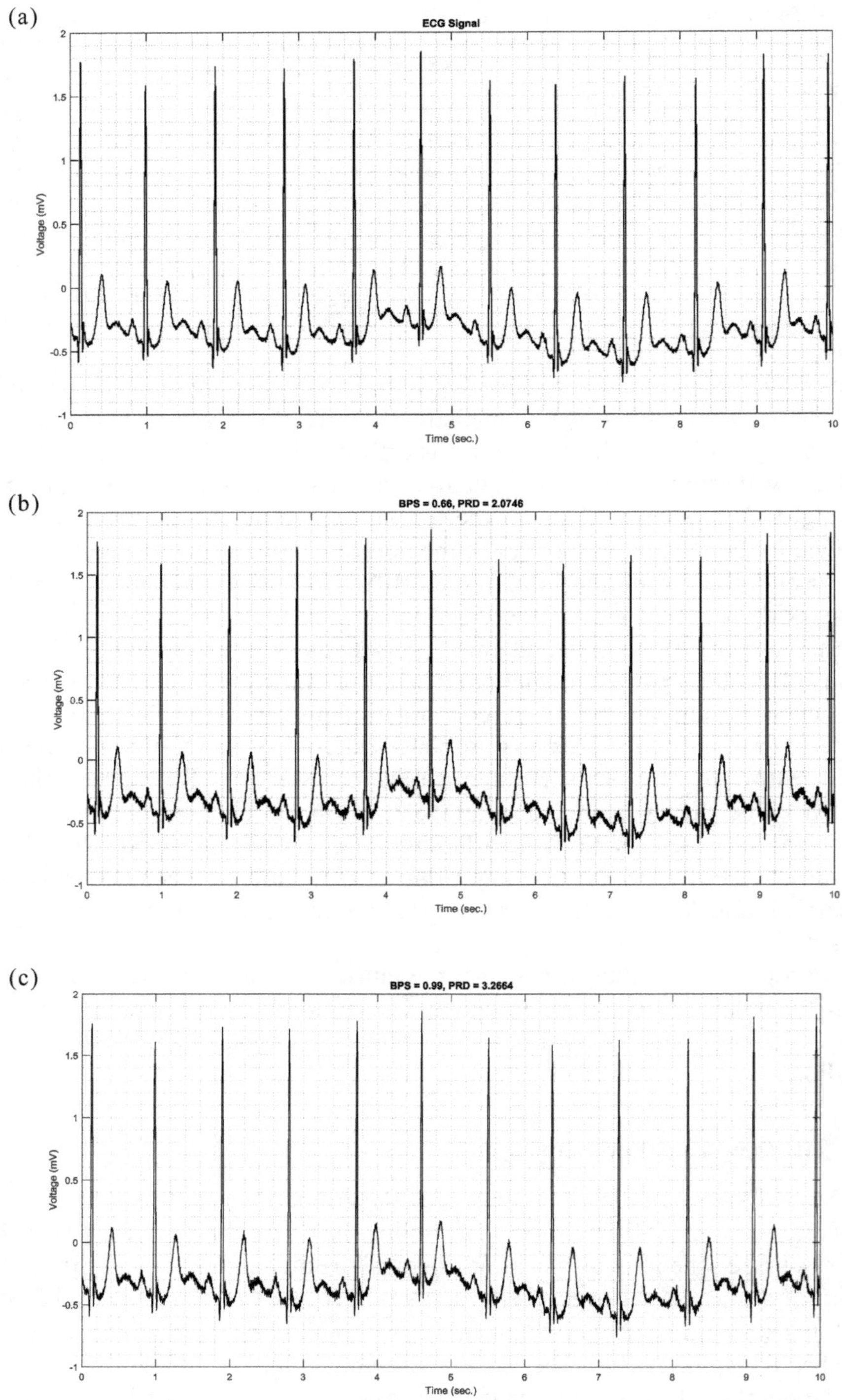

Figure 13.6. Embedding scheme 3: (a) host ECG signal, (b) watermarked ECG signal BPS = 0.66 and (c) watermarked ECG signal BPS = 0.99.

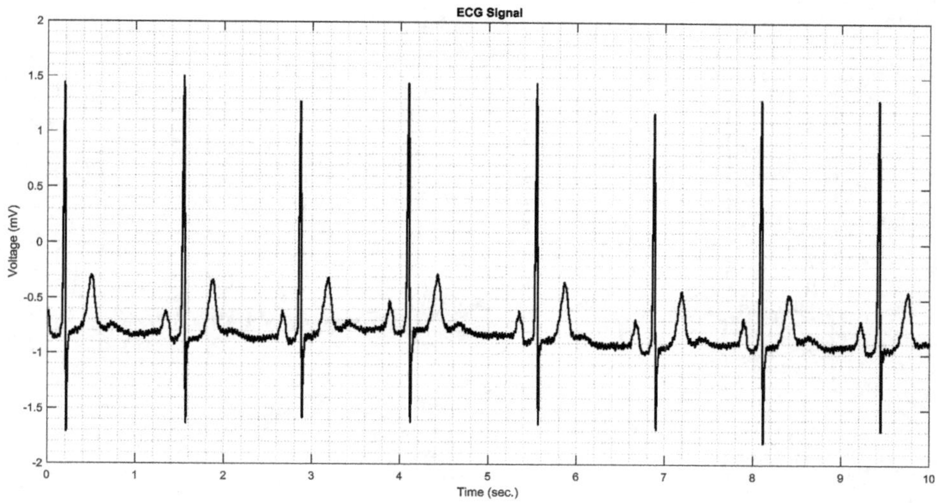

Figure 13.7. Sample host ECG signal for the improved scheme.

the embedding strength degrades the quality of the signal, but complete reversibility is guaranteed. Thus, this scheme doubles the embedding strength and the time taken by this scheme is also increased. The output for the increased embedding strength is presented in figures 13.7 and 13.8. In addition to increasing the embedding strength, this scheme updates the tampering detection ability compared to the scheme described in [35]. The tampering scheme developed in [35] struggles in detecting the tampering location when a specific attack is performed. The improved scheme inserts a fragment key in the tamper detection algorithm.

The performance of the improved scheme is presented in table 13.2. The embedded bits mentioned in the table are per 3600 ECG samples. Although the PRD and SNR are deteriorated, the reversibility is guaranteed. In figure 13.7 the host signal is present.

In figures 13.8(a), (b) and (c) watermarked output for embedding strengths 1.33, 1.66 and 1.99 BPS respectively are present. The problem with tampering detection of [35] is discussed next along with the improvement proposed over it.

The tampering detection watermark is generated using SHA-1 (the secure hashing algorithm), as discussed in [35], but with the introduction of the fragment key the scheme becomes more secure here. The process presented in [35] is illustrated in figure 13.9(a). The problem with this method is the self-authentication of the watermark. In [35], watermarking is performed in three phases and in each phase around 1200 bits can be embedded in the ECG signal. The tampering detection watermark is part of the watermark bits embedded in phase 1. To generate the tampering detection watermark, the ECG signal is first divided into 20 fragments of the same size. All the computations are performed on 10 s snippets of ECG signal with 3600 samples. The SHA-1 hash is generated for each fragment, and it is combined with ownership data. Now there are 20 fragments of the watermark, where each watermark fragment is a combination of ownership data and a tamper

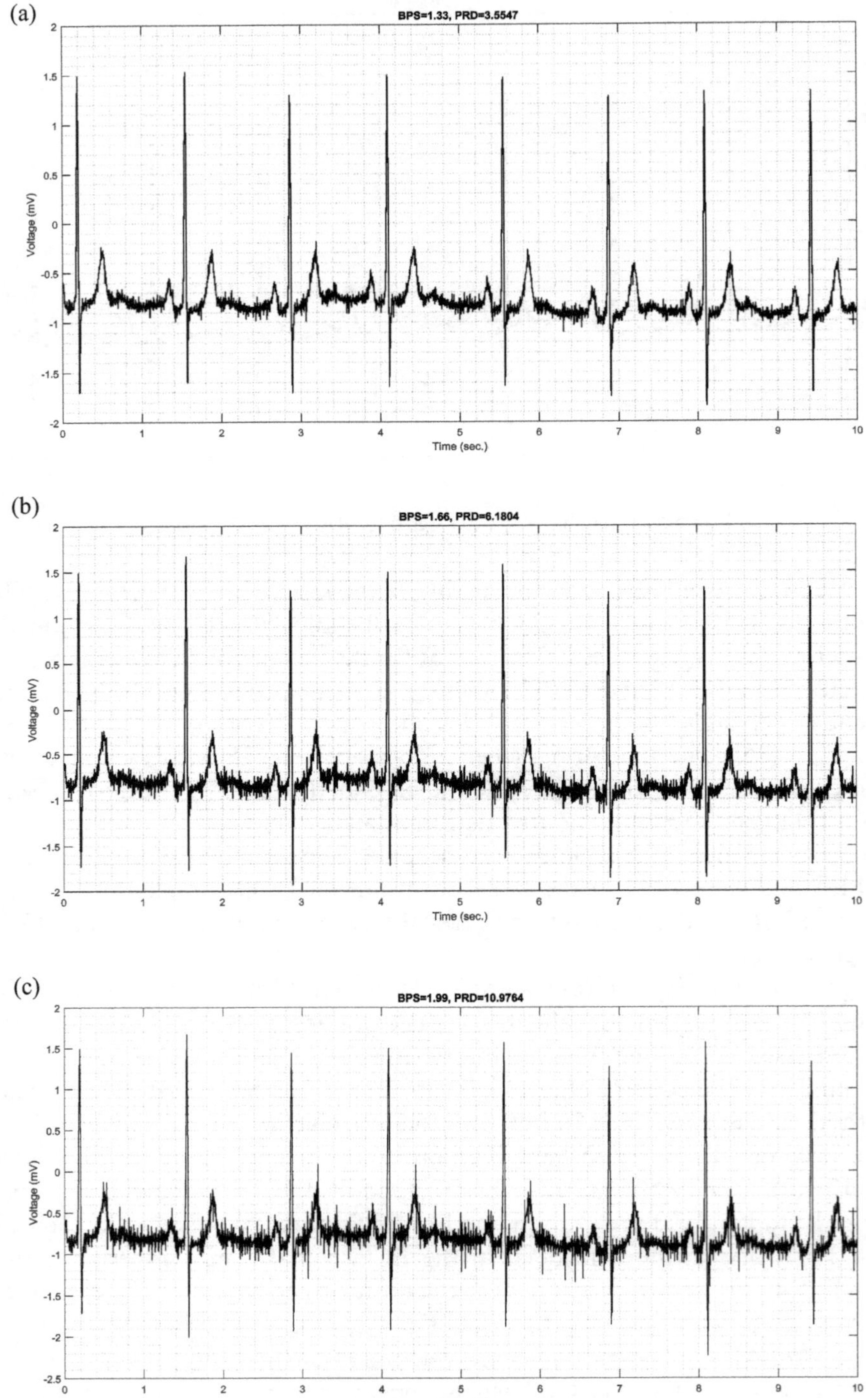

Figure 13.8. Embedding scheme 3: (a) watermarked ECG signal BPS = 1.33, (b) watermarked ECG signal BPS = 1.66 and (c) watermarked ECG signal BPS = 1.99.

Table 13.2. Performance of improved embedding scheme.

BPS	PRD	SNR	NCC	Bits embedded	Time (s)
1.33	7.15	23.96	0.99	4795	0.15
1.66	13.07	18.78	0.98	5994	0.19
1.99	24.85	13.33	0.94	7192	0.22

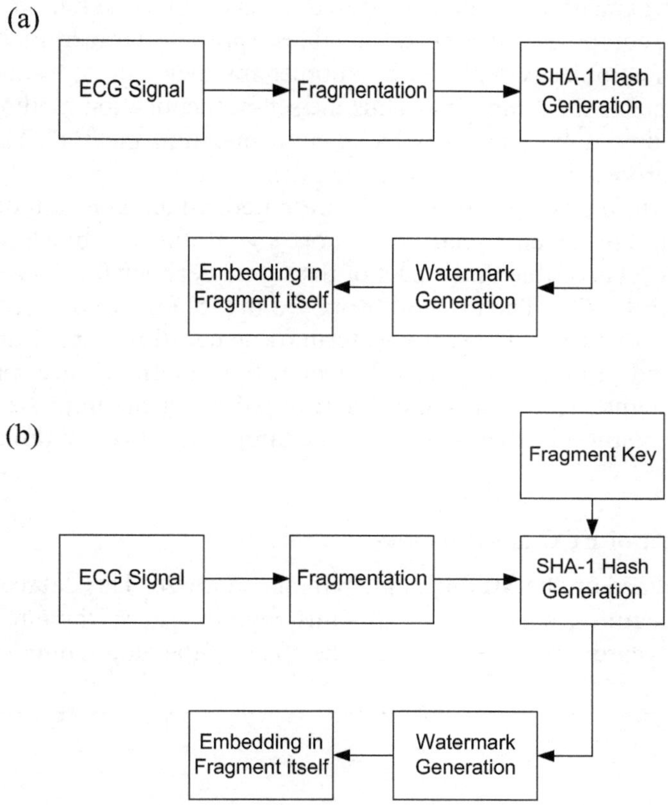

Figure 13.9. (a) The hash watermark generation used in [35] and (b) the proposed method of watermark generation.

detection watermark in the form of a hash. This watermark is embedded in the signal.

If an attack is carried out on the signal in such a manner that a fragment is completely replaced with another fragment of the same signal or from another signal watermarked by the same algorithm, the method presented in [35] fails to detect the tampering location in most of the signals. This happens due to the self-authentication of the fragment, because to test the tampering the hash value extracted from the fragment is compared with fragment itself. Therefore, when a fragment is completely replaced by another fragment, the self-authentication method will not find any problem with it. Although it detects that the signal has been tampered with, it cannot find its location. Therefore, in this chapter, in place of simple watermark embedding, a key-based embedding is implemented.

A unique security key is used to generate a random key for each fragment, called the fragment key. The hashed data are XORed with this fragment key to generate the tamper detection watermark. The same security key is present at the destination which generates the same fragment key for each fragment. During extraction, to validate the data hashing is performed again and the computed hash values are XORed with fragment keys and compared to the embedded hash value. Any difference in the computed and embedded values represents tampering. Now, even if any fragment is completely replaced by another fragment, the introduction of the fragment key detects the tampering. Thus the self-authentication method present in [35] is improved to a fragment key-based authentication method. The improved watermarking process is shown in figure 13.9(b).

In figure 13.10, the tampered signal is presented, which goes unnoticed by the method in [35]. The green signal is the host signal and the black signal is the tampered signal, it is visible that a part of the signal between 0 s–1 s is replaced by a fragment from 1 s–2 s. The method proposed in [35] fails to detect the location of tampering in this case. Thus, the watermark generation part is updated, and the existing and updated watermark generation methods are presented in figure 13.9. In figure 13.11 the same signal tested using the improved scheme is shown and the method is able to identify the tamper location, highlighted by the black rectangle.

13.6.1 The effect of ECG abnormalities

The DNN is trained on the MIT-BIH arrhythmia database. This database has ECG with different features, such as normal sinus rhythm, paced rhythm, fibrillation, ventricular tachycardia beats, etc. To test the effect of the data hiding algorithm on

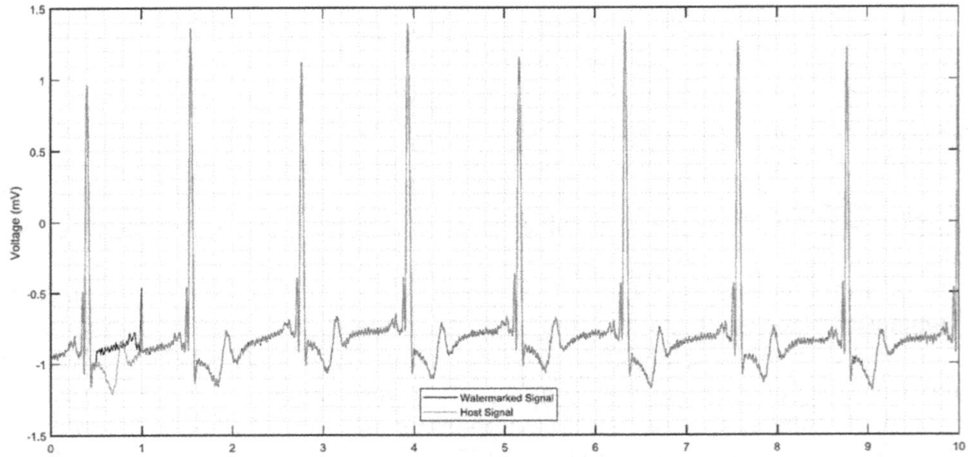

Figure 13.10. Tampering goes undetected in [35].

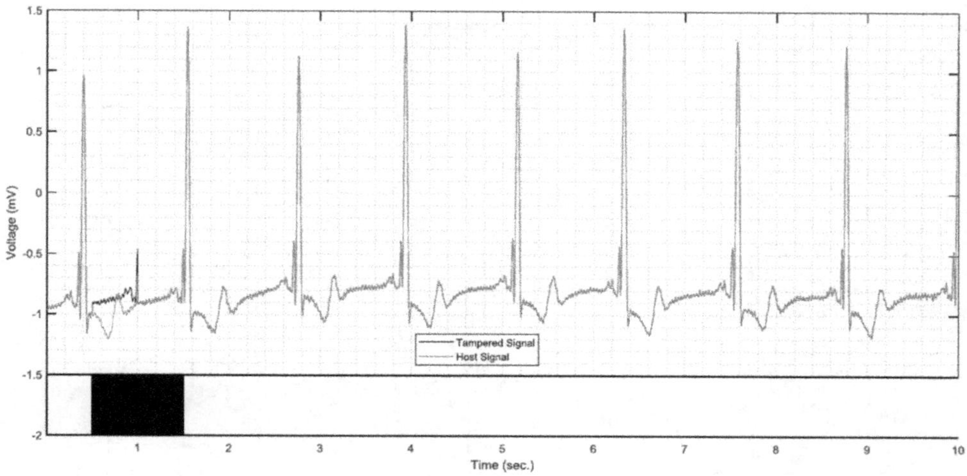

Figure 13.11. Tampering detected using the improved scheme.

ECGs with different types of heart conditions, the ECG signal database was divided into two datasets. In the two datasets, the first set consists of signals having normal or nearly normal ECG shapes. The second group contains ECG signals deformed due to different heart conditions. In the first dataset, 17 signals from the MIT-BIH arrhythmia database were present and in the second dataset 21 signals were present. The algorithm was separately tested on both datasets and different evaluation parameters were computed. To provide an idea of the types of ECG signals in each group, figure 13.12(a) shows the signal from the first dataset. In figure 13.12(b), signals which are deformed (from the second dataset) are present. The deformed ECG signal has hardly any resemblance to the normal ECG signal. To provide a conclusive idea of the performance of the watermarking algorithm on different types of ECG signals, the parameters are presented in table 13.3.

It is evident from figure 13.12 and table 13.3 that the watermarking algorithm performs better on normal ECG signals. The values present in the table are average values obtained for different datasets.

13.6.2 Performance on the ECG-ID database

Finally, it is important to examine the performance of the data hiding technique on a separate database. All the results discussed until now have been specific to the MIT-BIH arrhythmia database and the DNN was also trained on the same database. To test the performance of the DNN on a separate database, signals from the ECG-ID database [47, 48] are used. The ECG identification database contains ECG signals from 90 individuals. All signals are digitized with 12 bits and 500 Hz sampling frequency. To test the performance of the current algorithm, 35 signals are randomly selected from the ECG-ID database. In figure 13.13(a) the host signal from the ECG-ID database is presented and in figure 13.13(b) the output of signal watermarking on the ECG-ID database is presented with 0.99 BPS. The performance of the algorithm on the ECG-ID database for different BPS is tabulated in table 13.4.

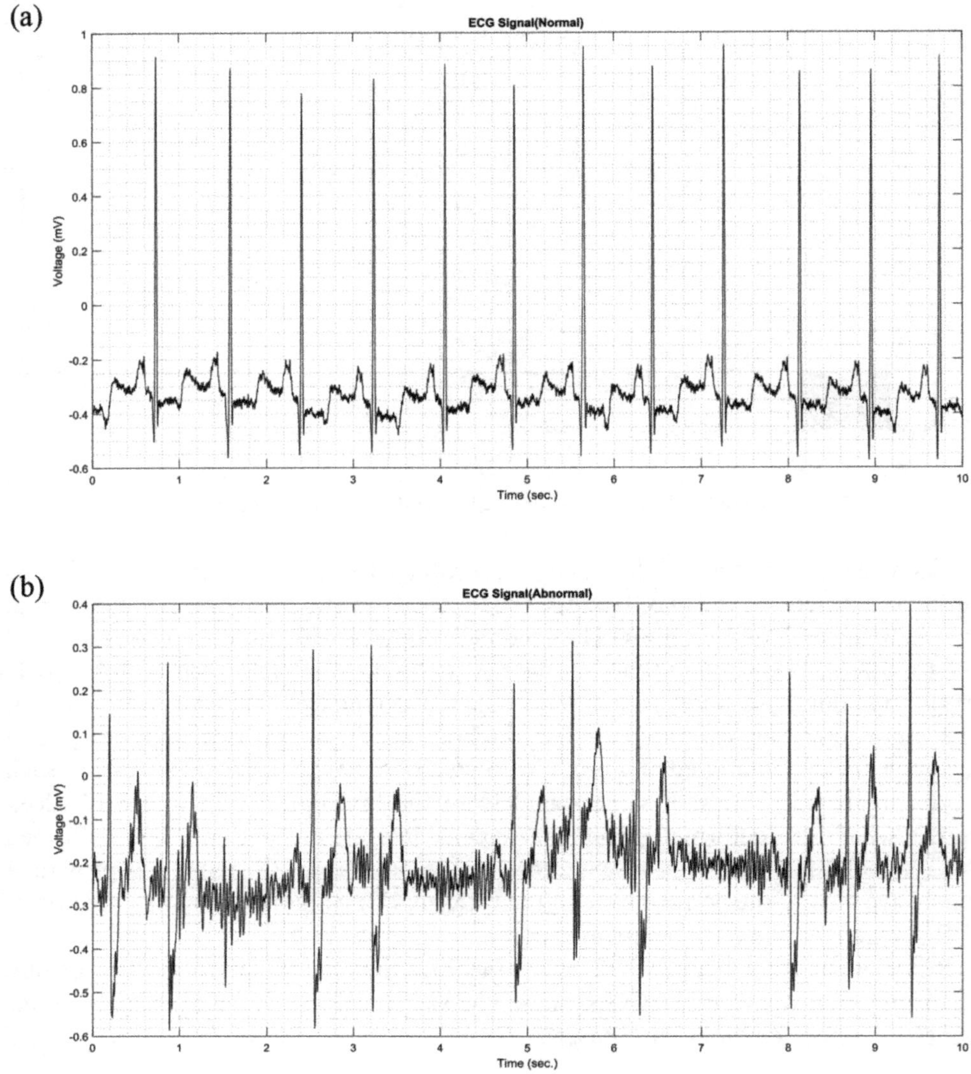

Figure 13.12. (a) Normal ECG signal and (b) deformed ECG signal.

The ECG-ID database is sampled at a higher frequency and is digitized at a higher bit resolution then the MIT-BIH database, but the performance of the DNN trained on the MIT-BIH database is satisfactory. In figure 13.13(b), the noise introduced in ECG signal at 0.99 BPS is shown.

13.7 Conclusion

To combat the problem of data security in telecardiology applications, a reversible data hiding algorithm for ECG signals was discussed in this chapter. A comprehensive survey of current watermarking and data hiding algorithms was also

Table 13.3. The effect of ECG shapes on the proposed algorithm.

BPS	Normal ECG			Deformed ECG		
	PRD	SNR	NCC	PRD	SNR	NCC
0.08	0.466	47.425	1.000	0.532	46.556	1.000
0.15	0.640	44.663	1.000	0.728	43.849	1.000
0.21	0.762	43.136	1.000	0.872	42.256	1.000
0.32	1.028	40.530	1.000	1.177	39.657	1.000
0.39	1.203	39.168	1.000	1.387	38.244	1.000
0.44	1.346	38.186	1.000	1.552	37.271	1.000
0.47	1.417	37.739	1.000	1.637	36.812	1.000
0.66	2.080	34.358	0.999	2.389	33.577	0.999
0.99	3.368	30.234	0.999	3.872	29.399	0.998

(a)

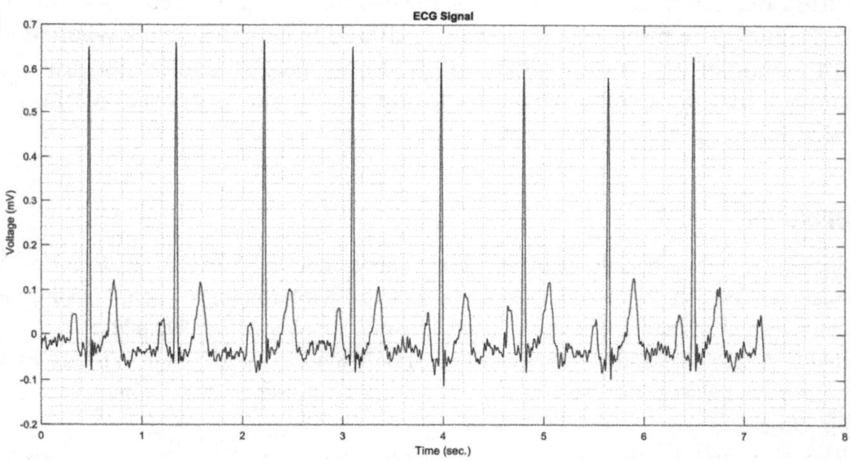

(b)

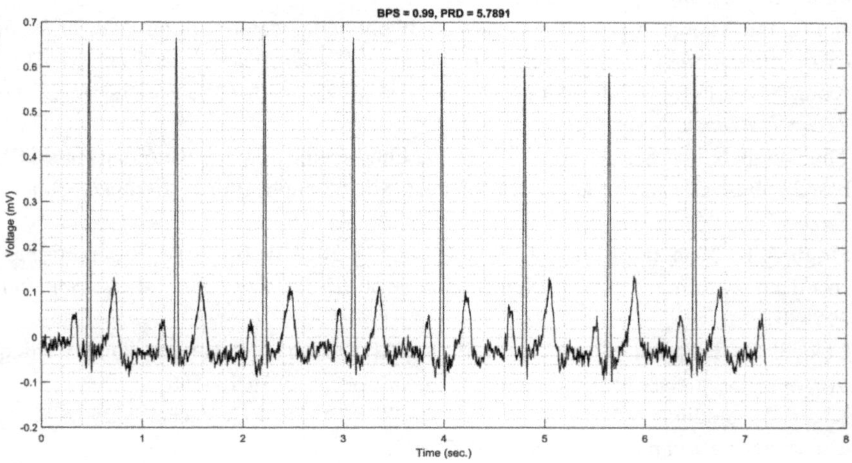

Figure 13.13. ECG-ID database: (a) host signal and (b) watermarked signal BPS = 0.99.

Table 13.4. Performance of the proposed algorithm on the ECG-ID database.

BPS	PRD	SNR	NCC	Time (s)
0.25	1.85	35.19	1.00	0.04
0.33	2.29	33.31	1.00	0.05
0.42	2.46	32.72	1.00	0.08
0.55	2.88	31.33	1.00	0.11
0.66	3.41	29.87	1.00	0.08
0.99	4.95	26.65	1.00	0.12

provided. The ECG data hiding technique was proposed based on prediction error expansion (PEE) using a DNN based prediction scheme. The performance of the algorithm was tested on ECG signals based on different criteria, such as embedding strength, signal abnormalities, etc. The chapter also investigated a hashing based tampering detection that helped in localization of the tampered region in a watermarked signal. The previous method [35] suffered from a drawback where it failed to detect the tampering location when attacked in a specific manner and this weakness is removed in the current work. In future work, the proposed method will be extended to other physiological signals.

References

[1] Shimizu K 1999 Telemedicine by mobile communication *IEEE Eng. Med. Biol. Mag.* **18** 32–44

[2] Backman W, Bendel D and Rakhit R 2010 The telecardiology revolution: improving the management of cardiac disease in primary care *J. R. Soc. Med.* **103** 442–6

[3] Ibaida A, Khalil I and Van Schyndel R 2011 A low complexity high capacity ECG signal watermark for wearable sensor-net health monitoring system *2011 Computing in Cardiology* pp 393–6

[4] Rao N V and Kumari V M 2011 Watermarking in medical imaging for security and authentication *Inf. Secur. J.* **20** 148–55

[5] Jian L and Xiangjian H 2005 A review study on digital watermarking *Proc. of 1st Int. Conf. on Information and Communication Technology, ICICT 2005* vol 2005 pp 337–41

[6] Kong X and Feng R 2001 Watermarking medical signals for telemedicine *IEEE Trans. Inf. Technol. Biomed.* **5** 195–201

[7] Thakkar F N and Srivastava V K 2017 A blind medical image watermarking: DWT-SVD based robust and secure approach for telemedicine applications *Multimed. Tools Appl.* **76** 3669–97

[8] Singh A K, Kumar B, Dave M and Mohan A 2015 Robust and imperceptible spread-spectrum watermarking for telemedicine applications *Proc. Natl Acad. Sci. India* A **85** 295–301

[9] Caldelli R, Filippini F and Becarelli R 2010 Reversible watermarking techniques: an overview and a classification *EURASIP J. Inf. Secur.* **2010** 134546

[10] Khan A, Siddiqa A, Munib S and Malik S A 2014 A recent survey of reversible watermarking techniques *Inf. Sci.* **279** 251–72

[11] Engin M, Çidam O and Engin E Z 2005 Wavelet transformation based watermarking technique for human electrocardiogram (ECG) *J. Med. Syst.* **29** 589–94
[12] Kaur S, Farooq O, Singhal R and Ahuja B S 2010 Digital watermarking of ECG data for secure wireless communication *ITC 2010–2010 Int. Conf. Recent Trends Information, Telecommun. Comput.* 140–4
[13] Ibaida A and Khalil I 2013 Wavelet-based ECG steganography for protecting patient confidential information in point-of-care systems *IEEE Trans. Biomed. Eng.* **60** 3322–30
[14] Awasarmol S P, Ashtekar S and Chintawar A 2018 Securely data hiding and transmission in an ECG signal using DWT *2017 Int. Conf. Energy, Commun. Data Anal. Soft Comput. ICECDS 2017* pp 2850–4
[15] Chen S T, Guo Y J, Huang H N, Kung W M, Tseng K K and Tu S Y 2014 Hiding patients confidential data in the ECG signal via transform-domain quantization scheme topical collection on mobile systems *J. Med. Syst.* **38** 1–8
[16] Dey N, Roy A B, Das A and Chaudhuri S S 2012 Stationary wavelet transformation based self-recovery of blind-watermark from electrocardiogram signal in wireless telecardiology *Int. Conf. on Security in Computer Networks and Distributed Systems* pp 347–57
[17] Edward Jero S and Ramu P 2016 Curvelets-based ECG steganography for data security *Electron. Lett.* **52** 283–5
[18] Jero S E, Ramu P and Ramakrishnan S 2015 ECG steganography using curvelet transform *Biomed. Signal Process. Control* **22** 161–9
[19] Mathivanan P, Edward Jero S, Ramu P and Balaji Ganesh A 2018 QR code based patient data protection in ECG steganography *Australas. Phys. Eng. Sci. Med.* **41** 1057–68
[20] Sanivarapu P V, Rajesh K N V P S, Reddy N V R and Reddy N C S 2020 Patient data hiding into ECG signal using watermarking in transform domain *Australas. Phys. Eng. Sci. Med.* **43** 213–26
[21] Patil V and Patil M 2018 Curvelet based ECG steganography for protection of data *Computational Vision and Bio Inspired Computing* (Lecture Notes in Computational Vision and Biomechanics vol 28) (Dordrecht: Springer) pp 238–48
[22] Soni N, Saini I and Singh B 2020 Integer wavelet transform-based ECG steganography for hiding patients' confidential information in e-healthcare systems *Advances in Intelligent Systems and Computing* vol 1053 (Berlin: Springer) pp 513–25
[23] Ni Z, Shi Y Q, Ansari N and Su W 2006 Reversible data hiding *IEEE Trans. Circuits Syst. Video Technol.* **16** 354–61
[24] Tian J 2003 Reversible data embedding using a difference expansion *IEEE Trans. Circuits Syst. Video Technol.* **13** 890–6
[25] Thodi D M and Rodríguez J J 2007 Expansion embedding techniques for reversible watermarking *IEEE Trans. Image Process.* **16** 721–30
[26] Tsai P, Hu Y C and Yeh H L 2009 Reversible image hiding scheme using predictive coding and histogram shifting *Signal Process.* **89** 1129–43
[27] Li X, Li B, Yang B and Zeng T 2013 General framework to histogram-shifting-based reversible data hiding *IEEE Trans. Image Process.* **22** 2181–91
[28] Pakdaman Z, Saryazdi S and Nezamabadi-pour H 2017 A prediction based reversible image watermarking in Hadamard domain *Multimed. Tools Appl.* **76** 8517–45
[29] Ou B, Li X, Zhao Y, Ni R and Shi Y Q 2013 Pairwise prediction-error expansion for efficient reversible data hiding *IEEE Trans. Image Process.* **22** 5010–21

[30] Li X, Yang B and Zeng T 2011 Efficient reversible watermarking based on adaptive prediction-error expansion and pixel selection *IEEE Trans. Image Process.* **20** 3524–33

[31] Jaiswal S P, Au O, Jakhetiya V, Guo A Y and Tiwari A K 2014 Adaptive predictor structure based interpolation for reversible data hiding *Int. Workshop on Digital Watermarking* pp 276–88

[32] Jaiswal S P, Au O C, Jakhetiya V, Guo Y, Tiwari A K and Yue K 2013 Efficient adaptive prediction based reversible image watermarking *Proc. of 2013 IEEE Int. Conf. Image Process. ICIP 2013* pp 4540–4

[33] Wu W, Liu B, Zhang W and Chen C 2015 Reversible data hiding in ECG signals based on histogram shifting and thresholding *2015 2nd Int. Symp. Future Inf. Commun. Technol. Ubiquitous Heal. Ubi-HealthTech 2015* pp 45–9

[34] Wang H, Zhang W and Yu N 2016 Protecting patient confidential information based on ECG reversible data hiding *Multimed. Tools Appl.* **75** 13733–47

[35] Bhalerao S, Ansari I A, Kumar A and Jain D K 2019 A reversible and multipurpose ECG data hiding technique for telemedicine applications *Pattern Recognit. Lett.* **125** 463–73

[36] Zheng K M and Qian X 2008 Reversible data hiding for electrocardiogram signal based on wavelet transforms *Proc. of 2008 Int. Conf. Comput. Intell. Secur. CIS 2008* vol 1 pp 295–9

[37] Dragoi I C and Coltuc D 2014 Local-prediction-based difference expansion reversible watermarking *IEEE Trans. Image Process.* **23** 1779–90

[38] Shiu H J, Lin B S, Huang C H, Chiang P Y and Lei C L 2017 Preserving privacy of online digital physiological signals using blind and reversible steganography *Comput. Methods Programs Biomed.* **151** 159–70

[39] Yang C Y, Cheng L T and Wang W F 2019 Effective reversible data hiding in electrocardiogram based on fast discrete cosine transform *Advances in Intelligent Systems and Computing* vol 880 (Berlin: Springer) pp 640–8

[40] Dey N, Dey M, Mahata S K, Das A and Chaudhuri S S 2015 Tamper detection of electrocardiographic signal using watermarked bio-hash code in wireless cardiology *Int. J. Signal Imaging Syst. Eng.* **8** 46–58

[41] Tabassum N and Izharuddin M 2018 Reversible data hiding scheme for ECG signal *Int. J. Rough Sets Data Anal.* **5** 42–54

[42] Li K, Pan W, Li Y, Jiang Q and Liu G 2018 A method to detect sleep apnea based on deep neural network and hidden Markov model using single-lead ECG signal *Neurocomputing* **294** 94–101

[43] Yu H, Tan Z-H, Ma Z, Martin R and Guo J 2017 Spoofing detection in automatic speaker verification systems using DNN classifiers and dynamic acoustic features *IEEE Trans. Neural Networks Learn. Syst.* **29** 4633–44

[44] Gong K, Guan J, Liu C-C and Qi J 2018 PET image denoising using a deep neural network through fine tuning *IEEE Trans. Radiat. Plasma Med. Sci.* **3** 153–61

[45] Moody G B and Mark R G 2001 The impact of the MIT-BIH arrhythmia database *IEEE Eng. Med. Biol. Mag.* **20** 45–50

[46] Moody G B, Mark R G and Goldberger A L 2001 PhysioNet: a web-based resource for the study of physiologic signals *IEEE Eng. Med. Biol. Mag.* **20** 70–5

[47] Goldberger A L, Amaral L A, Glass L, Hausdorff J M, Ivanov P C, Mark R G, Mietus J E, Moody G B, Peng C K and Stanley H E 2000 PhysioBank, PhysioToolkit, and PhysioNet: components of a new research resource for complex physiologic signals *Circulation* **101** E215–20

[48] Lugovaya T S 2005 Biometric human identification based on electrocardiogram *PhD Thesis* Electrotechnical University, St Petersburg

IOP Publishing

Modelling and Analysis of Active Biopotential Signals in Healthcare, Volume 1

Varun Bajaj and G R Sinha

Chapter 14

Statistical measures and analysis in electrocardiogram (ECG) signal processing

Ranjeet Kumar

School of Electronics Engineering (SENSE), VIT University, Chennai Campus, Chennai, Tamilnadu - 600 127, India

Electrocardiogram (ECG) signals are popular biopotential signals collected from prescribed locations on the human body using electrodes. The acquisition of signal or continuous recording/monitoring is affected by various types of noise and artifacts due to various sources and also requires a lot of storage in the device. The ECG signal is the primary signature of cardiovascular health or the condition of the heart. The heart is functionally rhythmic and this is reflected in the ECG signal as a quasi-periodic signal. Here, deviation in the rhythmic or quasi-periodic nature of the signal is interpreted as disease or abnormalities. In the healthcare sector, the analysis of long-duration signals is a very demanding and time-consuming task that directly takes the time of doctors and experts. In this chapter, we examine the statistical measures and analyses in ECG signal processing which, through making and reporting meaningful interpretations, can help the medical team in healthcare planning. Here, a brief summary of the understanding of statistical measurement variables, and the measures of central tendency and deviation or dispersion from the rhythmic nature of the ECG signal, are presented. This study is very important in various applications such as feature estimation, signal/disease classification, and qualitative analysis of original signals and processed signals.

14.1 Introduction

Signal processing is considered as the information extraction, retrieval, estimation and compact representation from a raw signal or raw data. In the field of biomedical signal processing, in particular electrocardiogram (ECG) signals or other biomedical

doi:10.1088/978-0-7503-3279-8ch14

signal recorded from the human body, the signals having non-stationary characteristics. Therefore, the signal characteristics vary or change over time with different activities and health conditions [1].

The time characteristics of the signal are fundamental features that reflect the signal source behavior in ideal observation conditions. However, some preprocessing or pre-filtering is required in real-time applications before signal analysis for accurate estimation [2–17]. Here, the statistical methods help to analyse the huge volumes of data, and allow an interface with other observations in scientific and engineering applications [17–20]. The statistical methods require a suitable study design, a significant selection of samples and suitable statistical tests to estimate the behavior of a population or subject [21]. Adequate knowledge of statistics is necessary for the suitable design of an epidemiological study or clinical observations. However, inappropriate statistical methods may result in inaccurate conclusions, which may lead to inappropriate medical treatment.

The statistical measures are generally descriptive and inferential statistical variables for a large dataset [22]. In the case of an ECG signal, time-series data are recorded to analyse the activity of the heart over time, and sometimes continuous monitoring is required for critical conditions. Therefore, a huge amount of data are

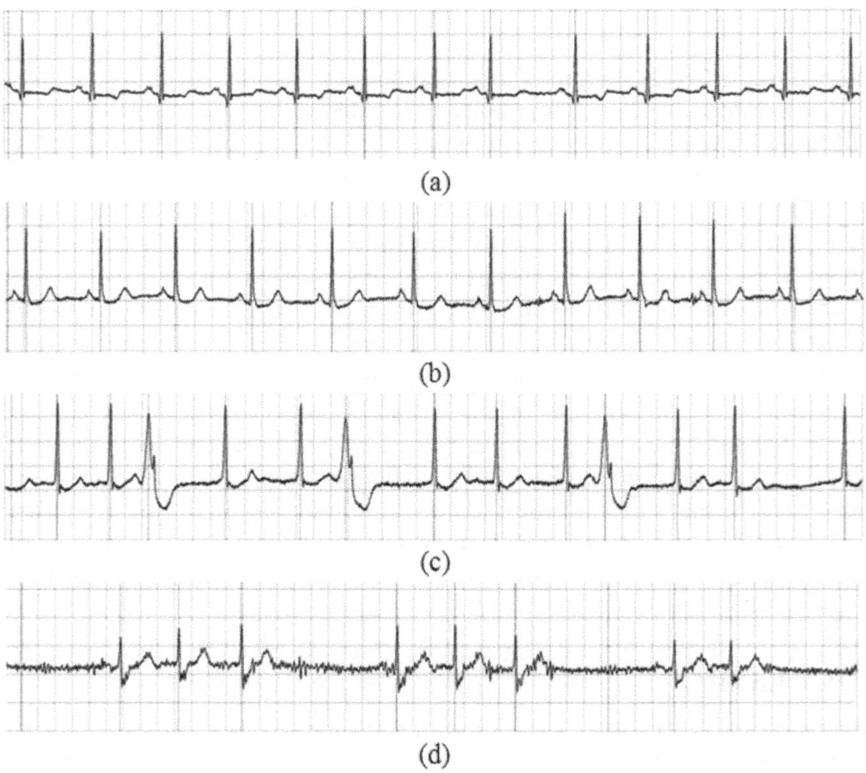

Figure 14.1. ECG waveforms from the MIT-BIH Arrhythmia Dataset [24]: (a) Rec. 100, (b) Rec. 101, (c) Rec. 221 and (d) Rec. 232. Courtesy of Physionet.org.

generated and stored for further review [23]. In figure 14.1, ECG signals are illustrated for different cases that reflect the importance of statistical measures.

Figure 14.1 illustrates different ECG waveforms of 10 s duration from the MIT-BIH Arrhythmia Dataset (recordings 100, 101, 221 and 232) which show different rhythmic behaviors. It is very easy to identify cardiac behavior with a very short signal duration. However, real-time monitoring or long-time recorded data characteristics may change compared to short time window data. Therefore, statistical measures, in particular descriptive and inferential variable analysis, are essential for healthcare management before the prescription of treatment [25, 26]. As is well known, central tendencies such as the mean, median and mode variables define a summary of the data, and these variables are considered as descriptive statistical measures [22]. The descriptive statistical measures are categorized as follows:

- *Measures of central tendency.* A group of data or a cluster is described by the central trend or position of distributed data. There are several statistical parameters that explain the central trend, such as mode, median and mean.
- *Measures of variability.* A group of data or a cluster is described or summarized by the distribution pattern or its spreadness, such as a deviation or variation score from its central tendency. The variability can be measured using several statistical parameters such as range, quartiles, absolute deviation, variance and standard deviation.

In contrast, inferential statistical measures are performed on a random dataset and then related to the complete dataset or observations. It is, therefore, important that the sample dataset accurately represents the complete observations.

In this chapter, a summary of statistical measures, including the measure of central tendency and the deviation from the rhythmic nature of the ECG signal, is presented. The organization of the chapter is as follows. An introduction has been provided in section 14.1 and in section 14.2 a detailed discussion of the electrocardiogram (ECG) and signal characteristics is presented. The statistical measures are discussed in section 14.3 and their use in analysis is illustrated using different applications for ECG signals in section 14.4. The chapter is concluded in section 14.5. This chapter explains various applications of statistical measures in feature estimation, signal/disease classification, and the qualitative analysis of original signals and processed signals.

14.2 The electrocardiogram (ECG) signal and its characteristics

During depolarization and repolarization a cardiac system generates tiny electrical signals through ionic activity of the heart muscles. These tiny signals are represented graphically as an electrocardiogram signal. The ECG is a physiological signal which can be exploited to diagnose heart disease and abnormalities. An electrocardiogram signal is sensed through electrodes that are placed at several prescribed locations on the human body. In this process, several electrode patterns can be used to form the lead representation, such as the single-, three-, six- and twelve-lead acquisition systems, and these leads are considered as channels that help to record the digital data or signal at various sampling rates as per the system configuration [1, 27].

Basically, an electrocardiogram signal contains the rhythmic behavior of the cardiac system in terms of the cardiac cycle or beats in the form of electrical waveforms such as the P, QRS, T and U waves. Normally, 72 beats are present in one minute of ECG signal for a healthy condition. It may vary with health conditions or abnormalities. The ECG is recorded from patients for both monitoring and diagnostic purposes, not only for immediate study but also for storage in a patient database for future retrieval and review.

The following subsections cover the generation of the ECG signal and its characteristics.

14.2.1 ECG signal generation

The ionic activity of the cardiac muscles during depolarization and repolarization can be measured or sensed from the human body surface and estimated as the potential difference with respect to the reference point. The beats of the ECG signal different waveforms, such as the P, QRS, T and U waves, that represents the different electrical phases of the cardiac cycle, as discussed above.

Typically, the ECG acquisition system uses three, six or 12 lead based instruments to record or monitor a real-time signal. The standard 12 lead based instrument uses ten electrodes placed in standard locations on the patient's skin, as shown in figure 14.2(a) [28]. An ECG waveform contains three prominent entities: the P wave, the QRS complex and the T wave. The P wave represents the *depolarization* of the atria with a duration of 80 ms. The rapid *depolarization* of the right and left ventricles is represented by the QRS complex with a duration of 80–100 ms, and the

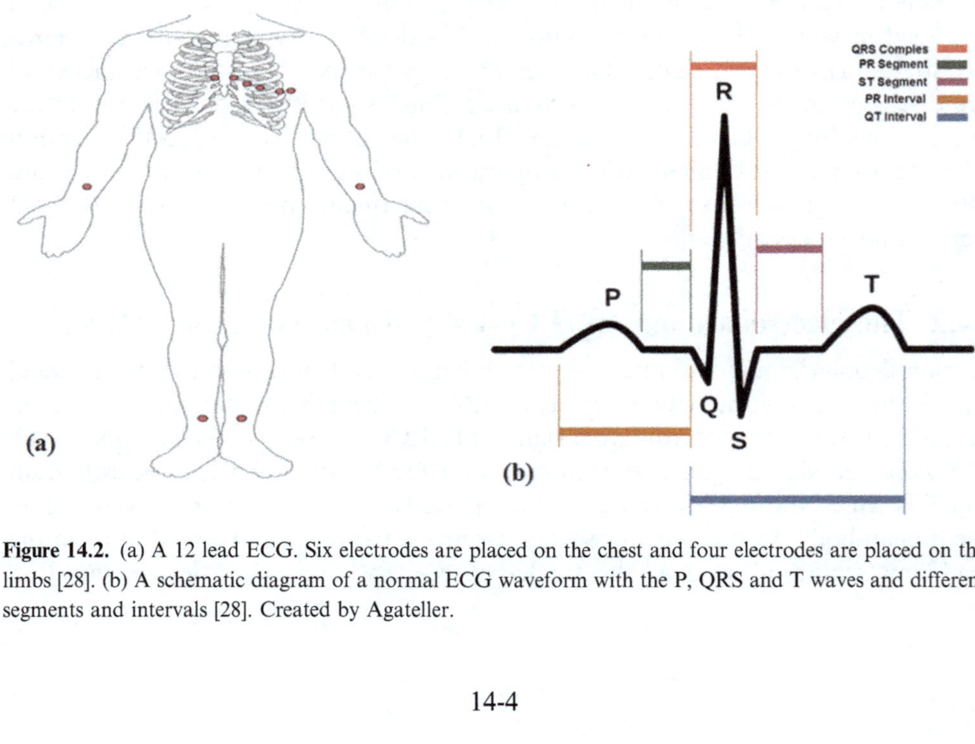

Figure 14.2. (a) A 12 lead ECG. Six electrodes are placed on the chest and four electrodes are placed on the limbs [28]. (b) A schematic diagram of a normal ECG waveform with the P, QRS and T waves and different segments and intervals [28]. Created by Agateller.

T wave represents the *repolarization* of the ventricles with a duration of 160 ms. There are several segments and intervals present between the P wave, QRS and T wave, which are known as the PR interval and segment, the ST segment and the QT interval, as shown in figure 14.2(b) [28], and are of different durations, namely 120–200 ms, 180 ms and 440 ms, respectively.

As shown in figure 14.2(a), a 12 lead ECG generation system generally uses the ten electrodes at prescribed points as listed in table 14.1.

Here, the 12 lead ECG generation/acquisition system is categorized into three sub-lead groups: limb leads, augmented limb leads and precordial leads. A virtual lead estimated by the average potential of the RA, LA and LL electrodes is known as the Wilson's central terminal (V_W):

$$V_W = (\mathrm{RA} + \mathrm{LA} + \mathrm{LL})/3. \tag{14.1}$$

In a 12 lead ECG acquisition system, all leads are unipolar (aVR, aVL, aVF, V1, V2, V3, V4, V5 and V6) except the limb leads (I, II and III). The measurement and estimation of a potential difference requires two terminal contacts or electrode points, therefore, the unipolar leads are measured as $+ve$ terminals with respect to common $-ve$ lead point. The details of the leads are as follows [28]:

- *Limb leads.* In a 12 lead acquisition system, leads I, II and III are called the limb leads. These leads are measured as the potential difference between each arm and the left leg. These lead representations, known as Einthoven's triangle, are shown in figure 14.3.
    - Lead I is estimated as the potential difference between the ($+ve$) left arm (LA) electrode and right arm (RA) electrode: I = LA − RA.

Table 14.1. List of electrodes and descriptions of their placement.

Electrode/lead name	Placement/position description
RA	On the right arm (avoiding thick muscle)
LA	On the left arm (in symmetry with RA)
RL	On the right leg (lower end of calf muscle)
LL	On the left leg (in symmetry with RL)
V1	Right of the sternum (at the fourth intercostal space between the fourth and fifth ribs)
V2	Left of the sternum (at the fourth intercostal space between the fourth and fifth ribs)
V3	In the center position between leads V2 and V4
V4	On the mid-clavicular line (at the fifth intercostal space between the fifth and sixth ribs)
V5	Horizontally level with V4
V6	Horizontally level with V4 and V5

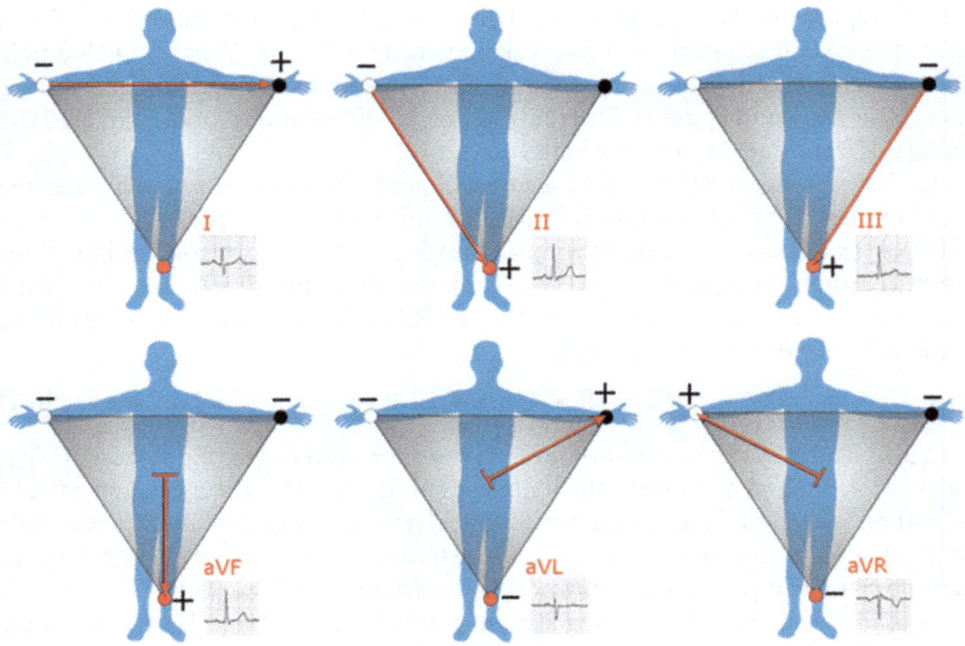

Figure 14.3. The limb leads and augmented limb leads [28]. Author Npatchett.

- ○ Lead II is estimated as the potential difference between the (+*ve*) left leg (LL) electrode and the right arm (RA) electrode: II = LL − RA.
- ○ Lead III is estimated as the potential difference between the (+*ve*) left leg (LL) electrode and the left arm (LA) electrode: III = LL − LA.

- *Augmented limb leads.* An augmented limb lead, such as aVR, aVL and aVF, is estimated and derived as the potential difference between the combination of the RA, LA and LL electrodes as defined below and shown in figure 14.4. Further, these leads are simplified using Wilson's central terminal (V_W). It is referred to immediately below as 'the negative pole'.
  - ○ *Lead augmented vector right* (aVR) is represented as the potential difference between the +*ve* electrode RA on the right arm and the −*ve* pole, a combination of the left arm electrode LA and the left leg electrode LL:

$$\mathrm{aVR} = \mathrm{RA} - \frac{1}{2}(\mathrm{LA} + \mathrm{LL}) = \frac{3}{2}(\mathrm{RA} - V_W).$$

  - ○ *Lead augmented vector left* (aVL) is represented as the potential difference between the +*ve* electrode LA on the left arm and the −*ve* pole, a combination of the right arm electrode RA and the left leg electrode LL:

$$\mathrm{aVL} = \mathrm{LA} - \frac{1}{2}(\mathrm{RA} + \mathrm{LL}) = \frac{3}{2}(\mathrm{LA} - V_W).$$

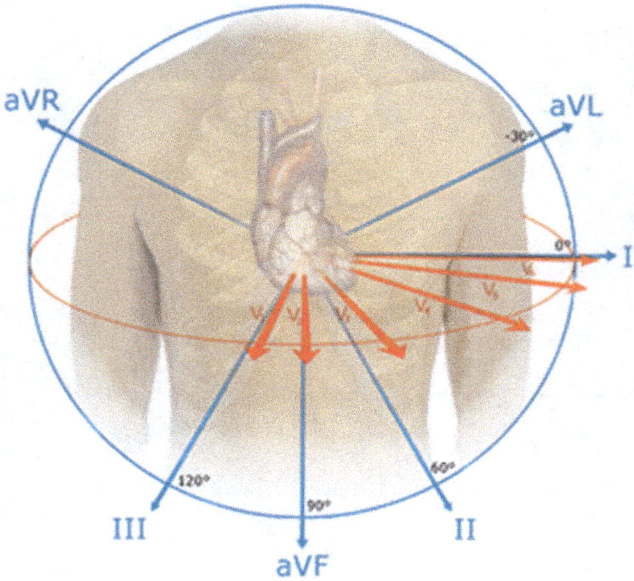

Figure 14.4. Precordial lead representation with limb and augmented limb leads [28]. Author Npatchett.

- ○ *Lead augmented vector foot* (aVF) is represented as the potential difference between the $+ve$ electrode on the left leg LL and $-ve$ pole, a combination of the right arm electrode RA and the left arm electrode LA:

$$\mathrm{aVF} = \mathrm{LL} - \frac{1}{2}(\mathrm{RA} + \mathrm{LA}) = \frac{3}{2}(\mathrm{LL} - V_W).$$

- *Precordial leads.* The precordial leads (V1, V2, V3, V4, V5 and V6) lie in the horizontal plane and perpendicular to the limb and augmented limb leads. These leads are considered as $+ve$ poles and potential measures with respect to Wilson's central terminal (V_W), which is considered as the $-ve$ pole or terminal.

Typically, these electrodes are placed at the prescribed positions and the leads are estimated as discussed above. The response of the leads is referred to as electrical waveforms or ECG signals. An ECG signal can provide a range of information related to heart health and disease such as rhythm, heart rate, variability and abnormality. As discussed above, the waveform is generated during cardiac action such as depolarization and repolarization. The waveform may deviate from its baseline which represents the resting sate of the cardiac muscle cells. These actions can be interpreted from the waveform and vice versa, as illustrated in figure 14.5 [28], where the potential or waveform is associated with different regions of the cardiac system, described as follows:

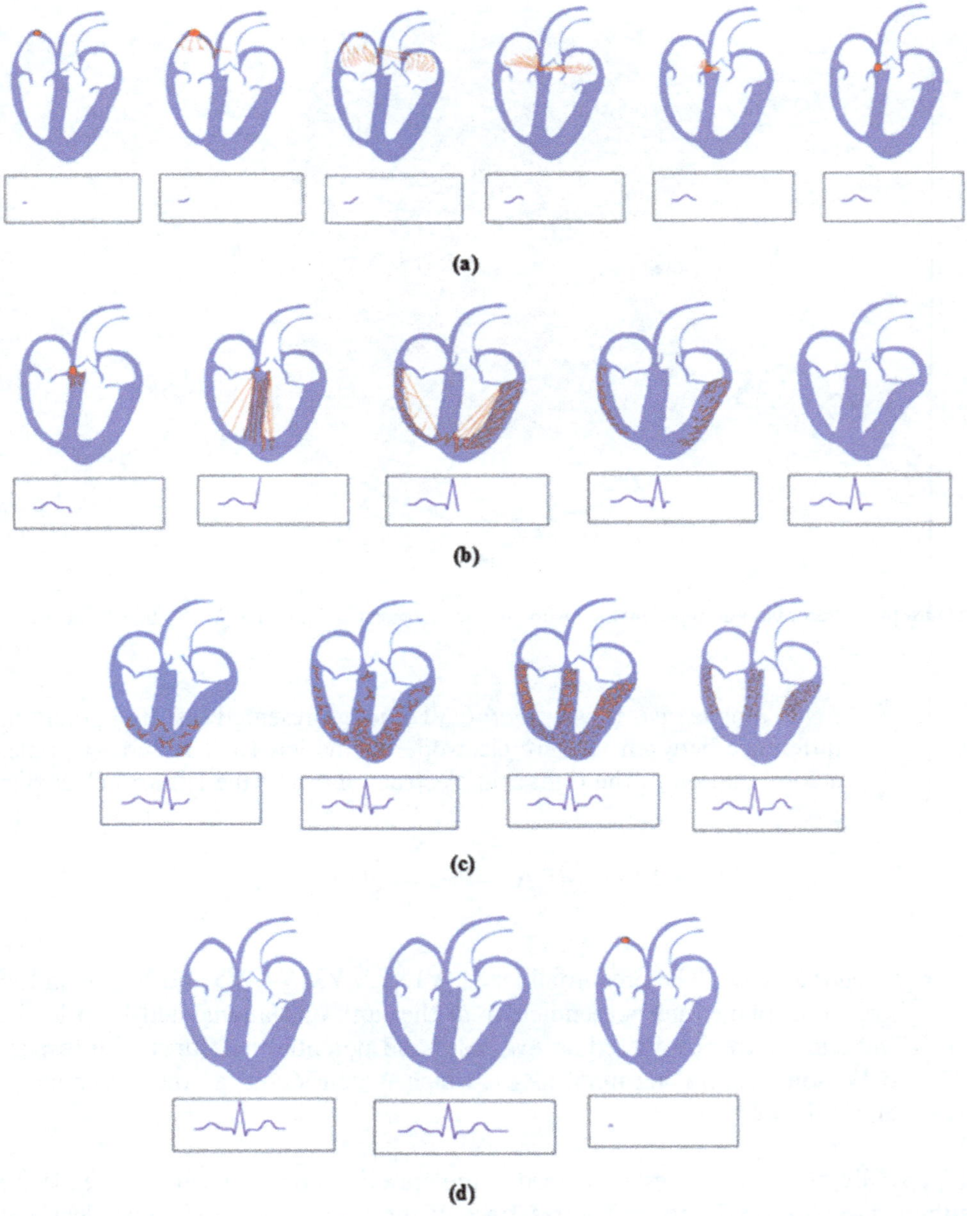

Figure 14.5. ECG signal generation: (a) generation of the P wave, (b) generation of the QRS complex, (c) generation of the T wave and (d) the heart in the rest position [28]. Author: Kalumet. Source: Selbst Erstellt.

- The *P wave* is generated by the normal *atrial depolarization* action, the key ionic potential vector is directed from the SA node towards the AV node, and moves from the right atrium to the left atrium, as shown in figure 14.5(a).
- The *QRS complex* reflects the rapid *depolarization* of the right and left *ventricle* action of the cardiac muscles. Ventricles have a large muscle mass

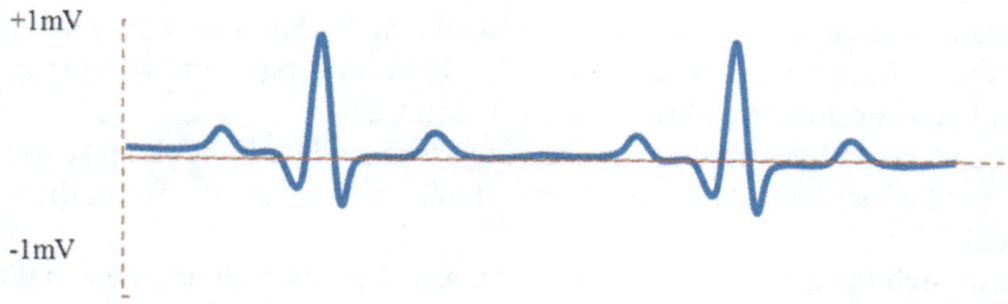

Figure 14.6. Two complete cycles of cardiac activity in an ECG signal (cardiac rhythm).

compared to the atria, therefore greater ionic activity is recognized to represent the QRS complex, with larger amplitude than that of the P wave, shown in figure 14.5(b).

- The *T wave* represents the *repolarization* action (or recovery to the rest state) of the ventricles, as shown in figure 14.5(c).
- At the end of the T wave the heart is in the rest position due to a *repolarization* (or recovery) of the ventricles, the rhythm follows with atrial depolarized and generates a new P wave, shown in figure 14.5(d).

In the recording of ECG signals, the clinical bandwidth is 0.05–100 Hz using the standard clinical 12 lead system. In contrast, it is restricted to 0.5–50 Hz in the case of monitoring and ambulatory purposes using a two-lead ECG system. The amplitude range of the ECG signal is also an important issue in monitoring systems because of the low amplitude of the ECG signal. Its maximum peak amplitude is in the range of 1 mV, so an amplifier is needed to step-up the amplitude of a signal with a gain of 1000 in order to obtain the peak signal in the range of about 1 V [27, 28].

In a recording and monitoring system, ECG signals are attenuated with higher frequency noise due to the power-line interference of the electrical system, muscles contraction noise and motion of the electrodes. Therefore, the filtering process is involved to eliminate such types of noise from the original ECG signal. With the increasing role of digital signal processing in biomedical signals, it has become necessary to convert the analog signal into a digital or discrete signal for further processing with a digital computers [1].

14.2.2 ECG signal characteristics

ECG signals are quasi-periodic in nature, where one cardiac cycle (P, QRS and T) of an ECG signal is known as a beat. These beats illustrate the heart rate of a cardiac system over time as beats per minute. Normally, the heart rate of a cardiac system lies between 60 to 100 beats per minute for a healthy heart [1, 27, 28].

The *rhythm* of the normal heart is represented by electrical impulses or a waveform originating from the SA node of the heart as shown by the simple heart rhythms in figure 14.6. A disturbance in the rhythm or a deviation from the normal sinus rhythm is known as arrhythmia. It occurs when the depolarization action is initiated by another pacemaker cell rather that the SA node, which results in the formation of an electrical impulse that deviates from the normal rhythm.

In general, the pacemaker responsible for the arrhythmia may be present in the atria or ventricles, or a combination [1, 27, 28]. Several types of arrhythmia can be classified based on the originating points or sources:

- The atrial and ventricular rhythms are associated with the same action rate in a cardiac system. Therefore, arrhythmia may originate from the atria or ventricles.
- Although the atrial and ventricular rhythms are associated, the atrial rhythm may be faster than the ventricular rate. Then the source of the arrhythmia may be in the atria muscles.
- If the atrial rhythm is faster than the atrial rate, the source of arrhythmia may be in the ventricles.

An ECG signal or arrhythmia is also categorized with respect to the action rate or heart rate as *bradycardic* and *tachycardic* arrhythmia for less than 60 beats/min and more than 100 beats/min heart rates, respectively. Similarly, several health conditions are recognized based on the ECG signal interpretation, such as sudden cardiac arrest, myocardial infarction (MI), stress tests and many more [27].

14.3 Statistical measures and analysis

The statistical measures and analysis, as discussed in section 14.1, can be defined as descriptive or inferential measures [22, 29, 30]. These are defined as follows:

- *Descriptive measures.* Descriptive statistical measures represent datasets or clusters with using a central location. The central measure or tendency is described by the degree of dispersion. These measures include the mean, median, mode, standard deviation, range, variance and other statistical variables.
  - *Mean.* The mean ($\bar{x}$, μ or average) is the statistical parameter which measures the central tendency of the distributed data or data clusters. The mean value of discrete or continuous data is equal to the sum of all the data values divided by the total number of data values in a dataset or cluster. It is defined as

$$\bar{x} = \frac{x_1 + x_2 + \ldots + x_n}{n}. \tag{14.2}$$

  - *Median.* The median represents the center value among the all observations or data values and is measured by arranging the values in ascending order and obtaining the center or middle value of the dataset.
  - *Mode.* The mode is represented by the data value with the highest frequency in the dataset. In other words, the mode is calculated by counting the number of times each value occurs.
  - *Range.* The easiest measure of the data spread is the range. It is simply the difference between the highest data value and the lowest data value in the observations.

- *Variance.* Variance is the representation of the average squared difference of the values from the mean value of the dataset. The measures of variability in dataset is its variance, and the variance includes all values in the calculation by comparing each value to the mean value of the dataset. In other words, it represents the spreadness of data from its mean value. It is defined as

$$\sigma^2 = \frac{\sum(x - \bar{x})^2}{n}. \tag{14.3}$$

- *Standard deviation.* The amount of variation or dispersion of a set of values is represented by the measure of standard deviation (SD). A lower SD indicates that the values tend to be closer to the mean of the dataset, while a higher SD indicates that the values are spread out over a wider range from the mean. The SD of an observation is the square root of variance, defined as

$$s = \sqrt{\frac{\sum(x - \bar{x})^2}{n}}. \tag{14.4}$$

The descriptive measures, such as central tendency and variability, are related to each other, in particular to the mean value of observations. Figure 14.7(a) illustrates the mean, median and mode variables for a normal frequency distribution of the data or observations. Similarly, figures 14.7(b) and (c) illustrate the mean, median and mode variables for positively and negatively skewed frequency distributions.

Similarly, figure 14.8 illustrates the relation between the mean and standard deviation (SD) for the variability measures. Here, the data are normally distributed and the variance of data from its mean value is defined. The distributions of the data are categorized as the first SD, second SD and third SD, which represent 68%, 95% and 99.7% of the spread of the total data.

- *Inferential measures.* Large datasets are analysed using a sample dataset to make inferences to the larger collection of observations based on inferential statistical measures. In this process, a hypothesis is set to explain the observations and their characteristics. This hypothesis and its test procedures make rational decisions on the reality of the observations or data of any event.

In inferential statistical measures, the term 'null hypothesis' (H_0, 'H-naught' or 'H-null') represents that there is no relationship between the population variables in the hypothesis [30].

Further, the P-value is the probability of the event occurring by chance if the null hypothesis is true. The P-value is a numerical value between 0 and 1 and is

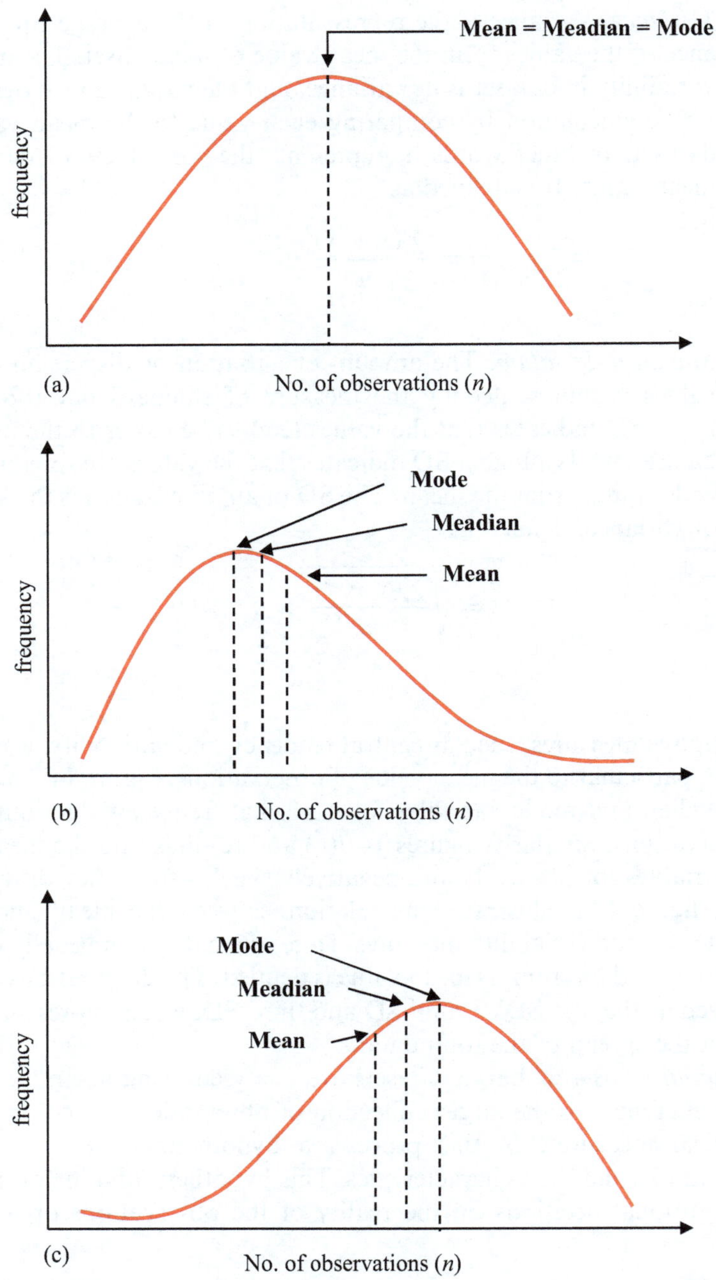

Figure 14.7. Central tendency variables for various frequency distributions of data.

interpreted by researchers to decide whether to reject or retain the null hypothesis as described in table 14.2.

In the field of signal processing, the statistical measures play a great role in distinguishing events over time. It also plays a role in summarizing the performance

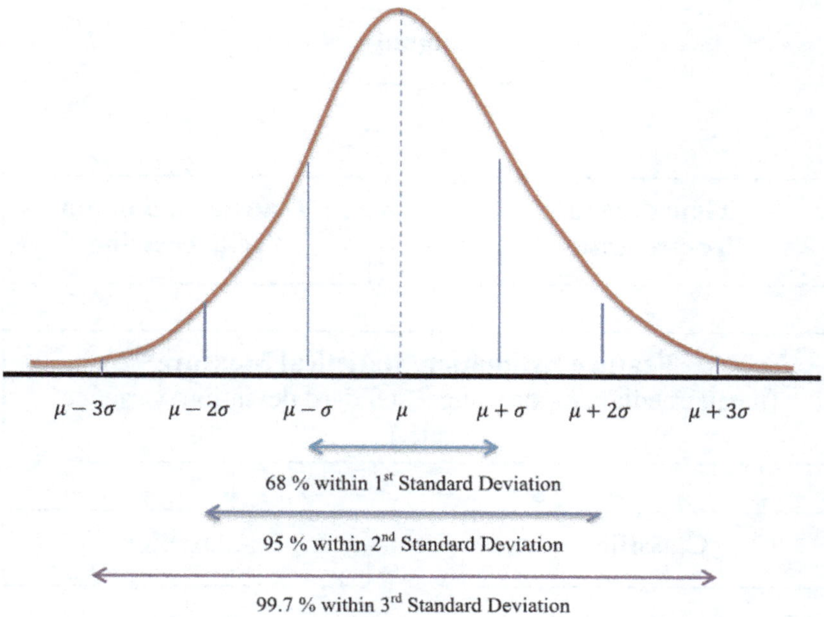

Figure 14.8. Illustration of variability measures in terms of SD and mean.

Table 14.2. The probability P values with interpretation of the null hypothesis [30].

P-value	Result	Null hypothesis
< 0.01	Result is highly significant	Reject H_0
$\geqslant 0.01$ but < 0.05	Result is significant	Reject H_0
$\geqslant 0.05$	Result is not significant	Do not reject H_0

of signal processing operations on long-duration time-series data, similar to other scientific and engineering applications.

14.4 Statistical analysis in ECG signal processing

An ECG signal is data collected using electrodes represented as a lead system. These data illustrate the rhythmic behavior of a cardiac system, i.e. muscle repolarization and depolarization corresponding to electric potential changes in the form of waves such as P, QRS, T and U, as discussed above. Several challenges are associated with the handling of big ECG data and its interpretation due to limited resources in terms of infrastructure as well as human healthcare resources. Therefore, an automatic system is needed to fill the gap and perform this task. As we know, an automatic system such as an artificially intelligent and machine learning-based system needs training using data, features and interpretation in many applications. The statistical measures help to summarize the data characteristics and features [18–20, 31]. There are several applications of statistical methods to validate the performance of signal

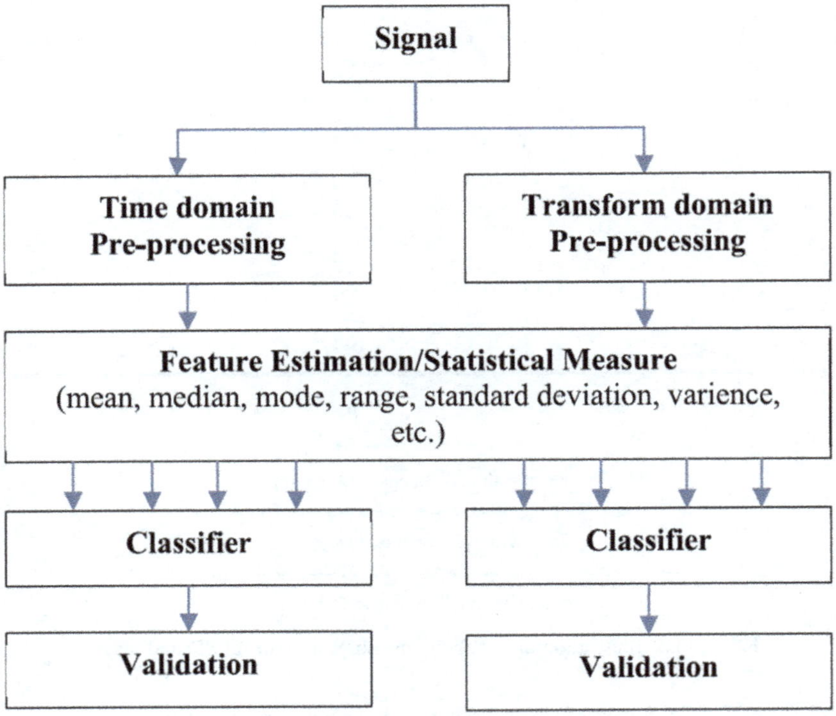

Figure 14.9. Signal classification function block representation.

processing operations. In the last two decades, several algorithms have been developed to classify signals, estimate features and interpret the performance in ECG signal processing.

In this context, many algorithms for classification, feature estimation, filtering and data compression have been developed and their performances are validated through statistical measures for different applications. The feature estimation and classification of the ECG signal completely depends on suitable statistical measures, as shown in figure 14.9. In the classification operations, accurate and significant features should be easily separable through classifiers to distinguish subjects. As we know, the ECG signal is non-stationary and quasi-periodic, with multiple time domain features and time–frequency domain features with suitable transformation. Time domain features are those such as the heart rate variability (HRV) signal extracted from consecutive differences between R-peaks, and the patterns of different waveforms within the signals [18]. These features are suitable for manual observation for very short durations of signals, but this is very complex for the long-duration signals. Therefore, statistical measures such as central tendency and variability features summarize the long-duration signal features into finite and discrete feature sets for classification problems. Generally, time domain feature classification efficiency is low compared to transform domain or time–frequency domain feature classification, as has been reported in many studies [18, 32–44].

The statistical measures discussed in the previous section are illustrated for the HRV data of ECG signals in figure 14.10. Here, these measures may be considered as features to classify the signal characteristics or its classes. In this study, a public database is utilized, namely the MIT-BIH Arrhythmia (MITA) ECG Database and Normal Sinus Rhythm Database (NSRDB) [45]. In the classification process, features play a key role in identifying the class of the signal that may indicate a particular health condition. The HRV signals represent variations of the heart rate, and can be measured in terms of R–R intervals. The variance of R–R intervals clearly represents the variability in cardiac rhythm; it may be normal or abnormal [26].

The MITA [24] ECG database contains 48 records of 30 min duration and the NSRDB [45] contains 18 records of 24 h duration. In this study, an initial 1 min of signal is extracted for feature calculation. In figure 14.10, the R–R interval duration is illustrated for 48 and 13 records from the respective database. The variation of the

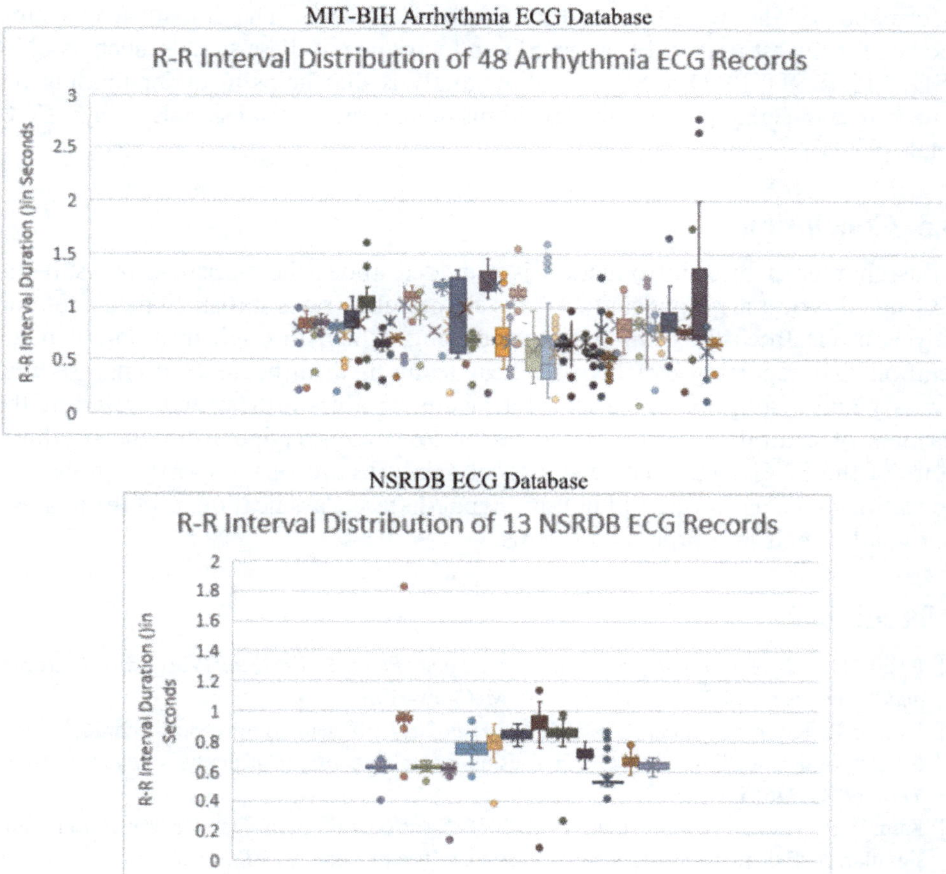

Figure 14.10. Box-plots and whisker diagrams of different of R–R intervals for different signals from the MITA-NSRDB. The box represents the values from the lower to the upper quartile and the central line is representative of the median.

R–R interval is clearly shown between the different records and databases using the statistical measures and analysis. In this analysis, box-plot and whisker diagrams [29] are used to understand the variability of the R–R time intervals for different signals. The descriptive statistical measures, such as mean, median, variance and standard deviation, help us understand the characteristics of datasets, and these measures are further considered as features for classification systems.

In other applications, such as filtering, segmentation and compression of ECG signals, statistical methods play an important role in measuring the cumulative performance for the different signals. In these applications, the performance is measured using fidelity parameters such as the signal-to-noise ratio (SNR), maximum error (ME), percentage-root-mean-square error (PRD), peak signal-to-noise ratio (PSNR) and compression ratio. These fidelity parameter values are different for different signals; in other words, performance is variable for different signals due to different characteristics. This variability is represented by the center tendency and variability statistical measures in many studies to represent the performance of the algorithm and its feasibility [46–96]. The statistical measures reported for the estimation of range, SD and variance of these signals, such as SNR ± SD, ME ± SD, PRD ± SD etc, allow analysis that helps in understanding the performance of signal processing algorithms on real biomedical signals such as ECG signals [93–96].

14.5 Conclusion

In this chapter, a brief introduction is provided about the statistical measures in ECG signal processing applications. This chapter illustrates statistical measures and analysis in electrocardiogram signal processing, which through meaningful interpretation and reporting can help medical teams in healthcare planning. A brief summary and description of statistical measure variables is presented, including the measures of central tendency and of deviation or dispersion from the rhythmic nature of the ECG signal. These statistical measures are very important in various applications such as feature estimation, signal/disease classification, and the analysis of original signals and processed signals.

References

[1] Reddy D C 2011 *Cardiological Signal Processing. Biomedical Signal Processing: Principles and Techniques* vol 2 (New Delhi: Tata/McGraw-Hill)

[2] Kumar R, Kumar A, Singh G K and Lee H-N 2017 Efficient compression technique based on temporal modelling of ECG signal using principle component analysis *IET Sci. Meas. Technol.* **11** 346–53

[3] Kumar R, Kumar A and Singh G K 2016 Electrocardiogram signal compression using singular coefficient truncation and wavelet coefficient coding *IET Sci. Meas. Technol.* **10** 266–74

[4] Kumar R, Kumar A and Singh G K 2016 Hybrid method based on singular value decomposition and embedded zero tree wavelet technique for ECG signal compression *Comput. Methods Programs Biomed.* **129** 135–48

[5] Kumar R, Kumar A and Singh G K 2015 Electrocardiogram signal compression based on singular value decomposition (SVD) and adaptive scanning wavelet difference reduction (ASWDR) technique *AEU—Int. J. Electron. Commun.* **69** 1810–22
[6] Kumar R, Kumar A and Pandey R K 2013 Beta wavelet-based ECG signal compression using lossless encoding with modified thresholding *Comput. Electr. Eng.* **39** 130–40
[7] Kumar R, Kumar A, Akhil G, Signh A and Jafri S N H 2014 Computationally efficient method for ECG signal compression based on modified SPIHT technique *Int. J. Biomed. Eng. Technol.* **15** 173–88
[8] Kumar R, Kumar A and Pandey R K 2012 Electrocardiogram signal compression using beta wavelets *J. Math. Model. Algorithms* **11** 235–48
[9] Kumar A and Ranjeet K 2011 ECG signal compression using the optimised wavelet filter banks *Int. J. Signal Imaging Syst. Eng.* **5** 187–95
[10] Ranjeet K, Kumar A and Pandey R K 2012 ECG signal compression using optimum wavelet filter bank based on Kaiser window *Procedia Eng.* **38** 2889–902
[11] Ranjeet K A 2011 Wavelet based electrocardiogram compression at different quantization levels *Information Technology and Mobile Communication* (Communications in Computer and Information Science vol 147) (Berlin: Springer) pp 392–8
[12] Ranjeet K, Kumar A and Pandey R K 2011 ECG signal compression using different techniques *Advances in Computing, Communication and Control* (Communications in Computer and Information Science vol 125) (Berlin: Springer) pp 231–41
[13] Farida R K 2011 Retained signal energy based optimal wavelet selection for denoising of ECG signal using modified thresholding *IEEE Int. Conf. on Multimedia, Signal Processing and Communication Technologies (IMPACT 2011)* pp 196–9.
[14] Kumar R and Kumar A 2015 Dual tree DWT analysis based electrocardiogram signal compression using zero coding technique *2nd IEEE Int. Conf. on Electronics and Communication Systems (ICECS 2015)* pp 624–9.
[15] Kumar R and Kumar A 2015 ECG signal compression algorithm based on joint-multi-resolution analysis (J-MRA) *2nd IEEE Int. Conf. on Electronics and Communication Systems (ICECS 2015)* 618–24
[16] Sedaaaghi M H 1998 ECG wave detection using morphological filters *Appl. Signal Process.* **5** 182–94
[17] Elghazzawi Z and Geheb F 1996 A knowledge-based system for arrhythmia detection *Comput. Cardiol.* **23** 541–4
[18] Acharya U R, Joseph K P, Kannathal N, Lim C M and Suri J S 2006 Heart rate variability: a review *Med. Biol. Eng. Comput.* **44** 1031–51
[19] Acharya U R, Min L C and Joseph P 2002 HRV analysis using correlation dimension and DFA *Innovations and Technology in Biology and Medicine (ITBM-RBM)* **vol 23** pp 333–9
[20] Acharya U R, Kannathal N, Seng O W, Ping L Y and Chua T 2004 Heart rate analysis in normal subjects of various age groups *Biomed. Online J. USA* **3** 24
[21] Akaike H 1974 A new look at statistical model identification *IEEE Trans. Autom. Control* **19** 716–23
[22] Trochim W M K 2020 *Research Methods Knowledge Base, Changes and additions by Conjoint.ly* https://conjointly.com/kb/ accessed in March 2020
[23] Kumar R, Kumar A and Singh G K 2016 Electrocardiogram signal compression based on 2D-transforms: a research overview *J. Med. Imaging Health Inform.* **6** 285–96

[24] Moody G B and Mark R G 2001 The impact of the MIT-BIH arrhythmia database *IEEE Eng. Med. Biol.* **20** 45–50

[25] Jahmunah V, Oh S L, Wei J K E, Ciaccio E J, Chua K, San T R and Acharya U R 2019 Computer-aided diagnosis of congestive heart failure using ECG signals—a review *Phys. Med.* **62** 95–104

[26] Kumar R, Verma A R, Panda M K and Kumar P 2020 HRV signal feature estimation and classification for healthcare system based on machine learning *Communications in Computer and Information Science* vol 1241 (Berlin: Springer) pp 437–448

[27] Sornmao L and Laguna P 2005 *The Electrocardiogram—A Brief Background* (Amsterdam: Elsevier)

[28] Electrocardiography https://en.wikipedia.org/wiki/Electrocardiography accessed March 2020

[29] Descriptive statistics https://en.wikipedia.org/wiki/Descriptive_statistics accessed Mar 2020

[30] Statistical inference https://en.wikipedia.org/wiki/Statistical_inference accessed March 2020

[31] Alekseev M and Alekseev A 2018 Statistical methods of ECG signal processing in diagnostics of coronary artery disease *Int. J. Cardiovasc. Res.* **7** 1

[32] Martis R J, Acharya U R and Min L C 2013 ECG beat classification using PCA, LDA, ICA and discrete wavelet transform *Biomed. Signal Process. Control* **8** 437–48

[33] Bušek P, Vaòková J, Opavský J, Salinger J and Nevšímalová S 2005 Spectral analysis of heart rate variability in sleep *Physiol. Res.* **54** 369–76

[34] Ceylan R and Özbay Y 2007 Comparison of FCM, PCA and WT techniques for classification ECG arrhythmias using artificial neural network *Expert Syst. Appl.* **33** 286–95

[35] Fujita H, Acharya U R, Sudarshan V K, Ghista D N, Sree S V, Eugene L W J and Koh J E W 2016 Sudden cardiac death (SCD) prediction based on nonlinear heart rate variability features and SCD index *Appl. Soft Comput.* **43** 510–9

[36] Bahadure N B, Ray A K and Thethi H P 2018 Comparative approach of MRI-based brain tumor segmentation and classification using genetic algorithm *J. Digit. Imaging* **3** 477–89

[37] Avola D, Cinque L and Placidi G 2013 Customized first and second order statistics-based operators to support advanced texture analysis of MRI images *Comput. Math. Methods Med.* **2013** 213901

[38] Banday S A and Mir A H 2017 Statistical textural feature and deformable model-based brain tumor segmentation and volume estimation *Multimed. Tools Appl.* **76** 3809–28

[39] Salem A M, Revett K and El-Dahshan E A 2009 Machine learning in electrocardiogram diagnosis *2009 Int. Multiconf. on Computer Science and Information Technology* 429–33

[40] Jambukia S H, Dabhi V K and Prajapati H B 2015 Classification of ECG signals using machine learning techniques: a survey *2015 Int. Conf. on Advances in Computer Engineering and Applications* pp 714–21

[41] Zubair M, Kim J and Yoon C 2016 An automated ECG beat classification system using convolutional neural networks *2016 6th Int. Conf. on IT Convergence and Security (ICITCS)* pp 1–5

[42] Hegde V N, Deekshit R and Satyanarayana P S 2011 Comparison of characterizing and data analysis methods for detecting abnormalities in ECG *2011 Asian Himalayas Int. Conf. on Internet (AH-ICI)* pp 1–5

[43] Yu S-N and Chen Y-H 2007 Electrocardiogram beat classification based on wavelet transformation and probabilistic neural network *Pattern Recognit. Lett.* **28** 1142–50

[44] Alarsan F I and Younes M 2019 Analysis and classification of heart diseases using heartbeat features and machine learning algorithms *J. Big Data* **6** 81

[45] Goldberger A L, Amaral L A N, Glass L, Hausdorff J M, Ivanov P C, Mark R G, Mietus J E, Moody G B, Peng C-K and Stanley H E 2003 PhysioBank, PhysioToolkit, and PhysioNet: components of a new research resource for complex physiologic signals *Circulation* **101** e215–20
[46] Raj S, Maurya K and Ray K C 2015 A knowledge-based real time embedded platform for arrhythmia beat classification *Biomed. Eng. Lett.* **5** 271–80
[47] Jalaleddine S M S, Hutchens C G, Strattan R D and Cobery W A 1999 ECG data compression techniques—a unified approach *IEEE Trans. Biomed. Eng.* **37** 329–43
[48] Anand R S and Kumar V 1995 Efficient and reliable detection of QRS-segment in ECG signals *14th IEEE/Engineering in Medicine and Biology Society Annual Conf.* vol 14 pp 256–7
[49] Li H 1999 Classification method of QRS waves by fractal dimension *Proc. 21st Annual Int. Conf. of the IEEE Engineering in Medicine and Biology* vol 2 p 907
[50] Tan K F, Chan K L and Choi K 2000 Detection of the QRS complex, P wave and T wave in electrocardiogram *Int. Conf. on Advances in Medical Signal and Information Processing (MEDSIP 2000)* vol 476 pp 41–7
[51] Hailer B and Van Leeuwen P 2004 Detection of coronary artery disease with MCG *Neurol. Clin. Neurophysiol.* **2004** 82
[52] Logan B and Healey J 2005 Robust detection of a trial fibrillation for a long term telemonitoring system *Comput. Cardiol.* **32** 619–22
[53] Chawla M P S, Verma H K and Kumar V 2006 ECG modeling and QRS detection using principal component analysis *IET 3rd Int. Conf. MEDSIP 2006* p 60
[54] Krasteva V and Jekova I 2007 QRS template matching for recognition of ventricular ectopic beats *Ann. Biomed. Eng.* **35** 2065–76
[55] Mehta S S and Lingayat N S 2008 Development of SVM based ECG pattern recognition technique *IETE J. Res.* **54** 5–11
[56] Sabouri S, SadAbadi H and Dabanloo N J 2010 Neural network classification of body surface potential contour map to detect myocardial infarction location *Comput. Cardiol.* **37** 301–4
[57] Dashouk M, Zhang Z, Lall C and Chen Y 2011 Performance challenges in current multi-lead QRS detection systems *Comput. Cardiol.* **38** 745–8
[58] Abo-Zahhad M, Ahmed S M and Zakaria A 2012 An efficient technique for compressing ECG signals using QRS detection, estimation, and 2D DWT coefficients thresholding *Model. Simul. Eng.* **2012** 1–10
[59] Craven D, McGinley B, Kilmatin L, Glavin M and Jones E 2016 Energy-efficient compressed sensing for ambulatory ECG monitoring *Comput. Biol. Med.* **71** 1–13
[60] Yochum M, Renaud C and Jacquir S 2016 Automatic detection of P, QRS and T patterns in 12 leads ECG signal based on CWT *Biomed. Signal Process. Control* **25** 46–52
[61] Sharma L D and Sunkaria R K 2016 A robust QRS detection using novel pre-processing techniques and kurtosis based enhanced efficiency *Measurement* **87** 194–204
[62] Nave G and Cohen A 1993 ECG compression using long-term prediction *IEEE Trans. Biomed. Eng.* **40** 877–85
[63] Mueller W C 1978 Arrhythmia detection program for an ambulatory ECG monitor *Biomed. Sci. Instrum.* **41** 81–5
[64] Manikandan M S and Dandapat S 2014 Wavelet-based electrocardiogram signal compression methods and their performances: a prospective review *Biomed. Signal Process. Control* **14** 73–107

[65] Abenstein P and Tompkins W J 1982 New data-reduction algorithm for real-time ECG analysis *IEEE Trans. Biomed. Eng.* **29** 43–8
[66] Wang L, Belina J, Vasinonta A, Berner M and Ramprashad S 1994 Compression of ECG using a code excited linear prediction (CELP) *IEEE Proc. Adv.: New Opportunities for Biomedical Engineers* vol 2 pp 1264–5
[67] Cardenas-Barrera J L and Lorenzo-Ginori J V 1999 Mean-shape vector quantizer for ECG signal compression *IEEE Trans. Biomed. Eng.* **46** 62–70
[68] Aydin A C, Cetin A E and Koymen H 1991 ECG data compression by sub-band coding *Electron. Lett.* **27** 359–60
[69] Chen J and Itoh S A 1998 Wavelet transform-based ECG compression method guaranteeing desired signal quality *IEEE Trans. Biomed. Eng.* **45** 1414–9
[70] Chen J, Wang F, Zhang Y and Shi X 2008 ECG compression using uniform scalar dead-zone quantization and conditional entropy coding *Med. Eng. Phys.* **30** 523–30
[71] Bilgin A, Marcellin M W and Altbach M I 2003 Compression of electrocardiogram signals using JPEG2000 *IEEE Trans. Consumer Electron.* **49** 833–40
[72] Chen D H and Yang S 2008 Compression of ECG signal using video coded technology-like scheme *J. Biomed. Sci. Eng.* **1** 22–6
[73] Miaou S, Yen H and Lin C 2002 Wavelet-based ECG compression using dynamic vector quantization with tree code vectors in single codebook *IEEE Trans. Biomed. Eng.* **49** 671–80
[74] Lu Z, Kim D Y and Pearlman W A 2000 Wavelet compression of ECG signals by the set partitioning in hierarchical trees algorithm *IEEE Trans. Biomed. Eng.* **47** 849–56
[75] Zigel Y, Cohen A and Katz A 2000 The weighted diagnostic distortion (WDD) measure for ECG signal compression *IEEE Trans. Biomed. Eng.* **47** 1422–30
[76] Gurukan H 2012 Compression of ECG signal using variable-length classified vectorsets and wavelet transforms *EURASIP J. Adv. Signal Process.* **1** 119–36
[77] Al-Shroud A, Abo-Zahhad M and Ahmed S M 2003 A novel compression algorithm for electrocardiogram signals based on the linear prediction of the wavelet coefficients *Digital Signal Process.* **13** 604–12
[78] Husoy J H and Gjerde T 1996 Computationally efficient sub-band coding for ECG signals *Med. Eng. Phys.* **18** 132–42
[79] Cetin A E, Koymen H and Aydin M C 1993 Multichannel ECG data compression by multirate signal processing and transform domain coding techniques *IEEE Trans. Biomed. Eng.* **40** 495–9
[80] Shaou-Gang M and Shu-Nien C 2005 Wavelet-based lossy-to-lossless ECG compression in a unified vector quantization framework *IEEE Trans. Biomed. Eng.* **52** 539–43
[81] Mohammad P, Ali T, Morteza M G and Saboori S 2005 Wavelet compression of ECG signal using SPIHT algorithm *Int. J. Inf. Commun. Eng.* **2** 219
[82] Hilton M L 1997 Wavelet and wavelet packet compression of electrocardiograms *IEEE Trans. Biomed. Eng.* **44** 394–402
[83] Lee H and Buckley K M 1999 ECG data compression using cut and align beats approach and 2-D transforms *IEEE Trans. Biomed. Eng.* **46** 556–64
[84] Alexandre E, Pena A and Sobreira M 2006 On the use of 2-D coding techniques for ECG signals *IEEE Trans. Inf. Technol. Biomed.* **10** 809–11
[85] Ahmad R A M and Kambiz N 2001 A two dimensional wavelet packet approach for ECG compression *Int. Symp. on Signal Processing and its Applications* pp 226–9

[86] Sahraeian S M E and Fatemizadeh E 2007 Wavelet-based 2-D ECG data compression method using SPIHT and VQ coding *EUROCON'2007* pp 133–7
[87] Tai S C, Sun C C and Yan W C 2005 A 2-D ECG compression method based on wavelet transform and modified SPIHT *IEEE Trans. Biomed. Eng.* **52** 999–1008
[88] Wang X and Meng J 2008 A 2-D compression algorithm based on wavelet transforms and vector quantization *Digit. Signal Process.* **18** 179–88
[89] Wei J, Chang C, Chou N and Jan G 2001 ECG data compression using truncated singular value decomposition *IEEE Trans. Inf. Technol. Biomed.* **5** 290–9
[90] Sharma L N, Dandapat S and Mahanta A 2012 Multichannel ECG data compression based on multiscale principal component analysis *IEEE Trans. Inf. Technol. Biomed.* **16** 730–6
[91] Manikandan M S and Dandapat S 2006 Wavelet threshold-based ECG compression using USZZQ and Huffman coding of DSM *Biomed. Signal Process. Control* **1** 261–70
[92] Blanco-Velasco M, Cruz-Roldán F, Moreno-Martínez E, Godino J I and Barner K E 2008 Embedded filter bank-based algorithm for ECG compression *Signal Process.* **88** 1402–12
[93] Tompkins W J 1999 Electrocardiography *Biomedical Digital Signal Processing* (Englewood Cliffs, NJ: Prentice Hall) ch 2
[94] Schooley B, Abed Y, Murad A, Horan T A and Roberts J 2013 Design and field test of a m-health system for emergency medical services *Health Technol.* **3** 327–40
[95] Han T S and Kobayashi K 2002 *Mathematics of Information and Coding* (Providence, RI: American Mathematical Society) pp 19–20
[96] Sharma L N, Dandapat S and Mahanta A 2010 ECG signal denoising using higher order statistics in wavelet subbands *Biomed. Signal Process. Control* **5** 214–22

Chapter 15

The impact of regional atrophy on Alzheimer's disease and its identification using 3D texture analysis

Shiwangi Mishra[1,2], Pritee Khanna[1]

[1]*PDPM-Indian Institute of Information Technology, Design and Manufacturing, Jabalpur, India*

[2]*Jabalpur Engineering College, Jabalpur, India*

Regional atrophy is a major concern in the identification of Alzheimer's disease (AD). It is evident that progressive deterioration of brain regions leads to severe dementia. Various neuroimaging studies have revealed that atrophy of the hippocampal region plays a critical role in worsening dementia and leading to Alzheimer's disease (AD). The atrophy of other brain regions is also part of this chronic disease. The recent advancements in neuroimaging techniques and image processing methods have resulted in the non-invasive investigation of each brain region at a microscopic level. In this chapter, the atlas-based segmentation of brain MRI from the Alzheimer's Disease Neuroimaging Initiative (ADNI) dataset is considered. These segmented brain MRIs are subjected to texture analysis using 3D discrete wavelet transformation (DWT) for identification of AD based on structural alterations. Also, a feature selection method, the support vector machine–recursive feature elimination (SVM-RFE), is applied to the features of the top discriminating regions. Further, comparative analysis is performed on the performance of various texture feature based methods applied on brain MRIs.

15.1 Introduction

Alzheimer's disease (AD) is a neurodegenerative disorder commonly observed in elderly people, which affects cognitive functions such as memory, learning and behavior [1]. It is responsible for about 60%–80% of dementia cases resulting in medical conditions of memory loss and other cognitive abilities. Alzheimer's disease deteriorates the neural connections present in the brain associated with memory,

doi:10.1088/978-0-7503-3279-8ch15

thereby creating hollow structures [2]. It depletes the functioning of the brain over time and is mostly found in people aged 65 and above. The affected people lose their ability to perform daily tasks and activities [3]. Although there are no treatments to completely cure this disease or stop its progression, an early stage diagnosis with current treatments can temporarily suppress the worsening of the AD symptoms and improve the quality of life through a slow transition. The World Health Organization (WHO) has projected that 82 million people will be affected by AD by 2030, which is further expected to increase to 152 million by 2050 [4]. Thus, the early stage diagnosis of AD in patients has become highly critical in deciding future courses of action. Although the causes of AD are not precisely known, it is attributed to the failure of the normal functions of brain proteins (beta-amyloid and tau protein) leading to a series of events eventually resulting in shrinkage of the brain [5]. When beta-amyloid clusters together to form amyloid plaques, they disrupt intercellular communication, while tau proteins organize themselves into structures called neurofibrillary tangles, which are toxic in nature and disrupt the transport functionality in the brain. Along with the protein failures, age, genetics, poor lifestyle, past trauma, gender and heart complications are other factors that can also cause this disease. The current absence of a cure makes this field an active area of research [6, 7]. Figure 15.1 shows the differences in a healthy and an AD brain, the latter showing clear shrinkage in brain regions.

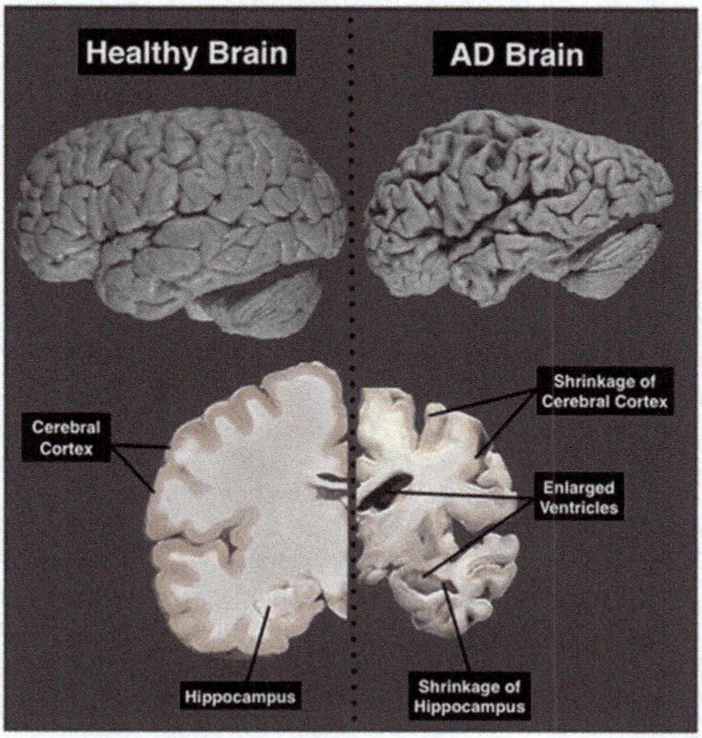

Figure 15.1. Healthy and AD brains of human (https://knowingneurons.com/brainatrophy_600/). Courtesy of Kate Fehlhaber, Knowing Neurones (knowingneurons.com)

The primary diagnosis of AD starts with the analysis of brain structural magnetic resonance images (sMRI) to identify distinctive biomarkers for the possible onset of AD. Even a small change in these biomarkers can signify the possibility of AD in the patient. These biomarkers are not very prominent in the early stages of AD and are thus difficult to detect. Recent advancements in fields such as image processing and artificial intelligence have significantly enhanced the accuracy and precision of object detection algorithms. Machine learning algorithms have an inherent ability to learn complex patterns and this ability to make data driven decisions enables their use in many fields [8]. The application of these techniques could prove to be a great aid for AD detection through anatomical sMRI scan classification as they can detect even the slightest change in the biomarkers.

Different imaging modalities used in the literature apart from sMRI [9–15] are resting state MRI (rs-MRI) [16], single photon emission computed tomography (SPECT) [17–22] and positron emission tomography (PET) [18, 21, 23–27]. However, sMRI (referred as MRI in the following text) is more widely used to analyse the presence of AD due to its non-invasive nature and good soft tissue contrast.

MRI based methods to diagnose AD are generally categorized as follows:

1. Multivariate based approaches apply reduction methods and brain morphometry methodologies on whole brain images to find patterns of AD. In the literature, various reduction methods are used on whole brain sMRIs [11, 28–30].
2. Region of interest (ROI) based approaches work using specific brain region/volume of interest methods to find patterns for AD. In this approach, a segmented part of the whole brain is used for performing analysis [31–33].

Several different types of features, such as gray matter density [34], morphometry [35], group comparison of cortical thickness [36] and texture measures [37] are extracted from MRI of the whole brain. In comparison to using these features separately, their fusion can improve AD diagnosis [38]. Minute details of MRIs can be harnessed by use of texture features. Therefore, these features have been used in AD diagnosis in recent works.

In this study, the aim is to perform texture analysis on the different regions of the brain to identify regional atrophy for AD classification. For completing this task, wavelet-based feature extraction is performed on 3D MRIs. A brain MRI can be divided into three brain regions, namely white matter (WM), gray matter (GM) and cerebrospinal fluid (CSF). The analysis of these segmented regions individually may help in determining the impact of regional atrophy in AD detection. For a better microscopic view, The MRIs are further segmented into sub-regions and this helps in identifying the structural alterations of these regions. Using segmented regions and performing wavelet analysis using a 3D wavelet helps to acquire detail and approximation coefficients, resulting in more detailed information. 3D discrete wavelet transform (3D-DWT) is performed on 116 regions obtained through atlas-based segmentation on 3D MRIs. A segmented MRI region gives 22 wavelet sub-bands of different dimensions. Five volumetric features are obtained from each sub-band. These fused features are used for classification. Results are also obtained using important dominant regions selected by a region ranking method. Also, a feature selection procedure, support vector machine–recursive feature elimination (SVM-RFE), is applied to features of the selected regions and these selected features are classified by the SVM.

The chapter is organized as follows. Regional atrophy on AD and its impact are discussed in section 15.2. State-of-the-art methods for AD detection using voxel-based morphometry (VBM), texture, shape and other methods are discussed in section 15.3. Section 15.4 gives the details of the dataset used in this work and explains the proposed methodology. Section 15.5 includes the experiments and results. Finally, section 15.6 concludes the work.

15.2 Regional atrophy and Alzheimer's disease

Regional atrophy is a major biomarker in the diagnosis of AD. Different sub-regions constituting the WM and GM regions are responsible for different functions of the brain from control to movement. In comparison to healthy controls (HCs), patients with AD show high levels of GM volume reduction in the several regions including the hippocampus, insula, amygdala, thalamus, cingulate gyrus, parietal lobule, middle occipital gyrus, caudate gyrus, para-hippocampal gyrus, temporal gyrus and frontal gyrus [39]. WM reductions were mainly apparent in the temporal lobe, including the para-hippocampal, superior temporal, sub gyrus and corpus callosum. In addition, some reductions are observed in the the WM regions including the precentral gyrus, superior/inferior frontal gyrus, cingulate fasciculus and the sub-gyral in the parietal lobe [31]. Based on neuroimaging studies of the functioning of important GM and WM regions, each brain region is responsible for different functions of the body and any damage or difference in the volume of these regions results in the loss of brain functionalities. The hippocampus region is most important for consolidation of information, which affects short- and long-term memory. The amygdala plays a primary role in decision-making, emotional reactions and consolidation of memory. The para-hippocampal cortex (PHC) is also an important region and is responsible for cognitive processes. Reduction in brain MRI volume of the temporal lobe along the olfactory regions, and the thalamus, insula and fusiform are also considered as biomarkers for AD detection.

Various automatic methods have been applied in the literature using hippocampal volumes and structural alterations [40]. Entorhinal cortex measures have been used in studies for AD and normal control discrimination [41]. Temporal lobe and para-hippocampal cortex measures have also been used for distinguishing AD and normal subjects [42]. It seems that a single region of the brain is not enough for detection of this disease. Some studies showed that simultaneous analysis of different brain regions may help in detecting AD patients [43, 44]. Considering this view, in our study brain MRIs are segmented into 116 sub-regions using atlas-based segmentation. These segmented regions are used for classifying AD patients. Features from different regions of the brain help in understanding the behavior of different regions for atrophy detection. Also, important dominant regions identified by region selection methods are used for the classification task.

15.3 Related works

Most of the works for distinguishing AD subjects and healthy subjects are based on volume-based changes in GM regions and other structural changes in various brain regions. The different proposed methods include various feature extraction methods, such as VBM, texture features, shape features and cortical thickness-based methods. Also, methods have been based on specific regional atrophy in addition to full brain 2D/3D images.

15.3.1 VBM based methods

In the literature AD detection is based on either ROI based measurements or VBM based methods for evaluating morphometric changes. VBM based studies capture sensitive information and are capable of identifying minute details and small-scale regional differences in GM regions. Thus, it is quite a popular technique for studying GM changes in AD. A few works discussing the use of VBM based studies are discussed below.

Beheshti and Demirel [45] evaluated the probability distribution function (PDF) from VBM generated features for performing feature selection and classification using a SVM. In another study [46], different feature selection methods were implemented, namely the statistical dependency (SD), mutual information (MI), information gain (IG), Gini index (GI), Pearson correlation coefficient (PCC), *t*-test-score (TS) and Fisher criterion (FC) were used for feature ranking using classification error. The performance of these feature selection techniques was evaluated using an SVM classifier. Mishra *et al* [47] proposed a CAD system which used the top regions selected by *t*-test statistics for volumetric feature extraction. The classification of these features was explored using multiple classifiers. Khedher *et al* [48] used partial least square (PLS) and principal component analysis (PCA) on WM and GM regions and performed classification using an SVM. Ahmed *et al* [49] computed circular harmonic functions (CHF) on the hippocampus and posterior cingulate cortex regions of the brain for feature extraction. The extracted features were quantized using the bag of visual words (BOVW) approach to create a signature as single feature. Krashenyi *et al* [50] applied a feature ranking procedure on the segmented 24 regions of the brain and performed classification using a fuzzy inference system.

15.3.2 Texture analysis based methods

Medical image analysis is very useful in the healthcare field as it can assist doctors/pathologists in diagnosing neurodegenerative diseases using non-invasive methodologies. Several methods have been proposed for the identification of AD using morphological, color and texture features [37]. The spatial arrangement of the gray levels and their statistical and structural relationships with other pixels results in important information known as texture. Texture is usually described using attributes such as coarse, random, granular, filiform, ordered, fine-grained, punctuating, etc. The characteristics of texture are evaluated using co-occurrence matrices, run-length matrices, randomness, entropy, contrast, homogeneity, etc. Many methods for feature extraction and texture analysis are based on statistics of different orders [51]. As texture analysis is useful for analysing small changes and structural patterns, it has been widely used in AD diagnosis. Sorensen *et al* [52] showed that texture can easily depict conversion to dementia independent of volume changes due to its ability to capture small details. Nanni *et al* [53] combined different texture descriptors and VBM features obtained from slices of 3D MRI images and ensemble classification was performed using an SVM classifier. Luk *et al* [54] performed texture analysis using 3D VGLCM-TOP features for extracting eight texture measures and combined those with the hippocampal volume for better results. Zhang *et al* [13] performed texture analysis on 3D MRI by evaluating the histogram, gradients, gray-level co-occurrence matrix (GLCM) and run-length matrix (RLM) as features from a circular ROI obtained from the hippocampus and entorhinal cortex region. Vaithinathan and Parthiban [55] performed 2D texture analysis on 3D MRI imaging considering three axes: axial, sagittal and coronal. The texture features used for analysis were the central moments, homogeneity, contrast, inverse different moment and entropy in segmented ROI acquired using a rough ROI (RROI) technique. These features were subjected to multiple feature selection methods and were classified using the random forest, SVM and *k*-nearest neighbors (*k*-NN) classifiers. Wang *et al* [56] used stationary wavelet transform (SWT) for extracting features from MR brain images and applied PCA for reduction of the SWT coefficients. Classification was performed using a feed-forward neural network (FNN) and the hybridization methods of particle swarm optimization (PSO) and artificial bee colony (ABC). Torabi *et al* [57] performed feature extraction using GLCM in 336 features and used a feed-forward artificial neural network (ANN) for classification of AD and HCs. Murcia *et al* [58] evaluated GLCM measures from different cortical and subcortical structures in the brain. Jha and Kwon [59] applied a curvelet and wavelet transform with PCA on MRI images. These features were classified using *k*-NN and ANN. El-Dahshan *et al* [60] applied DWT along with PCA and used ANN and *k*-NN classifiers. Zhang *et al* [61] performed discrete wavelet analysis with PCA and used an ANN as the classifier. Hu *et al* [62] used multiscale wavelet features implemented on GM images and the hippocampus region of the brain, and performed classification using an SVM classifier. Chaplot *et al* [63] detected AD patients from normal patients using DWT coefficients as the feature and performed classification using SVM and neural network classifiers. Zhang and Wang [64] designed a model for AD prediction which used displacement field estimation between AD and normal brains using an SVM, twin SVM (TWSVM) and generalized eigenvalue proximal SVM (GEPSVM). Zhang *et al* [28] used 3D wavelet features and triplet features. The features were reduced by PCA and classified using kernel SVM with particle swarm optimization with time varying acceleration coefficients (PSOTVAC). Alam *et al* [65] used dual-tree complex wavelet transforms (DTCWTs) and evaluated principal coefficients from trans-axial slices of MRIs. Simoes *et al* [66] evaluated local binary patterns (LBPs) for performing 3D texture analysis on the image patches obtained from the whole brain and used an ensemble classifier. Zhou *et al* [67] used wavelet entropy for feature extraction and a naïve Bayes classifier. Ortiz *et al* [68] proposed a texture-based method for selecting important ROIs using a self-organizing map (SOM) and performed classification on a reduced set of discriminative features for classification.

15.3.3 Shape analysis based methods

Shape descriptors are important for differentiating contour and region based methods. These methods evaluate specific features which give detailed information about the shape. Contour shape features define the shape based on contour lines in various representations, such as centroid distances, contour moments, shape signatures, spectral transforms, scale space methods and structural representation. In contrast, region based techniques describe the shape depending on all points in the shape and represent global features. Region based shape features are represented by moment invariants, orthogonal moment descriptors such as Zernike and Legendre descriptors, complex shape moments and generic Fourier descriptors. In AD detection, shape feature analysis is limited, as discussed here. To model the shape of the hippocampus, Gerardin *et al* [69] used spherical harmonics (SPHARM) coefficients as features which were classified using SVM. A bagging strategy was used for finding the most relevant features for classification. Gorji and Haddadnia [70] used pseudo-Zernike moments on structural MRI for the diagnosis of mild cognitive impairment (MCI) individuals from AD and normal people. Two types of ANNs are employed using learning vector quantization networks. Shakeri *et al* [71] computed triangulated surface meshes from segmented hippocampus regions and used a spectral matching method. A deep learning auto-encoder was computed on the vertex coordinates of the mesh models to create low-dimensional feature representation. Kim *et al* [72] used multiscale shape descriptors with non-Euclidean setting on wavelet features for the analysis of cortical thickness.

15.3.4 Other methods

Apart from voxel-based morphometry texture and shape features, other morphometry features have also been used in AD detection. Wolz *et al* [73] used four different feature extraction techniques: cortical thickness, hippocampal volume, tensor based morphometry (TBM) and manifold-based learning for evaluating features and then the extracted features were classified using linear discriminant analysis (LDA) and an SVM. The highest accuracy for AD versus HC classification was obtained by combining all extracted features and using the LDA classifier. Querbes *et al* [74] computed cortical thickness from brain MRI and divided it into 22 regions. Further, the normalized cortical index was evaluated from these regions for the classification of AD. Westman *et al* [9] used the FreeSurfer tool for generating regional volume and regional cortical thickness features. Further, classification was performed using orthogonal projections to latent structures (OPLS). Aguilar *et al* [11] used FreeSurfer on MRI data to evaluate regional cortical thickness and volumetric features, and applied four classifiers, i.e. OPLS, decision trees, ANN and SVM. Zhou *et al* [75] also used the FreeSurfer tool for evaluating 55 volumetric features. Features selected by incremental error analysis are combined with mini-mental score examination (MMSE) and classification is performed using an SVM. Papakostas *et al* [12] used VBM and deformation-based morphometry (DBM) with lattice computing (LC) based classifiers on brain MRI. Zhang and Wang [64] used displacement field (DF) estimation and PCA as feature reduction methods for distinguishing normal brains from AD brains. Three SVM variant classifiers were used for performing classification.

15.4 Materials and methods

15.4.1 Dataset

The dataset acquired for this chapter is available from the Alzheimer's Disease Neuroimaging Initiative (ADNI) (www.loni.ucla.edu/ADNI). These data were provided through a collaboration between the US National Institute on Aging (NIA), the US Food and Drug Administration (FDA), the US National Institute of Biomedical Imaging and Bioengineering (NIBIB), non-profit organizations and private pharmaceutical companies. The basic objective of ADNI is to combine modalities such as PET, sMRI and biological markers with neuropsychological and clinical analysis for measuring progression and performing identification of MCI and AD. Researchers and clinicians are trying to develop monitoring and treatment systems for determining the early progression of AD. Such systems can identify the sensitive information of specific markers, and also helps in reducing the time and cost of clinical trials. For further details, see http://www.adni-info.org.

The ADNI database contains 1.5 T and 3.0 T T1w MRI scans for HC and AD patients noted at different points of time. In this study 417 T1-weighted MRI images, comprising 229 HCs and 188 AD images, are used for the experiments [72]. A screening test of AD and HC subjects, given as ADNI 1, is used in this work. The demographics of the dataset used are given in table 15.1. General criteria (inclusion/exclusion) are as follows:

1. HC subjects have MMSE scores ranging from 24 to 30 (inclusive) and a clinical dementia rating (CDR) of 0. These patients are non-depressed, non-MCI and non-demented.

Table 15.1. Data for the patients with AD and the HC subjects used in this study.

Diagnosis	Number of subjects	Age group	Gender (M/F)	MMSE	CDR 0/0.5/1
AD	188	75.3 ± 7.5	99/89	23.28 ± 2	0.75 ± 0.25 0/95/93
HC	229	75.97 ± 5	119/110	29 ± 1	0.00 ± 0.00 229/0/0

2. AD subjects have MMSE scores ranging from 20 to 26 (inclusive), a CDR of 0.5 or 1.0, and meet the NINCDS/ADRDA [76] criteria for probable AD.

Selection criteria. The selection query for the database used is project = 'ADNI 1' AND Study = 'ADNI Screening' AND slice thickness = '1.2' AND weighing = 'T1'. It is noted that there are no significant differences in the age and gender ranges in the two groups (HC and AD). All geometric distortions caused due to gradient non-linearity, bias field and intensity inhomogeneity are removed using image corrections such as GradWarp and N3.

15.4.2 The proposed methodology

The pipeline of the proposed methodology given in figure 15.2 depicts five major steps: pre-processing of brain MRI, region segmentation, region selection, feature extraction, feature selection and classification. The results obtained using the selected features from the selected regions are compared to state-of-art studies, which reveals the impact of regional atrophy on AD disease identification.

15.4.2.1 MRI data pre-processing

It is important to identify structural changes in brain MRI for the study of neurodegenerative diseases such as AD. It is evident that various patterns of atrophy are related with specific diseases [77]. VBM is used to perform automated analysis. While performing statistical analysis on multiple MRI scans of different individuals, these images need to be registered with the other subjects' MRIs through spatial normalization.

Statistical parameter mapping (SPM) performs spatial normalization, inhomogeneity correction and segmentation of a brain image into important regions: GM, WM and CSF. These normalized images are registered with standard Montreal neurological imaging (MNI) templates using linear affine transformation and a non-linear deformation using high-dimensional DARTEL normalization. DARTEL processing is useful for performing inter-subject alignment by modeling brain shape. This high-dimensional warping process results in aligning of WM with WM regions or GM with GM regions among all the MRIs simultaneously. Smoothing is performed on these DARTEL warped GM images using an 8 mm full width half maximum (FWHM) Gaussian kernel [78].

Figure 15.3 shows the results of pre-processing steps performed on 3D MRI. Smoothing results in the normally distributed data moving closer to a Gaussian field model, which reduces inter-subject variability. Also, it increases the sensitivity and reduces variance across subjects, which results in increasing validity of parametric tests to detect structural variabilities. These pre-processing steps are employed on brain MRI from ADNI 1 using SPM version 12 (http://www.fil.ion.ucl.ac.uk/spm) and the VBM (http://dbm.neuro.uni-jena.de/vbm) toolbox with default settings [78]. The pre-processing steps on the 3D MRI are shown in figure 15.3. As a result of this step, the GM volumes are obtained with voxels of dimension $120 \times 145 \times 120$, where the voxel size is 1.5 mm $\times$ 1.5 mm $\times$ 1.5 mm.

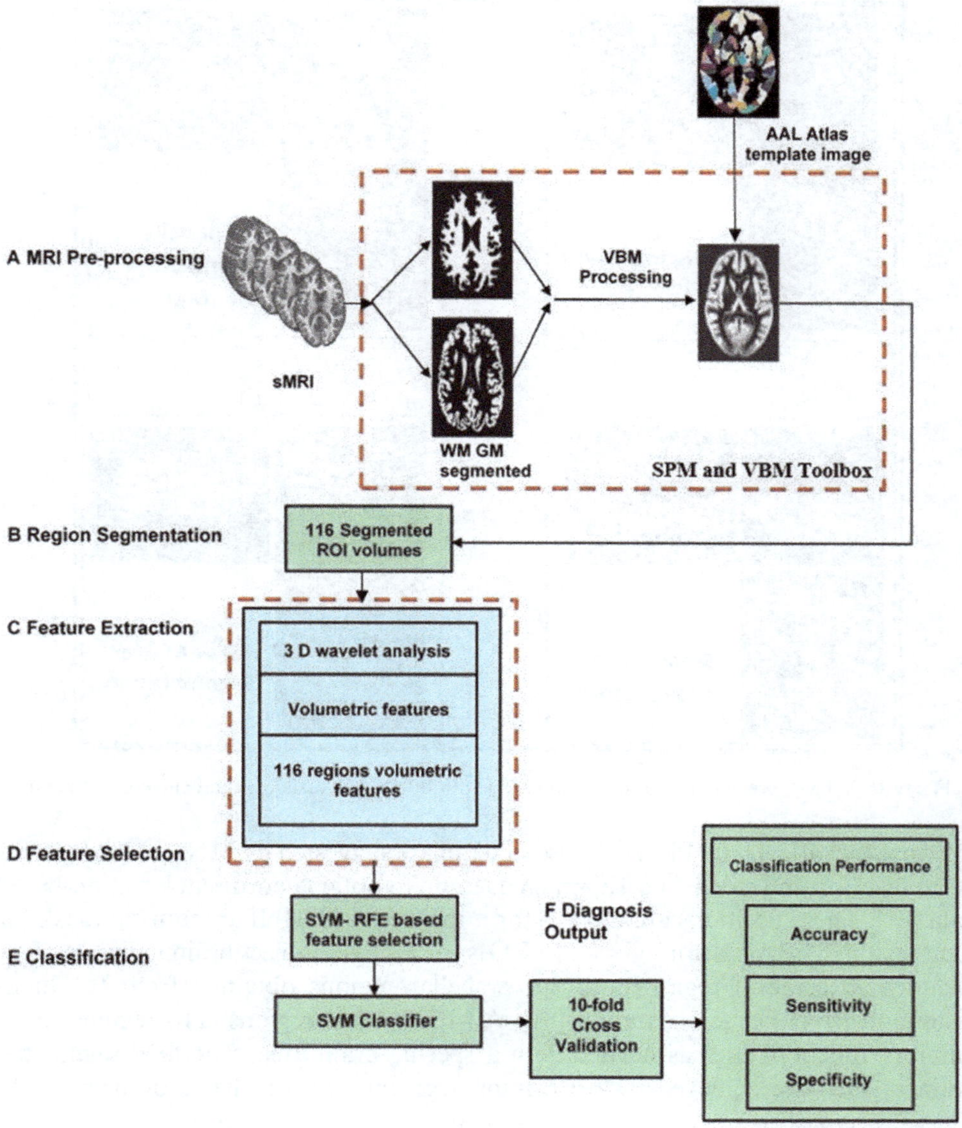

Figure 15.2. Pipeline of the proposed study for AD detection.

15.4.2.2 Region segmentation

Brain MRIs are segmented into different brain regions to analyse their roles for AD detection. Brain MRIs can be segmented into ROIs either through manual segmentation by expert neurologists or by atlas-based segmentation. In atlas-based segmentation, a given image is registered with an atlas image and anatomized labels are given to the target brain image [79]. For performing region segmentation, the WFU PickAtlas3 tool is used to anatomize GM images into respective ROIs using a standard automated anatomical labeling (AAL) atlas (http://fmri.wfubmc.edu/

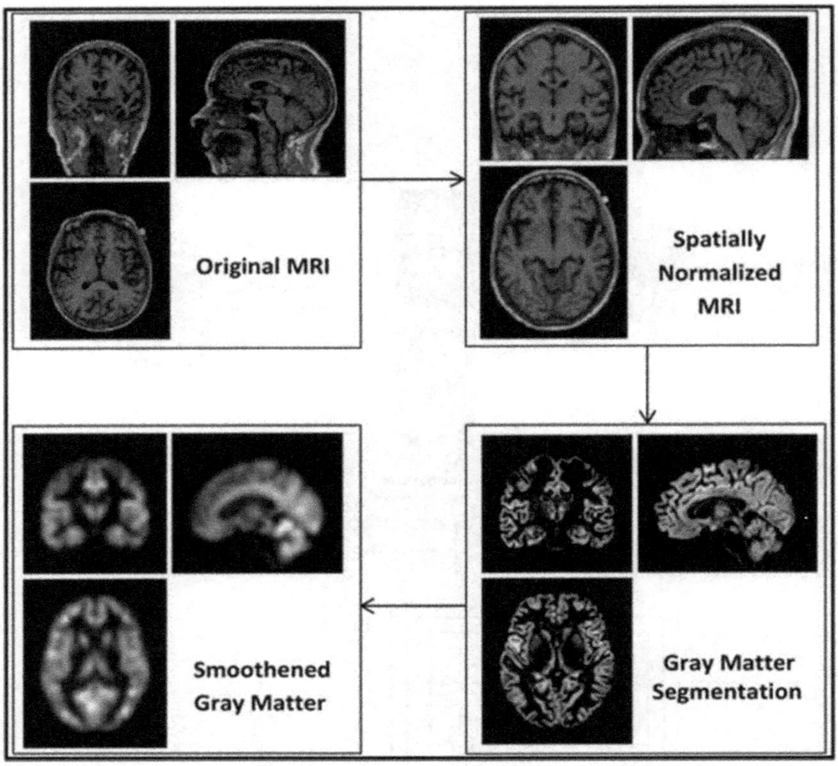

Figure 15.3. Pre-processing performed on 3D MRI with three views (axial, sagittal and coronal axes).

software/pickatlas) [80]. PickAtlas uses a template suggested by MNI for performing normalization and selects the Talairach daemon to obtain coordinate position-based data [71]. These positions are used to form the standard MNI anatomical mask for each region. The AAL atlas gives 116 ROIs for a single subject brain template. This includes 90 cerebral regions and 26 cerebellar regions obtained from the high-resolution MNI via segmentation [81]. All these brain regions (116 regions) have different functionalities associated with a specific brain area. The final segmented brain regions are achieved by performing segmentation on the smoothened GM density volumes.

15.4.2.3 Region selection

Region selection is an important process in the pipeline for the detection of AD subjects and healthy controls. The aim of this step is to rank 116 segmented regions. Region selection criteria help in finding informative/useful regions from which features can be extracted for further evaluation of AD subjects and HCs. Also, the differences in modality, dataset and morphometry results in the emergence of new biomarkers. Previously, we proposed a method based on statistical analysis for finding dominant biomarkers from 116 atlas-based segmented brain MRI regions [47]. This region ranking method is based on evaluating *t*-test statistics among different regions. The obtained ranking values are used for obtaining the most dominant biomarkers. These selected brain regions are expected to have the most discriminating features.

15.4.2.4 Feature extraction

Texture features are extracted to analyse the whole brain as well as its regions. Wavelets are powerful mathematical functions for performing feature extraction as they decompose data into different frequency components and then each component is studied with a resolution matched to its scale. It is used in the literature for performing texture analysis on MRI images. The main advantage of wavelets is that they provide localized frequency information about the function of a signal, which is very helpful in the classification task [6]. The wavelet transform has two important parameters, which are the basis function and the level of decomposition. In this work, 3D wavelet features are harnessed from the MRIs of different regions. Feature extraction using 3D wavelet transform, stationary wavelet transform, Haar wavelet, multiscale tight frames and dual-tree wavelet transform has been used previously in the literature, focusing on the whole brain or a specific region [28, 60–65]. In this work, texture analysis using 3D-DWT is performed on brain regions obtained through atlas-based segmentation. The volumetric features of all the sub-band coefficients of 3D-DWT are obtained as feature vectors. Volumetric features capture statistical information of the brain volume, thus providing insight on the 3D volume in the reduced form. The details of the 3D-DWT performed on brain regions and for volumetric feature extraction in this work are given in the following.

3D wavelet transform

Understanding that wavelet transforms [82] are able to extract features at different scales from signals and are useful for texture analysis in 3D images, 3D-DWT is performed on 3D MRI images in this work. DWT is a powerful implementation of the wavelet transform (WT) using dyadic scales and positions. It is seen that most of the AD classification methods reported in the literature have used 2D-DWT, which does not consider the information of all dimension of slices of 3D MRI [52]. It is observed in the literature that 3D-DWT outperforms 2D-DWT for volumetric data [84]. Figure 15.4 shows 3D-DWT performed on brain MRI.

In 1D-DWT, the signal is decomposed into approximation and detail coefficients using low-pass and high-pass filters, denoted as functions L and H, respectively, in figure 15.4. Down-sampling as represented by the downward arrow separates the low frequency and high frequency sub-bands. To summarize, first the low-pass and high-pass filters are applied along the x-dimension of the 3D image to obtain the sub-bands L and H. After that, these filters are applied along the y-dimension, resulting in LL, LH, HL and HH images. The approximation component of the image is represented by the LL sub-band and the detailed components are represented by the HH, HL and LH sub-bands. 3D-DWT is obtained by expanding 2D-DWT for a third spatial direction, i.e. the z-direction, to consider volumetric data, as shown in figure 15.4. 2D-DWT gives data in rows and columns, while 3D-DWT adds the data of slice information. Three-dimensional wavelet transform

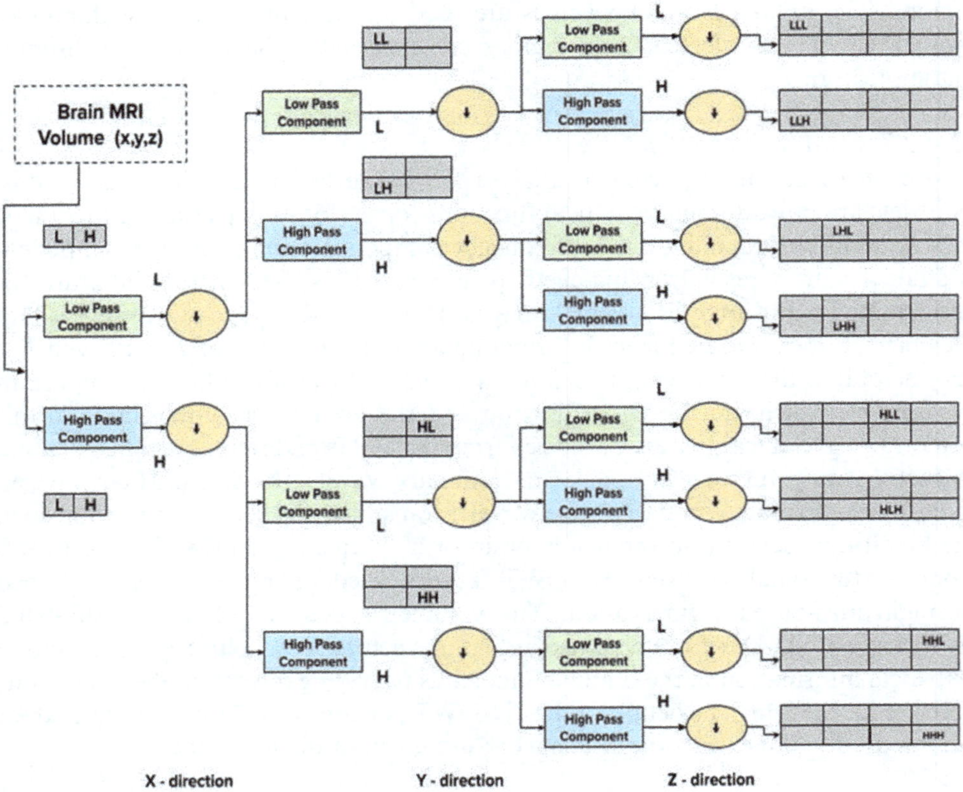

Figure 15.4. 3D-DWT. Reprinted from [83], copyright 2016, with permission from Elsevier.

results in eight sub-bands, named LLL, LLH, LHL, LHH, HLL, HLH, HHL and HHH. The naming convention is easy to follow as, for example, the LLH band is interpreted as a high-pass sub-band from directional filtering of low-pass along the *x*-direction, low-pass along the *y*-direction and high-pass along the *z*-direction. More compact and coarser approximation components of volumetric images can be obtained by increasing the level of decomposition of the 3D brain MRI images. Three-level 3D-DWT for a 3D image is shown in figure 15.5.

A three-level 3D-DWT with the Daubechies basis function is applied on each 3D brain MRI. Daubechies are compact orthogonal wavelets, that are compact in both the temporal and spectral domains. This gives 22 sub-bands, including eight sub-bands of level 3 (L3), seven sub-bands of level 2 (L2) and seven sub-bands of level 1 (L1) for each subject.

Volumetric features

Volumetric features consider statistical information of the region under consideration, thus resulting in reduction in dimensionality in comparison to the dimension of the extracted ROI [47]. By performing volumetric feature extraction, each ROI can be reduced to a smaller dimension and give information for each anatomical

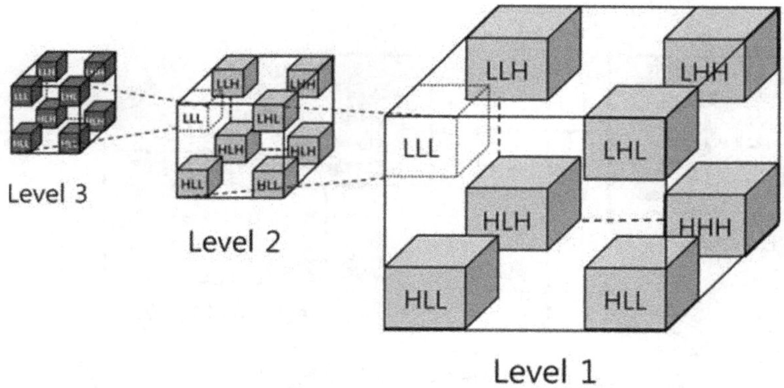

Figure 15.5. Three-level 3D-DWT. Reprinted from [85].

region. Statistics calculated on the wavelet coefficients have often been used in biomedical signal processing. Thus, feature extraction as shown in figure 15.6 is performed for each ROI wavelet feature, i.e. different brain regions, by considering all the wavelet features obtained by 22 sub-bands of that region. In the previous step, the 3D wavelet is evaluated on the MRI image which gives 22 sub-bands for each ROI. In order to retain the information of all the brain regions, volumetric features (standard deviation (SD), skewness (skw), kurtosis (kur), energy (en) and Shannon entropy (shen)) are evaluated from each of these sub-bands.

Let V_r (x, y, z) be the pixel value at location (x, y, z) for a region r, $\overline{m_r}$ be the mean of all gray values of region r and N_r be the number of the pixels present in the region r. These volumetric features are described as follows:

$$\mathrm{SD}_r = \sqrt{\frac{1}{N_r}\sum_x\sum_y\sum_z(V_r(x, y, z) - \overline{m_r})^2} \tag{15.1}$$

$$\mathrm{skw}_r = \frac{\frac{1}{N_r}\sum_x\sum_y\sum_z(V_r(x, y, z) - \overline{m_r})^3}{\mathrm{SD}_r^{\,3}} \tag{15.2}$$

$$\mathrm{kur}_r = \frac{\frac{1}{N_r}\sum_x\sum_y\sum_z(V_r(x, y, z) - \overline{m_r})^4}{\mathrm{SD}_r^{\,4}} \tag{15.3}$$

$$\mathrm{en}_r = \sum_x\sum_y\sum_z V_r^2(x, y, z) \tag{15.4}$$

$$\mathrm{shen}_r = -\sum_x\sum_y\sum_z V_r^2(x, y, z)\log(1 + V_r^2(x, y, z)). \tag{15.5}$$

The volumetric images of each subject's 3D image are of dimension 120 × 145 × 120 = 2 088 000. For each of the 22 sub-bands, the five defined statistical features (SD, skw, kur, en, shen) are extracted, resulting in a total of 110 volumetric feature. These features are evaluated for 116 regions, resulting in 127 600 features, and

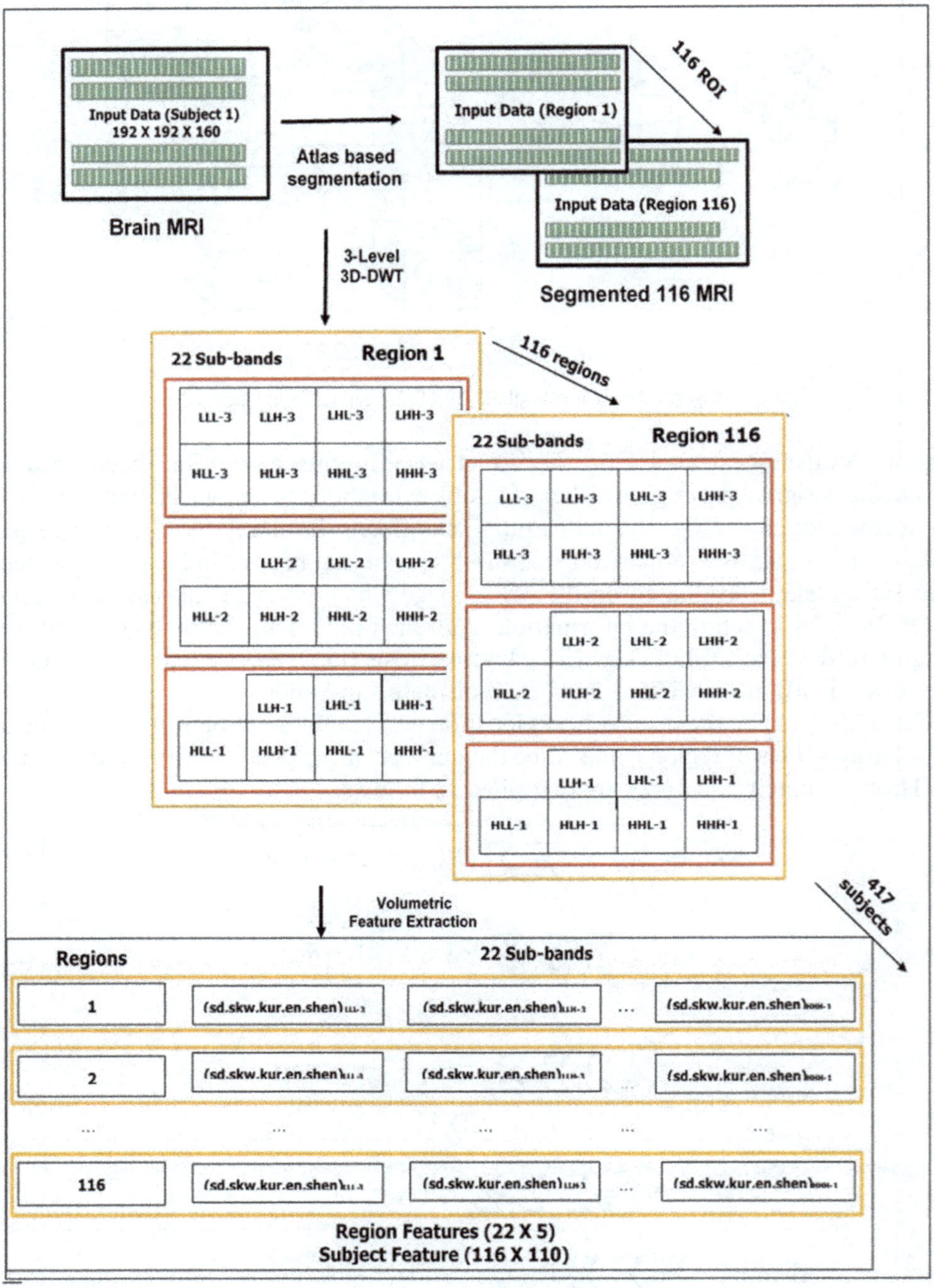

Figure 15.6. Feature extraction using 3D-DWT from 116 regions.

finally five important brain regions, resulting in a total of 550 features. Therefore, 550 volumetric features in total are obtained from the volume data for each subject.

15.4.2.5 Feature selection and classification

Feature selection is an important task which reduces the cardinality of the feature set based on predefined criteria to select a number of attributes based on their usefulness to complete a given purpose. SVM-RFE is one of the popular wrapper approaches for feature selection, proposed by Guyon *et al* [86]. SVM-RFE computes ranking weight vectors and uses weight vectors for ranking features. The algorithm is designed as follows. Initially, the dataset is used to train the SVM classifier. Then, based on this classification, ranking weights are allotted to each feature. Features with smaller weights are iteratively removed, resulting in a sorted list of all features based on their significance. Using a subset of selected ranked features to train the SVM classifier and using SVM classification accuracy as a criterion, an optimal feature set is obtained.

After obtaining an optimal set of features as the output of the feature selection method, the classification task is carried out using a supervised learning technique, the SVM. For completing this task, LIBSVM (www.csie.ntu.edu.tw/cjlin/libsvm/) is used. The settings used for performing the two-class classification are the defaults with a linear kernel.

15.5 Experiments and results

The experiments are carried out on the volumetric features obtained from 3D wavelet sub-bands of different brain regions obtained through VBM followed by atlas-based segmentation. SPM version 12 is used for performing VBM. For performing region segmentation WFU PickAtlas (v3.0.5b) is employed. Classification experiments are performed in MATLAB. The image processing and numerical computations are performed on a 64 bit system with a 2.30 GHz Intel®Xeon processor and 128 GB RAM.

First, a whole MRI is segmented into GM and WM regions. Further, the GM region is segmented into 116 regions obtained through atlas-based segmentation. Five experiments are performed to understand the importance of regional atrophy for the identification of AD. The experiments are performed on (i) voxel as features (VAFs) obtained from the GM and WM regions, (ii) volumetric features evaluated on 3D-DWT sub-bands obtained from all 116 regions, (iii) volumetric features evaluated on 3D-DWT sub-bands obtained from the top five selected regions and (iv) features obtained after applying feature selection on the features of the top five selected regions. Finally, the obtained results are compared to those obtained with related methods. The performance parameters for classification as given in the literature and used here in the context of AD subjects and HC classification are defined as follows:

$$\text{Accuracy (ACC)} = (\text{TP} + \text{TN})/(\text{TP} + \text{TN} + \text{FP} + \text{FN}) \tag{15.6}$$

$$\text{Sensitivity (SEN)} = \text{TP}/(\text{TP} + \text{FN}) \tag{15.7}$$

Table 15.2. Results obtained with the five experiments performed for AD classification.

Methods	Accuracy	Sensitivity	Specificity
GM features (VAF)	76.80%	75.20%	78.50%
WM features (VAF)	64.20%	66.80%	61.70%
3D-DWT + volumetric features from 116 regions	79.75%	85.18%	75.30%
3D-DWT + volumetric features from top five regions	88.19%	84.01%	91.64%
3D-DWT + volumetric features from top five regions with SVM-RFE feature selection	90.06%	87.00%	92.00%

$$\text{Specificity (SPE)} = \text{TN}/(\text{TN} + \text{FP}), \tag{15.8}$$

where true positive (TP) is the number of AD subjects correctly classified as AD, true negative (TN) is the number of HC subjects correctly classified as HC, false negative (FN) is the number of AD subjects incorrectly classified as HC and false positive (FP) is the number of HC subjects incorrectly classified as AD. These performance parameters are evaluated through ten-fold cross validation to ensure reliable results and prevent over-fitting [87]. The data are conveniently divided into training and testing sets by applying a 9:1 ratio randomly. The training set consists of 373 subjects and the test set contains randomly selected 41 subjects. The average of these performance parameters is reported in the results. The results obtained from all the experiments are summarized in table 15.2 and discussed below.

15.5.1 Experiment 1: Voxel as features (VAF) obtained from GM and WM regions

The whole MRI is initially segmented into GM and WM regions using VBM based segmentation. These pre-processed GM and WM images are smoothed images obtain using DARTEL based segmentation. A feature vector is obtained by using all voxels from the GM and WM regions. In other words, the feature sets used in this experiment are VAFs with an image size of $120 \times 145 \times 120 = 2\ 122\ 945$. This resulted in a classification accuracy of 76.8% and 64.2% for GM and WM, respectively.

15.5.2 Experiment 2: Volumetric features evaluated on the 3D-DWT sub-bands obtained from all 116 regions

Atlas-based segmentation of the GM image gives 116 regions parcellated from the brain MRI. Initially, feature extraction is performed on all 116 regions. Five volumetric features evaluated on 22 wavelet sub-bands of a region results in 110 features for one region. Following the same method, 110 features are evaluated from each of the 116 regions. This gives $110 \times 116 = 12\ 760$ features for each subject in the dataset. The accuracy obtained with these features is 79.75%, which is low. Therefore, the obtained classification accuracy and other performance parameters need to be improved.

15.5.3 Experiment 3: Volumetric features evaluated on 3D-DWT sub-bands obtained from the top five regions

In order to understand the impact of regional atrophy in the detection of AD, 3D-DWT is performed on the important brain regions and then the volumetric features from these regions are extracted. The region ranking algorithm [47] employed in this work gives important brain regions based on their statistical analysis. The top regions of the ranked regions are verified by performing individual classification of these regions. The selected top regions are the hippocampus (left and right), amygdala (left), para-hippocampal (left) and temporal lobe (left). These five regions are also important according to medical theory. The classification accuracies achieved for these regions are shown in table 15.3. It is seen that the top-most region hippocampus-left gives the highest accuracy of 85.71% followed by the other regions. This is done to validate the performance of these regions in comparison to the other brain regions ranked by *t*-test analysis and is the reason to select these five features. Figure 15.7 shows a boxplot representing the accuracy achieved by these regions.

However, fusion of the features of these regions is employed to achieve higher classification accuracy as it can integrate information from the identified important regions. The observed performance using these top five regions is shown in table 15.2. The observed accuracy is 88.19% with a sensitivity of 84.01% and a specificity of 91.64%. It is to be noted that the feature size in this case is $5 \times 22 \times 5 = 550$, which is much less in comparison to the feature size in the previous experiment, but with improved performance.

15.5.4 Experiment 4: Features obtained after applying feature selection on the features of the top five selected regions

In order to use optimal features in comparison to all the obtained features, an SVM-RFE based feature selection algorithm is applied on the 550 features obtained from the top-most five regions. This results in the selection of optimal features by performing a recursive feature elimination method. The importance of the selected top regions hippocampus (left and right), amygdala (left), para-hippocampal (left) and temporal lobe (left) has already been discussed. The proposed CAD system gives a better performance for volumetric features evaluated on 3D-DWT sub-bands

Table 15.3. Classification accuracy for the top five regions.

Brain regions	Accuracy	Sensitivity	Specificity
Hippocampus-left	85.71	84.21	86.96
Hippocampus-right	80.95	78.94	82.60
Amygdala-left	80.43	73.83	85.93
Para-hippocampus-left	82.14	78.07	85.53
Temporal lobe-left	78.07	74.82	80.65

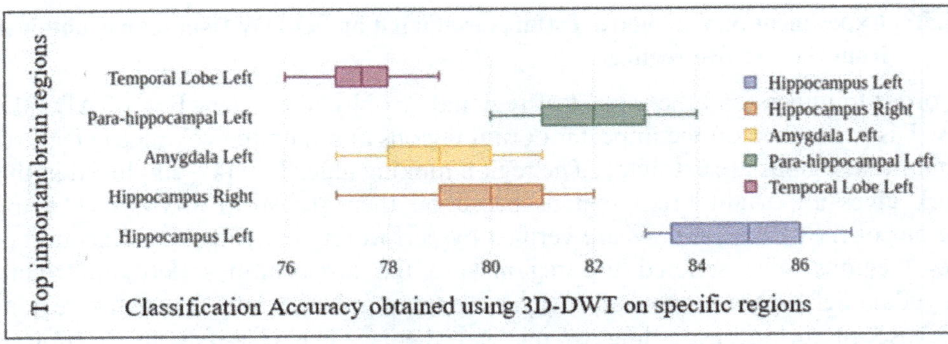

Figure 15.7. Boxplots showing classification accuracy obtained with the five selected regions.

Table 15.4. Performance comparison with existing methods.

Methods and reference	Accuracy	Sensitivity	Specificity
DWT + PCA + *k*-NN (El-Dahshan *et al* 2010) [60]	88.42%	92.16%	84.09%
LBP-TOP (Simoes *et al* 2014) [66]	84.00%	81.00%	89%
HCF + SVM (Ahmed *et al* 2014) [88]	83.71%	77.09%	88.2%
PCA + PLS + SVM (Khedher *et al* 2015) [48]	88.49%	85.11%	91.27%
PDF + SVM (Beheshti and Demirel 2015) [45]	89.65%	87.73%	91.57%
Mean + SD + FI (Krashenyi *et al* 2015) [50]	86.11%	81.39%	89.06%
Curvelet + PCA + K-NN (Jha and Kwon 2016) [59]	89.47%	94.12%	84.09%
Wavelet frame + SVM (Hu *et al* 2016) [62]	85.10%	84.57%	85.53%
3D GLCM (Murcia *et al* 2017) [58]	81.30%	77.50%	84.31%
VGLCM-TOP + hippocampal volume (Luk *et al* 2018) [54]	—	83.10%	92%
Volumetric features + SVM (Mishra *et al* 2018) [47]	89.15%	85.06%	92.53%
Fisher + RF (Vaithinathan and Parthiban 2019) [55]	87.39%	85.42%	88.81%
Texture descriptors + SVM (Nanni *et al* 2019) [53]	87.60%	84.10%	90.3%
Proposed method	**90.06%**	**87.00%**	**92.00%**

obtained from the top five regions. This gives 90.06% for accuracy, 87% for sensitivity and 92%for specificity. On comparing the results given in table 15.2, it can be seen that the features of the selected regions along with feature selection gives higher accuracy in comparison to the basic system using features from all 116 regions.

15.5.5 Performance comparison with state-of-art methods

Texture analysis based research for AD classification includes the use of GLCM, statistical features, voxel-based GLCM on three orthogonal planes (VGLCM-TOP), LBP and wavelet features. The current results are compared with some of these methods using similar datasets or similar features, as shown in table 15.4.

El-Dahshan *et al* [60] applied the DWT along with PCA and used the ANN and *k*-NN classifiers. They obtained a highest accuracy of 88.42% with the *k*-NN classifier. Simoes *et al* [66] evaluated LBP texture features on local image patches obtained from a 3D whole brain image. These features were classified using an ensemble classifier resulting in a classification accuracy of 84%. Ahmed *et al* [88] used the hippocampus and PCC regions of the MRI to extract HCF as features. The observed accuracy was 83.71% with the SVM classifier using the RBF kernel. Khedher *et al* [48] reported 88.49% accuracy with fused features extracted from the GM and WM regions. These features are extracted through VBM based segmentation and are reduced using PCA and PLS.

Beheshti and Demirel [45] performed probability distribution-based feature selection on VBM based features. Classification of these selected features was performed using an SVM showing an accuracy of 89.65%. Krashenyi *et al* [50] evaluated the mean and standard deviation from 24 brain regions and used a fuzzy inference system for classification to achieve 86.11% accuracy. Jha and Kwon [59] applied a curvelet and wavelet transform with PCA on the MRI images. These features are classified using the *k*-NN and ANN classifiers for diagnosis of AD and achieved a classification accuracy of 89.47%. Hu *et al* [62] used multiscale wavelet feature extraction on the MRI data from the hippocampus region and achieved 84.13% accuracy. Murcia *et al* [58] evaluated GLCM measures from different cortical and subcortical structures in the brain. Voxel features extracted from a whole MRI image are selected through a *t*-test, resulting in 81.3% classification accuracy for AD. At the same time, voxel features extracted from the hippocampus region only resulted in 75.1% accuracy. Luk *et al* [54] evaluated eight texture measures using 3D GLCM features and hippocampal volume. For AD and HC classification, this method achieved 83.1% and 92% sensitivity and specificity, respectively. Mishra *et al* [47] used volumetric features obtained from the top regions selected by *t*-test statistics and performed classification using multiple classifiers. The highest accuracy reported was 89.15% using an SVM classifier. Vaithinathan and Parthiban [55] evaluated central moments, homogeneity, contrast, inverse different moment and entropy features extracted from a segmented ROI acquired using the RROI technique. This method reported a highest accuracy of 87.39% with Fisher features classified by using the random forest as well as ANN classifier. Nanni *et al* [53] combined texture descriptors and VBM features using an ensemble of the SVM classifier. They achieved a highest accuracy of 87.60%.

Compared to these approaches, the proposed automated system uses 3D-DWT followed by extraction of volumetric features. The highest accuracy is achieved using dominant atrophy regions obtained though region ranking. The features extracted from these selected regions are subjected to feature selection using SVM-RFE and classified through SVM. With a highest accuracy of 90.06%, the results obtained in this work are highly competitive compared to the state-of-art methods. The proposed method performs better than the other methods using texture features, as shown in table 15.4.

15.6 Conclusions

In this chapter, the chronic neurodegenerative disease AD and its impact on different brain regions is discussed. Efforts have been made by various researchers to automate AD detection by using image processing and machine learning techniques. Different image modalities such as MRI, rs-MRI, PET and SPECT are available for brain images. Brain MRI based approaches are more popular due to the availability of MRI and its ability to provide good soft tissue contrast. A whole MRI can be segmented into different regions. A detail discussion of the importance of these region is presented. Each brain region performs specific tasks and is responsible for proper functioning of the human body. Structural alterations to these regions cause different brain disorders/diseases. Brain segmentation is performed to harness structural changes in the brain. This chapter also discusses the techniques and pipelines developed for AD detection using different types of feature extraction techniques. The feature extraction methods for AD detection are mainly categorized into VBM methods, texture-based methods and shape-based methods.

In this chapter, the focus has been to perform texture analysis on different regions of the brain for AD detection. It is evident that using 3D-DWT provides minute details of the MRI and it is found to be useful in combination with volumetric features. This work focuses on using important imaging biomarkers and using selected region features for AD and HC classification. To perform this task 3D wavelet analysis is performed on 3D MRI. In order to achieve this, brain MRI images are subjected to pre-processing and finally segmenting into the WM and GM brain regions. For attaining a microscopic view of regional atrophy, the 3D wavelet transform followed by volumetric features are evaluated on the atlas-based segmentation of GM regions, which results in 116 regions obtained for analysis. 3D-DWT performed on a region results in 22 sub-bands. Five volumetric features are obtained from each band and the resulting features are fused. These 116 regions are observed and the top five regions with high regional atrophy are found to be significant for AD detection. The regions used in the proposed method are important biomarkers for the identification of AD and it this evident through biological research and medical science as well [78]. Further, instead of considering five volumetric features of all 22 bands of these selected regions, feature selection is performed to obtain higher accuracy. Finally, features extracted from the top five regions and selected through SVM-RFE are classified through an SVM. The standard ADNI dataset (188 AD/229 normal) is used for experimentation. A performance comparison shows that the proposed method outperforms the state-of-art methods based on texture analysis.

Acknowledgments

This research is supported by the Collaborative Research Scheme (CRS) of TEQIP III, a Government of India project assisted by the World Bank, with CRS Project ID 1-5739421811.

The data used in this study (http://www.adni-info.org) are a collection funded by the Alzheimer's Disease Neuroimaging Initiative (ADNI) (National Institutes of

Health Grant U01 AG024904) and DOD ADNI (Department of Defense award number W81XWH-12-2-0012).

We also acknowledge the support provided by Dr Iman Beheshti, Research Fellow at CERVO Brain Research Center, Mental Health University Institute of Quebec, Canada.

References

[1] Mohs R C, Rosen W G and Davis K L 1983 The Alzheimer's disease assessment scale: an instrument for assessing treatment efficacy *Psychopharmacol. Bull.* **19** 448

[2] DeKosky S T and Marek K 2003 Looking backward to move forward: early detection of neurodegenerative disorders *Science* **302** 830–4

[3] Fan Y, Batmanghelich N, Clark C M and Davatzikos C 2008 Spatial patterns of brain atrophy in MCI patients, identified via high-dimensional pattern classification, predict subsequent cognitive decline *Neuroimage* **39** 1731–43

[4] Alzheimer's Association 2015 Alzheimer's Disease and Dementia http://alz.org/ (Accessed: 5 Apr 2015)

[5] Spillantini M G and Goedert M 1998 Tau protein pathology in neurodegenerative diseases *Trends Neurosci.* **21** 428–33

[6] Weiner M W *et al* 2015 The Alzheimer's disease neuroimaging initiative: a review of papers published since its inception *Alzheimer's Dement.* **11** S1–68

[7] Bron E E *et al* 2015 Standardized evaluation of algorithms for computer-aided diagnosis of dementia based on structural MRI: the CADDementia challenge *Neuroimage* **111** 562–79

[8] Wei T Y, Fu Y, Chang K H, Lin K J, Lu Y J and Cheng C M 2018 Point-of-care devices using disease biomarkers to diagnose neurodegenerative disorders *Trends Biotechnol.* **36** 290–303

[9] Westman E, Muehlboeck J S and Simmons A 2012 Combining MRI and CSF measures for classification of Alzheimer's disease and prediction of mild cognitive impairment conversion *Neuroimage* **62** 229–38

[10] Davatzikos C, Bhatt P, Shaw L M, Batmanghelich K N and Trojanowski J Q 2011 Prediction of MCI to AD conversion, via MRI, CSF biomarkers, and pattern classification *Neurobiol. Aging* **32** 2322.e19–27

[11] Aguilar C *et al* 2013 Different multivariate techniques for automated classification of MRI data in Alzheimer's disease and mild cognitive impairment *Psychiatry Res. Neuroimaging* **212** 89–98

[12] Papakostas G A, Savio A, Graña M and Kaburlasos V G 2015 A lattice computing approach to Alzheimer's disease computer assisted diagnosis based on MRI data *Neurocomputing* **150** 37–42

[13] Zhang J, Yu C, Jiang G, Liu W and Tong L 2012 3D texture analysis on MRI images of Alzheimer's disease *Brain Imaging Behav.* **6** 61–9

[14] Lai Z and Deng H 2018 Medical image classification based on deep features extracted by deep model and statistic feature fusion with multilayer perceptron *Comput. Intell. Neurosci.* **2018** 2061516

[15] Xiao Z, Ding Y, Lan T, Zhang C, Luo C and Qin Z 2017 Brain MR image classification for Alzheimer's disease diagnosis based on multifeature fusion *Comput. Math. Methods Med.* **2017** 1952373

[16] Remes J J *et al* 2013 On applicability of PCA, voxel-wise variance normalization and dimensionality assumptions for sliding temporal window sICA in resting-state fMRI *Magn. Reson. Imaging* **31** 1338–48
[17] Chaves R *et al* 2009 SPECT image classification based on NMSE feature correlation weighting and SVM *IEEE Nucl. Sci. Symp. Conf. Record* vol 1 pp 2715–9
[18] Wang Y *et al* 2004 Development of a PET/SPECT agent for amyloid imaging in Alzheimer's disease *J. Mol. Neurosci.* **24** 55–62
[19] Stoeckel J *et al* 2001 Classification of SPECT images of normal subjects versus images of Alzheimer's disease patients *Medical Image Computing and Computer-Assisted Intervention —MICCAI 2001 Lecture Notes in Computer Science* vol 2208 (Berlin: Springer) pp 666–74
[20] Fung G and Stoeckel J 2007 SVM feature selection for classification of SPECT images of Alzheimer's disease using spatial information *Knowl. Inf. Syst.* **11** 243–58
[21] Dinesh E, Kumar M S, Vigneshwar M and Mohanraj T 2013 Instinctive classification of Alzheimer's disease using FMRI, PET and SPECT images *2013 7th Int. Conf. on Intelligent Systems and Control (ISCO),* pp 405–9
[22] Hanyu H, Sato T, Hirao K, Kanetaka H, Iwamoto T and Koizumi K 2010 The progression of cognitive deterioration and regional cerebral blood flow patterns in Alzheimer's disease: a longitudinal SPECT study *J. Neurol. Sci.* **290** 96–101
[23] Dukart J, Mueller K, Barthel H, Villringer A, Sabri O and Schroeter M L 2013 Meta-analysis based SVM classification enables accurate detection of Alzheimer's disease across different clinical centers using FDG-PET and MRI *Psychiatry Res. Neuroimaging* **212** 230–6
[24] Kuba R *et al* 2011 MRI-negative PET-positive temporal lobe epilepsy: invasive EEG findings, histopathology, and postoperative outcomes *Epilepsy Behav.* **22** 537–41
[25] Silveira M and Marques J 2010 Boosting Alzheimer disease diagnosis using PET images *Proc. of Int. Conf. on Pattern Recognition* pp 2556–9
[26] Colliot O *et al* 2018 Reproducible evaluation of classification methods in Alzheimer's disease: framework and application to MRI and PET data *Neuroimage* **183** 504–21
[27] Gray K R *et al* 2011 Regional analysis of FDG-PET for use in the classification of Alzheimer's disease *Proc. of Int. Symp. on Biomedical Imaging* pp 1082–5
[28] Zhang Y, Wang S, Phillips P, Dong Z, Ji G and Yang J 2015 Detection of Alzheimer's disease and mild cognitive impairment based on structural volumetric MR images using 3D-DWT and WTA-KSVM trained by PSOTVAC *Biomed. Signal Process. Control* **21** 58–73
[29] Cova T F G G, Pereira J L G F S C and Pais A A C C 2013 Is standard multivariate analysis sufficient in clinical and epidemiological studies? *J. Biomed. Inform.* **46** 75–86
[30] López M *et al* 2009 Multivariate approaches for Alzheimer's disease diagnosis using Bayesian classifiers *IEEE Nucl. Sci. Symp. Conf. Record* pp 3190–3
[31] Ziegler D A, Piguet O, Salat D H, Prince K, Connally E and Corkin S 2010 Cognition in healthy aging is related to regional white matter integrity, but not cortical thickness *Neurobiol. Aging* **31** 1912–26
[32] Kennedy K M *et al* 2009 Age-related differences in regional brain volumes: a comparison of optimized voxel-based morphometry to manual volumetry *Neurobiol. Aging* **30** 1657–76
[33] Smits L L *et al* 2014 Regional atrophy is associated with impairment in distinct cognitive domains in Alzheimer's disease *Alzheimer's Dement.* **10** S299–305
[34] Lemaître H, Crivello F, Grassiot B, Alpérovitch A, Tzourio C and Mazoyer B 2005 Age- and sex-related effects on the neuroanatomy of healthy elderly *Neuroimage* **26** 900–11

[35] Whitwell J L *et al* 2008 MRI correlates of neurofibrillary tangle pathology at autopsy: a voxel-based morphometry study *Neurology* **71** 743–9

[36] Dale A, Fischl B and Sereno M I 1999 Cortical surface-based analysis: I. segmentation and surface reconstruction *Neuroimage* **9** 179–94

[37] Humeau-Heurtier A 2019 Texture feature extraction methods: a survey *IEEE Access* **7** 8975–9000

[38] Groves A R *et al* 2012 Benefits of multi-modal fusion analysis on a large-scale dataset: life-span patterns of inter-subject variability in cortical morphometry and white matter microstructure *Neuroimage* **63** 365–80

[39] Taki Y *et al* 2011 Correlations among brain gray matter volumes, age, gender, and hemisphere in healthy individuals *PLoS One* **6** e22734

[40] Chupin M *et al* 2009 Fully automatic hippocampus segmentation and classification in Alzheimer's disease and mild cognitive impairment applied on data from ADNI *Hippocampus* **19** 579

[41] Matsuda H *et al* 2012 Automatic voxel-based morphometry of structural MRI by SPM8 plus diffeomorphic anatomic registration through exponentiated lie algebra improves the diagnosis of probable Alzheimer disease *Am. J. Neuroradiol.* **33** 1109–14

[42] Son J H, Han D H, Min K J and Kee B S 2013 Correlation between gray matter volume in the temporal lobe and depressive symptoms in patients with Alzheimer's disease *Neurosci. Lett.* **548** 15–20

[43] Gray K R *et al* 2012 Multi-region analysis of longitudinal FDG-PET for the classification of Alzheimer's disease *Neuroimage* **60** 221–9

[44] Hinrichs C *et al* 2011 Predictive markers for AD in a multi-modality framework: an analysis of MCI progression in the ADNI population *Neuroimage* **55** 574–89

[45] Beheshti I and Demirel H 2015 Probability distribution function-based classification of structural MRI for the detection of Alzheimer's disease *Comput. Biol. Med.* **64** 208–16

[46] Beheshti I and Demirel H 2016 Feature-ranking-based Alzheimer's disease classification from structural MRI *Magn. Reson. Imaging* **34** 252–63

[47] Mishra S, Beheshti I and Khanna P 2018 A statistical region selection and randomized volumetric features selection framework for early detection of Alzheimer's disease *Int. J. Imaging Syst. Technol.* **28** 302–14

[48] Khedher L, Ramírez J, Górriz J M, Brahim A and Segovia F 2015 Early diagnosis of Alzheimer's disease based on partial least squares, principal component analysis and support vector machine using segmented MRI images *Neurocomputing* **151** 139–50

[49] Ben Ahmed O, Mizotin M, Benois-Pineau J, Allard M, Catheline G and Ben Amar C 2015 Alzheimer's disease diagnosis on structural MR images using circular harmonic functions descriptors on hippocampus and posterior cingulate cortex *Comput. Med. Imaging Graph.* **44** 13–25

[50] Krashenyi I, Popov A, Ramirez J and Gorriz J M 2015 Fuzzy classification of Alzheimer's disease using statistical moments *2015 IEEE 35th Int. Conf. Electronics Nanotechnology, ELNANO 2015* pp 409–12

[51] Cai J H *et al* 2020 Magnetic resonance texture analysis in Alzheimer's disease *Acad. Radiol.* https://doi.org/10.1016/j.acra.2020.01.006

[52] Sørensen L *et al* 2016 Early detection of Alzheimer's disease using MRI hippocampal texture *Hum. Brain Mapp.* **37** 1148–61

[53] Nanni L, Brahnam S, Salvatore C and Castiglioni I 2019 Texture descriptors and voxels for the early diagnosis of Alzheimer's disease *Artif. Intell. Med.* **97** 19–26

[54] Luk C C *et al* 2018 Alzheimer's disease: 3-dimensional MRI texture for prediction of conversion from mild cognitive impairment *Alzheimer's Dement. Diagnosis Assess. Dis. Monit.* **10** 755–63

[55] Vaithinathan K and Parthiban L 2019 A novel texture extraction technique with T1 weighted MRI for the classification of Alzheimer's disease *J. Neurosci. Methods* **318** 84–99

[56] Zhang L, Song M, Liu X, Bu J and Chen C 2013 Fast multi-view segment graph kernel for object classification *Signal Process.* **93** 1597–607

[57] Torabi M, Ardekani R D and Fatemizadeh E 2006 Discrimination between Alzheimer's disease and control group in MR-images based on texture analysis using artificial neural network *2006 Int. Conf. on Biomedical and Pharmaceutical Engineering* pp 79–83

[58] Martinez-Murcia F J, Górriz J M, Ramírez J and Ortiz A(Alzheimer's Disease Neuroimaging Initiative) 2016 A spherical brain mapping of MR images for the detection of Alzheimer's disease *Curr. Alzheimer Res.* **13** 575–88

[59] Jha D and Kwon G-R 2016 Alzheimer disease detection in MRI using curvelet transform with *k*-NN *J. Korean Inst. Inf. Technol.* **14** 121

[60] El-Dahshan E S A, Hosny T and Salem A B M 2010 Hybrid intelligent techniques for MRI brain images classification *Digit. Signal Process.* **20** 433–41

[61] Zhang Y, Dong Z, Wu L and Wang S 2011 A hybrid method for MRI brain image classification *Expert Syst. Appl.* **38** 10049–53

[62] Hu K, Wang Y, Chen K, Hou L and Zhang X 2016 Multi-scale features extraction from baseline structure MRI for MCI patient classification and AD early diagnosis *Neurocomputing* **175** 132–45

[63] Chaplot S, Patnaik L M and Jagannathan N R 2006 Classification of magnetic resonance brain images using wavelets as input to support vector machine and neural network *Biomed. Signal Process. Control* **1** 86–92

[64] Zhang Y and Wang S 2015 Detection of Alzheimer's disease by displacement field and machine learning *PeerJ* **3** e1251

[65] Alam S, Kwon G R, Kim J I and Park C S 2017 Twin SVM-based classification of Alzheimer's disease using complex dual-tree wavelet principal coefficients and LDA *J. Healthc. Eng.* **2017** 8750506

[66] Simões R, van Cappellen van Walsum A M and Slump C H 2014 Classification and localization of early-stage Alzheimer's disease in magnetic resonance images using a patch-based classifier ensemble *Neuroradiology* **56** 709–21

[67] Zhou X *et al* 2015 Detection of pathological brain in MRI scanning based on wavelet-entropy and naive Bayes classifier *International Conference on Bioinformatics and Biomedical Engineering Lecture Notes in Computer Science* vol 9043 (Berlin: Springer) pp 201–9

[68] Ortiz A, Górriz J M, Ramírez J and Martínez-Murcia F J 2013 LVQ-SVM based CAD tool applied to structural MRI for the diagnosis of the Alzheimer's disease *Pattern Recognit. Lett.* **34** 1725–33

[69] Gerardin E *et al* 2009 Multidimensional classification of hippocampal shape features discriminates Alzheimer's disease and mild cognitive impairment from normal aging *Neuroimage* **47** 1476–86

[70] Gorji H T and Haddadnia J 2015 A novel method for early diagnosis of Alzheimer's disease based on pseudo Zernike moment from structural MRI *Neuroscience* **305** 361–71

[71] Shakeri M *et al* 2016 Statistical shape analysis of subcortical structures using spectral matching *Comput. Med. Imaging Graph.* **52** 58–71
[72] Kim W H *et al* 2014 Multi-resolutional shape features via non-Euclidean wavelets: applications to statistical analysis of cortical thickness *Neuroimage* **93** 107–23
[73] Wolz R *et al* 2011 Multi-method analysis of MRI images in early diagnostics of Alzheimer's disease *PLoS One* **6** 1–9
[74] Querbes O *et al* 2009 Early diagnosis of Alzheimer's disease using cortical thickness: impact of cognitive reserve *Brain* **132** 2036–47
[75] Zhou Q *et al* 2014 An optimal decisional space for the classification of Alzheimer's disease and mild cognitive impairment *IEEE Trans. Biomed. Eng.* **61** 2245–53
[76] McKhann G, Drachman D, Folstein M, Katzman R, Price D and Stadlan E M 1984 Clinical diagnosis of Alzheimer's disease report of the NINCDS-ADRDA work group under the auspices of department of health and human services task force on Alzheimer's disease *Neurology* **34** 939
[77] Ahmed M R, Zhang Y, Feng Z, Lo B, Inan O T and Liao H 2019 Neuroimaging and machine learning for dementia diagnosis: recent advancements and future prospects *IEEE Rev. Biomed. Eng.* **12** 19–33
[78] Ashburner J and Friston K J 2000 Voxel-based morphometry—the methods *Neuroimage* **11** 805–21
[79] Evans A C, Janke A L, Collins D L and Baillet S 2012 Brain templates and atlases *Neuroimage* **62** 911–22
[80] Lancaster J L *et al* 2000 Automated Talairach Atlas labels for functional brain mapping *Hum. Brain Mapp.* **10** 120–31
[81] Maldjian J A, Laurienti P J, Kraft R A and Burdette J H 2003 An automated method for neuroanatomic and cytoarchitectonic atlas-based interrogation of fMRI data sets *Neuroimage* **19** 1233–9
[82] Torrence C and Compo G P 1998 A practical guide to wavelet analysis *Bull. Am. Meteorol. Soc.* **79** 61–78
[83] Kumar C A, Madhavi B K and Lalkishore K 2016 Pipeline and parallel processor architecture for fast computation of 3D-DWT using modified lifting scheme *Proc. of the 2016 IEEE Int. Conf. on Wireless Communications, Signal Processing and Networking, WiSPNET 2016* pp 2123–8
[84] Chen X, Yang W, He Z and Zhong W 2012 Adaptive acquisition footprint suppression based on a 3D stationary wavelet transform: a case study from China *J. Appl. Geophys.* **77** 1–6
[85] Hong Y K *et al* 2010 Three-dimensional discrete wavelet transform (DWT) based on unary arithmetic *WMSCI 2010–Proc. of 14th World Multi-Conf. on Systemics, Cybernetics and Informatics* vol 1 pp 30–3
[86] Guyon I, Weston J, Barnhill S and Vapnik V 2002 Gene selection for cancer classification using support vector machines *Mach. Learn.* **46** 389–422
[87] Kohavi R 2016 A study of cross-validation and bootstrap for accuracy estimation and model selection *Learning* 1137–43
[88] Ben Ahmed O *et al* 2014 Classification of Alzheimer's disease subjects from MRI using hippocampal visual features *Multimed. Tools Appl.* **74** 1249–66

Lightning Source UK Ltd.
Milton Keynes UK
UKHW050712270421
382700UK00003B/138

9 780750 332774